Celestial Mechanics and Astrodynamics: Theory and Practice

Astrophysics and Space Science Library

More information about this series at http://www.springer.com/series/5664

Pini Gurfil • P. Kenneth Seidelmann

Celestial Mechanics and Astrodynamics: Theory and Practice

Springer

Pini Gurfil
Faculty of Aerospace Engineering
Technion-Israel Institute of Technology
Haifa, Israel

P. Kenneth Seidelmann
Department of Astronomy
The University of Virginia
Charlottesville, USA

ISSN 0067-0057 ISSN 2214-7985 (electronic)
Astrophysics and Space Science Library
ISBN 978-3-662-50368-3 ISBN 978-3-662-50370-6 (eBook)
DOI 10.1007/978-3-662-50370-6

Library of Congress Control Number: 2016943837

Cover illustration: Technion's Space Autonomous Mission for Swarming and Geo-locating Nanosatellites (SAMSON). Credit: Asher Space Research Institute, Technion

Printed on acid-free paper

This Springer imprint is published by Springer Nature
The registered company is Springer-Verlag GmbH Berlin Heidelberg

I dedicate this book to my children, Eytam, Oshri, and Ohav, and to my parents, Arie and Sara.

Pini Gurfil

This book is dedicated to Bobbie Seidelmann and our family, Holly, Kent, Jutta, Alan, Karen, and Sarah.

P. Kenneth Seidelmann

Also we dedicate the book to the scientists that preceded us and taught, mentored, and inspired us.

Foreword

The early contributions to artificial satellite orbit theory were mostly made by the celestial mechanicians, e.g., Brouwer, Garfinkel, Vinti and Kozai. Then, as aerospace engineering curricula emerged, their astrodynamics graduates began to make contributions. Most of the recent astrodynamics books have been written by engineering graduates. This book, co-authored by a celestial mechanician, Ken Seidelmann, and an astrodynamicist, Pini Gurfil, is a welcome addition to the aerospace community as it merges the two backgrounds.

Chapter 1 begins with a short history of celestial mechanics and then transitions to introductions to some of the key topics covered in the book. Topics included that are not usually seen in astrodynamics books are stability, chaos, Poincaré sections, KAM (Kolmogorov-Arnold-Moser) theory, and observation systems. Chapter 2 covers the basic math and physics concepts needed for the subjects in the book. Chapter 3 provides an excellent discussion of coordinate systems and introduces relativity, a subject not usually included in astrodynamics books but certainly present in celestial mechanics, e.g., the precession of Mercury's perihelion. Chapters 4 and 5 provide a thorough discussion of the central force and two-body problems. Included is a section on Einstein's modification of the orbit equation. The focus of Chap. 6 is initial orbit determination. Chapter 7 provides a thorough discussion of the N-body problem and the integrals associated with this problem. Chapter 8 then addresses the special case of the circular restricted 3-body problem (CR3BP). The coverage of the CR3BP is more comprehensive than found in most astrodynamics books and includes a discussion of families of periodic orbits. Chapter 9 is an introduction to numerical procedures used in astrodynamics and celestial mechanics. This chapter is not a comprehensive coverage and comparison of numerical integration methods but an introduction to the methods needed to understand numerical methods and error computation.

Chapter 10 begins a group of five chapters that this writer considers very important for astrodynamics and celestial mechanics but is often not found in astrodynamics books. I believe that the motion under the influence of conservative perturbations, those derivable from a potential, is best addressed and understood using Hamiltonian mechanics and perturbation methods such as Lie series. Chapter 10 discusses the basics of Hamiltonian mechanics, canonical transformations, generating functions, and Jacobi's theorem and applies these to the two-body problem. The focus of Chap. 11 is perturbation methods, and it begins with an excellent discussion of the variation of parameters (VOP), which leads to Lagrange's planetary equations. Then, with the perturbations expressed as specific disturbing accelerations instead of the accelerations obtained from a potential, Gauss' variational equations are derived for the accelerations in the radial, transverse, and orbit normal directions and the tangential, normal, and orbit normal directions. Included is a discussion of Lagrange brackets, which are needed for the VOP. Also in this chapter is the presentation of the Kustaanheimo-Stiefel variables. Using the foundations developed in Chap. 10, Chap. 11 addresses the solution for the 3rd body perturbations, atmospheric drag, and gravitational potential. Then Chap. 12 focuses on the solution for motion about an oblate planet. There are many such solutions beginning with Brouwer's 1959 paper, and presenting even a few solutions would be prohibitive. The solution presented here is the Cid-Lahulla radial intermediary. Special perturbation (numerical integration) methods are the most accurate and the general perturbation analytical methods, e.g, Brouwer's solution, are the most efficient. Chapter 13 presents the semianalytical approach, which is more efficient than numerical integration and more accurate than the analytical solution. The method is then applied to the four problems, a LEO satellite perturbed by drag, frozen orbits, sun-synchronous and repeat ground track orbits, and the motion of a geosynchronous satellite.

Chapters 10–13 address the problem of the motion of a space object under the influence of forces derivable from a potential except for the section on the effects of atmospheric drag. Chapters 14 and 15 consider the problem of the control of a space object using both continuous and impulsive control. Chapter 14 considers the control of specific types of orbits such as sun-synchronous orbits, frozen orbits, and geosynchronous orbits, as well as gravity assists. Both impulsive and continuous thrust control are addressed. Chapter 15 provides a very thorough coverage of the well-known problem of optimal impulsive orbit transfers.

Chapter 16 addresses the problem of orbit data processing and presents batch least squares and recursive filtering. Also discussed is the use of polynomials for the compression/representation of ephemerides. Chapter 17 provides a summary of the problem of space debris including probability of collision and collision avoidance maneuvers. The book concludes with another discussion of main contributors to celestial mechanics and the early pioneers of astrodynamics.

Entire books have been written on the subjects presented in many of the chapters in this book. Thus, when writing a book on astrodynamics, there has to be a balance between the amount of material presented and the necessary balance of mathematical rigor and its application to the problem at hand. I believe this book has achieved such a balance. There is a breadth of topics and each one is presented with the necessary depth needed for the reader to understand the topic. The book can be used for a senior/1st-year graduate class in astrodynamics and also for a 2nd-year graduate class in astrodynamics. It is a pleasure for me to write this Foreword and recommend this book to the astrodynamics community.

Texas A&M University, College Station, TX, USA Kyle T. Alfriend

Preface

While astrodynamics is a relatively new science, celestial mechanics, dealing with the motion of planets, satellites, comets, stars, and galaxies, is over three centuries old, dating back to Kepler's laws and Newton's *Principia*. Celestial mechanics has evolved into a myriad of approaches, methods, and results, some of which are the bases for astrodynamics. Indeed, celestial mechanics and astrodynamics share some fundamental tools, ranging from analytical dynamics to computer programs, used for the calculation of spacecraft and planetary orbits.

In recent years, an unprecedented interest in celestial mechanics and astrodynamics has risen due to new space programs. Astrophysicists, astronomers, space systems engineers, mathematicians, and scientists have been cooperating to develop and implement groundbreaking space missions. Progress in the theory of dynamical systems and computational methods has enabled development of low-energy spacecraft orbits; significant progress in the research and development of electric propulsion systems promises revolutionary, energy-efficient spacecraft trajectories; and the idea of flying several spacecraft in formation may break the boundaries of mass and size by creating virtual spaceborne platforms. The problems with debris have been recognized and studied. All of these factors have generated a growing interest in astrodynamics, a science devoted to understanding and controlling the interaction between a spacecraft and the space environment.

Whereas there are many books dealing separately with celestial mechanics and astrodynamics, one rarely finds a book dealing with these two topics in a unified manner. The juxtaposition of celestial mechanics and astrodynamics is a unique approach that is expected to be a refreshing attempt to discuss both the dynamics of celestial objects and the mechanics of space flight. The purpose of this book is to holistically describe methods and applications common to celestial mechanics and astrodynamics. The book includes classical and emerging topics, manifesting the state of the art and beyond. The book contains homogenous and fluent discussion of the key problems, rendering a portrayal of recent advances in the field together with some basic concepts and essential infrastructure in orbital mechanics. The text contains introductory material followed by a gradual development of ideas interweaved to yield a coherent presentation of advanced topics. The book presents the main challenges and future prospects for the two fields, in an elaborate, comprehensive, and mathematically rigorous manner.

This book is designed as an introductory text and reference book for graduate students, researchers, and practitioners in the fields of astronomy, celestial mechanics, astrodynamics, satellite systems, space sciences, and astrophysics. The purpose of the book is to emphasize the similarities between celestial mechanics and astrodynamics and to present recent advances in these two fields, so that the reader can understand the interrelations and mutual influences.

This book is of value to graduate students and academic researchers, for its introduction of concepts in the field for future work and its comprehensive discussion of the scientific and engineering state of the art; to university professors teaching courses on orbital and/or celestial mechanics; to aerospace engineers, for its discussion of advanced trajectory analysis and control techniques; to mathematicians, for its discussion of nonlinear dynamics and mechanics; and to astronomers, for its presentation of perturbation methods and orbit determination schemes. It is also of value for commercial, economic, and space policymakers, as it presents the forefront of space technology and science from a broad and innovative perspective.

Some of the developments in the book are based on the classical books by Danby, McCuskey, Brouwer and Clemence, Kovalevsky, and Hildebrand. We cite these authors throughout the text.

Haifa, Israel Pini Gurfil
Charlottesville, VA, USA P. Kenneth Seidelmann

Acknowledgments

The authors wish to acknowledge all the celestial mechanicians and astrodynamicists that preceded them in written works, in research, and in teaching and mentoring.

Pini Gurfil wishes to acknowledge his collaborators and students, who contributed to various chapters of this book: Dr. Vladimir Martinusi, Dr. Martin Lara, Dr. David Mishne, Dr. Dmitry Pisarevsky, Ohad Ben-Yaacov, Alex Galperin, Elad Denenberg, Ariel Vaknin, Gali Nir, Anton Jigalin, Sofia Belyanin, Changxuan Wen, and Weichao Zhong. Pini Gurfil's contribution to this book was supported by grants from the European Research Council, the German-Israeli Foundation, and the Israeli Ministry of Science and Technology.

Special thanks go to Dr. Michael Efroimsky, whose far-reaching vision and true commitment to science paved the way for the creation of this book.

Contents

List of Figures

List of Tables

Notation and Acronyms

Throughout this book, we denote a vector **w** and a matrix **W** by a boldface font. By default, $w = \|\mathbf{w}\| \equiv \|\mathbf{w}\|_2$ denotes the Euclidean norm. $\hat{\mathbf{w}}$ denotes a unit vector, i.e., $\hat{\mathbf{w}} = \mathbf{w}/w$. We will use the notation $\delta/\delta t$ for differentiation with respect to a rotating frame and d/dt for differentiation with respect to a fixed frame.

The Lagrangian equilibrium points will be numbered according to the older convention, in which L_1 is the point exterior to the secondary body and L_2 is the interior point. In recent years, a different convention for numbering the Lagrangian points has been widespread. Under this convention, L_2 is the exterior point and L_1 is the interior point.

We mostly use the following convention for denoting conservative specific forces. For a position vector **r** and a potential function R, we choose the sign of R such that $\ddot{\mathbf{r}} = \nabla R$. This is consistent with, e.g., Battin's notation and will be used in most parts of the book, unless specifically noted otherwise.

Upon a first occurrence of a term, we emphasize it using an italic font, for example, the term *Lagrangian point*.

Finally, we elaborate below the main acronyms used throughout this book:

2MASS	= 2-Micron Astronomical Sky Survey
ACPL	= Accepted Collision Probability Level
AD	= Anno Domini (Common Era)
AGI	= Analytical Graphics Inc.
AMMD	= Allowed Minimum Miss Distance
ANCAS	= Alfano-Negron Close Approach Software
BC	= Before Christ (Before Common Era)
BCRF	= Barycentric Celestial Reference Frame
BCRS	= Barycentric Celestial Reference System
BDT	= BeiDou System Time
BIH	= Bureau Internationale de l'Heure
BIPM	= Bureau International des Poids et Mesures
c	= Speed of Light
CCD	= Charge-Coupled Device

CfA	= Center for Astrophysics
CGCS 2000	= China Geodetic Coordinate System 2000
CIO	= Celestial Intermediate Origin
CIP	= Celestial Intermediate Pole
CIRS	= Celestial Intermediate Reference System
CRS	= Celestial Reference System
CSSI	= Center for Space Standards and Innovation (AGI)
DE	= Development Ephemeris
DSST	= Draper Semianalytical Satellite Theory
ΔT	= TT – UT1
ΔUT1	= UT1 – UTC
ECEF	= Earth-Centered Earth-Fixed
ECI	= Earth-Centered Inertial
ECR	= Earth-Centered Rotating
EGM	= Earth Gravitational Model
EKF	= Extended Kalman Filter
ERA	= Earth Rotation Angle
ESA	= European Space Agency
ET	= Ephemeris Time
FK5	= Fundamental Catalog 5
FLI	= Fast Lyapunov Indicator
GAM	= Gravity Assist Maneuver
GAST	= Greenwich Apparent Sidereal Time
GCRF	= Geocentric Celestial Reference Frame
GCRS	= Geocentric Celestial Reference System
GCT	= Greenwich Civil Time
GEO	= Geostationary Orbit
GHA	= Greenwich Hour Angle
GLONASS	= GLObal NAvigation Satellite System
GMST	= Greenwich Mean Sidereal Time
GMT	= Greenwich Mean Time
GNSS	= Global Navigation Satellite System
GPS	= Global Positioning System
GRACE	= Gravity Recovery and Climate Experiment
GST	= Greenwich Sidereal Time
GTRF	= Galileo Terrestrial Reference Frame
GVE	= Gauss Variational Equations
HCRF	= Hipparcos Catalog Reference Frame
HCW	= Hill-Clohessy-Wiltshire
IAG	= International Association for Geodesy
IAU	= International Astronomical Union
IAU2000A	= IAU Precession-Nutation Model of 2000
IC	= Initial Conditions
ICRF	= International Celestial Reference Frame

ICRS = International Celestial Reference System
IERS = International Earth Rotation and Reference Systems Service
ILRS = International Laser Ranging Service
IRP = IERS Reference Pole
IRM = IERS Reference Meridian
IRS = International Reference Star Catalog
ISS = International Space Station
ITRF = International Terrestrial Reference Frame
ITRF2 = 2nd ITRF
ITRS = International Terrestrial Reference System
ITU-R = International Telecommunications Union-Radio communications
JD = Julian Date
JDN = Julian Day Number
JPL = Jet Propulsion Laboratory
KAM = Kolmogorov-Arnold-Moser
KF = Kalman Filter
KS = Kustaanheimo-Stiefel
LAGEOS = LAser GEOdynamics Satellites
LBT = Large Binocular Telescope
LCE = Lyapunov Characteristic Exponent
LE = Lunar Ephemeris
LEO = Low-Earth Orbit
LHA = Local Hour Angle
LLR = Lunar Laser Ranging
LPE = Lagrange Planetary Equations
LTAN = Local Time of the Ascending Node
LVLH = Local Vertical, Local Horizontal
μas = Microarcsecond
mas = Milliarcsecond
MJD = Modified Julian Day
mm = Millimeter
nas = Nanoarcsecond
NASA = National Aeronautics and Space Administration
NEO = Near-Earth Object
nm = Nanometer
NOMAD = Naval Observatory Merged Astrometric Database
NORAD = North American Aerospace Defense Command
NOVAS = Naval Observatory Vector Astrometry Subroutines
NPOI = Navy Precise Optical Interferometer
ns = Nanosecond
OGLE = Optical Gravitational Lensing Experiment
PPN = Parameterized Post-Newtonian
PVT = Position Velocity Time
PZT = Photographic Zenith Tube

RAAN = Right Ascension of Ascending Node
RCS = Radar Cross Section
RGT = Repeat Ground Track
RMS = Root Mean Square
RVD = Rendezvous and Docking
s = Second
SAO = Smithsonian Astrophysical Observatory
SAOC = Smithsonian Astrophysical Observatory Catalog
SGP = Simplified General Perturbations
SI = Systeme International
SIM = Space Interferometry Mission
SOFA = Standards of Fundamental Astronomy
SOI = Sphere of Influence
SRP = Solar Radiation Pressure
SSN = Space Surveillance Network
SSO = Sun-Synchronous Orbit
ST = Sidereal Time
STK = Satellite Tool Kit
TAI = International Atomic Time
TCA = Time of Critical Approach
TCB = Barycentric Coordinate Time
TCG = Geocentric Coordinate Time
TDB = Barycentric Dynamical Time
TDV = Transit Duration Variations
Teph = Barycentric Ephemeris Time
TIO = Terrestrial Intermediate Origin
TIRS = Terrestrial Intermediate Reference System
TLE = Two-Line Elements
TOD = True-of-Date
TRS = Terrestrial Reference System
TT = Terrestrial Time
TTV = Transit Timing Variation
UCAC = USNO CCD Astrographic Catalog
USNO = US Naval Observatory
USNO B = US Naval Observatory Large Catalog
UT = Universal Time
UTC = Universal Time Coordinated
UTM = Universal Transverse Mercator
UT0 = Universal Time 0
UT1 = Universal Time 1
UT1R = UT1 Regularized

UT2 = Universal Time 2
VLBI = Very Long Baseline Interferometer
VOP = Variation of Parameters
WGS-84 = World Geodetic System 84
yr = Year

Chapter 1
Introduction

1.1 Definitions

Celestial mechanics embraces the dynamical and mathematical theories describing the motions of planets, satellites, one member of a double star pair around another, and similar phenomena.

Dynamical astronomy includes celestial mechanics, plus galactic, stellar, and all other studies of astronomical motions.

Astrodynamics is generally defined as the study of motion of artificial satellites, where thrust is used to create and/or correct orbits (the term *astrodynamics* is often attributed to R.M.L. Baker in the late 1950s). In particular, astrodynamics constitutes the application of celestial mechanics, astroballistics, propulsion theory and allied fields to the problem of planning and directing the trajectories of space vehicles. Astrodynamics is now a well-recognized, stand-alone field of discipline comprising many auxiliary fields including dynamical systems analysis, optimization, control, estimation, numerical analysis, perturbation methods, Lagrangian and Hamiltonian dynamics, geometric mechanics, and chaos.

The natural objects are composed of stars, which provide their own emissions of energy and generally are more massive than 0.09 the mass of the Sun; planets, which orbit around a star, reflect light, and are 0.001 the mass of the Sun or less (planets are also now categorized as dwarf, minor, and extrasolar); moons, which orbit around a planet; and comets, which generally have tails, can be long or short periodic, or can make a first appearance.

The artificial bodies can be in orbit around the Earth, interplanetary objects going to orbits around other bodies, or into orbit around the Sun. Objects revolve around a point, so planets revolve around the Sun. Objects rotate about an axis, so the Earth rotates. Each object creates a gravitational force on all other objects, which causes perturbations in the motions of the objects.

P. Gurfil, P.K. Seidelmann, *Celestial Mechanics and Astrodynamics: Theory and Practice*, Astrophysics and Space Science Library 436,
DOI 10.1007/978-3-662-50370-6_1

In this chapter, we will present some background material, meant to provide the reader with the context of this book. The topics discussed encompass fundamental concepts in astrodynamics and celestial mechanics; orbital stability of celestial bodies and artificial satellites; and observational data.

1.2 History

Celestial mechanics has a long and distinguished history. Many mathematical developments were due to problems in celestial mechanics. Many advances can be identified with specific people (IAU 2015). This will become apparent in the different chapters of this book.

The ancient understanding of the solar system was based on the Earth being at the center of the celestial motions, with the motions of the bodies being on epicycles. Nicholas Copernicus proposed that the motions of the planets were around the Sun, but still with epicycles to describe the motions.

Johannes Kepler used the observations of Tycho Brahe from the sixteenth century to determine his laws of motion. In practice, the observations of Mars were the only ones accurate enough for him to determine his laws. Kepler enunciated two laws in 1609, and the third law in 1619. The laws were the following:

1. The orbit of each planet is an *ellipse* with the Sun at one focus.
2. Each planet revolves so that the line joining it to the Sun sweeps out equal areas in equal intervals of time (*law of areas*).
3. The ratio of the squares of the periods of any two planets is equal to the ratio of the cubes of their mean distances from the Sun (this law is approximate, its validity depends on the masses of the planets being small compared to the Sun's mass.)

Motion following these laws is called *Keplerian motion*. This is two-body motion, and it is not accurate in the presence of other planets perturbing the motion.

These laws led Isaac Newton, and others, to the conclusion that the force that keeps a planet in its orbit around the Sun varies inversely as the square of the distance. Once Newton proved that gravitational attraction between two homogeneous spheres could be calculated as if the masses of the spheres were concentrated at their centers, progress in dynamical astronomy was clear and rapid. The *Law of Universal Gravitation* by Newton is:

If two particles of masses, m_1 and m_2, are situated at a distance r apart, each particle attracts the other with a force Gm_1m_2/r^2, where G is a universal constant. The forces act along the line joining the masses.

Newton's Principia (Newton 1713) contains three dynamic postulates on which subsequent analyses rests:

Newton's 1st Law: A particle of constant mass remains at rest, or moves with constant speed in a straight line, unless acted upon by some force.

Newton's 2nd Law: A particle subjected to the action of a force moves in such a way that the time rate of change of the linear momentum equals the force.
Newton's 3rd Law: When two particles act upon each other, the force exerted on the second particle by the first is equal in magnitude and opposite in direction to the force exerted by the second on the first.

Astrodynamics really started with the launch of Sputnik in 1957 and US preparations for the launch of satellites. Paul Herget prepared the programs for orbit calculations for the Vanguard project (Green and Lomask 1970) and those programs were used for computations of the orbit of Sputnik from US observations. Herget's programs were continued for the Mercury and Apollo missions and continue today in some NASA and military applications. Artificial satellite theories were developed by Brouwer (1959), Vinti (1960), Musen (1961), Kozai (1962), and Garfinkel (1964) among others. With time, the concepts of transfer orbits and gravity assists were developed and will be described in the chapters of this book.

1.3 Properties of Conics

Conic, from a *conic section*, means any plane cross-section through a right circular cone. Any plane cuts a cone in a conic and any conic can be so obtained. Also, the general equation of the second degree in Cartesian coordinates is a conic, and any conic can be represented by a Cartesian equation of the second degree. By a suitable change of axes, we can usually reduce the general quadratic equation to two variables in the form

$$ax^2 + by^2 + c = 0 \tag{1.1}$$

The origin of the coordinates is called the *center* of the conic. The axes of the coordinates are the axes of the conic, also the axes of symmetry.

Another definition of the conic leads to the general equation in polar coordinates. The conic is the locus of points such that there is a constant ratio of its distance from a fixed point to its distance from a fixed line. The fixed point is the *focus*; the fixed line is the *directrix*; and the fixed ratio is the *eccentricity*.

For the polar coordinates equation, choose the origin at the focus, let the polar angle, v, be measured from a line, OD, perpendicular to the directrix, which is at a distance k from the directrix, as shown in Fig. 1.1.
Let $P(r, v)$ be a point on the conic; then, by definition,

$$r = e(k - r\cos v) \quad \text{or} \quad \frac{p}{r} = 1 + e\cos v \tag{1.2}$$

where e is a constant ratio, eccentricity, and $p = ek$ is the *parameter* of the conic; e determines the shape, p determines the size, r has a maximum and minimum along the line OD; the chord through O, perpendicular to OD, is the *latus rectum*, with

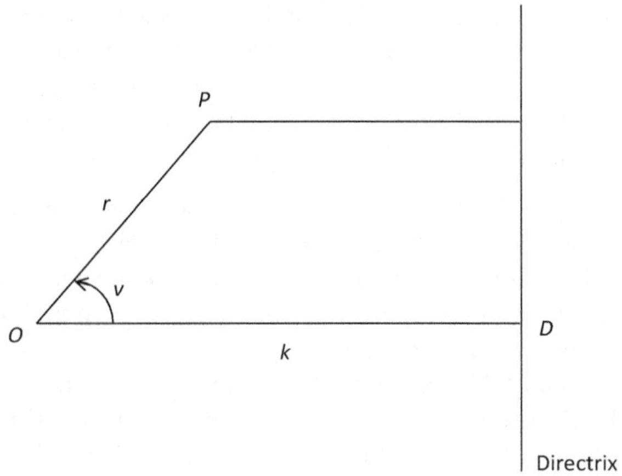

Fig. 1.1 A conic

a length $2p$. The parameter p is referred to as the *semilatus rectum* (Danby 1962, pp. 319–320).

1.3.1 The Ellipse, $0 < e < 1$

The Cartesian equation of an ellipse can be written as

$$\frac{x^2}{a^2} + \frac{y^2}{b^2} = 1 \tag{1.3}$$

where a and b are positive, $a > b$, x is the major axis, y the minor axis, a the semimajor axis, and b the semiminor axis. The origin is on the x axis, the directrix is parallel to the y axis, and the foci are S and S', as shown in Fig. 1.2. We have the relations

$$CA = CA' = a, \ CB = CB' = b, \ SA = q = a(1-e), \ SA' = q' = a(1+e) \tag{1.4}$$

$$CS = CS' = a\,e, \ p = a(1-e^2), \ b^2 = a^2(1-e^2), \ SB = a \tag{1.5}$$

The polar equation is

$$\frac{a(1-e^2)}{r} = 1 + e\cos v \tag{1.6}$$

where v is called the *true anomaly*, and is also commonly denoted by f.

One can define an ellipse as the locus of P such that $SP + S'P = constant = 2a$. SP and $S'P$ make equal angles with the tangent at P.

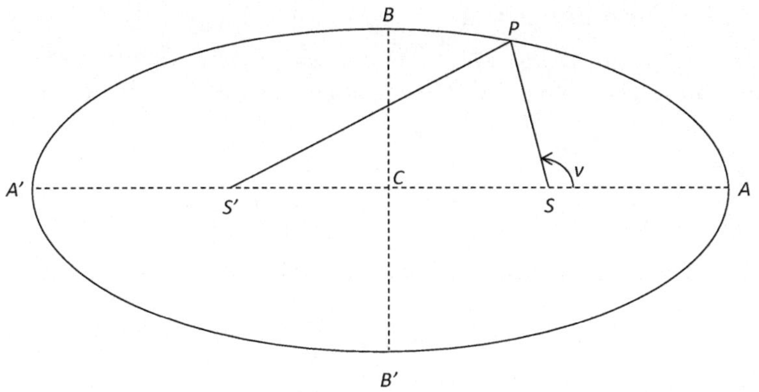

Fig. 1.2 An ellipse

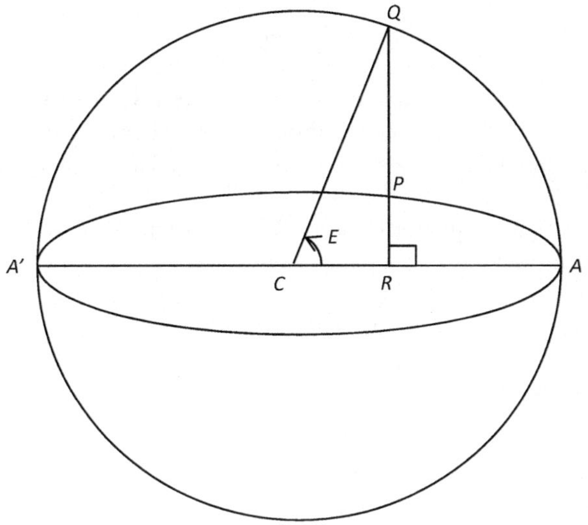

Fig. 1.3 A circle and an ellipse

The ellipse is the orthogonal projection of an auxiliary circle of radius a. Draw a circle of radius a and choose a diameter AA'. From Q, draw a perpendicular QR, as seen in Fig. 1.3, and make P such that $\frac{PR}{QR} = \frac{b}{a}$. P traces the ellipse with semiaxes a and b. The area of the ellipse is πab. Let $\angle QCA = E$. The coordinates of P are $x = a \cos E$ and $y = b \sin E$. The angle E is called the *eccentric anomaly*. If $e = 0$, the ellipse becomes a circle (Danby 1962, pp. 321–22).

1.3.2 The Parabola, $e = 1$

As e goes to 1, a goes to infinity. Remembering that $q = SA$ remains finite, we can write

$$\frac{q(1+e)}{r} = 1 + e \cos v \tag{1.7}$$

Putting $e = 1$, the polar equation of the *parabola* is

$$r = q \sec^2 \frac{v}{2} \tag{1.8}$$

The size of the parabola is q; the semilatus rectum is $2q$. The Cartesian equation is

$$y^2 = 4qx \tag{1.9}$$

There is one focus at $(q, 0)$ and one axis, the x axis. Consider the ellipse and let e tend to 1. The second focus tends to infinity; the tangent at P makes equal angles with SP and the parabola axis, as shown in Fig. 1.4. This explains the focusing of light by a parabolic mirror.

If we wish to have equations based on the variation of a single variable, or parameter, the simplest case for a parabola is

$$x = qt, \; y = qt^2 \tag{1.10}$$

(Danby 1962, pp. 323–24).

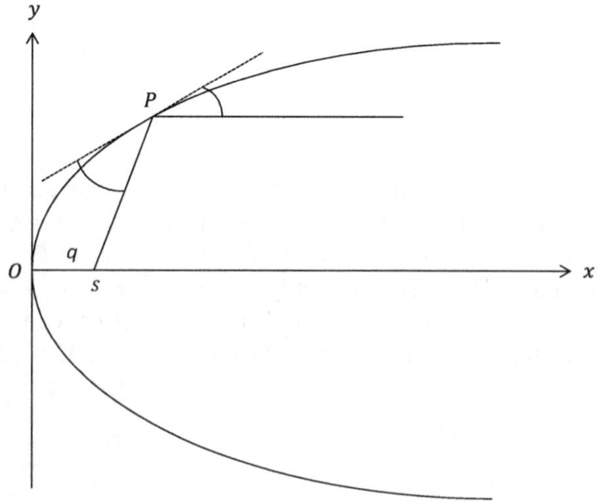

Fig. 1.4 A parabola

1.3.3 *The Hyperbola, e > 1*

The curve is not bounded, since r can become large. When r becomes large, v tends to a value from

$$v_a = \cos^{-1}\left(-\frac{1}{e}\right) \tag{1.11}$$

and we obtain a *hyperbola*, whose Cartesian equation is

$$\frac{x^2}{a^2} - \frac{y^2}{b^2} = 1 \tag{1.12}$$

In this case, a and b are taken as positive; a can also be taken as negative. When x and y become large, Eq. (1.12) is nearly the same as

$$\left(\frac{x}{a} + \frac{y}{b}\right)\left(\frac{x}{a} - \frac{y}{b}\right) = 0 \tag{1.13}$$

which is the equation of two lines through the origin. As r becomes larger, the closer the curve resembles these lines; they are called *asymptotes*, and are shown in Fig. 1.5. The directions given by v_a must be parallel to these asymptotes, so the angle between the asymptotes is $2\cos^{-1}(-1/e)$ or $\cos^{-1}\left(2/e^2 - 1\right)$.

The hyperbola has two branches, the polar equation gives only one. Comparing Eq. (1.13) and Eq. (1.11), we find

$$\tan^{-1}\left(\frac{b}{a}\right) = \cos^{-1}\left(\frac{1}{e}\right) \tag{1.14}$$

from which

$$b^2 = a^2(e^2 - 1) \tag{1.15}$$

so that $CA = a,\ AS = q = a(e - 1)$.

The tangent at A cuts the asymptotes at points $(-a, \pm b)$. The semilatus rectum is $p = a(e^2 - 1)$. The hyperbola is defined as the locus of P such that there is a constant difference, $2a$, between SP and $S'P$, so $SS' = 2c = 2ae$. S and S' are fixed points, foci, as shown in Fig. 1.5. The Cartesian coordinates of any point on the hyperbola are given by the parametric equations

$$x = \pm a\cosh F,\ y = b\sinh F \tag{1.16}$$

(Danby 1962, pp. 324–26).

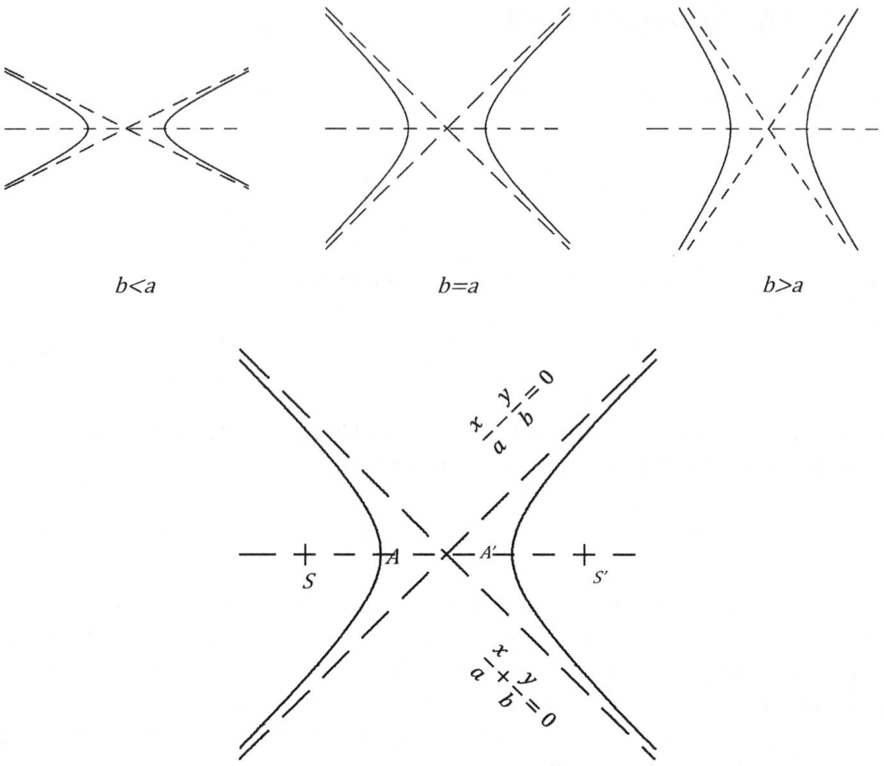

Fig. 1.5 A hyperbola. The case $a = b$ yields a rectangular hyperbola

1.4 Astronomical Background

In an orbit of one body about another, the least and greatest separations are indicated by applying prefixes *peri-* and *apo-*, respectively, to the Greek word for the more massive bodies. For a satellite around the Earth, the words are *perigee* and *apogee*. For a planet around the Sun, it is *perihelion* and *aphelion*. For motion around any star, the words are *periastron* and *apastron*; and for motion around a center of force, *pericentron* and *apocentron*.

Time for a complete revolution in an orbit with respect to the stars is a *sidereal period*. Due to the revolution of the orbit itself, many different periods can arise. The positions with respect to the Sun and Earth in orbits of inner and outer planets, as seen from the north pole of the *ecliptic* (the orbital plane of Earth about the Sun; see Sect. 3.5.1 for an exact definition), are shown in Fig. 1.6. As inner planets move around the Sun, their positions, as seen from the Earth, change from

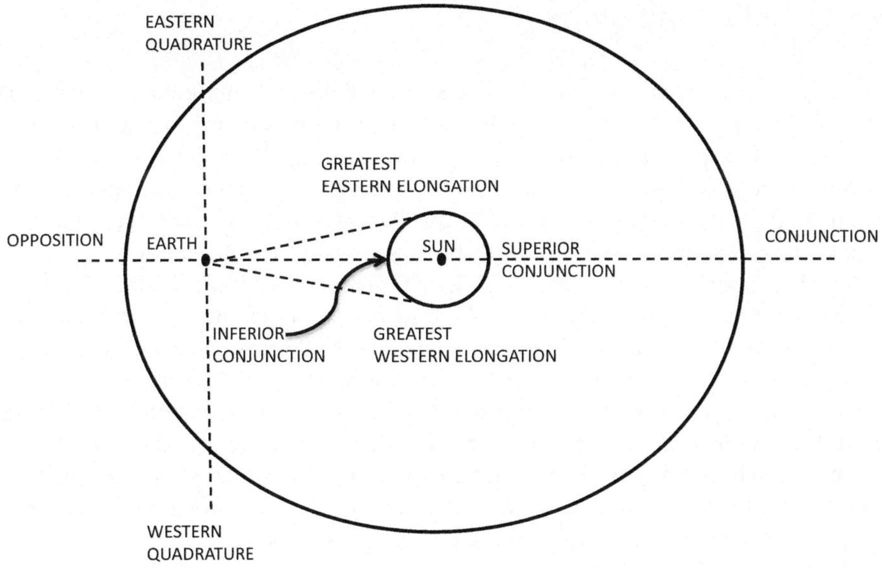

Fig. 1.6 The positions with respect to the Earth in orbits of planets as seen from the north pole of the ecliptic

superior conjunction, when they are behind the Sun, to greatest eastern elongation, to inferior conjunction, when they are between the Sun and the Earth, and to greatest western elongation. Outer planets move from conjunction, when they are farthest away beyond the Sun, to eastern quadrature, to opposition, when they are in the opposite direction from the Sun, to western quadrature. Inner planets are visible at elongations, and outer planets are visible for some time before and after quadratures. The average time for a planet to return to the same position relative to Earth is the *mean synodic period*. For a superior planet

$$\frac{1}{sidereal\ period} + \frac{1}{synodic\ perid} = 1 \tag{1.17}$$

For an inferior planet

$$\frac{1}{sidereal\ period} - \frac{1}{synodic\ period} = 1 \tag{1.18}$$

The unit being the sidereal year (the sidereal period of Earth).

We next discuss the stability of orbits, as well as notions related to the periodic motion mentioned above.

1.5 Stability and Chaos

The concepts of *stability* and *chaos* are of fundamental importance in celestial mechanics and astrodynamics. There are formal definitions for different types of stability and chaos (Belbruno 2004). However, the determination of the lack of stability, or of the presence of chaos, does not mean that the objects cannot exist. For example, our solar system is stable for some period of time and is in computational chaos after an extended period of time, but our solar system exists now.

We can describe systems as being deterministic, if we can calculate their past and future motions, when we know all the forces acting on them. Systems are not deterministic, if their past or future motion cannot be determined unequivocally.

Physical systems can be described by continuous or discrete models. The variation of a set of coordinates can be provided as a continuous function of time or at discrete intervals of time. A physical system can be described by differential equations, whose solution is a function of time, which varies continuously. A discrete model of a system can be written using a mapping, which selects time according to an iteration index over a set of integer numbers.

A definition of chaos is that, given a set of initial conditions, a small change in the initial conditions can lead to an indeterminate change in the past or future motion. The *three-body problem*, modelling the motion of three mutually-attracting bodies, has led to many studies of stability. The *restricted three-body problem*, which is a three-body problem wherein one of the masses is infinitesimally small compared to the other two, attracted the attention of mathematicians such as Poincaré, who was an early investigator of the restricted three-body problem and the related stability issues, in particular those concerning the *circular restricted three-body problem*, in which the two massive orbiting masses follow circular orbits about their common center of mass.

1.5.1 Three-Body Problem

The three-body problem has led to many studies of stability and chaos. Sundman developed a series solution, but its convergence is so slow that it is not useable.

Many determinations of stability are for the three-body problem. Euler and Lagrange found specific solutions to the three-body problem, which are now referred to as the *Lagrangian points*. Three of the solutions are for positions of the small body on a straight line with the two massive bodies, exterior to the two bodies or between them. These are the L_1, L_2, and L_3 points, and have been shown to be unstable. The other two points are at the vertices of an equilateral triangle and labeled the L_4 and L_5 points. These are stable orbits, and bodies have been discovered at these points for three-body systems involving the Sun and planets in the solar system; these are referred to as *Trojan orbits*.

1.5.2 Solar System

The solar system has led to many studies of stability and chaos. Some of this is based on n-body studies and others specifically on our solar system. Very long term integrations show a lack of determination of positions in the solar system, so the system is chaotic. Attempts to determine the history of the solar system and how the bodies move within the solar system, for example from the *Oort cloud* into the *Kuiper belt* and into the inner solar system, show that relatively stable orbits can change quickly, primarily through the change in the eccentricity of the orbit.

Until the later part of the twentieth century, our solar system was the only one known. With the discovery of *extrasolar planets*, first individual planets around other stars became known, and then multiple planets around individual stars—other solar systems—became known. The variety of the sizes and distributions of the planets has led to questions concerning how solar systems are formed and how they evolve with time. The existence of a solar system alone does not prove its stability. There is, of course, particular interest in planets that exist in the habitable zone around a star.

1.5.3 Resonances, Singularities and Regularization

There are many *resonances* between the motions of celestial bodies. These are cases where the mean motion of one body, multiplied by a small integer, is almost equal to the mean motion of another body, multiplied by another small integer. The small integers are usually single digits. In the main asteroid belt, these resonances cause what are called the *Kirkwood gaps*. In the rings of Saturn, there are the *Cassini divisions*. Among the planets, the resonances lead to long-period, large-amplitude perturbations. The resonances also lead to *tidal locking*, such that, for example, the same side of the Moon always faces the Earth. In perturbation theory, these resonances lead to small divisors in the denominator of the series of the perturbation theory and prevent convergence of the series.

In the three-body problem the source of instability is *singularities*, which are based on collisions between bodies. These singularities can be eliminated by *regularization*. So regularizations have been developed to avoid instabilities. Regularization aims to reduce singular differential equations to regular differential equations. The regularizing transformations are used when two or more bodies approach a collision. The elementary regularizing methods are Levi-Civita, Kustaanheimo-Stiefel, and Birkhoff transformations (Stiefel and Scheifele 1971; Kustaanheimo and Stiefel 1965; Celletti 2010).

1.6 Stability Determination

Various methods of calculating measures of stability have been developed and computer applications of these methods have led to faster calculation techniques. Poincaré was one of the first to formalize such techniques, and he inspired a number of people to follow his lead.

One way to investigate the dynamical behavior of a system is by mapping techniques. A number of these have been developed, such as the *logistic map*, the *standard map*, the *dissipative standard map*, *Hénon's map*, and the *Poincaré map* (Celletti 2010).

1.6.1 Poincaré Surface of Section

When dealing with a set of equations of motion consisting of two simultaneous, nonlinear, second-order differential equations, the solution consists of values of $x, y, \dot{x}$ and $\dot{y}$ at a sequence of times, where the quantities denote the position and velocity vectors in the rotating reference frame. The values of $x, y, \dot{x}$ and $\dot{y}$ at any given time correspond to a single point in a four dimensional phase space. The path of the particle is confined to a surface. Thus, we only require three of the quantities to define the orbit at that time uniquely. If we define a plane, for example $y = 0$, in the three dimensional space, the values of x and $\dot{x}$ can be plotted every time $y = 0$. This is the *Poincaré surface of section*, or the Poincaré map. This technique can be used to show the regular and chaotic regions in the restricted three-body problem. The points are only plotted when the trajectory intersects the surface in a particular direction. For regular orbits, a pattern develops with distinct islands, which are a characteristic of resonant motion. For chaotic motion, the points cover a larger region of the phase space. Also, the points do not fall on a smooth curve; they fill in an area of the phase space (Murray and Dermott 1999).

1.6.2 Hill Stability

The concept of *Hill stability* is expressed in terms of the *Hill sphere*. An astronomical body's Hill sphere is the region in which it dominates the attraction of satellites. To be retained by a planet, a moon must have an orbit that lies within the planet's Hill sphere. That moon would, in turn, have a Hill sphere of its own. Any object within that distance would tend to become a satellite of the moon, rather than that of the planet itself. One simple view of the extent of the Solar System is the Hill sphere of the Sun with respect to local stars and the galactic nucleus.

In more precise terms, the Hill sphere approximates the gravitational *sphere of influence* (SOI) of a smaller body in the face of perturbations from a more massive body. It was defined by the American astronomer George William Hill, based upon the work of the French astronomer Édouard Roche. For this reason, it is also known as the *Roche sphere*. The sphere of influence is a two-body concept, which geometrically defines the region where a celestial body would be an attractor in the presence of solar gravitation. In a sense, it defines the border after which an object departing a planet would be considered having reached infinity. The sphere of influence radius can be approximated using the formula

$$r_{SOI} = \left(\frac{M_{planet}}{M_{sun}}\right)^{\frac{2}{5}} r_{planet} \tag{1.19}$$

where M denotes mass and r_{planet} is the distance between the planet and the Sun.

The Hill sphere extends between the Lagrangian points L_1 and L_2, which lie along the line of centers of the two bodies. The region of influence of the second body is shortest in that direction, and so it acts as the limiting factor for the size of the Hill sphere. Beyond that distance, a third object in orbit around the second (e.g. Jupiter) would spend at least part of its orbit outside the Hill sphere, and would be progressively perturbed by the tidal forces of the central body (e.g. the Sun), eventually ending up orbiting the latter.

The Hill sphere is only an approximation, and other forces (such as radiation pressure or the *Yarkovsky effect*) can eventually perturb an object out of the sphere. This third object should also be of small enough mass that it introduces no additional complications through its own gravity. Detailed numerical calculations show that orbits at, or just within, the Hill sphere are not stable in the long term; it appears that stable satellite orbits exist only inside 1/2 to 1/3 of the Hill radius. The region of stability for retrograde orbits at a large distance from the primary, is larger than the region for prograde orbits at a large distance from the primary. This was thought to explain the preponderance of retrograde moons around Jupiter; however, Saturn has a more even mix of retrograde/prograde moons, so the reasons are more complicated.

1.6.3 Lyapunov

One measure of stability is the *Lyapunov time*, which is the time period on which a dynamical system is chaotic. Thus, the Lyapunov time indicates the limits of the predictability of the system. Formally, it is the time for the distance between nearby trajectories of the system to increase by e. The Lyapunov time, named after the Russian mathematician Aleksandr Lyapunov, is used in celestial mechanics for the stability of solar systems and elsewhere for computational uncertainties.

The rate of separation of infinitesimally close trajectories can be characterized by the *Lyapunov characteristic exponent* (LCE). There is a spectrum of Lyapunov exponents, equal in number to the dimensionality of the phase space. It is common to refer to the largest one as the maximal Lyapunov exponent (MLE), because it determines the predictability of a dynamical system. The *fast Lyapunov indicator* (FLI) obtains the value of the MLE at a fixed time. Comparing the FLIs as the initial conditions are varied allows distinguishing between different kinds of motion; for example, regular, resonant, or chaotic (Froeschlé et al. 1997).

1.6.4 Kolmogorov-Arnold-Moser Theorem

Kolmogorov (1954), and then Möser (1962) and Arnold (1963), provided the breakthroughs that have led to what is called *KAM theory*. It allows the proof of persistent *invariant tori*. The implementation of computer-assisted KAM proofs has permitted obtaining realistic results in simple models, like the planar circular restricted three-body problem. KAM theory provides a lower bound on the persistence of invariant tori, and the *converse KAM theory* gives an upper bound on the non-existence of invariant tori. Just above the critical breakdown threshold, the invariant tori transform into *invariant Cantor sets* or *Cantori* (Percival 1982; Celletti 2010).

The *KAM theorem* is a result in dynamical systems about the persistence of quasiperiodic motion under small perturbations. The theorem partly resolves the small-divisor problem that arises in the perturbation theory of classical mechanics. The problem is whether or not a small perturbation of a conservative dynamical system results in a lasting quasiperiodic orbit. The original breakthrough to this problem was given by Andrey Kolmogorov in 1954. This was rigorously proven and extended by Vladimir Arnold (in 1963 for analytic Hamiltonian systems) and Jurgen Möser (in 1962 for smooth twist maps), and the general result is known as the KAM theorem. The KAM theorem, as it was originally stated, could not be applied directly as a whole to the motions of the solar system. However, it is useful in generating corrections of astronomical models, and to prove long-term stability and the avoidance of orbital resonance in the solar system. Arnold used the methods of KAM to prove the stability of elliptical orbits in the planar three-body problem.

The KAM theorem is usually stated in terms of trajectories in phase space of an *integrable Hamiltonian system*. In general, we say that a dynamical system is *integrable* if the number of independent *constants of motion*, or *integrals of motion*, is equal to the number of degrees of freedom. The system will be *super-integrable* if the number of constants of motion exceeds the number of degrees of freedom. The motion of an integrable system is confined to a doughnut-shaped surface, an invariant torus. Different initial conditions of the integrable Hamiltonian system will trace different invariant tori in phase space. Plotting the coordinates of an integrable system would show that they are quasiperiodic.

The KAM theorem states that if the system is subjected to a weak nonlinear perturbation, some of the invariant tori are deformed and survive, while others are destroyed. The ones that survive are those that meet the non-resonance condition, i.e., they have "sufficiently irrational" frequencies. This implies that the motion continues to be quasiperiodic, with the independent periods changed (as a consequence of the non-degeneracy condition). The KAM theorem specifies quantitatively what level of perturbation can be applied for this to be true. An important consequence of the KAM theorem is that for a large set of initial conditions, the motion remains perpetually quasiperiodic.

The methods introduced by Kolmogorov, Arnold, and Moser have developed into a large body of results related to quasiperiodic motions. Notably, it has been extended to non-Hamiltonian systems, to non-perturbative situations, and to systems with fast and slow frequencies.

The non-resonance and non-degeneracy conditions of the KAM theorem become increasingly difficult to satisfy for systems with more degrees of freedom. As the number of dimensions of the system increases, the volume occupied by the tori decreases.

As the perturbation increases and the smooth curves disintegrate, we move from KAM theory to *Aubry-Mather theory*, which requires less stringent hypotheses and works with the Cantor-like sets (Celletti 2010).

1.6.5 Spacecraft Orbit Stability

Artificial satellites and interplanetary missions introduced a new category of stability studies, with the added feature that the spacecraft could have a propulsion system on board to alter the orbit.

For satellites around the Earth, there are a number of forces acting to change the orbits. These include *atmospheric drag*, *solar radiation pressure*, *oblateness*, and the gravitational pull of the Moon and the Sun, referred to as the *lunisolar perturbation*. Also there are resonances to consider.

For interplanetary missions, the perturbations of the solar system bodies have to be considered, and the shapes and gravity anomalies of the bodies need to be considered as well. The gravity assists of bodies can be used to alter the trajectory and achieve a different orbit without the use of propellant.

1.7 Chaos Determination

The availability of computers and the development of numerical experiments led to tests of long time-period orbital evolutions and the identification of chaotic orbits. Chaos can be studied in the circular restricted three-body problem or the solar system. For the circular three-body problem, the different mappings and tests of chaos indicators are applicable. Examples of chaos indicators are the LCE and FLI mentioned previously, the *mean experimental growth factor in nearby orbits*, the *smaller alignment index*, and the *spectral analysis method* (Dvorak and Lhotka 2013).

For the solar system, a map of the n-body problem is needed. Wisdom and Holman (1991) developed a map that is quick and *symplectic* in nature, which results in no secular changes in the energy of the system. This map is the Hamiltonian of the gravitational n-body problem, where the solution for a system of n mutually-perturbed bodies is found. To derive the map, the Hamiltonian is separated into a Keplerian part and an interaction part. In order to achieve the separation, a new system of variables, called the *Jacobian coordinates*, are chosen (Murray and Dermott 1999, pp. 440–448). The use of *symplectic integrators*, preserving the symplectic structure of the integrated dynamical system, has become widespread.

Long-period integrations of the solar system were done by Sussman and Wisdom (1988) and Wisdom and Holman (1991). By generating a system of averaged equations, using terms from the disturbing function, which approximate the real system, the equations can be integrated using a much longer time step; a step size of 40 days for the outer planets can be replaced with a 500-year step. Laskar (1994) investigated the planetary orbits for 10 billion years in the past and 15 billion years in the future. This showed couplings of Earth and Venus in eccentricities and inclinations, and chaotic variations in Mercury and Mars. Laskar (1994) also carried out further integrations by shifting the initial position of the Earth. These studies show that the planets, except possibly for Mercury, remain close to their current orbits for about a billion years or more, although they are technically chaotic.

In summary, these studies lead to the concept of *marginal stability* of the solar system. The solar system is unstable, but catastrophic phenomena leading to the destruction of the solar system can take place only in a time comparable to the age of the solar system, about 5 billion years. Thus, it is possible to suppose that the chaotic evolution of the orbits was a part of the process for structuring the system. There could have been additional bodies, but the system would have been more unstable, and a collision or ejection could have taken place. Thus, the system evolved into a more stable system. In the outer planets, the direct gravitational short period perturbations are more significant. So particles among the outer planets do not remain beyond a few hundred million years.

Applying these results to exoplanet systems, if the planetary formation from planetesimals is correct, planetary systems will be in a state of marginal stability. So at the end of the formation phase, a great number of bodies may remain, but the system will be unstable, leading to collisions or ejections. If a system exists with one or two planets, a multitude of small bodies will also remain, which have not been evacuated by gravitational instabilities (Laskar 1996).

1.8 Observational Data

Observational techniques have evolved over the years. We will primarily deal with positional and distance measurements necessary to be able to determine the positions and motions of objects.

Prior to about 1830, positional observations were of limited accuracy. With the advent of improved star catalogs, the observations of the planets improved significantly. The earliest observations were made with transit or meridian instruments, which were used to determine star and planetary positions. Optical positions of faint objects were observed visually. Circa 1900, photographic techniques were introduced and the plates were measured with measuring machines. In the 1960s, radar became available for measuring the distances to the Moon, Venus, and Mars. It was also the technique of choice for observing uncooperative artificial satellites. Artificial satellites with transmitters can be observed directly, and their velocities determined from the *Doppler effect*.

In the mid 1960s, *laser ranging* became available and retroreflectors were placed on satellites and the Moon to provide very accurate distance measures between specific points. In the late 1970s, *Charge Coupled Devices* (CCDs) were developed and could be used for astronomical observations. A CCD camera, the *Widefield/Planetary Camera*, was selected for the *Hubble Space Telescope*, and this promoted the development of the CCDs for astronomy. In the 1980s, *Very Long Baseline Interferometry* (VLBI) was developed for observations of the positions of distant radio sources. Also optical interferometry was developed as a method of obtaining much more accurate positional measures of optical sources (Kovalevsky and Seidelmann 2004). Over the years, infrared sensitive detectors have been developed that permit observations of natural and artificial objects. The IR observations have an advantage in being more sensitive to warm and redish objects.

The use of artificial satellites, as a means of determining accurate positions and motions on the Earth or around the Earth, led to the *Global Navigation Satellite Systems* (GNSS), and most specifically the *Global Positioning System* (GPS), which is now used for navigation and positioning worldwide, for the determination of the kinematics of the Earth, and for measuring terrestrial coordinates.

1.8.1 Transit Circle

Transit or meridian circles were used to develop star catalogs, observe solar system body positions, and determine the local time. With the availability of the telegraph and time signals from national services, the use of local transit instruments to determine local time declined and most positional observations were made by national observatories. International cooperation in the standardization of ephemerides, astronomical constants, star catalogs, and time scales developed in the beginning of the twentieth century. The transit circles were the source of star positions, the celestial reference frame, and solar system observations, and remained the primary sources of observations of the outer planets until the 1990s (Urban and Seidelmann 2012, pp. 316–319). The observations of star positions with the transit circles ended with the availability of the *Hipparcos Star Catalog*, whose positional observations were much more accurate than could be achieved with transit circle observations.

1.8.2 Photographic

With the availability of photographic observations, the capability to observe fainter objects made it possible to achieve astrometric studies for parallaxes, of faint satellites, and fainter star catalogs. Full sky surveys were now possible. There are now archives of plates that permit searches for historical observations of newly discovered objects to improve their orbits.

1.8.3 Radar Observations

After World War II, radar became available. After Sputnik, means of observing satellites were introduced. As larger antennas were developed, the ability to use radar for observing distance measures to the nearby planets was possible. This provided both the means of greatly improving the ephemerides of these planets, but also the ephemeris of the Earth. The accuracy of the measures was limited by the variations in the terrain of the targeted planets (Urban and Seidelmann 2012, pp. 317–325).

With the launch of Sputnik, the initial satellite observations were all optical, as radar equipment was not available for such observations. The US Navy developed the capability of a radar fence across the southern United States, so that any satellite passing over that fence would be detected. That *Naval Space Surveillance Fence* operated, with upgrades, from about 1960 until 2013. A new, more capable fence is being developed in 2015. A group of radars were developed both to detect possible launches of intercontinental missiles, and also to observe artificial satellites and maintain catalogs of these satellites. Both the US and Russia developed systems for the observation and cataloging of artificial satellites.

1.8.4 Laser Ranging

The advent of lasers led to the development of corner cubes that would reflect the laser beam directly back to its source.

The US Apollo missions to the Moon and the Russian lunar missions were opportunities to place scientific instruments on the Moon. A logical instrument to place there was a laser retroreflector that permitted measuring the precise distance between an observatory on the Earth and that reflector on the Moon. The problem was that the reflected signal would be very weak and the photons would be mixed in with photon noise. So sophisticated means had to be developed to sort out the reflected photons from the many other photons being detected. With time the lasers improved, the detection methods improved, and the accuracies currently are at the centimeter level. The lunar laser ranging measures have been used to greatly improve the lunar ephemeris, study the kinematics of both the Moon and the Earth, determine the parameters of general relativity, and determine the values of the *Universal Time* UT1 (Kovalevsky and Seidelmann 2004), to be discussed in Chap. 3.

The lasers and corner cubes led to the development of artificial satellites that were round with corner cubes on the surface and in stable orbits. Observations of these satellites, such as LAGEOS, were excellent for geodetic studies.[1]

With increased laser power, it has become possible to observe large objects in Earth orbit with lasers. These observations can contribute significantly for determination of the orbits of these objects.

1.8.5 VLBI

In the 1980s, VLBI, mentioned above, was developed. Organizations with a large radio antenna obtained the VLBI recording equipment and correlators were established at some organizations. A schedule of observations was developed, so several antennas would be observing the same radio source at the same time. By correlating the observational data recorded on large magnetic tapes, the difference in the times of the observations could be determined and, hence, the angle to the object with respect to the baseline between the antennas was determined very accurately. By this means the locations of a number of radio sources could be determined. The radio sources were very distant from the Earth and, hence, not subject to detectible motions. The one problem was to determine whether there were any changes in the source structure of the radio sources. The catalog of these objects was independent of time, since they were not moving (Urban and Seidelmann 2012, pp. 175–177). From the observations, the kinematics of the Earth could be solved and a celestial reference frame, independent of epoch, established.

The Earth orientation parameters were determined from optical observations, mostly *Photographic Zenith Tubes* (PZTs), since the discovery of the variability in the rotation of the Earth. Laser ranging to satellites and the Moon contributed to the determination of the parameters. When VLBI observations became available, they provided a much more accurate source of the Earth orientation parameters. All of the parameters, Earth rotation, polar motion, precession, and nutation, which we discuss in Chap. 3, could be determined from the VLBI observations and more accurately than from other observations (Urban and Seidelmann 2012, pp. 175–186).

As the sensitivity and accuracy of the VLBI observations improved, it became possible to make observations of spacecraft at planets and achieve more accurate observations than could be made with optical means. These observations could be added to other spacecraft observations to improve the ephemerides (Urban and Seidelmann 2012, pp. 320).

---

[1]See http://ilrs.gsfc.nasa.gov/missions/satellite_missions/current_missions/lag1_general.html.

Recently, the Chinese have sent spacecraft to the Moon and have used VLBI observations of the spacecraft signals to determine the spacecraft orbit. By transmitting the observational data from the antennas directly to the correlator, the data could be reduced and positional observations determined in 5 min.

1.8.6 CCDs

CCDs, mentioned above, started to become available in the 1970s. Their availability was initially very limited, but increased during the 1980s. The use of CCDs in cameras for everyone, and their many other applications, pushed the technical developments of CCDs. The CCDs are much more efficient with 60 %–90 % of the photons being recorded, as opposed to less than 10 % for photographic plates. Also, the increase in the detection is linear with time for CCDs, while photographic plates do not have that characteristic. Eventually, the availability of CCDs greatly reduced the use of photographic plates, such that in the 1990s it became impossible to obtain photographic plates for astronomy.

The quality of the CCDs improved significantly and the costs were greatly reduced. Now, CCD cameras for astronomy are readily available, most observatories have CCD cameras, and many amateurs use them for astrometric observations and astronomical photographs. Recent developments in the CCD technology have led to CMOS detectors, which are similar to CCDs, but permit reading out individual pixels, rather than the CCD sequential reading out of the columns and rows of the entire CCD. Hence, a pixel that is observing a bright star can be read out before it saturates, and the observation of the field of stars continues uninterrupted.

By observational techniques of combining the tracking rate of the telescope and the charge transfer rate within the CCD, observations of moving artificial satellites and star backgrounds can be obtained with stellar images of the satellite and the stars, and tracking images of the objects combined on a single image. These observations can be repeated to get multiple measured positions of the satellite during its track across the sky.

The astrometric observations with CCDs are much faster, reach much fainter stars, and more accurate. The need for a measuring machine is eliminated as the CCD locations can be determined directly for the observational data. A dynamical range of magnitudes can be achieved by obscuring masks, multiple exposures of different lengths, and other techniques. The accuracies of the measurements can be optimized by the centroiding techniques in the measurements of the images.

The use of CCDs permitted making much better images of astronomical phenomena and combining images taken with different filters to produce outstanding color images of objects. The combined astrometric accuracy with faint images permits studies of astrophysical activity.

1.8.7 Optical Interferometry

While radio interferometry is done over long distances with the observations recorded and then sent to a correlator for processing, optical interferometry is at optical wavelengths and done over connected optical paths with a delay line to combine the two images temporally, and achieve optical fringes that indicate the time delay is correct. Optical interferometers have been developed over the past 30 years for imaging and astrometric measurements. Some interferometers are arrays of small telescopes, others are large telescopes combined with smaller telescopes (Kovalevsky and Seidelmann 2004, pp. 18–21).

The *Naval Prototype Optical Interferometer* (NPOI), now the *Ken Johnston Naval Precise Optical Interferometer*, was developed on Anderson Messa outside Flagstaff, Arizona. It is a joint project of the US Naval Observatory (USNO) and the Naval Research Laboratory. It is an array of telescopes on three legs designed to achieve both imaging and astrometric observations. The astrometric accuracy is limited by the atmospheric effects on optical observations.

The *Large Binocular Telescope* (LBT) is designed to do both imaging and optical interferometry, by combining the images from the two mirrors through a delay line to achieve both images and astrometric measures.

The *Space Interferometry Mission* (SIM) was a planned optical interferometer to be launched into space. It went through a development process at JPL to prove that it could meet all the technical specifications required by the mission. It was designed to achieve microarcsecond accuracies for 20th magnitude stars and achieve a variety of research observations. It went through various reductions due to funding limitations, but finally was terminated due to a lack of funding at NASA.[2]

1.8.8 Surveys

Historically, surveys have been made to achieve positional star catalogs, usually of all the sky. Some have been limited to the sky coverage possible from an individual telescope. The *Palomar Sky Survey* is an example of a photographic survey from a single telescope. The USNO B catalogs were achieved by combining survey observations from different telescopes that together covered the entire sky, going down to 20th magnitude. The *Hipparcos catalog* was achieved from an astrometric satellite. The *2 Micron Astronomical Sky Survey* (2MASS) infrared catalog was achieved by observations from both hemispheres in the near infrared. The *USNO CCD Astrograph Catalog* (UCAC) is based on a CCD camera on an 8 inch telescope that observed in both the southern and northern hemispheres. The Gaia astrometric

[2] See http://science.nasa.gov/missions/sim/.

satellite is beginning an all-sky survey to achieve much better accuracies at faint magnitudes.

The interest in *Near Earth Objects* (NEO), space debris, and astronomical variability has led to the development of wide field-of-view telescopes for search capability. These telescopes can search for objects and identify where there is a need for follow up observations. They can identify moving objects that should be re-observed immediately. They also produce sky survey observations that can be used to produce a catalog of objects and positions. They discover and observe a large number of asteroids, artificial satellites, and debris.

The *Hipparcos astrometric satellite* was launched in 1989. It did not achieve its planned orbit, but through introduction of special efforts to overcome the difficulties of the real orbit, its scientific objectives were achieved. It observed for three years and produced a catalog of positions, proper motions, and parallaxes of 118,000 stars down to 12th magnitude with milliarcsecond accuracies. With time, the positional accuracies have degraded due to the limitations of the accuracies of the proper motions. However, Hipparcos set a new level of astrometric accuracies, which could not be achieved by telescopes on Earth (ESA 1997).

The UCAC was developed from observations with an 8 inch telescope with a CCD camera. Observations were made from Cerra Tololo in Chile and Flagstaff, Arizona, to obtain a full sky coverage. The resulting catalog of 100 million stars goes down to 16th magnitude with accuracies of 15–100 mas dependent on magnitude (Zacharias et al. 2010). The USNO NOMAD catalog is a combination of Hipparcos, UCAC, and USNO B to achieve the best accuracy available for every star down to 20th magnitude (Urban and Seidelmann 2012, pp. 530).

The 2MASS project was an all sky survey of the sky in J, H, and K near infrared bands going down to 14th magnitude. There are 471 million sources with accuracies of about 80 mas (Cutri et al. 2003; Zacharias et al. 2005).

Gaia is a European Space Agency (ESA) astrometric satellite designed to obtain astrometric positions and radio velocities of stars down to the 20th magnitude. It has two optical systems with a fixed angle between them that are scanning the sky in a rotational pattern and projecting the images onto an array of CCD detectors. It has been launched to the Earth-Moon L_2 point for an orbit. It is to observe for three years.[3]

1.8.9 GNSS

There are a family of GNSS, which we briefly mentioned above, in existence and in development. The GPS was developed by the US Defense Department and has at least 24 satellites in about 12 hour orbits transmitting a coded signal that receivers, accessing at least four satellites at a time, can use to measure their distance from the

[3]See http://sci.esa.int/gaia/ and http://www.esa.int/Our_Activities/Space_Science/Gaia_overview.

satellites. Originally, there were eight satellites in three orbital planes. Currently, there are four satellites in six orbital planes. The GPS was designed for military navigation purposes, but now the GPS receivers are in all kinds of equipment. It is used for both military and civilian navigation on the seas, in the air, and on land. It can be used for tracking objects anywhere. GPS can be used for scientific purposes, such as measuring Earth tectonics and determining the Earth rotation and polar motion. It can also be used to determine the positions of satellites in orbits around the Earth, both below and above the GPS orbits. The Global Navigation Satellite System GLONASS was developed by the Russian military and is in operation. The *Galileo satellite system* is being developed by ESA. The BeiDou/Compass is being developed by China (Urban and Seidelmann 2012, pp. 177).

1.8.10 Satellite Observations

Active satellites with transmission capabilities can be tracked and measurements made to be able to determine their orbits to the required accuracy. These methods include two-way distance measures, time of signals from the satellite, Doppler determinations of velocities, optical observations, radar observations, and so on. The concern for active satellites is possible impacts from other objects. As the ability is developed to observe fainter and smaller objects around the Earth, the number of objects to be cataloged and maintained in the catalog increases dramatically. This requires improved observing capabilities and computational abilities. For safety reasons, there is a need to observe smaller objects that could damage active satellites, particularly the *International Space Station* (ISS). We shall discuss this issue in Chap. 17.

References

Arnold, V.I.: Small denominators and problems of stability of motion in classical and celestial mechanics. Russ. Math. Surv. **18**(6), 85–191 (1963)

Belbruno, E.: Capture Dynamics and Chaotic Motions in Celestial Mechanics: With Applications to the Construction of Low Energy Transfers. Princeton University Press, Princeton (2004)

Brouwer, D.: Solution of the problem of artificial satellite theory without drag. Astron. J. **64**, 378–396 (1959)

Celletti, A.: Stability and Chaos in Celestial Mechanics. Springer Science and Business Media, New York (2010)

Cutri, R., Skrutskie, M., Van Dyk, S., Beichman, C., Carpenter, J., Chester, T., Cambresy, L., Evans, T., Fowler, J., Gizis, J., et al.: 2MASS all sky catalog of point sources. NASA/IPAC Infrared Science Archive (2003)

Danby, J.: Fundamentals of Celestial Mechanics. The Macmillan Company, New York (1962)

Dvorak, R., Lhotka, C.: Celestial Dynamics: Chaoticity and Dynamics of Celestial Systems. Wiley, Weinheim (2013)

ESA: The Hipparcos and Tycho Catalogues. European Space Agency SP-1200, Noordwijk, Netherlands (1997)

Froeschlé, C., Lega, E., Gonczi, R.: Fast Lyapunov indicators. application to asteroidal motion. Celest. Mech. Dyn. Astron. **67**(1), 41–62 (1997)

Garfinkel, B.: An improved theory of motion of an artificial satellite. Astron. J. **69**(3), 223–229 (1964)

Green, C.M., Lomask, M.: Vanguard: a history. NASA Special Publication 4202, Washington, DC (1970)

IAU: IAU Commission 7: Celestial mechanics and dynamical astronomy. In: Montmerle, T. (ed.) Proceedings of the XXVIII IAU General Assembly, vol. XXIXA, pp. 1–19 (2015)

Kolmogorov, A.N.: On conservation of conditionally periodic motions under small perturbations of the Hamiltonian. Dokl. Akad. Nauk SSSR **98**(4), 527–530 (1954)

Kovalevsky, J., Seidelmann, P.K.: Fundamentals of Astrometry. Cambridge University Press, Cambridge (2004)

Kozai, Y.: Second-order solution of artificial satellite theory without air drag. Astron. J. **67**(7), 446–461 (1962)

Kustaanheimo, P., Stiefel, E.: Perturbation theory of Kepler motion based on spinor regularization. J. Reine. Angew. Math. **218**, 204 (1965)

Laskar, J.: Large-scale chaos in the solar system. Astron. Astrophys. **287**, L9–L12 (1994)

Laskar, J.: Marginal stability and chaos in the solar system. In: Dynamics, Ephemerides, and Astrometry of the Solar System, IAU Symposium, vol. 172, pp. 75–88 (1996)

Möser, J.: On invariant curves of area-preserving mappings of an annulus. Nachr Akad Wiss Göttingen, II, pp. 1–20 (1962)

Murray, C.D., Dermott, S.F.: Solar System Dynamics. Cambridge University Press, Cambridge (1999)

Musen, P.: The theory of artificial satellites in terms of the orbital true longitude. J. Geophys. Res. **66**(2), 403–409 (1961)

Newton, I.: Philosophia Naturalis Principia Mathematica, 2nd edn. S. Pepys, London (1713)

Percival, I.C.: Chaotic boundary of a Hamiltonian map. Phys. D Nonlinear Phenomena **6**(1), 67–77 (1982)

Stiefel, E., Scheifele, G.: Linear and Regular Celestial Mechanics. Springer, Berlin, Heidelberg, New York (1971)

Sussman, G.J., Wisdom, J.: Numerical evidence that the motion of Pluto is chaotic. Science **241**(4864), 433–437 (1988)

Urban, S.E., Seidelmann, P.K. (eds.): Explanatory Supplement to the Astronomical Almanac. University Science Books, Mill Valley, California (2012)

Vinti, J.P.: Theory of the orbit of an artificial satellite with use of spheroidal coordinates. Astron. J. **65**, 353–354 (1960)

Wisdom, J., Holman, M.: Symplectic maps for the n-body problem. Astron. J. **102**, 1528–1538 (1991)

Zacharias, N., McCallon, H.L., Kopan, E., Cutri, R.M.: Extending the ICRF into the infrared: 2MASS-UCAC astrometry. Highlights Astron. **13**, 603–604 (2005)

Zacharias, N., Finch, C., Girard, T., Hambly, N., Wycoff, G., Zacharias, M., Castillo, D., Corbin, T., DiVittorio, M., Dutta, S., et al.: The Third US Naval Observatory CCD Astrograph Catalog (UCAC3). Astron. J. **139**(6), 2184 (2010)

Chapter 2
Vectors

2.1 Introduction

This chapter is meant to provide the basic information concerning vectors, hopefully as a review, but to ensure that the reader is familiar with the basics.

A scalar has a magnitude and a sign. A vector is a scalar with direction. This can be given the notation from the ends of the vector or as a symbol; we will use boldface notation for vectors. Thus, AB, shown in Fig. 2.1, would be equal the vector $\mathbf{a}$. Similarly, $BC = \mathbf{b}$ and $AC = \mathbf{c}$.

A scalar multiples a vector to change its length, $k\,\mathbf{a} = \mathbf{a}\,k$.

In vector addition, $\mathbf{a} + \mathbf{b} = \mathbf{c}$.

The commutative law means $\mathbf{a} + \mathbf{b} = \mathbf{b} + \mathbf{a}$.

The associative law is $(\mathbf{a} + \mathbf{b}) + \mathbf{c} = \mathbf{a} + (\mathbf{b} + \mathbf{c})$.

Vectors can be given in components. If $\mathbf{i}$ and $\mathbf{j}$ are not parallel, and $\mathbf{i}, \mathbf{j}$, and $\mathbf{r}$ are coplanar, then unique scalars x and y exist such that

$$\mathbf{r} = x\mathbf{i} + y\mathbf{j} \tag{2.1}$$

If we add $\mathbf{k}$ not in the plane, then $\mathbf{r}$ can be three dimensional and

$$\mathbf{r} = x\mathbf{i} + y\mathbf{j} + z\mathbf{k} \tag{2.2}$$

$\mathbf{r}$ is resolved along directions $\hat{\mathbf{i}}, \hat{\mathbf{j}}, \hat{\mathbf{k}}$ and x, y, z are its components. Normally, $\mathbf{i}, \mathbf{j}$, and $\mathbf{k}$ are mutually perpendicular in a right-hand triad, with the positive directions as shown in Fig. 2.2.

The shorthand notation for $\mathbf{r}$ in Eq. (2.2) may be written as $\mathbf{r} = [x,\ y,\ z]$. Written this way, $\mathbf{r}$ is a *row vector*. To obtain a *column vector*, we use the *transpose operator*, denoted by $(\cdot)^T$. In the following chapters, where appropriate, we will usually assume that a vector $\mathbf{v}$ is a column vector , unless written otherwise.

P. Gurfil, P.K. Seidelmann, *Celestial Mechanics and Astrodynamics: Theory and Practice*, Astrophysics and Space Science Library 436,
DOI 10.1007/978-3-662-50370-6_2

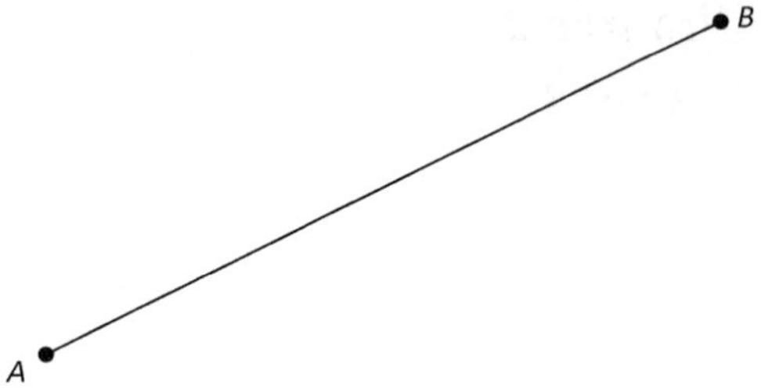

Fig. 2.1 Ends of a vector

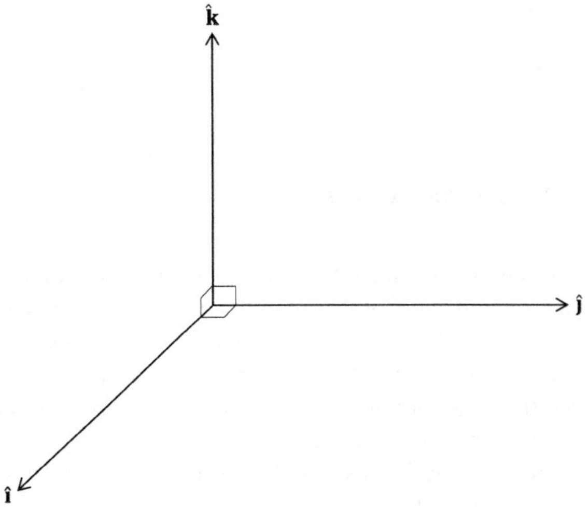

Fig. 2.2 A mutually-perpendicular right-hand triad with positive directions

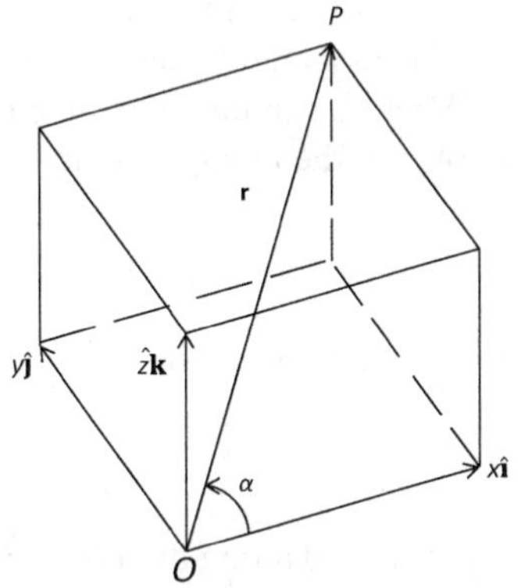

Fig. 2.3 Components of a vector

Thus, the components can be written without ambiguity as $\mathbf{r} = [x, y, z]$. If α is the angle between OP and $x\mathbf{i}$, then $\cos \alpha = x/r = l$, as shown in Fig. 2.3.

If m and n are defined similar to l, then l, m, n are *direction cosines* of $\mathbf{r}$ with respect to the triad, and $l^2 + m^2 + n^2 = 1$.

If we denote $\hat{\mathbf{r}}$ as a unit vector, then our vector $\mathbf{r}$ can be written as

$$OP = \mathbf{r} = x\mathbf{i} + y\mathbf{j} + z\mathbf{k} = [x, y, z] = [rl, rm, rn] = r[l, m, n] = r\hat{\mathbf{r}} \tag{2.3}$$

If $\mathbf{r_1} = [x_1, y_1, z_1]$ and $\mathbf{r_2} = [x_2, y_2, z_2]$, then

$$\mathbf{r_1} + \mathbf{r_2} = [x_1 + x_2, y_1 + y_2, z_1 + z_2]$$

(Danby 1962, pp. 14–19).

2.2 Scalar Product

Take vectors **a** and **b** and the angle θ between them as in Fig. 2.4. The *scalar product* or *dot product* is

$$\mathbf{a} \cdot \mathbf{b} = ab \cos \theta \tag{2.4}$$

We see by the commutative law that

$$\mathbf{a} \cdot \mathbf{b} = ab \cos \theta = ba \cos \theta = \mathbf{b} \cdot \mathbf{a} \tag{2.5}$$

We can interpret this as **a** times the projected length of **b** on **a**, which is $b \cos \theta$. Since $\mathbf{b} + \mathbf{c}$ is the projection of the sum of separate projections on **a**, it follows that

$$\mathbf{a} \cdot (\mathbf{b} + \mathbf{c}) = \mathbf{a} \cdot \mathbf{b} + \mathbf{a} \cdot \mathbf{c} \tag{2.6}$$

Also

$$m(\mathbf{a} \cdot \mathbf{b}) = (m\mathbf{a}) \cdot \mathbf{b} = \mathbf{a} \cdot (m\mathbf{b}) \tag{2.7}$$

If $\{\hat{\mathbf{i}}, \hat{\mathbf{j}}, \hat{\mathbf{k}}\}$ are a triad, then the components of $\mathbf{r}$ on the triad are

$$\mathbf{r} \cdot \hat{\mathbf{i}},\ \mathbf{r} \cdot \hat{\mathbf{j}},\ \mathbf{r} \cdot \hat{\mathbf{k}}$$

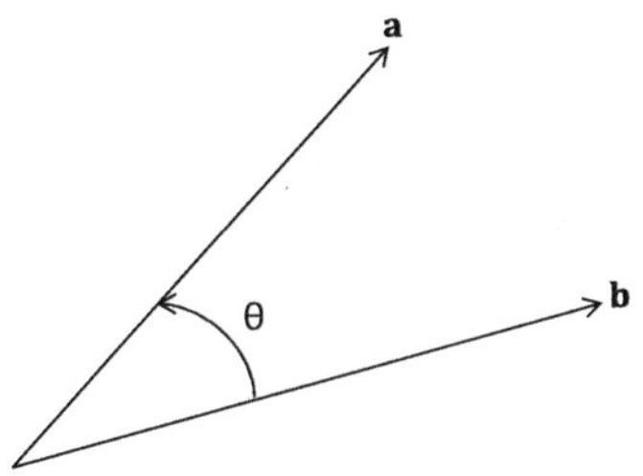

Fig. 2.4 An angle between two vectors

and in addition

$$\hat{\mathbf{i}} \cdot \hat{\mathbf{j}} = \hat{\mathbf{i}} \cdot \hat{\mathbf{k}} = \hat{\mathbf{j}} \cdot \hat{\mathbf{k}} = 0$$

The square of the vector $\mathbf{r} = x\hat{\mathbf{i}} + y\hat{\mathbf{j}} + z\hat{\mathbf{k}}$ is written as $\mathbf{r^2} \equiv \mathbf{r} \cdot \mathbf{r} = rr = r^2$, $r^2 = x^2 + y^2 + z^2$ (Danby 1962, pp. 20–21).

2.3 Vector Product

The right-hand convention defines a unique direction (American screws for mechanics), as illustrated in Fig. 2.5.

Then the *vector cross-product* is

$$\mathbf{a} \times \mathbf{b} = (ab \sin \theta)\mathbf{i} \tag{2.8}$$

This is not commutative, since $\mathbf{a} \times \mathbf{b} = -\mathbf{b} \times \mathbf{a}$. It is distributive, however, so that $\mathbf{a} \times (\mathbf{b} + \mathbf{c}) = \mathbf{a} \times \mathbf{b} + \mathbf{a} \times \mathbf{c}$. Note that $\mathbf{r} \times \mathbf{r} = 0$. In the right-hand triad $\{\hat{\mathbf{i}}, \hat{\mathbf{j}}, \hat{\mathbf{k}}\}$,

$$\hat{\mathbf{i}} \times \hat{\mathbf{j}} = \hat{\mathbf{k}},\ \hat{\mathbf{j}} \times \hat{\mathbf{k}} = \hat{\mathbf{i}},\ \hat{\mathbf{k}} \times \hat{\mathbf{i}} = \hat{\mathbf{j}},\ \hat{\mathbf{k}} \times \hat{\mathbf{j}} = -\hat{\mathbf{i}}$$

If $\mathbf{b}$ and $\mathbf{c}$ are $[b_x, b_y, b_z]$ and $[c_x, c_y, c_z]$, then

$$\mathbf{b} \times \mathbf{c} = [b_y c_z - b_z c_y, b_z c_x - b_x c_z, b_x c_y - b_y c_x]$$

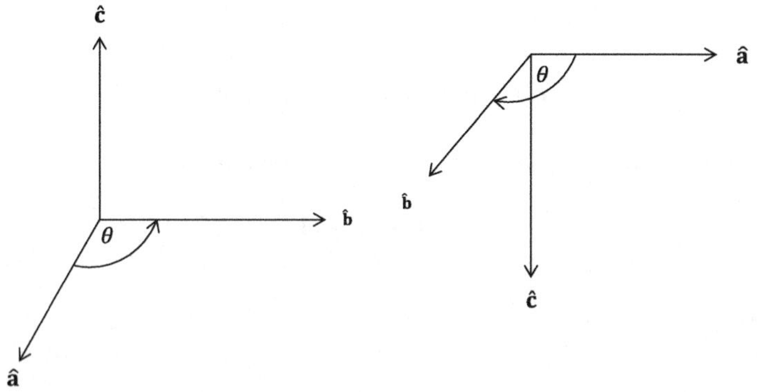

Fig. 2.5 Illustration of the right-hand convention

In the determinant form,

$$\mathbf{b} \times \mathbf{c} = \begin{vmatrix} \hat{\mathbf{i}} & \hat{\mathbf{j}} & \hat{\mathbf{k}} \\ b_x & b_y & b_z \\ c_x & c_y & c_z \end{vmatrix}$$

(Danby 1962, pp. 22–23).

2.4 Triple Scalar and Vector Products

The *triple scalar product* is defined as $\mathbf{a} \cdot (\mathbf{b} \times \mathbf{c})$ and can be written $[\mathbf{a}, \mathbf{b}, \mathbf{c}]$. If $\mathbf{a} = [a_x, a_y, a_z]$, then

$$\mathbf{a} \cdot (\mathbf{b} \times \mathbf{c}) = \begin{vmatrix} a_x & a_y & a_z \\ b_x & b_y & b_z \\ c_x & c_y & c_z \end{vmatrix}$$

Note that

$$[\mathbf{a}, \mathbf{b}, \mathbf{c}] = [\mathbf{b}, \mathbf{c}, \mathbf{a}] = [\mathbf{c}, \mathbf{a}, \mathbf{b}] = -[\mathbf{c}, \mathbf{b}, \mathbf{a}]$$

The *triple vector product* is defined by

$$\mathbf{a} \times (\mathbf{b} \times \mathbf{c}) = (\mathbf{a} \cdot \mathbf{c})\mathbf{b} - (\mathbf{a} \cdot \mathbf{b})\mathbf{c}$$

If the vector equation has terms you do not like, cross multiply the equation by its direction to eliminate it. For example, the equation of motion for a central orbit is

$$\frac{d^2\mathbf{r}}{dt^2} = -f(r)\hat{\mathbf{r}} \tag{2.9}$$

where $f(r)$ is a scalar function. Vectorial multiplication by $\mathbf{r}$ results in

$$\mathbf{r} \times \frac{d^2\mathbf{r}}{dt^2} = 0 \tag{2.10}$$

(Danby 1962, pp. 23–25).

2.5 Velocity of Vector

The velocity can vary as a function of a scalar, such as time, or as a function of another vector, such as position. We assume the variation is continuous, so it can be differentiated. We shall consider vectors varying with time, as shown in Fig. 2.6, so it is particle motion. In Fig. 2.6, AB is the path of P. P is the location at time t. P' is the location at time $t + \delta t$. δt is a small time interval. PP' is $\delta \mathbf{r}$. $\hat{\boldsymbol{\tau}}$ is the unit vector along PP', in the limit it will be tangent at P. $\delta \mathbf{r} = \hat{\boldsymbol{\tau}} \delta s$, where s is the distance along the curve. We have

$$\lim_{\delta t \to 0} \frac{PP'}{\delta t} = \lim_{\delta t \to 0} \frac{\delta \mathbf{r}}{\delta t} = \frac{d\mathbf{r}}{dt} = \hat{\boldsymbol{\tau}} \frac{ds}{dt} \tag{2.11}$$

This is the particle velocity at P, also commonly written as $\dot{\mathbf{r}}$ or $\mathbf{v}$. There is a magnitude and a direction, the properties of a vector. The components of $\dot{\mathbf{r}}$ are $(\dot{x}, \dot{y}, \dot{z})$ and those of $\hat{\tau}$ are $(dx/ds, dy/ds, dz/ds)$. Note that in modern convention *velocity* is a vector $\mathbf{v}$ or $\dot{\mathbf{r}}$; *speed* is a scalar value of velocity. The speed of P is ds/dt.

Also, $\|d\mathbf{r}/dt\|$, the speed of P, is not the same as $d\|\mathbf{r}\|/dt$, the component of velocity along the radius vector, or dr/dt.

Because have $\mathbf{r} \cdot \mathbf{r} = r^2$, then differentiating gives $\mathbf{r} \cdot \dot{\mathbf{r}} = r\dot{r}$. Here $\mathbf{r}$ is a position vector, defining the path traced by P and the locus of P; $\dot{\mathbf{r}}$ is a vector, if $OQ = \dot{\mathbf{r}}$, then Q follows a path called the *hodograph* of motion. The rate of change of OQ is the acceleration of P, which is a vector written as

$$\frac{d^2\mathbf{r}}{dt^2} = \frac{d\mathbf{v}}{dt} = \ddot{\mathbf{r}} \tag{2.12}$$

In Fig. 2.6, $OP \times OP' = \mathbf{r} \times \delta \mathbf{r}$. The modulus is twice the area of triangle OPP', the rate of change of the area is the *areal velocity* of OP, whose magnitude is $\|\mathbf{r} \times \dot{\mathbf{r}}\|/2$. The direction of $\mathbf{r} \times \dot{\mathbf{r}}$ is perpendicular to the plane of $\mathbf{r}$ and $\dot{\mathbf{r}}$. This direction is constant when the motion is in the plane.

Consider Kepler's first two laws, mentioned in Sect. 1.2: (i) Motion is in a plane for a planet; and (ii) the areal velocity is constant. So for Keplerian motion, $\mathbf{r} \times \dot{\mathbf{r}} = \mathbf{h}$,

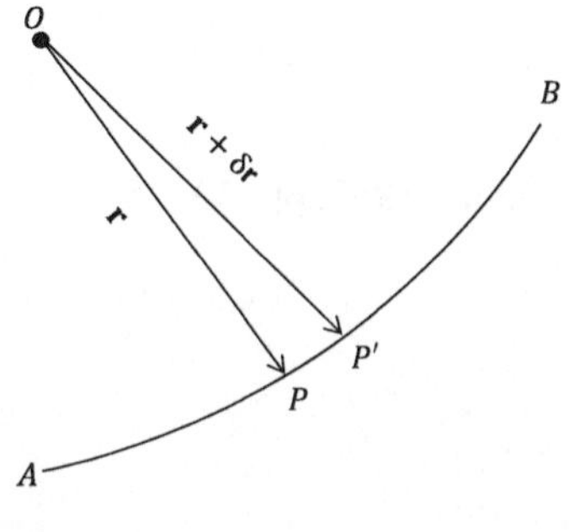

Fig. 2.6 Time-varying path of a particle

where **h** is a constant vector. Note that this is the first integral of $\mathbf{r} \times d^2\mathbf{r}/dt^2 = 0$ (Danby 1962, pp. 26–28).

2.6 Rotation of Axes

Given a system of right-handed rectangular axes as in Fig. 2.7, we can perform a *rotation of axes* through specified angles. A rotation about the x-axis positively through θ, as shown in Fig. 2.7, the new axes $[x', y', z']$ are related to the old axes $[x, y, z]$ by

$$[\mathbf{x}', \mathbf{y}', \mathbf{z}'] = [\mathbf{x}, \mathbf{y}, \mathbf{z}] \begin{bmatrix} 1 & 0 & 0 \\ 0 & \cos\theta & -\sin\theta \\ 0 & \sin\theta & \cos\theta \end{bmatrix}$$

A rotation about the y-axis positively through ω, as shown in Fig. 2.8, the new axes $[x', y', z']$ are related to the old axes $[x, y, z]$ by

$$[\mathbf{x}', \mathbf{y}', \mathbf{z}'] = [\mathbf{x}, \mathbf{y}, \mathbf{z}] \begin{bmatrix} \cos\omega & 0 & \sin\omega \\ 0 & 1 & 0 \\ -\sin\omega & 0 & \cos\omega \end{bmatrix}$$

A rotation about the z-axis positively through i, as shown in Fig. 2.9, the new axes $[x', y', z']$ are related to the old axes $[x, y, z]$ by

$$[\mathbf{x}', \mathbf{y}', \mathbf{z}'] = [\mathbf{x}, \mathbf{y}, \mathbf{z}] \begin{bmatrix} \cos i & -\sin i & 0 \\ \sin i & \cos i & 0 \\ 0 & 0 & 1 \end{bmatrix}$$

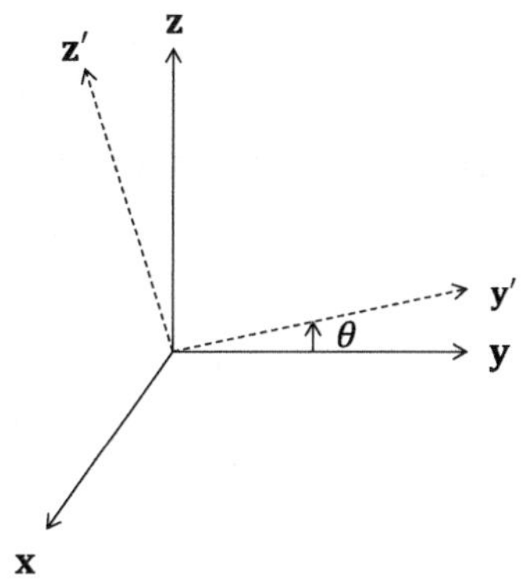

Fig. 2.7 A rotation about the x-axis by an angle θ

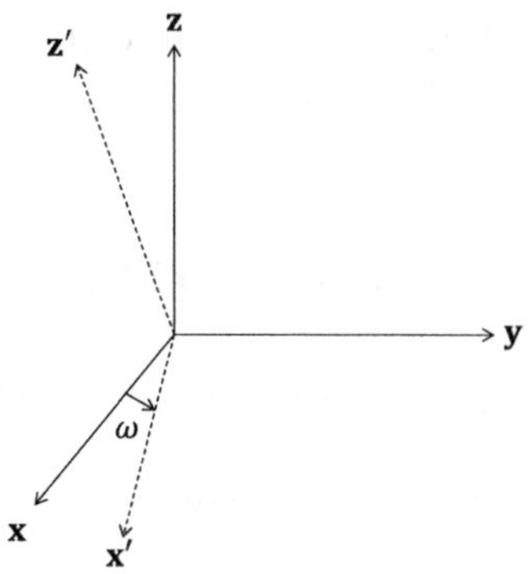

Fig. 2.8 A rotation about the y-axis by an angle ω

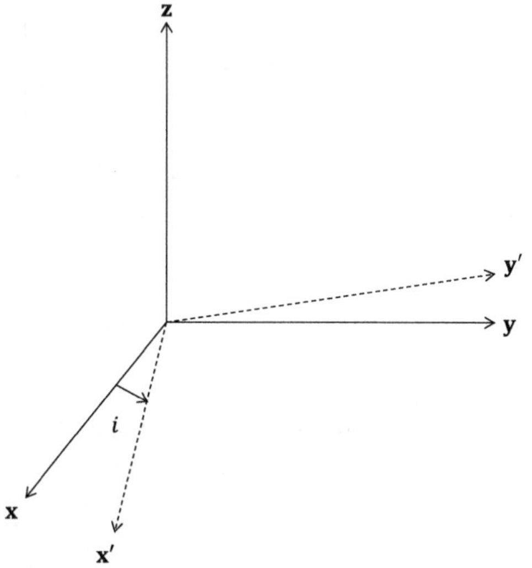

Fig. 2.9 A rotation about the z-axis by an angle i

Combinations of rotations can be made by matrix multiplication. Care must be taken of the signs of the rotations.

2.7 Angular Velocity

A vector can change both direction and length, so we can differentiate a unit vector, $\hat{\mathbf{i}}$. Let $\hat{\mathbf{i}}$ rotate through a small angle $\delta\theta$, as shown in Fig. 2.10; the new vector is $\hat{\mathbf{i}} + \delta\hat{\mathbf{i}}$. This is still a unit vector, so $\hat{\mathbf{i}} \cdot \delta\hat{\mathbf{i}} = 0$.

Let the rotation be in a right-handed sense. Then the direction of $\delta\hat{\mathbf{i}}$ is $\hat{\mathbf{k}} \times \hat{\mathbf{i}}$ and the length is $\delta\theta$, so

$$\delta\hat{\mathbf{i}} = \delta\theta\, \hat{\mathbf{k}} \times \hat{\mathbf{i}} \tag{2.13}$$

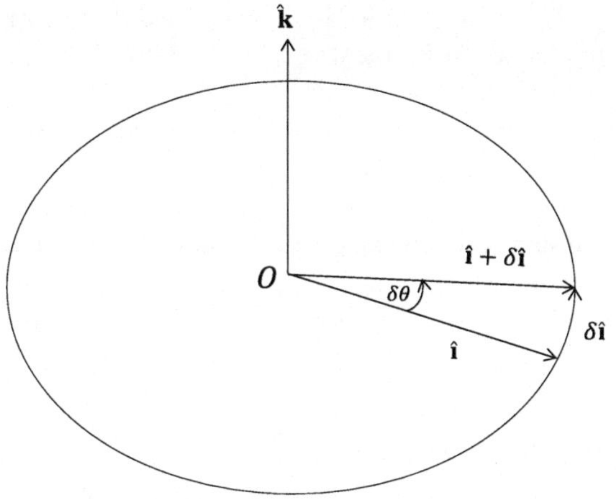

Fig. 2.10 A rotation by a small angle $\delta\theta$

and

$$\frac{d\hat{\mathbf{i}}}{dt} = \left(\hat{\mathbf{k}}\frac{d\theta}{dt}\right) \times \hat{\mathbf{i}} = \boldsymbol{\omega} \times \hat{\mathbf{i}} \tag{2.14}$$

where $\boldsymbol{\omega}$ is an *angular velocity*. Note, the derivative of a unit vector is normal to that vector (Danby 1962, pp. 29–30).

2.8 Rotating Axes

Write $\mathbf{r} = r\,\hat{\mathbf{r}}$ and differentiate,

$$\dot{\mathbf{r}} = \dot{r}\,\hat{\mathbf{r}} + r\frac{d\hat{\mathbf{r}}}{dt} \tag{2.15}$$

The components on the right side are (i) the change in the length of the radius along the radius; and (ii) the change perpendicular to $\mathbf{r}$ due to rotation.

Rotating axes are a common phenomena. Suppose a vector $\mathbf{r}$ has a rate of change $d\mathbf{r}/dt$ with respect to a fixed frame F_1 and $\delta\mathbf{r}/\delta t$ with respect to frame F_2, rotating with respect to F_1 with an angular velocity $\boldsymbol{\omega}$, which is not necessarily constant. Equation (2.15) applies where the axes of F_2 are rotating with $\mathbf{r}$, so $\delta\mathbf{r}/\delta t$ is written instead of $\dot{r}\mathbf{r}$. We will use the notation $\delta(\cdot)/\delta t$ for differentiation with respect to a rotating frame and $d(\cdot)/dt$ for differentiation with respect to a fixed frame. Then, Eq. (2.15) can be written as

$$\frac{d\mathbf{r}}{dt} = \frac{\delta\mathbf{r}}{\delta t} + \boldsymbol{\omega} \times \mathbf{r} \tag{2.16}$$

This is a general result, which we demonstrate. Let $\{\hat{\mathbf{i}}, \hat{\mathbf{j}}, \hat{\mathbf{k}}\}$ be an orthogonal triad, fixed to F_2, then

$$\mathbf{r} = \sum_{\hat{\mathbf{i}},\hat{\mathbf{j}},\hat{\mathbf{k}}} (\mathbf{r} \cdot \hat{\mathbf{i}})\hat{\mathbf{i}} \tag{2.17}$$

the summation being over $\hat{\mathbf{i}}, \hat{\mathbf{j}}$, and $\hat{\mathbf{k}}$. Then as in Eq. (2.15),

$$\frac{d\mathbf{r}}{dt} = \sum \left(\frac{d(\mathbf{r} \cdot \hat{\mathbf{i}})}{dt} \hat{\mathbf{i}} \right) + \sum \left((\mathbf{r} \cdot \hat{\mathbf{i}}) \frac{d\hat{\mathbf{i}}}{dt} \right) \tag{2.18}$$

But $d\hat{\mathbf{i}}/dt = \boldsymbol{\omega} \times \hat{\mathbf{i}}$, so

$$\sum \left((\mathbf{r} \cdot \hat{\mathbf{i}}) \frac{d\mathbf{i}}{dt} \right) = \boldsymbol{\omega} \times \sum (\mathbf{r} \cdot \hat{\mathbf{i}})\hat{\mathbf{i}} = \boldsymbol{\omega} \times \mathbf{r} \tag{2.19}$$

For $\delta\mathbf{r}/\delta t$, take $\hat{\mathbf{i}}, \hat{\mathbf{j}}, \hat{\mathbf{k}}$ as non-rotating constant vectors, so

$$\frac{\delta\mathbf{r}}{\delta t} = \sum \left(\frac{d(\mathbf{r} \cdot \hat{\mathbf{i}})}{dt} \hat{\mathbf{i}} \right) \tag{2.20}$$

Written in notation of operators,

$$\frac{d}{dt} = \frac{\delta}{\delta t} + \boldsymbol{\omega} \times \tag{2.21}$$

To illustrate, let us find components of velocity and acceleration in polar coordinates. $OP = \mathbf{r}$ makes the angle θ with a fixed direction, as shown in Fig. 2.11. Let $\hat{\mathbf{i}}$ point in the radial direction, $\hat{\mathbf{j}}$ in the transverse direction (increasing θ), and

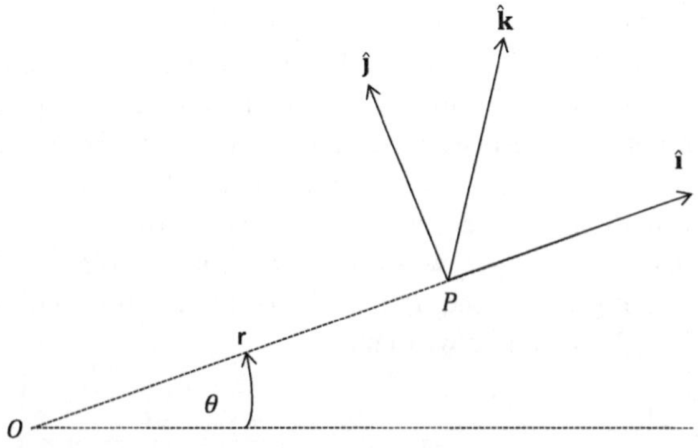

Fig. 2.11 Finding the components of velocity in polar coordinates

$\hat{\mathbf{k}} = \hat{\mathbf{i}} \times \hat{\mathbf{j}}$. Then the angular velocity of $\mathbf{r}$ is $\dot{\theta}\,\hat{\mathbf{j}}$, and

$$\frac{d\mathbf{r}}{dt} = \dot{r}\,\hat{\mathbf{i}} + r\dot{\theta}\,\hat{\mathbf{j}} \tag{2.22}$$

The first term on the right is the radial velocity, and the second term is the transverse component of the velocity. It is assumed that there is no velocity in the $\hat{k}$ direction. Similarly,

$$\frac{d^2\mathbf{r}}{dt^2} = \frac{d}{dt}(\dot{r}\hat{\mathbf{i}} + r\dot{\theta}\hat{\mathbf{j}}) = \left(\frac{\delta}{\delta t} + \dot{\theta}\hat{\mathbf{k}}\times\right)(\dot{r}\hat{\mathbf{i}} + r\dot{\theta}\hat{\mathbf{j}})$$
$$= \ddot{r}\hat{\mathbf{i}} + (\dot{r}\dot{\theta} + r\ddot{\theta})\hat{\mathbf{j}} + \dot{\theta}\hat{\mathbf{k}} \times (\dot{r}\hat{\mathbf{i}} + r\dot{\theta}\hat{\mathbf{j}}) \tag{2.23}$$

which reduces to

$$\frac{d^2\mathbf{r}}{dt^2} = (\ddot{r} - r\dot{\theta}^2)\,\hat{\mathbf{i}} + (r\ddot{\theta} + 2\dot{r}\dot{\theta})\,\hat{\mathbf{j}} \tag{2.24}$$

The first term on the right is the radial acceleration, and the second term is the traverse component of the acceleration, which can be written as

$$\frac{1}{r}\frac{d}{dt}(r^2\dot{\theta}) \tag{2.25}$$

where $r^2\dot{\theta}$ is twice the areal velocity of P.

Consider a particle traveling in a circle with a constant speed. Then $\ddot{r} = 0$. The traverse component of acceleration is zero, so

$$\frac{d^2\mathbf{r}}{dt^2} = -r\dot{\theta}^2\hat{\mathbf{i}} \tag{2.26}$$

Acceleration is the result of kinematics of a particle. If an observer is moving with a particle, and wishes to work with respect to rotating axes, the observer will be unable to account for the observations, unless "centrifugal acceleration" of $-r\theta^2\hat{\mathbf{i}}$ is applied to every particle. Alternatively, a fictitious force, the "centrifugal force", can be introduced, which would produce this acceleration.

Consider a conical pendulum with a mass m at P, suspended from O by a string with length r, moving with a constant angular velocity, $\dot{\theta}$, about the vertical, as shown in Fig. 2.12. OP makes constant angle, ϕ, from the vertical. The radius of the circle is $r \sin\phi$, and the centrifugal force magnitude is $C = mr \sin\phi\dot{\theta}^2$.

Resolving forces at right angles to the string to avoid the force $\mathbf{T}$,

$$mg \sin\phi = C \cos\phi = mr \sin\phi \cos\phi\dot{\theta}^2 \tag{2.27}$$

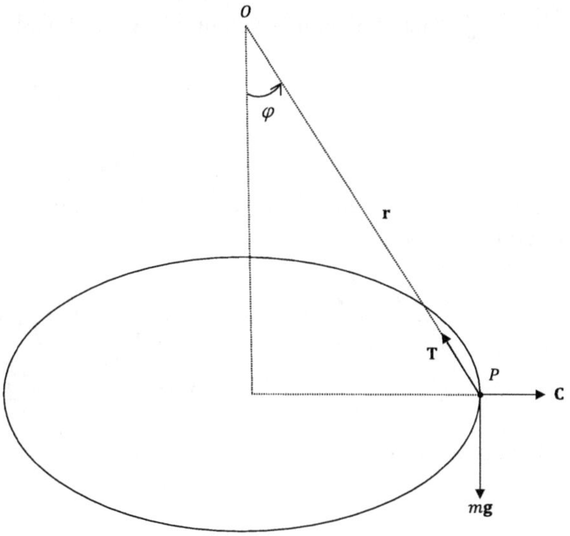

Fig. 2.12 Illustration of a conical pendulum

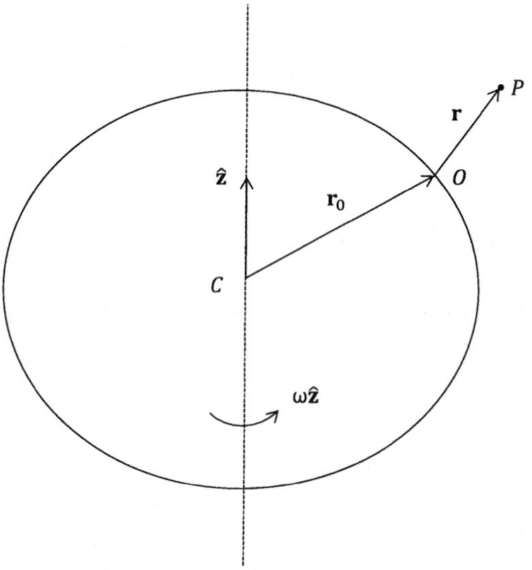

Fig. 2.13 Finding the acceleration of a point above the surface of the Earth

So,

$$\dot{\theta}^2 = \frac{g}{r\cos\phi} \tag{2.28}$$

Also, consider motion observed from a fixed point on the Earth. C is the center of the Earth. Let a person at O observe the motion of a point P, as shown in Fig. 2.13. We have $CO = \mathbf{r}_0$, $OP = \mathbf{r}$, r is small compared to r_0, and the angular velocity of the Earth is $\omega\hat{\mathbf{z}}$. The velocity of P with respect to the non-rotating axes is

$$\frac{d}{dt}(\mathbf{r_0}+\mathbf{r}) = \frac{\delta\mathbf{r}}{\delta t} + \omega\hat{\mathbf{z}}\times\mathbf{r_0} + \omega\hat{\mathbf{z}}\times\mathbf{r} \tag{2.29}$$

where $d/dt = \delta/\delta t + \boldsymbol{\omega}\times$ and $\delta\mathbf{r_0}/\delta t = 0$. The acceleration of P is

$$\frac{d^2(\mathbf{r}_0+\mathbf{r})}{dt^2} = \left(\frac{\delta}{\delta t}+\omega\mathbf{z}\times\right)\frac{d(\mathbf{r}_0+\mathbf{r})}{dt} = \frac{\delta^2\mathbf{r}}{\delta t^2} + 2\omega\hat{\mathbf{z}}\times\frac{\delta\mathbf{r}}{\delta t} + \omega^2\mathbf{z}\times(\hat{\mathbf{z}}\times\mathbf{r_0})$$
$$+\,\omega^2\hat{\mathbf{z}}\times(\hat{\mathbf{z}}\times\mathbf{r}) \tag{2.30}$$

The equation of motion of P contains $d^2(\mathbf{r}_0+\mathbf{r})/dt^2$, and terms due to the forces acting on P. If the equation is in terms of $\delta^2(\mathbf{r}_0+\mathbf{r})/\delta t^2$, or equivalently $\delta^2\mathbf{r}/\delta t^2$, which is in terms of what the person at O observes, $d^2\mathbf{r}/dt^2$ can be replaced with $\delta^2\mathbf{r}/\delta t^2$, if the following terms are added

$$-\,2\omega\hat{\mathbf{z}}\times\frac{\delta\mathbf{r}}{\delta t} - \omega^2\hat{\mathbf{z}}\times(\hat{\mathbf{z}}\times\mathbf{r_0}) - \omega^2\hat{\mathbf{z}}\times(\hat{\mathbf{z}}\times\mathbf{r}) \tag{2.31}$$

to the other side, i.e. with the terms dealing with forces acting on P. Since ω is small, the last two terms are normally neglected. These forces are Coriolis forces (we will see them again later).

The Coriolis force is named for a nineteenth century French mathematician. It is the inertial force caused by the Earth's rotation that deflects a moving body. This deflection is produced by the acceleration of any body moving at a constant speed above the Earth with respect to the surface of the rotating Earth (Danby 1962, p. 31–35).

2.9 Gradient of a Scalar

Take a scalar function of a position, written as $f(x,y,z)$ or $f(\mathbf{r})$. Assume that it is defined and continuous in space. To move from $f(x,y,z)$ to $f(x+\delta x,y,z)$, the change in f can be written as $(\partial f/\partial x)\,\delta x$, where $\partial/\partial x$ is conventional partial differentiation with respect to x, so y and z are assumed to be constant. This could be written

$$\frac{\partial f}{\partial x}\delta x = \left(\frac{\partial f}{\partial x}\hat{\mathbf{i}} + \frac{\partial f}{\partial y}\hat{\mathbf{j}} + \frac{\partial f}{\partial z}\hat{\mathbf{k}}\right)\cdot\hat{\mathbf{i}}\,\delta x = \nabla f\cdot\hat{\mathbf{i}}\,\delta x \tag{2.32}$$

Likewise, a change from (x,y,z) to $(x,y+\delta y,z)$ is $\nabla f\cdot\hat{\mathbf{j}}\,\delta y$, and from $\mathbf{r}$ to $\mathbf{r}+\delta\mathbf{r}$ is $\nabla f\cdot\delta\mathbf{r}$.

The vector components $(\partial f/\partial x, \partial f/\partial y, \partial f/\partial z)$ are called the *gradient* of f, and are denoted by grad$f$, or ∇f. The operator ∇ is called the *Nabla operator* or *Del operator*. Obviously, ∇f is a vector.

Now, consider a curve C and the values of f along the curve. The rate of change of f along C with respect to the arc length is

$$\frac{df}{ds} = \frac{\partial f}{\partial x}\frac{dx}{ds} + \frac{\partial f}{\partial y}\frac{dy}{ds} + \frac{\partial f}{\partial z}\frac{dz}{ds} = \hat{\boldsymbol{\tau}} \cdot \nabla f \tag{2.33}$$

Consider a surface with $f(x, y, z) = constant$. Along a line on the surface $df/ds = 0$, so $\hat{\boldsymbol{\tau}} \cdot \nabla f = 0$ and ∇f is perpendicular to the surface. ∇f is a *field vector*, a vector that is a function of position (Danby 1962, pp. 35–36).

2.10 Momentum and Energy

Starting with some definitions, mass, m, is the measure of material in a body independent of the object's location on the Earth or Moon. Weight is mg, where g is a localized constant depending on the location, on Earth or elsewhere. A point mass is when all the mass of an object is treated as being localized at the center of the object. This is valid for planetary theory; it is not valid for lunar theory, Earth satellites, or close natural satellites. Linear momentum is $m\mathbf{v}$, the mass times the velocity of a body; it is a vector. The vector $\mathbf{r} \times m\mathbf{v}$ about O is *angular momentum* (see Fig. 2.14). This is two times the areal velocity about O.

The *kinetic energy* is $\frac{1}{2}mv^2$; this is a scalar. The *work* is $\mathbf{P} \cdot \mathbf{r}$ where the force $\mathbf{P}$ on a particle moves it through $\mathbf{r}$. If $\mathbf{P}$ is not constant, we integrate over small displacements $d\mathbf{r}$,

$$W_{AB} = \int_A^B \mathbf{P} \cdot d\mathbf{r} \tag{2.34}$$

The rate of doing work, $\mathbf{P} \cdot d\mathbf{r}/dt = \mathbf{P} \cdot \mathbf{v}$, is termed *power*; this is a scalar.

Using these definitions, the angular momentum of a mass m moving in a curved path C, as shown in Fig. 2.15, is $\mathbf{L} = \mathbf{r} \times m\mathbf{v}$, where $\mathbf{r}$ and $\mathbf{v}$ are position and velocity vectors at time t. Differentiation with respect to time with m constant,

$$\dot{\mathbf{L}} = \dot{\mathbf{r}} \times m\mathbf{v} + \mathbf{r} \times m\dot{\mathbf{v}} = \mathbf{r} \times \mathbf{F} = \mathbf{N} \tag{2.35}$$

where $\dot{\mathbf{r}} = \mathbf{v}$ so the first term is zero.

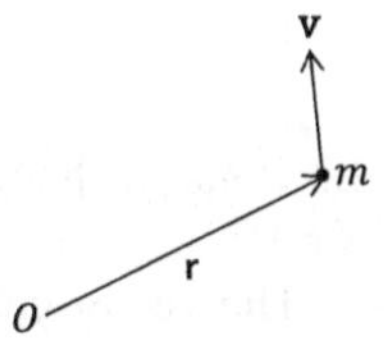

Fig. 2.14 Mass, position and velocity

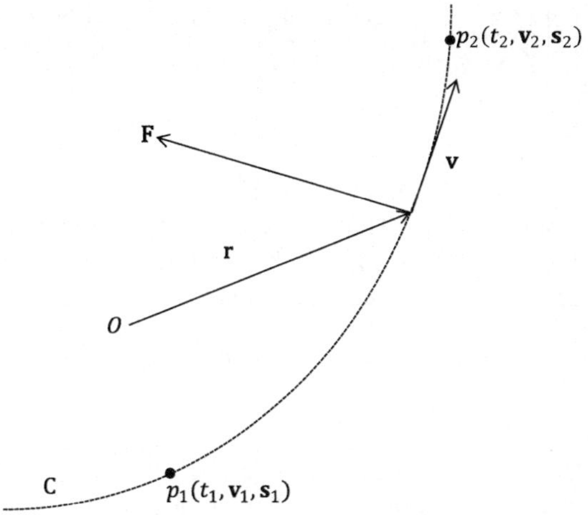

Fig. 2.15 Velocity and position of a mass m along a path C

By Newton's second law of motion, $\mathbf{F} = d(m\mathbf{v})/dt$. The particle m is caused to move by the force $\mathbf{F}$, and $\mathbf{N}$ is the *torque* or *moment* of $\mathbf{F}$ about O. It is a vector perpendicular to the plane defined by $\mathbf{r}$ and $\mathbf{F}$. Equation (2.35) expresses Newton's second law for rotational motion, namely the time rate of change of angular momentum equals the torque.

Note that, in a plane, if $\mathbf{N}$ is zero, the forces on the particle pass through the origin, and angular momentum is constant. Conversely, if angular momentum is constant, the resultant forces pass through the origin. So forces acting on planets causing curved paths, pass through the Sun.

In general, $\mathbf{F}$ may be a function of position, velocity, or time, or any combination thereof. Normally, however, $\mathbf{F}$ will be independent of velocity. Suppose that two masses m_1 and m_2 interact with each other. Let the force on m_1 due to m_2 be $\mathbf{F}_{12}$ and let the force on m_2 due to m_1 be $\mathbf{F}_{21}$. Newton's third law is expressed mathematically as $\mathbf{F}_{12} = -\mathbf{F}_{21}$.

Consider a constant mass m moving under the action of force $\mathbf{F}$ along C. Then $\mathbf{F} = m\dot{\mathbf{v}}$. Take the scalar product by $\mathbf{v}$, then

$$m\mathbf{v} \cdot \dot{\mathbf{v}} = \mathbf{v} \cdot \mathbf{F} \tag{2.36}$$

Since $d(\mathbf{v} \cdot \mathbf{v})/dt = 2\mathbf{v} \cdot \dot{\mathbf{v}}$ and $\mathbf{v} \cdot \mathbf{v} = v^2$, then

$$\frac{d}{dt}\left(\frac{1}{2}mv^2\right) = \mathbf{v} \cdot \mathbf{F} \tag{2.37}$$

Note that $\frac{1}{2}mv^2$ is kinetic energy and $\mathbf{v} \cdot \mathbf{F}$ is the rate of working by the force $\mathbf{F}$ on the mass m. Multiply both sides of Eq. (2.37) by dt and integrate,

$$\frac{1}{2}m(v_2^2 - v_1^2) = \int_{t_1}^{t_2} \mathbf{v} \cdot \mathbf{F}dt \tag{2.38}$$

$\mathbf{v}dt$ is an element of arc $d\mathbf{s}$, so the integral becomes

$$\int_{s_1}^{s_2} \mathbf{F} \cdot d\mathbf{s} \tag{2.39}$$

This is a line integral along C and represents work done by $\mathbf{F}$ in motion of m from P_1 to P_2. Thus, the change in kinetic energy in the motion of m from P_1 to P_2 equals the work done by the force in that interval. If $\int_{s_1}^{s_2} \mathbf{F} \cdot d\mathbf{s}$ is independent of path, then there is a scalar function of position, V, defined by

$$\int_{s_1}^{s_2} \mathbf{F} \cdot d\mathbf{s} = V(s_1) - V(s_2) \tag{2.40}$$

This is called the *potential energy* function. $V(s)$ denotes the potential energy at s. Then substituting Eq. (2.40) into Eq. (2.38),

$$\frac{1}{2}mv_2^2 + V(s_2) = \frac{1}{2}mv_1^2 + V(s_1) \tag{2.41}$$

This is *conservation of energy* for moving mass, or the *energy integral*. The sum of kinetic and potential energies of the system remains constant.

In Cartesian coordinates, let $\mathbf{F} = X\hat{\mathbf{i}} + Y\hat{\mathbf{j}} + Z\hat{\mathbf{k}}$ and $d\mathbf{s} = dx\hat{\mathbf{i}} + dy\hat{\mathbf{j}} + dz\hat{\mathbf{k}}$. X, Y, Z are functions of x, y, z. Then Eq. (2.40) becomes

$$\int_{(x_1,y_1,z_1)}^{(x_2,y_2,z_2)} (X\,dx + Y\,dy + Z\,dz) = V(x_1, y_1, z_1) - V(x_2, y_2, z_2) \tag{2.42}$$

This implies that $X = -\frac{\partial V}{\partial x}$, $Y = -\frac{\partial V}{\partial y}$, $Z = \frac{\partial V}{\partial z}$. The gradient of $V(x, y, z)$ is defined by

$$\nabla V = \frac{\partial V}{\partial x}\hat{\mathbf{i}} + \frac{\partial V}{\partial y}\hat{\mathbf{j}} + \frac{\partial V}{\partial z}\hat{\mathbf{k}} \tag{2.43}$$

Hence,

$$\nabla V = -X\hat{\mathbf{i}} - Y\hat{\mathbf{j}} - Z\hat{\mathbf{k}} = -\mathbf{F} \tag{2.44}$$

When this relationship between a force $\mathbf{F}$ at all points of a space and the corresponding $V(x, y, z)$, called the *potential*, exists, we say that the force field is a *conservative force field*.

A necessary and sufficient condition for the existence of V is

$$\frac{\partial Z}{\partial y} - \frac{\partial Y}{\partial z} = 0, \quad \frac{\partial Z}{\partial x} - \frac{\partial X}{\partial z} = 0, \quad \frac{\partial Y}{\partial x} - \frac{\partial X}{\partial y} = 0 \tag{2.45}$$

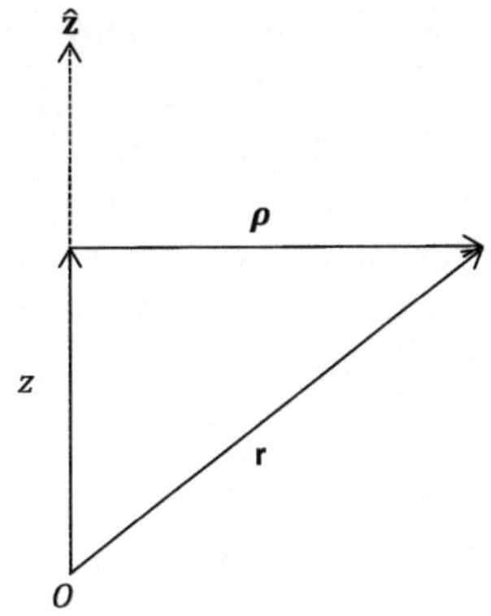

Fig. 2.16 Rotating motion with respect to the z axis

For illustration, consider motion with respect to axes rotating with a constant angular velocity ω about the z axis, as shown in Fig 2.16. If a mass m is acted on by force $\mathbf{F}$, its motion with respect to the rotating axes will be

$$\mathbf{F} = m\left[\ddot{r} + 2\,\omega\,\hat{\mathbf{z}} \times \dot{r} + \omega^2\,\hat{\mathbf{z}} \times (\hat{\mathbf{z}} \times \mathbf{r})\right] \tag{2.46}$$

where

$$\frac{d\mathbf{r}}{dt} = \frac{\partial \mathbf{r}}{\partial t} + \boldsymbol{\omega} \times \mathbf{r} \tag{2.47}$$

is applied twice. Let $\mathbf{r}$ have components z along $\hat{\mathbf{z}}$ and $\boldsymbol{\rho}$ be at right angles to $\hat{\mathbf{z}}$. Then $\mathbf{r} = z\hat{\mathbf{z}} + \boldsymbol{\rho}$ and

$$\mathbf{F} = m(\ddot{r} + 2\,\omega\hat{\mathbf{z}} \times \dot{\mathbf{r}} - \omega^2\boldsymbol{\rho}) \tag{2.48}$$

Multiple scalarly by $\dot{\mathbf{r}}$ and, since the component of $\dot{\mathbf{r}}$ in the z direction dotted on $\boldsymbol{\rho}$ is 0, and $\boldsymbol{\rho} \cdot \dot{\mathbf{r}} = \boldsymbol{\rho} \cdot \dot{\boldsymbol{\rho}}$,

$$\mathbf{F} \cdot \dot{\mathbf{r}} = m(\ddot{r} \cdot \dot{\mathbf{r}} - \omega^2\boldsymbol{\rho} \cdot \dot{\boldsymbol{\rho}}) \tag{2.49}$$

If $\mathbf{F}$ arises from a potential V, described with respect to a rotating axis, then the modified energy integral from integrating the above equation is, with $m = 1$,

$$V - \frac{1}{2}\omega^2\rho^2 + \frac{1}{2}\dot{\mathbf{r}}^2 = constant \tag{2.50}$$

which is the same as the usual energy integral for motion of a particle in a field of potential $V - \frac{1}{2}\omega^2\rho^2$. This is the modified potential; $\frac{1}{2}\omega^2\rho^2$ is the rotational potential.

Now we consider applications of the equations of motion to some simple problems (Danby 1962, pp. 41–42), (McCuskey 1963, pp. 21–22).

2.10.1 Simple Harmonic Motion

Assume that a particle is moving in a straight line with a force varying with distance from a point. The point is the origin and x is the distance from it. The force per unit mass is k^2x and $m = 1$. The equation of motion is $\ddot{x} = -k^2x$. The field is conservative with the potential $\frac{1}{2}k^2x^2$, the energy integral is

$$\frac{1}{2}\dot{x}^2 + \frac{1}{2}k^2\,x^2 = constant \tag{2.51}$$

The constant must be positive, so write it as $\frac{1}{2}k^2a^2$, then

$$\frac{dx}{dt} = k(a^2 - x^2)^{\frac{1}{2}} \tag{2.52}$$

A solution is $x = a\cos(kt + b)$, where a and b are arbitrary constants. Note the motion is symmetrical about the origin determined by the speed at origin. There are finite oscillations whatever the initial speed is (Danby 1962, pp. 47–48).

2.10.2 Linear Motion in an Inverse Square Field

The approximation of a uniform field is only valid near the Earth's surface. A force of $-\frac{k^2}{r^2}\hat{\mathbf{z}}$ per unit mass, where r is distance from the Earth's center, and k is a gravitational constant, must be used further away. We will only consider motion in the z direction, so use of vectors is unnecessary.

The field is conservative and the potential $-k^2/r$. Thus, the energy integral is

$$\dot{z}^2 = \frac{2k^2}{z} - \frac{2k^2}{z_0} + v_0^2 \tag{2.53}$$

where z_0 and v_0 are the position and velocity at $t = 0$. If we impose the condition that the right side must not be negative, there are three possible cases:

1. If $v_0^2 < 2k^2/z_0$, that is a negative constant in the equation which $2k^2/z$ will equal for some z_1, so $\dot{z}$ will become zero and the particle will start to descend.
2. If $v_0^2 > 2k^2/z_0$, $\dot{z}$ will never be zero, but will approach a finite number $\sqrt{v_0^2 - 2k^2/z_0}$ as z approaches infinity, i.e. the particle just continues to move away.
3. If $v_0^2 = 2k^2/z_0$, then $\dot{z} \to 0$ as $z \to \infty$. This critical value is known as the *escape velocity*. Each planet has such a velocity; it is the necessary speed for a rocket to escape completely from the Earth. To calculate this k^2 must be known.

k^2 is the product of mass of the body and the constant of gravitation. If we know the gravity acceleration, g, at the Earth's surface, $g = k^2/r^2$ where r is the Earth's radius,

$$v_0^2 = \frac{2k^2}{r_0} = 2gr_0 \tag{2.54}$$

(Danby 1962, pp. 49–50).

2.10.3 *Foucoult's Pendulum*

In 1851, Foucoult (1819–1868) devised an experiment, where a pendulum was suspended from the dome of the Pantheon in Paris. The plane in which it swung was slowly rotating due to the Earth's rotation under the pendulum. This phenomenon is termed *Foucoult's pendulum*.

Let the suspension of the pendulum be at O, and the bob be at P, as shown in Fig. 2.17. Let $OP = \mathbf{r}$, where r is constant. Let $\hat{\mathbf{k}}$ be along CO. We note that since $r \lll r_0$, which is the Earth's radius, whether the gravity mg acts along $\hat{\mathbf{k}}$ or PC makes no difference.

Thus, the forces on the bob of mass m are tension in the string T along PO and gravity mg along $-\hat{\mathbf{k}}$, i.e. $-T\hat{\mathbf{r}}$ and $-mg\hat{\mathbf{k}}$. The equation of motion with respect to the non-rotating axes fixed in space is

$$m\frac{d^2\mathbf{r}}{dt^2} = -T\hat{\mathbf{r}} - mg\hat{\mathbf{k}} \tag{2.55}$$

Fig. 2.17 Foucoult's pendulum

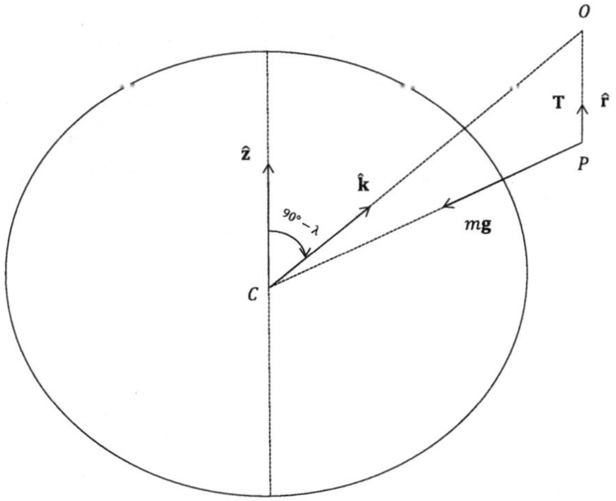

But the equation of motion with respect to the axes rotating about O requires adding in the Coriolis forces. In this case ω^2 terms are so small compared to ω that they can be neglected. Then

$$\frac{d^2\mathbf{r}}{dt^2} + 2\,\omega\hat{\mathbf{z}} \times \frac{d\mathbf{r}}{dt} = -g\hat{\mathbf{k}} - \frac{T}{m}\hat{\mathbf{r}} \tag{2.56}$$

The equation of motion of a simple pendulum is derived by putting $\omega = 0$. Multiply vectorially by $\hat{\mathbf{r}}$ to eliminate T. Then

$$\hat{\mathbf{r}} \times \frac{d^2\mathbf{r}}{dt^2} + 2\,\omega\hat{\mathbf{r}} \times \left(\hat{\mathbf{z}} \times \frac{d\mathbf{r}}{dt}\right) = -g\hat{\mathbf{r}} \times \hat{\mathbf{k}} \tag{2.57}$$

Since $dr/dt = 0$, and $\hat{\mathbf{r}} \cdot \frac{d\mathbf{r}}{dt} = r\frac{dr}{dt} = 0$, the second term on the left reduces to $-2\omega\frac{d\mathbf{r}}{dt}\sin\lambda$, where λ is the latitude of O,

$$\hat{\mathbf{r}} \times \frac{d^2\mathbf{r}}{dt^2} - 2\omega\frac{d\mathbf{r}}{dt}\sin\lambda = -g\hat{\mathbf{r}} \times \hat{\mathbf{k}} \tag{2.58}$$

We expect a pendulum to have a uniform rotation about $\hat{\mathbf{k}}$. Assume axes rotating with an angular velocity, $\omega'\hat{\mathbf{k}}$, and set $\delta/\delta t$ equal the rate of change with respect to these axes. Assume ω' is constant and so small that $\omega'^2 = 0$ and $\omega'\omega = 0$. Then

$$\hat{\mathbf{r}} \times \left(\frac{\delta^2\mathbf{r}}{\delta t^2} + 2\omega'\hat{\mathbf{k}} \times \frac{\delta\mathbf{r}}{\delta t}\right) - 2\omega\frac{\delta\mathbf{r}}{\delta t}\sin\lambda = -g\hat{\mathbf{r}} \times \hat{\mathbf{k}} \tag{2.59}$$

Since $\hat{\mathbf{r}} \cdot \frac{\delta\mathbf{r}}{\delta t} = 0$ and $\hat{\mathbf{r}}$ is perpendicular to $\frac{\delta\mathbf{r}}{\delta t}$, rearranging yields

$$\hat{\mathbf{r}} \times \frac{\delta^2\mathbf{r}}{\delta t^2} - 2(\hat{\mathbf{r}} \cdot \hat{k}\,\omega' + \omega\sin\lambda)\frac{\delta\mathbf{r}}{\delta t} = -g\hat{\mathbf{r}} \times \hat{\mathbf{k}} \tag{2.60}$$

Assume that the pendulum is always close to the vertical, so $\hat{\mathbf{r}} \cdot \hat{\mathbf{k}} \approx 1$. Then, if $\omega' = -\omega\sin\lambda$, the equation is that of a simple pendulum. So the observed motion is a simple pendulum rotating with an angular velocity $-\hat{\mathbf{k}}\omega\sin\lambda$ (Danby 1962, pp. 50–52).

References

Danby, J.: Fundamentals of Celestial Mechanics. The Macmillan Company, New York (1962)
McCuskey, S.W.: Introduction to Celestial Mechanics. Addison-Wesley, Reading (1963)

Chapter 3
Reference Systems and Relativity

3.1 Reference Systems

A reference system is a theoretical concept of coordinates, and includes the time and the standards necessary to specify the bases for giving positions and motions in the system. There are celestial and terrestrial reference systems. Previously, the celestial reference system was based on reference planes aligned to the orbital plane of the Earth—the ecliptic—and the Earth's equatorial plane, both of which are in motion. An intersection of these planes, the *vernal equinox*, was the origin of the coordinates, longitude in the ecliptic plane and right ascension in the equatorial plane. Latitude is measured perpendicular to the ecliptic plane and declination is perpendicular to the equatorial plane (see Fig. 3.1). The celestial reference system used to be based on the Newtonian dynamics of the solar system, star catalogs based on observations of nearby stars with proper motions, and the dynamical mean equator and equinox of the Julian date J2000.0. The equinox moved with time due to precession and nutation. Star catalogs used the equinox as the origin of right ascensions, but that catalog equinox did not necessarily agree with the dynamical equinox and could differ with declination. The kinematics of the Earth result in motions of the equator and equinox, variations in the pole of rotation, polar motion, and variations in the rotation rate of the Earth, given by Universal Time. So all these measurements are epoch dependent.

Now, the *International Celestial Reference System* (ICRS) is based on the theory of relativity, observations of distant extragalactic radio sources, and a fixed origin. The ICRS is basically fixed in space, determined from very distant sources, which do not have apparent motion, and, thus, it is not epoch dependent. There are a *Barycentric Celestial Reference System* (BCRS), centered at the barycenter of the solar system, and a *Geocentric Celestial Reference System* (GCRS), centered at the geocenter, both defined by the metric tensor of the *International Astronomical Union* (IAU) 2000 resolutions, and both are global space-fixed reference systems

P. Gurfil, P.K. Seidelmann, *Celestial Mechanics and Astrodynamics: Theory and Practice*, Astrophysics and Space Science Library 436,
DOI 10.1007/978-3-662-50370-6_3

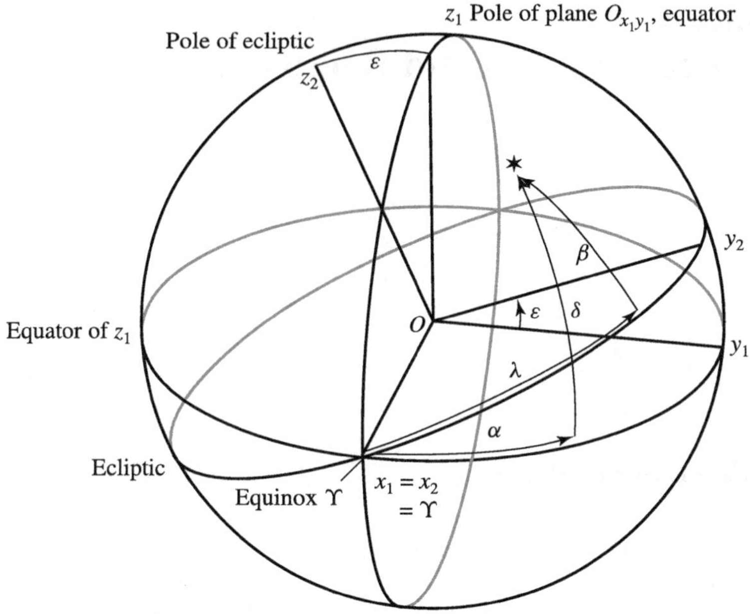

Fig. 3.1 Equatorial and ecliptic reference planes (reproduced from Urban and Seidelmann (2012), with permission)

(IAU-Resolutions 2000). The GCRS is defined such that its spatial coordinates are not kinematically rotating with respect to the BCRS.

3.2 Relativistic Coordinate Systems

Although relativity theory had existed for half a century, the observational effects of both special and general relativity were small, when compared to observational accuracies. Except for the advance of the perihelion of Mercury, deviations from Newtonian physics did not need to be taken into account in the solar system until the advent of highly precise techniques in the 1960s and 1970s. These include numerical integrations of orbits, planetary radar ranging, spacecraft ranging, very long baseline interferometry (VLBI), pulsar timing, and lunar laser ranging (LLR). More recently, optical astrometry has improved with Hipparcos satellite measurements at the milliarcsecond (mas) level.

Hence, a relativistically consistent approach is required for coordinate systems, time scales, and units of measurement. The IAU 2000 resolutions promote a consistency, even though relativistic effects can be ignored, or handled as corrections to Newtonian formulations, for some applications. The establishment of a self-consistent relativistic framework has the benefit, where the effects of relativity are often treated as small corrections to Newtonian developments, as it allows physical assumptions and errors to be more clearly understood (Seidelmann and Seago 2005).

3.2.1 Newtonian Coordinates

The inertial space-time coordinates of Newton's theory become unique once the origin, scales, and orientation of the three spatial axes are prescribed, together with the (constant) origin velocity. Once introduced, such inertial coordinates can cover the universe. There can be two conceptually relevant celestial systems serving different purposes under Newtonian theory: a geocentric system and a barycentric system. Neglecting external (galactic and extragalactic) gravity, the barycentric celestial system is useful for solar system ephemerides and interplanetary spacecraft navigation, and for defining concepts such as the ecliptic. In the fundamental system, the astrometric remote-object positions and other concepts can be defined, such as radial velocity or proper motion. The spatial axes of the geocentric celestial system are considered non-rotating in the Newtonian absolute sense, but the geocenter is accelerated within the solar system. This geocentric system might be called *quasi-inertial*. This system is useful for concepts like the equator, and is most convenient for describing dynamic processes near the Earth, including artificial satellite theory and Earth rotation.

Denoting space and time coordinates of the barycentric celestial system as (X, T), and those of the geocentric celestial system as (x, t), the trivial relationship between these two sets of coordinates is

$$x = X - X_E(t), \ t = T \tag{3.1}$$

where $X_E(t)$ denotes the position of the geocenter in the barycentric system. For this reason, the barycentric and the geocentric celestial systems are not always clearly distinguished in the Newtonian system, and it has become common to view ensembles of calculations as being carried out entirely in either a single reference system, or two reference systems (barycentric and geocentric), having parallel axes that differ only in the origin of coordinates (i.e., they are connected via a *Galilean transformation*).

In Newtonian coordinates, relativistic effects are interpreted as corrections to an otherwise classical result. However, at high levels of accuracy, and with some types of observations, careful consideration of relativity theory will lead to a more correct interpretation for the same calculations.

3.2.2 Relativistic Coordinates

Einstein defined an inertial frame under general relativity as a coordinate system freely falling in accordance with the local gravitational field due to all matter in the Universe. Newtonian-inertial frames are then replaced under general relativity by those that are locally inertial. The geometry of space-time in these freely falling frames is defined by a metric tensor, a 4×4 matrix of mathematical expressions, that serves as an operator on two 4-vectors (Soffel et al. 2003). The metric tensor directly yields the generalized distance between two neighboring events in space-time and effectively determines the equations through which physics is described in the frame. In such locally inertial frames, gravitational forces are expressed in terms of tidal potentials that appear in the metric; by construction, potentials due to masses external to the system are zero at the origin.

Time in general relativity comes in two varieties: *proper time* and *coordinate time*. Proper time is simply the time kept by a clock co-moving with the observer, in whatever trajectory and gravity field the observer exists. Coordinate time is one of the four independent variables used to characterize a locally inertial coordinate frame in general relativity; that is, its value assigns chronological order to sequentially occurring events within the coordinate frame. In general, coordinate time will not be kept by any physically real clock; rather, it is the independent argument of the equations of motion of bodies in its frame (Nelson 2000; Soffel et al. 2003).

Barycentric and geocentric coordinates are related by a 4-dimensional space-time (generalized Lorentz) transformation, complicated by acceleration terms and gravitational potentials in the actual solar system. That these two astronomical reference sub-systems are relativistically quite different has profound consequences for many classical concepts in astronomy. For example, if the solar system is gravitationally isolated, rays of light from some extremely distant source might be traced back to a region of space-time described as the *celestial sphere*, because the distance of a relatively close star will affect the appearance of its light rays.

One might also associate stellar location with a corresponding barycentric direction-cosine vector

$$\frac{\mathbf{X}}{\|\mathbf{X}\|} = \begin{bmatrix} \cos\alpha\cos\delta \\ \sin\alpha\cos\delta \\ \sin\delta \end{bmatrix} \tag{3.2}$$

where $\mathbf{X}$ is the *barycentric coordinate* position as a function of barycentric coordinate time. From this vector, spherical angles (α, δ) can be simply introduced as catalogued values on the celestial sphere. As the coordinate distance $\|\mathbf{X}\|$ of a source tends to infinity, the two constructions for astrometric position become coincident. Classical concepts such as proper motion, or radial velocity can then be employed as coordinate quantities in the barycentric system.

3.2.3 ICRS, BCRS, GCRS

The classical definitions of right ascension and declination (α, δ) use concepts from both systems; the Conventional Intermediate Pole (CIP) with its corresponding equator coming from the geocentric celestial system, and the ecliptic coming from the barycentric celestial system. In astrometry, one distinguishes between the two celestial systems; i.e., true (barycentric) places from the apparent places of stars; annual aberration and parallax are seen as corrections needed to realize the quasi-inertial celestial system. The coordinate system defined by the equator and equinox of J2000.0 may be either barycentric or geocentric.

The conversion between the ICRS and the dynamical mean equator and equinox, in either a barycentric or geocentric frame, is based on a frame bias matrix, **B**, which corresponds to a fixed set of small rotations. In the barycentric case it is

$$\mathbf{r}_{mean(2000)} = \mathbf{B}\,\mathbf{r}_{ICRS} \tag{3.3}$$

where $\mathbf{r}_{ICRS}$ is a vector with respect to the ICRS and $\mathbf{r}_{mean(2000)}$ is a vector with respect to the dynamical mean equator and equinox. The $\mathbf{r}'s$ are column vectors for directions on the sky in the form

$$\mathbf{r} = r \begin{bmatrix} \cos\delta\cos\alpha \\ \cos\delta\sin\alpha \\ \sin\delta \end{bmatrix} \tag{3.4}$$

where α is the right ascension and δ is the declination with respect to the appropriate reference system. In the geocentric case, the vectors are $\mathbf{r}_{GCRS}$ and $\mathbf{r}_{mean(2000)}$, which is a geocentric vector. The matrix **B** is, to first order,

$$\mathbf{B} = \begin{bmatrix} 1 & d\alpha_0 & \xi_0 \\ -d\alpha_0 & 1 & -\eta_0 \\ \xi_0 & \eta_0 & 1 \end{bmatrix} \tag{3.5}$$

where, in milliarcseconds, $d\alpha_0 = -14.6$, $\xi_0 = -16.6170$, and $\eta_0 = -6.8192$ (IERS-Conventions 2010). The angles ξ_0 and η_0 are the ICRS pole offsets and $d\alpha_0$ is the offset in the ICRS right ascension with respect to the dynamical equinox of J2000.0, as measured in an inertial (non-rotating) system. A more accurate second-order matrix is given in *The Explanatory Supplement* (Urban and Seidelmann 2012, p. 118).

The terrestrial reference system used to be based on the Greenwich meridian determined from the Airy transit instrument and the equator determined from astronomical observations of nearby stars. Now measurements by multiple means of the locations of many stations on Earth, and considering the motions of those stations, are the basis of the terrestrial reference system. As a result the terrestrial reference system has changed from astronomical position coordinates to geodetic position coordinates. The difference between these coordinates is the deflection of the vertical, due to local gravitational anomalies. The *International Terrestrial Reference System* (ITRS), to which positions on the Earth are referred, is geocentric with no global residual rotation with respect to horizontal motions at the Earth surface (IERS-Conventions-2003 2004; Altamini and Boucher 2002).

At the basic level, the transformation of a vector from the *Terrestrial Reference System* (TRS), $\mathbf{x}_{TRS}$, to the *Celestial Reference System* (CRS), $\mathbf{x}_{CRS}$, is by the transformation

$$\mathbf{x}_{CRS} = \mathbf{PNTW}\,\mathbf{x}_{TRS} \tag{3.6}$$

where $\mathbf{P}, \mathbf{N}, \mathbf{T}, \mathbf{W}$ are rotational matrices that account for precession, nutation, Earth rotation, and polar motion, respectively. Details are given in Sect. 3.5.

Each of these rotation matrices is the product of three basic rotation matrices,

$$\mathbf{R}_1(\boldsymbol{\theta}) = \begin{bmatrix} 1 & 0 & 0 \\ 0 & \cos\theta & \sin\theta \\ 0 & -\sin\theta & \cos\theta \end{bmatrix} \tag{3.7}$$

$$\mathbf{R}_2(\boldsymbol{\theta}) = \begin{bmatrix} \cos\theta & 0 & \sin\theta \\ 0 & 1 & 0 \\ -\sin\theta & 0 & \cos\theta \end{bmatrix} \tag{3.8}$$

$$\mathbf{R}_3(\boldsymbol{\theta}) = \begin{bmatrix} \cos\theta & \sin\theta & 0 \\ -\sin\theta & \cos\theta & 0 \\ 0 & 0 & 1 \end{bmatrix} \tag{3.9}$$

where θ is a rotation angle and the subscripts refer to positive rotations about the x, y, z axes, respectively, of a right-handed Cartesian coordinate system.

3.2.4 Geodesic Precession and Nutation

A dynamically non-rotating, freely falling, locally inertial, geocentric reference system would slowly precess with respect to the BCRS, the largest component being called *geodesic precession*. Geodesic precession amounts to 19.2 mas/yr and *geodesic nutation* is dominated by an annual term with amplitude 0.15 mas. By imposing the constraint of kinematical non-rotation to the GCRS, these Coriolis-type perturbations must be added (via the tidal potential in the metric) to the equations of motion of bodies referred to the GCRS. The motion of the celestial pole is defined within the GCRS, and geodesic precession, therefore, appears in the precession-nutation theory rather than in the transformation between the GCRS and BCRS, per IAU Resolution B1.6 (IAU-Resolutions 2000). Other barycentric-geocentric transformation terms that affect the equations of motion of bodies in the constrained GCRS are described by Soffel et al. (2003), and Kopeikin and Vlasov (2004).

3.3 Reference Frames

A reference frame is the practical realization of a reference system and is based on observational data.

3.3.1 *Celestial Reference Frames*

The *International Celestial Reference Frame* (ICRF) is the realization of the barycentric, fixed, celestial reference system (ICRS) based on observations of extragalactic radio sources. The sources are selected for stable source structure and a lack of apparent motion, so the ICRF is fixed and stable. The different versions of the ICRF, which are based on improved observations and more sources, are indicated by dates (IERS 2009). There are other realizations of the ICRS at different wavelengths and magnitude ranges, such as the *Hipparcos Celestial Reference Frame* (HCRF) (ESA 1997) and the 2 MASS catalog (Cutri et al. 2003).

The *Geocentric Celestial Reference Frame* (GCRF) is the result of a transformation of the *Barycentric Celestial Reference Frame* (BCRF) to the geocenter and, thus, a realization of the GCRS by extragalactic objects. It is a geocentric ICRF.

3.3.2 *CIP and CIO*

The *Celestial Intermediate Pole* (CIP) is the geocentric equatorial pole determined by the IAU precession-nutation model for the transformation from the ICRF to the GCRF. The *Celestial Intermediate Origin* (CIO) is an origin for right ascensions on the instantaneous celestial true equator of date. Its motion has no component along the true instantaneous equator around the z-axis, and at J2000.0 it has the position of the mean equinox of J2000.0. Since the original determination, a slight misalignment (about 20 mas) between the pole of the GCRS and the CIP, and between the CIO and the mean equinox of J2000.0 has been determined, see Fig. 3.2, which depicts the relationship between the CIO, σ, the origin of right ascension of the GCRS, Σ, the true equinox of J2000.0, Υ_0, the mean equinox of date, Υ_M, and the true equinox of date, Υ_T. The true equator of date is identical with the equator of the CIP. The intersection of the equator of the GCRS and the true equator of date occurs at N'. Since the GCRS is nearly aligned with the Earth's mean equator of J2000.0, there is a significant separation between Υ_0 and Σ.

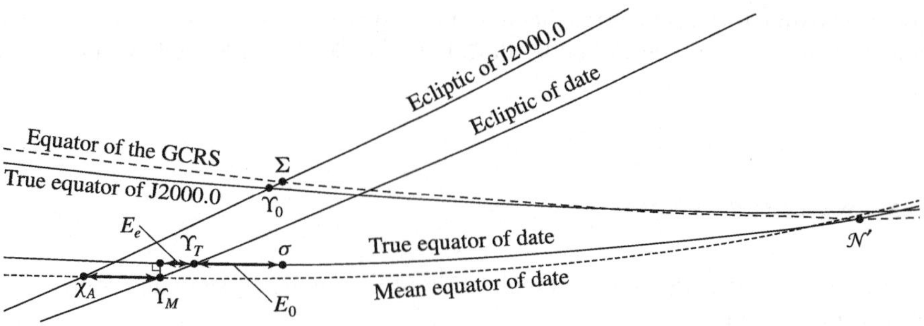

Fig. 3.2 Ecliptic and equatorial planes (reproduced from Urban and Seidelmann (2012), with permission)

The CIO is to be used with a combined bias, precession, nutation matrix. This parametrization is in two angles, X in the direction of the origin of right ascension in GCRS, and Y in the direction 90° to the east of X. The third parameter is the CIO locator, s. s is the difference between the length of the arcs from Σ and N', and σ and N'. The point Σ is the origin of right ascensions of the GCRS, σ is the CIO, and N' is the node of the equator of the CIP on the equator of the GCRS. The value of s is given by

$$s = -\int_{t_0}^{t} \frac{X\dot{Y} - Y\dot{X}}{1+Z} dt - (\sigma N' - \sigma_0 N') \tag{3.10}$$

where σ_0 is the position of σ at the Julian date JD 2451545.0 and $Z = (1 - X^2 - Y^2)^{1/2}$. The precession-nutation bias matrix is

$$\begin{aligned} \mathbf{PNB}_{CIO} &= \mathbf{R}_3(-s)\mathbf{R} \\ &= \mathbf{R}_3(-s) \begin{bmatrix} 1 - aX^2 & -aXY & X \\ -aXY & 1 - aY^2 & Y \\ -X & -Y & 1 - a(X^2 + Y^2) \end{bmatrix} \end{aligned} \tag{3.11}$$

where

$$a \approx \frac{1}{2} + \frac{X^2 + Y^2}{8} \tag{3.12}$$

$$X = \sum_{i=1}^{5} x_i T^i + \sum_j \sum_{k=0}^{3} T^k (b_{sjk} \sin d_k + b_{cjk} \cos d_k) \tag{3.13}$$

$$Y = \sum_{i=1}^{5} y_i T^i + \sum_j \sum_{k=0}^{3} T^k (e_{sjk} \sin f_k + e_{cjk} \cos f_k) \tag{3.14}$$

The variable T is the time in Julian centuries from J2000.0 *Terrestrial Time* (TT),[1] x_i and y_i are the coefficients for the frame bias and precession in X and Y, respectively, $b_{sjk}, b_{cjk}, e_{sjk}$, and e_{cjk} are the coefficients for the nutation and coupling between precession and nutation, and d_k and f_k are the fundamental angular arguments for the nutation and coupling terms.

The parameter s may also be represented by a series, but it can be derived from the series for $s+\frac{1}{2}XY$. The non-periodic parts of Eq. (3.7) and $s+\frac{1}{2}XY$ are in seconds of arc,

$$\begin{aligned} X = & -0.016617 + 2004.191898T - 0.42978297T^2 \\ & - 0.19861834T^3 + 7.578 \times 10^{-6}T^4 + 5.9285 \times 10^{-6}T^5 \end{aligned} \tag{3.15}$$

$$\begin{aligned} Y = & -0.006951 - 0.025896T - 22.4072747T^2 \\ & + 0.00190059T^3 + 0.0011112526T^4 + 0.1358 \times 10^{-6}t^5 \end{aligned} \tag{3.16}$$

$$\begin{aligned} s = & -\frac{1}{2}XY + 9.4 \times 10^{-5} + 0.00380865T - 0.00012268T^2 \\ & - 0.7257411T^3 + 2.798 \times 10^{-5}t^4 + 1.562 \times 10^{-5}T^5 \end{aligned} \tag{3.17}$$

Approximate formulae for the position of the CIP and CIO, accurate to about 0.001", from 1975 to 2025, are given on pp. B46-B47 of *The Astronomical Almanac*.[2] More information is given elsewhere (IERS-Conventions-2003 2004).

The elements of the matrix in Eq. (3.7) are dimensionless, so the values of X, Y, s, and the coefficients for the nutation and coupling between precession and nutation should be in radians. However, they are usually given in arcseconds and must be converted.

The *Celestial Intermediate Reference System* (CIRS) is a geocentric reference system related to the GCRS by a time-dependent rotation taking into account precession-nutation. It is defined by the intermediate equator of the CIP and the CIO on a specific date.

3.3.3 Equation of Equinoxes

The *equation of equinoxes*, E_e, is the difference between the position of the true and mean equinoxes of date, and equivalent to the difference between the apparent and mean sidereal time,

$$E_e(T) = \Upsilon_T - \Upsilon_M = GAST - GMST \tag{3.18}$$

[1] A detailed discussion of time scales will be given in Sect. 3.4.

[2] The Astronomical Almanac Online, http://asa.usno.navy.mil or http://asa.hmnao.com.

where Υ_T is the true equinox of date and Υ_M is the mean equinox of date. E_e can be expressed in series form as

$$E_e(T) = \Delta\psi \cos \varepsilon_A + \sum_k (C'_k \sin A_k + S'_k \cos A_k) = 0.''00000087\, T \sin \Omega \qquad (3.19)$$

where $\Delta\psi$ is the total nutation in longitude, ε_A is the mean obliquity of the ecliptic, Ω is the mean longitude of the ascending node of the Moon, and the coefficients, C'_k and S'_k, and the angular arguments, A_k, are used to describe the complementary terms arising from coupling between precession and nutation (Capitaine and Wallace 2006). The values of the coefficients are given in The Explanatory Supplement (Urban and Seidelmann 2012, p. 207).

3.3.4 Equation of Origins

The IAU has recommended the use of the CIO in place of the equinox. If the CIO is used, the equation of the equinoxes is not needed. Instead, determining the true equinox requires the *equation of the origins*, $E_0(T)$, which determines the difference between the CIO and the equinox of date. This difference is the same as the difference between the *Earth Rotation Angle* (ERA) and the *Greenwich Apparent Sidereal Time* (GAST). Thus,

$$E_0(T) = CIO - \Upsilon_T = \theta - GAST \qquad (3.20)$$

A celestial object can have an equinox right ascension, α_e, with respect to the equinox and an intermediate right ascension, α_i, with respect to the CIO. The equinox right ascension is related to the intermediate right ascension by

$$\alpha_e = \alpha_i - E_0(T) \qquad (3.21)$$

E_0 is the difference between the CIO and the true equinox of date, so it is a function of the position of both. Thus,

$$E_0(T) = s - \tan \frac{\mathbf{NPB}_1^T \cdot \mathbf{R}_1}{\mathbf{NPB}_2^T \cdot \mathbf{R}_1} \qquad (3.22)$$

where s is the CIO locator, $\mathbf{NPB}_1$ and $\mathbf{NPB}_2$ are the row matrices consisting of the first and second rows, respectively, of the matrix product of the frame bias, $\mathbf{B}$, precession $\mathbf{P}$, and nutation $\mathbf{N}$ with respect to the equinox, and $\mathbf{R}_1$ is the row matrix of the first row of the bias precession nutation matrix, Eq. (3.11), for rotation from the intermediate reference frame to the GCRS. See The Explanatory Supplement (Urban and Seidelmann 2012, p. 208).

3.3.5 Terrestrial Reference Frames

The *International Terrestrial Reference Frame* (ITRF) is the realization of the ITRS based on a set of coordinates and velocities of fiducial points on the Earth. The Earth is not really a sphere, but rather an oblate ellipsoid of revolution, or spheroid. So the height above mean sea level is measured from an irregular surface called a *geoid*. These heights are such that water flows downhill under gravity. There are multiple systems of terrestrial coordinates, so precise coordinates need to be specified as to their basis.

The oblateness of the Earth is due to the rotation of the Earth. The axis of rotation coincides on average with the axis of the principal moment of inertia, referred to as the *axis of figure*. The axis of rotation moves in the Earth around the axis of figure in a quasiperiodic motion, which is referred to as *polar motion* (see Sect. 3.6.4). The Earth is not a rigid body, has secular and periodic changes in shape and distribution of mass. The crust of the Earth consists of plates that move slowly over the mantle—continental drifts. So the terrestrial coordinate frame is realized by adoptions of positions and motions of primary reference points on the surface of the Earth. The values of these reference points are chosen so there is no net rotation of the primary points, and so the prime meridian is near Greenwich. In practice, the observations and analyses can lead to differences in the terrestrial reference frames. This frame, fixed in a mean sense with respect to the Earth's surface, has a prime meridian that is slightly offset from the historical Greenwich meridian through the *Airy transit instrument*.

3.3.6 Terrestrial Intermediate Origin

The *Terrestrial Intermediate Origin* (TIO) is the non-rotating origin of the *Terrestrial Intermediate Reference System* (TIRS). The TIO was originally set at the ITRF origin of longitude and throughout 1900–2100 stays within 0.1 milliarcsecond of the ITRF zero meridian. The TIRS is a geocentric reference system defined by the intermediate equator of the CIP and the TIO on a specific date. It is related to the CIRS by the ERA. The TIO locator denoted by s', is an arc used in the location of the TIO. Considering the node of the ITRS equator on the instantaneous equator (orthogonal to the CIO), s' is the difference in longitudes of that node measured from the (1) ITRS origin and (2) the TIO (see Fig. 3.3). As a consequence of polar motion, the TIO moves, and the s' changes by approximately 50 microarcseconds (1.5 mm) per century. The TIO meridian is the moving plane through the geocenter, the CIP, and the TIO. The TIO is determined by successive rotations though two small angles x and y, and the TIO locator, s'. The angles are the polar motion, the angular coordinates of the CIP with respect to the terrestrial pole measured along

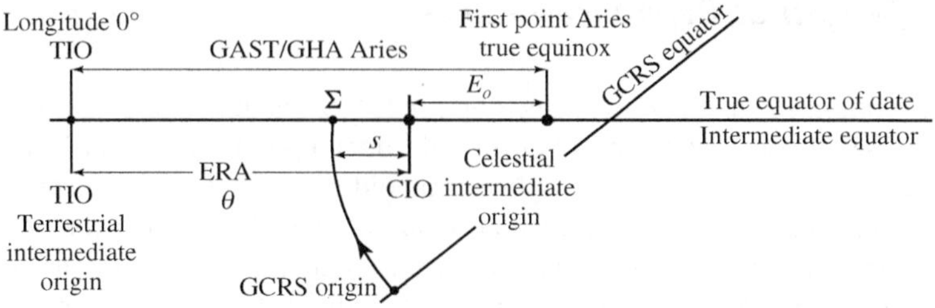

Fig. 3.3 Relationships of origins (reproduced from Urban and Seidelmann (2012), with permission)

the meridians at longitude 0° and 270°(90° W). The TIO locator is determined by

$$s' = -\int_{t_0}^{t} \frac{x\dot{y} - y\dot{x}}{1+z} dt \tag{3.23}$$

where $t_0 = JD\ 2451545.0$ and

$$z = (1 - x^2 - y^2)^{1/2} \tag{3.24}$$

The Earth Rotation Angle (ERA) is the angle, θ, measured along the equator of the CIP between the direction of the CIO and the TIO. It is a linear function of Universal Time 1 (UT1),[3] and its time derivative is the Earth's angular velocity (see Fig. 3.4). In 2001, the IAU redefined UT1 with respect to the Earth Rotation Angle (ERA or θ) beginning on January 1, 2003 as

$$\begin{aligned}\theta(UT1) = {} & 2\pi[0.7790572732640 \\ & + 1.00273781191135448(JD\ UT1 - 2451545.0)]\end{aligned} \tag{3.25}$$

in radians.

3.3.7 ECEF, ECI, ECR

Earth-Centered, Earth-Fixed (ECEF), also known as *Earth Centered Rotating* (ECR), is a Cartesian coordinate system, and is sometimes known as a "conventional terrestrial" system. It represents positions as X, Y, and Z coordinates. The point $(0, 0, 0)$ is defined as the center of mass of the Earth, hence the name Earth-Centered. Its axes are aligned with the *International Reference Pole* (IRP) and

[3] We will elaborate on this time scale in Sect. 3.4.

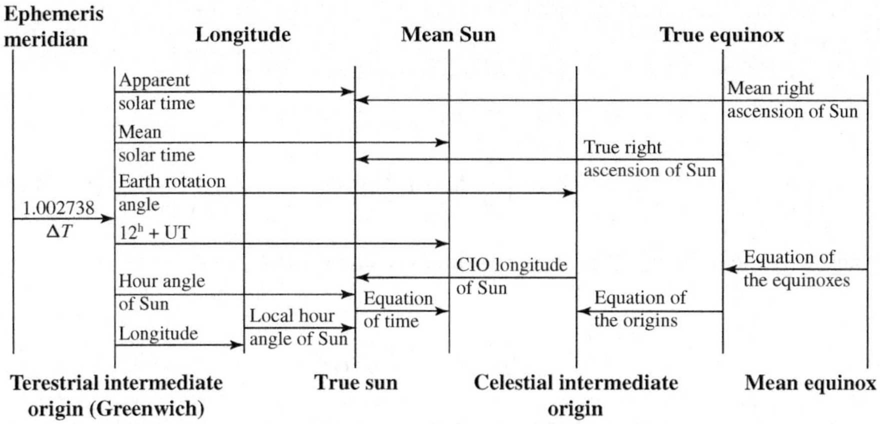

Fig. 3.4 Schematic representation of the relationships between the different fiducial meridians encountered in the celestial and terrestrial coordinate systems (reproduced from Urban and Seidelmann (2012), with permission)

International Reference Meridian (IRM), which are fixed with respect to the surface of the Earth, hence the name Earth-Fixed. This term can cause confusion, since the Earth does rotate with the axis as opposed to an inertial system, such as *Earth Centered Inertial* (ECI), and is, therefore, alternatively called *Earth Centered Rotating* (ECR). The Z-axis is pointing towards the north, but it does not coincide exactly with the instantaneous Earth rotational axis. The slight "wobbling" of the rotational axis is known as polar motion, mentioned above. The X-axis intersects the sphere of the Earth at 0° latitude and 0° longitude. This means that ECEF rotates with the Earth and, therefore, coordinates of a point fixed on the surface of the Earth do not change.

3.3.8 Satellite Geodesy

In satellite geodesy, two reference systems are required: (i) a space-fixed, inertial reference system for the satellite motions, and (ii) an Earth-fixed, terrestrial reference system for positions on the surface of the Earth. These reference systems and the transformations between them have been described previously. The terrestrial reference systems used, WGS-84 and ITRF, differ by a few centimeters due to the different reference stations used by each system.

The representation of geocentric positions can be in Cartesian coordinates, X, Y, Z, or ellipsoidal representation, longitude (λ), latitude (ϕ), and height above the ellipsoid (h), which are more commonly used in navigation. The relation between Cartesian and ellipsoidal coordinates is given by the formulae

$$\begin{aligned} X &= (N + h)\cos\lambda\cos\phi \\ Y &= (N + h)\sin\lambda\cos\phi \\ Z &= \left(\frac{b^2}{a^2}N + h\right)\sin\phi \end{aligned} \tag{3.26}$$

where N is the radius of curvature in prime vertical given by

$$N = \frac{a^2}{\sqrt{a^2\cos^2\phi + b^2\sin^2\phi}} \tag{3.27}$$

and a, b are the semiaxes of the ellipsoid. A set of parameters, called a *geodetic datum*, defines the relationship between a specific reference ellipsoid and a terrestrial reference system. A number of ellipsoids and geodetic datum have been used over the years. A list of these and their values are given in The Explanatory Supplement (Urban and Seidelmann 2012, pp. 143–146).

The *Helmert transformation* is a means of a distortion-free transformation between datum. With $\mathbf{x}_T$ as the transformed vector and $\mathbf{x}$ as the initial vector

$$\mathbf{x}_T = \mathbf{C} + \mu\mathbf{R}\mathbf{x} \tag{3.28}$$

where $\mathbf{C}$ contains the three translations along the coordinate axes, μ is a unitless scale factor, and $\mathbf{R}$ is the rotation matrix with rotations about the x, y, and z axes. The values of the parameters of the rotations between the datum must be determined based on the datum involved. These are available from the IERS.

3.3.9 GNSS Reference Systems

The *World Geodetic System* (WGS-84) is a unified terrestrial reference system for position and velocity reference, based on the BIH-84 reference frame used for UT1. The WGS-84 has been refined by more accurate coordinates, WGS-84(G1150) was introduced in 2002 and agrees with ITRF2000 at the centimeter level (Merrigan and Saffel 2002). The origin of WGS is meant to be at the Earth's center of mass, and the error is thought to be less than 2 cm. The meridian of zero longitude is the ITRF reference meridian, which is 5.31 arcsec, or 102.5 m (336.3 feet) east of the Greenwich meridian of the Airy meridian circle. The WGS-84 datum surface is an oblate spheroid with major equatorial radius $a = 6378137$ m and flattening $f = 1/298.257223563$. The polar semiminor axis $b = a(1-f) = 6356752.3142$ m. The Global Positioning System (GPS) uses the WGS-84.

The GLONASS ephemeris is given in the *Parametry Zemli 1990* (Parameters of the Earth 1990) (PZ-90) reference frame, which is basically an ECEF system

(Boykov et al. 1993; Mitrikas et al. 1998). From 2007, PZ-90.02 is used and only requires an origin shift to match ITRF2000 (Boucher and Altamimi 2001). The transformation parameters between PZ90 and WGS-84 are published by the IERS (Leick 2003).

The *Galileo Terrestrial Reference Frame* (GTRF) will be a dedicated terrestrial reference frame, which will be an independent realization of the ITRS. The Galileo requirements are that the differences of positions from the ITRF shall not exceed 3 cm.

The Chinese BeiDou-2 (formerly called COMPASS) will be a constellation of 35 satellites, 5 in geostationary orbits and 30 non-geostationary satellites (27 in medium Earth orbit and 3 in inclined geosynchronous orbits) that will give complete global coverage. The coordinate system is the *China Geodetic Coordinate System* 2000 (CGCS 2000), which is aligned with the ITRS. The time system is the BeiDou System time (BDT), without leap seconds. BDT is linked to the national UTC(k), which is consistent with UTC (Yang and Han 2012); these time scales are discussed in the next section.

3.4 Time Scales

Historically, mean solar time was the basis of time keeping. It is based on the concept of the diurnal motion of a fictitious mean sun, assuming that the Earth's rate of rotation is constant. This differs from the apparent solar time, based on the motion of the true Sun, by the *equation of time* (see Fig. 3.5). With the discovery of the variability of the rate of rotation of the Earth, *Ephemeris Time* (ET) was introduced as the independent variable in gravitational theories of the solar system from 1960 to 1984 with an ephemeris second defined as 1/31556925.9747 of the tropical year for 1900 January 0 12 hours ET. For a uniform time scale prior to atomic time scales (1956), Ephemeris Time must still be used. With the development of atomic clocks, the *Systeme International* (SI) second and *International Atomic Time* (TAI) were adopted, based on transitions of the cesium atom.

Universal Time (UT) is a measure of time that closely conforms to the mean diurnal motion of the Sun. UT1 is the angle of the Earth's rotation about the CIP axis defined by its conventional linear relation to the Earth Rotation Angle (ERA) and regarded as a time determined by the rotation of the Earth. UT1 is related to the Greenwich Apparent Sidereal Time (GAST) through the ERA and determined from observations. The rotation of the Earth is measured, primarily by Very Long Baseline Interferometry (VLBI), and by GPS and satellite and lunar laser ranging. Sidereal time is the apparent diurnal motion of the equinox, so it is a measure of the rotation of the Earth with respect to the stars, rather than the Sun.

Coordinated Universal Time (UTC) is a time scale differing from TAI by an integral number of seconds and maintained within ± 0.9 s of UT1 by the introduction of *leap seconds*. Thus, UTC is the international basis for legal and civil time with local times differing due to time zones.

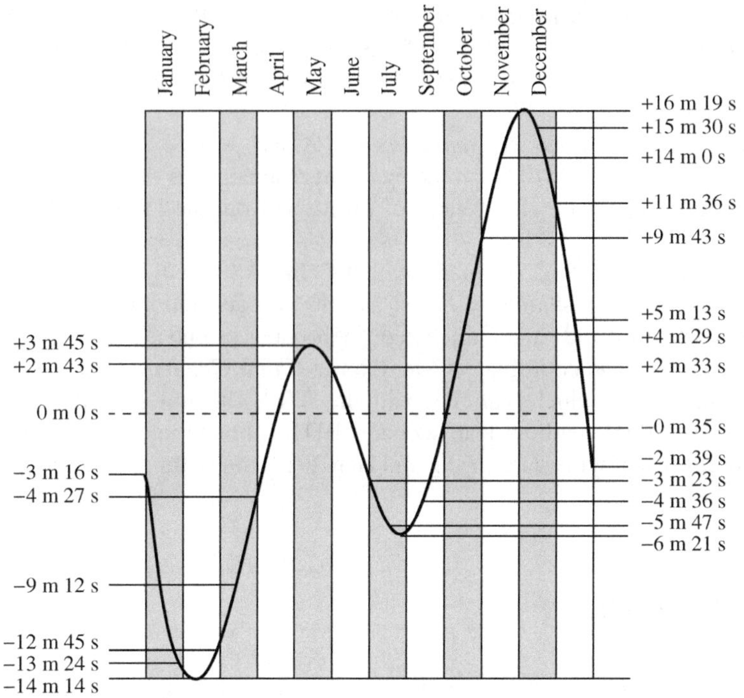

Fig. 3.5 Variation in the equation of time through the year (reproduced from Urban and Seidelmann (2012), with permission)

There are a family of dynamical time scales, as necessary for the theory of relativity. *Terrestrial Time* (TT) is a coordinate time whose mean rate is close to the mean rate of the proper time of an observer located on the rotating geoid. At 1977 January 1.0 TAI, the value of TT was 1977 January 1.0003725 exactly. An accurate realization of TT is TT(TAI) = TAI +32.184 s. TT is related to *Geocentric Coordinate Time* (TCG) by a conventional linear transformation.

$$TCG - TT = L_G \times (MJD(TAI) - 43144.0) \times 86400s \tag{3.29}$$

$L_G = 6.969290134 \times 10^{-10}$, which provides continuity with TT, whose time unit agrees with the SI second on the geoid (IERS-Conventions 2010). TCG is the coordinate time of the GCRS based on the SI second. MJD is the Modified Julian Date, JD−2400000.5.

The *Barycentric Coordinate Time* (TCB) is the coordinate time of the BCRS. TCB is related to TCG by relativistic transformations that include *secular terms*,

$$TCB - TCG = c^{-2}\left[\int_{t_0}^{t}\left(\frac{\mathbf{v}_e^2}{2} + U_{ext}(\mathbf{x}_e)\right)dt + \mathbf{v}_e \cdot (\mathbf{x} - \mathbf{x}_e)\right] + \mathcal{O}\left(c^{-4}\right) \tag{3.30}$$

where c is the speed of light, and $\mathbf{x}_e$ and $\mathbf{v}_e$ denote the barycentric position and velocity of the Earth's center of mass, and $\mathbf{x}$ is the barycentric position of the observer. The external potential U_{ext} is the Newtonian potential of all solar system bodies, apart from the Earth, and is evaluated at the geocenter. In the integral, $t = TCB$ and t_0 is chosen to agree with the epoch in Terrestrial Time. $TCB - TCG$ can be approximated in seconds by

$$TCB - TCG = \frac{L_C \times (TT - TT_0) + P(TT) - P(TT_0)}{1 - L_B} + c^{-2}\mathbf{v}_e \cdot (\mathbf{x} - \mathbf{x}_e)$$

where $L_B = 1.550519768 \times 10^{-8}$, and the current estimate of L_C is $1.48082686741 \times 10^{-8} (\pm 2 \times 10^{-17})$. $TT_0 = JD2443144.5$ TAI (1977 January 1, 0 h). The periodic terms $P(TT)$ have a maximum amplitude of about 1.7 ms and can be evaluated by an analytical model.[4] Also $P(TT) - P(TT_0)$ can be determined from a numerical time ephemeris (Irwin and Fukushima 1999). The values of $L_C(TT - TT_0) + P(TT) - P(TT_0)$ as a function of TT for 1600–2200 are available (Harada and Fukushima 2003). The last term of Eq. (3.31) is diurnal at the surface of the Earth with an amplitude of less than 2.1 μ s. See The Explanatory Supplement (Urban and Seidelmann 2012, p. 89).

Barycentric Dynamical Time (TDB) was originally intended to serve as an independent time argument of barycentric ephemerides and equations of motion. In 1991 the IAU resolutions noted that TDB is a linear function of TCB and in 2006 TDB was redefined through a linear transformation of TCB,

$$TDB = TCB - L_B \times (JD_{TCB} - 2443144.5003725) \times 86400s + TDB_0 \tag{3.31}$$

where $TDB_0 = -6.55 \times 10^{-5}$ and $L_B = 1.550519768 \times 10^{-8}$ by definition.

Barycentric Ephemeris Time Teph is the independent time argument of the JPL and MIT/CfA solar system ephemerides. It differs from TCB by an offset and constant rate. The rate of Teph is as close as possible to that of TT for the time span covered by the particular ephemeris. Each ephemeris defines its own version of Teph. For practical purposes the Teph of JPL ephemeris DE405 is the same as TDB.

The International Telecommunications Union-Radio (ITU-R) has introduced a resolution to redefine UTC without leap seconds. This would be a time scale strictly based on TAI and differing from UT1 by an increasing amount. The Radio Communication Assembly in 2015 again postponed a decision on the proposed UTC resolution for more study, with consideration of the future of UTC to be considered no sooner than 2023.

[4]Bretagnon, 2001, private communication.

The GNSS use specific time scales for their systems. GPS uses an atomic clock-based system with clocks on the satellites and ground stations. The time scale is steered by the US Naval Observatory (USNO) to match the anticipated rate of TAI with a constant offset of -19 s. So the GPS time scale itself is not affected by leap seconds. The Galileo time scale will use the same constant offset from TAI as GPS, but the time scale will be determined by the Bureau International des Poids et Mesures (BIPM). The GLONASS system uses Moscow time, so it is adjusted by leap seconds when they occur.

3.5 Coordinate Systems

A Cartesian coordinate system specifies each location uniquely in a three-dimensional space by three Cartesian coordinates, its signed distances to three mutually perpendicular planes (or, equivalently, by its perpendicular projection onto three mutually perpendicular lines). Each reference line is called a *coordinate axis*, or just axis, of the system, and the point where they meet is its origin, $(0, 0, 0)$. The invention of *Cartesian coordinates* in the seventeenth century by René Descartes (Latinized name: Cartesius) revolutionized mathematics by providing the first systematic link between Euclidean geometry and algebra.

3.5.1 Origins and Planes

The origins of celestial coordinate frames are designated as follows: *topocentric* are measured from the surface of the Earth, *geocentric* are measured from the center of the Earth, *barycentric* are measured from the center of the mass of the solar system, *heliocentric* are measured from the center of the Sun, *planetocentric* are measured from the center of a planet, and *selenocentric* are measured from the center of the Moon.

The principal celestial reference planes through appropriate origins are: *astronomical horizon* is normal to the local vertical and passes through the observer; *local meridian* contains the local vertical and the direction of the axis of rotation of the Earth; *celestial equator* is normal to the axis of rotation of the Earth and passes through the Earth's center; *ecliptic* is the mean plane of the orbit of the Earth-Moon barycenter around the solar system barycenter; *invariable plane*, or *Laplacian plane*, is normal to the axis of angular momentum of a system and passes through its center; a *planet meridian* contains the axis of rotation of a planet and passes through the observer; a *planet equator* is normal to the axis of rotation of the planet and passes through the planet's center; *orbital plane* is the plane of the orbit of a body around another body; the *galactic equator* is the mean plane of the Milky Way normal to the North Galactic Pole, which is in the constellation Coma at $\alpha(1950) = 12\text{h}49\text{m}, \delta(1950) = 27°27'.4$ and passes through the Galactic Center, which is in the constellation Sagitarius at ICRS(J2000) $\alpha = 17\text{h}45.6\text{m}, \delta = -26.56°$.

3.5.2 *Horizon Reference Frame*

The intersections of the plane of the meridian with the planes of the horizon and equator define the directions for measuring azimuth and local hour angle. The azimuth is measured in degrees in the plane of the horizon from the north, increasing in positive value toward the east. Local hour angle is measured in units of time positive to the west with respect to the local meridian. The latitudinal angles with respect to the horizon and equator are altitude, which is measured positively toward the zenith from the horizon, and declination, which is measured from the equator, positive toward the north pole of rotation (see Fig. 3.6).

3.5.3 *Geocentric Coordinates*

The *geocentric coordinates* are the longitude, λ', and latitude, ϕ', of a point on the Earth's surface relative to the center of the Earth, with the equator and a prime meridian as reference planes. Coordinates can be given in rectangular coordinates or spherical coordinates, longitude, latitude, and radial distance. Longitude is measured from the prime meridian positively to the east, usually in degrees to $\pm 180°$. Latitude is measured from the equator to 90°, positively to the north.

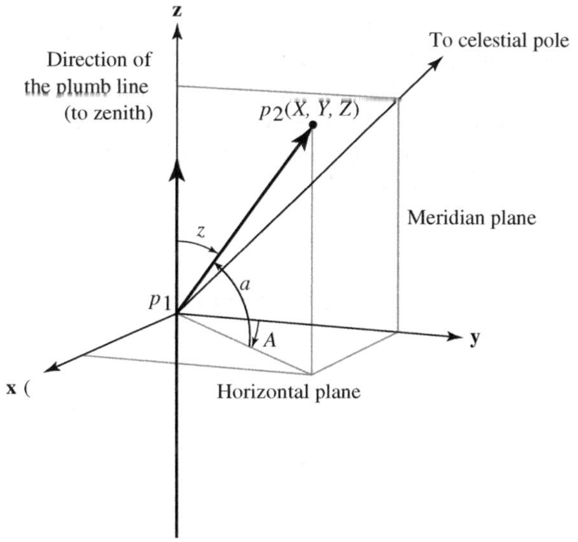

Fig. 3.6 Horizon system altitude and azimuth (reproduced from Urban and Seidelmann (2012), with permission)

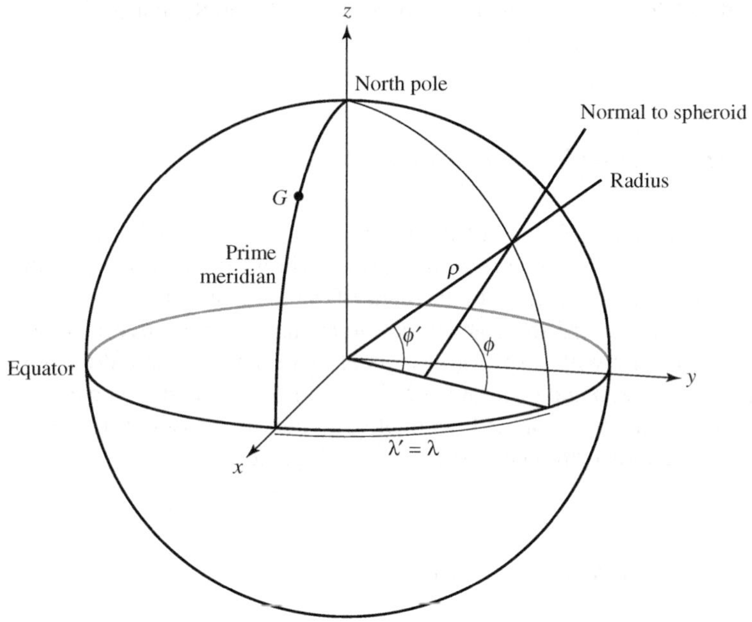

Fig. 3.7 Geocentric and geodetic coordinates (reproduced from Urban and Seidelmann (2012), with permission)

3.5.4 Geodetic Coordinates

The *geodetic coordinates* are the longitude and latitude of a point on the Earth's surface determined from the geodetic vertical, which is normal to the reference ellipsoid. The equatorial radius and flattening of the adopted spheroid must be specified. Geodetic and geocentric longitude $\lambda = \lambda'$ are defined the same way and agree. Geodetic latitude, ϕ, is the inclination to the equatorial plane of the normal to the spheroid. Geodetic latitude may differ from geocentric latitude by up to 10 min of arc in mid-latitudes. Geodetic height is the distance above the spheroid along the normal to the spheroid (see Fig. 3.7).

3.5.5 Geographic Coordinates

The *geographic coordinates* refer to terrestrial longitude and latitude, when determined by astronomical observations with respect to the celestial pole and local meridian through the local vertical, and a height above the geoid, also called the height above mean sea level (see Figs. 3.8 and 3.9). These coordinates are also called *astronomical coordinates* and *terrestrial coordinates*. The geographic longitude is

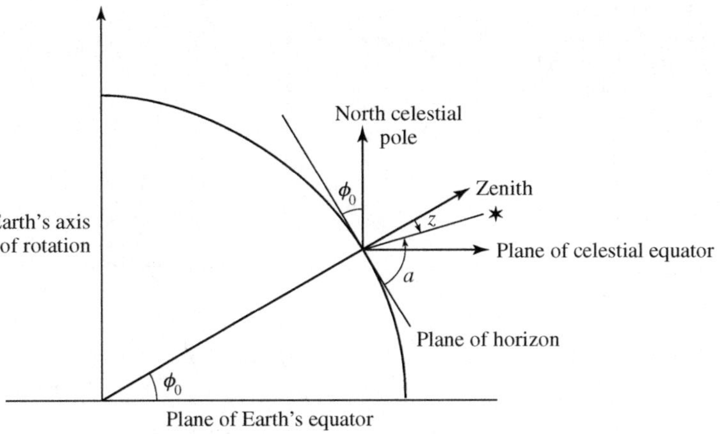

Fig. 3.8 Relation between geographic latitude and the latitude of the celestial pole (reproduced from Urban and Seidelmann (2012), with permission)

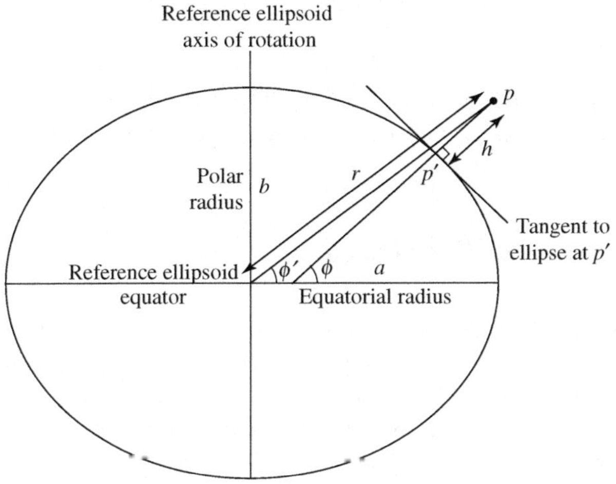

Fig. 3.9 Geocentric (ϕ') and geodetic (ϕ) latitude (reproduced from Urban and Seidelmann (2012), with permission)

the angle between the plane of the astronomical meridian through a point and the plane of the ITRF prime meridian. The astronomical meridian contains the direction of the local vertical and the direction of the line through the point that is parallel to the axis of rotation of the Earth. The local vertical is affected by local gravity anomalies and by the varying gravity fields of the Sun, Moon, and oceans (see Fig. 3.10). Thus, the astronomical meridian is not precisely the same as the geodetic meridian through a point. So geographic and geodetic longitude of a point differ slightly. Geographic latitude is also defined by the local vertical and axis of rotation,

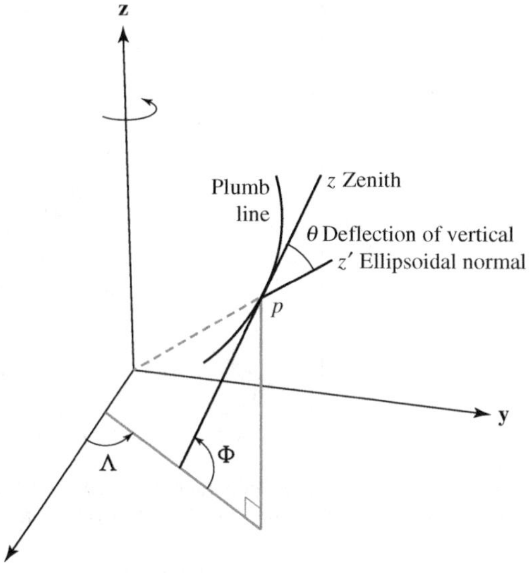

Fig. 3.10 Astronomical latitude and longitude (reproduced from Urban and Seidelmann (2012), with permission)

and so is not equal to the geodetic latitude. The inclination of the local vertical to the normal to the reference spheroid is known as the *deflection of the vertical.*

3.5.6 Astronomical Coordinates

The longitude and latitude of a point on the Earth relative to the geoid are provided by astronomical coordinates. The coordinates are influenced by local gravity anomalies. A proper, or virtual, place is the direction of an object in the GCRS that takes into account orbital or space motion and light-time, light deflection, and annual aberration. Thus, the geocentric right ascension and declination are the position where the object would be seen from the center of the Earth, if the Earth were transparent, non-refracting, and massless. Apparent place is the proper place of an object expressed with respect to the true, or intermediate, equator and equinox of date.

The *mean place* is the coordinates of an object at a specific date in the BCRS. The coordinates represent the direction of the object as it would be observed from the solar system barycenter at the specific time with respect to a fixed coordinate system, e.g the ICRF, if the masses of the Sun and other solar system bodies were negligible. The *local place* is the topocentric place in the coordinate system of the reference system. The local place represents the position of an object as it would be seen from some specific location on the Earth at some date and time in the coordinate system of the GCRS, assuming the atmosphere was non-refracting. An astrometric place is the direction of a solar system body formed by applying the correction for the barycentric motion of this body during the light-time to the geometric position referred to the ICRS. For more information see The Explanatory Supplement (Urban and Seidelmann 2012, pp. 281–283).

3.6 Kinematics of the Earth

3.6.1 Earth Orientation

The connection between an observation from the surface of the Earth in a local coordinate system and a celestial coordinate system is complicated by three aspects of the kinematics of the Earth: precession and nutation of the axes of rotation, the motion of the pole of rotation within the Earth (polar motion), and the variability of the rate of rotation of the Earth. Precession and nutation are the only ones of these that are accurately predictable, the others have to be observed. The variability of the rotation of the Earth is due to a number of causes, primarily a long term tidal friction effect, and interchange of angular momentum between the mantle and a number of elements, specifically the core, atmosphere, oceans, and others. The changes result in variations in the length of the day. The observed determination of the rotation of the mantle of the Earth is given by Universal Time, with versions UT0, UT1, and UT2. UT1 is the measure used. This is available as observed and in predicted values from the IERS. For observations and pointing from the surface of the Earth, UT1 must be known. The difference between UT1 and UTC is currently maintained at less than 0.9 s.

3.6.2 Precession

The main sources of forced precession for the Earth's rotation are the torques caused by the attraction of the Sun and Moon on the Earth's equatorial bulge, called *precession of the equator* (formerly called *lunisolar precession*). The slow change in the orientation of the Earth's orbital plane is called *precession of the ecliptic* (formerly called *planetary precession*). The combination of both motions, the motion of the equator with respect to the ecliptic, is called *general precession* (see Fig. 3.11). The precession matrix from mean equator and equinox of epoch to mean equator and equinox of date is made up of the rotations

$$\mathbf{P} = \mathbf{R}_3(-z_A)\mathbf{R}_2(+\theta_A)\mathbf{R}_3(-\zeta_A) \tag{3.32}$$

An expansion of the equation and the values of the constants are given in The Explanatory Supplement (Urban and Seidelmann 2012, pp. 216–217).

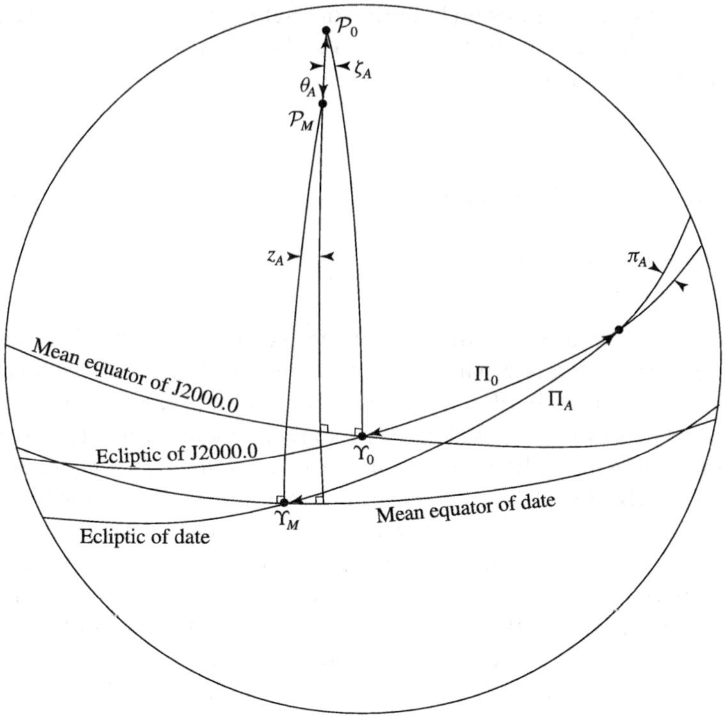

Fig. 3.11 The general precession connects the mean equinox of epoch, Υ_0, to the mean equinox of date, Υ_M (reproduced from Urban and Seidelmann (2012), with permission)

3.6.3 Nutation

Nutation is the oscillations in the motion of the Earth's pole due to torques from external gravitational forces and is specified in terms of components in obliquity and longitude. It is limited to motions with periods longer than two days. The IAU 2000A Theory of Nutation was computed for the celestial intermediate pole by determining the nutations in longitude and obliquity of a rigid Earth and making modifications for the non-rigid Earth, so there are no nearly diurnal motions of the celestial pole with respect to either space-fixed or body-fixed coordinates. See Kaplan (2005) for a table of the fundamental arguments and nutation amplitudes. The nutation matrix, **N**, is a sequence of three rotations

$$\mathbf{N} = \mathbf{R}_1(-\varepsilon)\mathbf{R}_3(-\Delta\psi)\mathbf{R}_1(+\varepsilon_A) \tag{3.33}$$

where $\Delta\psi$ is the nutation in longitude, $\varepsilon = \varepsilon_A + \Delta\varepsilon$, $\Delta\varepsilon$ is the nutation in obliquity, and ε_A is the obliquity of the ecliptic of date. Software implementing the nutation theory is available from the IERS,[5] USNO NOVAS subroutines,[6] and SOFA.[7] These software routines implement the IAU 2000 nutation in slightly different ways. More information about nutation is given in The Explanatory Supplement (Urban and Seidelmann 2012, pp. 223–227).

3.6.4 Polar Motion

Polar motion, mentioned in Sect. 3.3.5, refers to the quasiperiodic motion of the Earth's pole of rotation with respect to the Earth's solid body. This is the angular excursion of the CIP from the ITRS z axis. There is a secular polar motion, which is the non-periodic motion of the Earth's pole toward the direction of approximately 75° west longitude. For more information about polar motion see The Explanatory Supplement (Urban and Seidelmann 2012, pp. 240–244).

3.7 Observation Effects

3.7.1 Aberration

The *aberration* phenomenon is the relativistic apparent angular displacement of the observed position of a celestial object from its geometric position caused by the motion of the observer and the object in the reference system, in which the trajectories of the observed object and the observer are described. The displacement due to the motion of the object may be considered a correction for light time. In Fig. 3.12, E denotes the stationary observer at time t and P is the position of the moving object at time t. The dotted line is the motion of the moving object and P' is the location of the object when the light was emitted at $(t - \tau)$, where τ is the light time, the time for the light to travel from P' to E. The light time is calculated iteratively.

The correction for the motion of the observer is called *stellar aberration* and is due to the following effects: the diurnal rotation of the Earth, *diurnal aberration*; the orbital motion of the Earth around the barycenter of the solar system, *annual aberration*; and the motion of the solar system in space, *secular aberration*. In Fig. 3.13, $\mathbf{p}$ is the direction of EP', the direction of the object allowing for light time, the observer is moving with a velocity $\mathbf{V}$ relative to a fixed reference frame, and at time t observes the object at P'' in the direction $\mathbf{p}_1$. $P'EB$ is the angle θ between the

---

[5] The IERS, http://www.iers.org.

[6] Naval Observatory Vector Astrometry Software (NOVAS), http://www.usno.navy.mil/USNO/astronomical-applications/software-products/novas/.

[7] IAU Standards of Fundamental Astronomy (SOFA) Software Collection, http://www.iausofa.org/.

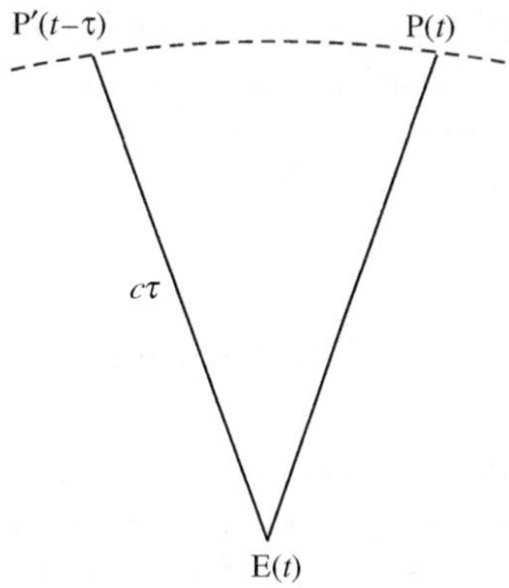

Fig. 3.12 Light-time displacement (reproduced from Urban and Seidelmann (2012), with permission)

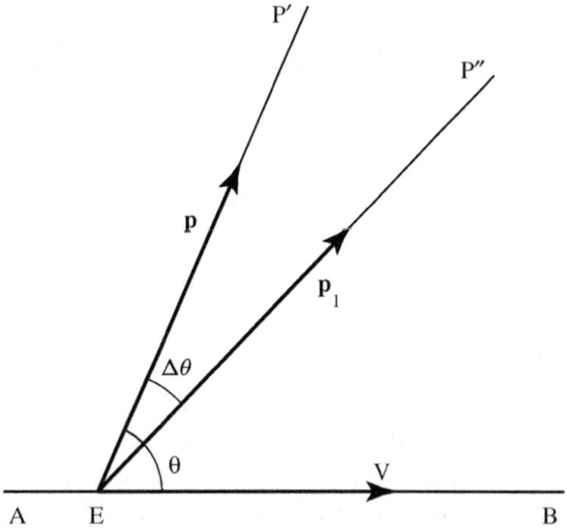

Fig. 3.13 Stellar aberration (reproduced from Urban and Seidelmann (2012), with permission)

motion and **p**, and $P'EP''$, the angle $\Delta\theta$, is the displacement due to aberration in the moving frame, which is in the direction of the motion. The correction for stellar aberration can be determined from

$$\sin\Delta\theta = \frac{V}{c}\sin\theta - \frac{1}{4}\left(\frac{V}{c}\right)^2 \sin 2\theta + \ldots \tag{3.34}$$

The corrections for annual aberration in the sense of apparent place minus mean place are

$$\begin{aligned}\cos\delta\Delta\alpha &= -\frac{\dot{X}}{c}\sin\alpha + \frac{\dot{Y}}{c}\cos\alpha \\ &\quad + \frac{1}{c^2}(\dot{X}\sin\alpha - \dot{Y}\cos\alpha)(\dot{X}\cos\alpha + \dot{Y}\sin\alpha)\sec\delta + \ldots\end{aligned} \tag{3.35}$$

$$\begin{aligned}\Delta\delta = &-\frac{\dot{X}}{c}\cos\alpha\sin\delta - \frac{\dot{Y}}{c}\sin\alpha\sin\delta + \frac{\dot{Z}}{c}\cos\delta \\ &- \frac{1}{2c^2}(\dot{X}\sin\alpha - \dot{Y}\cos\alpha)^2\tan\delta \\ &+ \frac{1}{c^2}(\dot{X}\cos\delta\cos\alpha + \dot{Y}\cos\delta\sin\alpha + \dot{Z}\sin\delta) \\ &\times (\dot{X}\sin\delta\cos\alpha + \dot{Y}\sin\delta\sin\alpha - \dot{Z}\cos\delta) + \ldots \end{aligned} \tag{3.36}$$

where $\dot{X}, \dot{Y}, \dot{Z}$ are the components of the Earth's velocity parallel to equatorial rectangular axes.

The diurnal aberration corrections in the sense of apparent place minus mean place are

$$\Delta\alpha = 0.02133^s \frac{\rho}{a}\cos\phi'\cos h\sec\delta \tag{3.37}$$

$$\Delta\delta = 0.3200'' \frac{\rho}{a}\cos\phi'\sin h\sin\delta \tag{3.38}$$

where ρ and ϕ' are the geocentric distance and latitude of the observer and h is the hour angle. The superscript s denotes seconds of time, and $''$ denotes seconds of arc.

For more information about aberration corrections see The Explanatory Supplement (Urban and Seidelmann 2012, pp. 263–270).

3.7.2 *Proper Motion*

The apparent space motion of a star in two dimensions relative to the celestial reference frame is called *proper motion*. Thus, it is usually tabulated as changes in right ascension and declination. Care must be taken to ensure that the units are properly used. Modern catalogs tend to tabulate $\mu_\alpha = 15\mu_\alpha\cos\delta$ and μ_δ, where μ_α and μ_δ are the measurements on the celestial sphere in seconds of time and of arc/year, respectively. The third dimension of the space motion is given as radial velocity.

3.7.3 *Radial Velocities*

The *radial velocity* is the rate of change of the distance to an object, usually corrected for the Earth's motion, with respect to the solar system barycenter. For more information about radial velocities see The Explanatory Supplement (Urban and Seidelmann 2012, pp. 258–260).

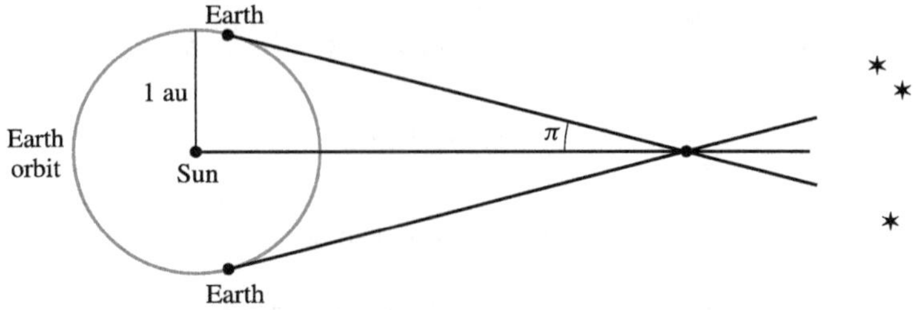

Fig. 3.14 Parallax of an object (reproduced from Urban and Seidelmann (2012), with permission)

3.7.4 Parallax

The term *parallax* denotes the difference in the apparent direction of an object as seen from two different locations; conversely, the angle at the object that is subtended by the line joining two designated points. Thus, there are different types of parallaxes due to different locations for observations. Annual, or heliocentric, parallax is the difference between the geocentric and heliocentric directions toward an object (see Fig. 3.14). Geocentric, or diurnal, parallax is the difference between the topocentric and geocentric directions toward an object. Horizontal parallax is the difference between the topocentric and geocentric directions toward an object, when the object is on the astronomical horizon. For more information concerning parallax see The Explanatory Supplement (Urban and Seidelmann 2012, pp. 261–263).

3.7.5 Refraction

The phenomenon of *refraction* is the change in direction, or bending, of a light ray as it passes obliquely from a medium of lesser/greater density to a medium of greater/lesser density. Astronomical refraction is the change in direction as a light ray passes obliquely through the atmosphere. The result of refraction is that the observed altitude of an object is greater than the geometric altitude. The magnitude of refraction is dependent on the altitude of the object and atmospheric conditions. At the horizon a value of $34'$ is frequently used in computations for sea level observations. *Saastamoinen's refraction formula* can be used for zenith distances down to 70°. Given the observed zenith distance, z_0, the temperature, pressure, and partial pressure of water vapor, $(T_0, P_0, P_{\omega 0})$, the refraction for a wavelength of 0.574 μm, visible light, and an observer at sea level is

$$\xi = 16.271 Q \tan z_0 (1 + 0.0000394 Q \tan^2 z_0) - 0.0000749 P_0 (\tan z_0 + \tan^3 z_0) \tag{3.39}$$

where $Q = (P_0 - 0.156 P_{\omega 0})/T_0$. For more information on refraction see The Explanatory Supplement (Urban and Seidelmann 2012, pp. 277–280).

3.7.6 Relativistic Light Deflection

The phenomenon of *relativistic light deflection* is the bending of the path of light towards the body in a gravitational field of a massive body. In Fig. 3.15, S is the Sun, P the body being observed, and E is the Earth. The unit vectors $\mathbf{e}$ and $\mathbf{q}$ are the heliocentric directions to the Earth and the body, respectively. The heliocentric angle of the Earth from P is ψ, where $\cos\psi = \mathbf{q}\cdot\mathbf{e}$. The unit vector $\mathbf{p}$ is the geocentric direction of the body P when the light left it. The dotted arc AEB is the light path as it passes the Earth. As the light travels along path AB, it is deflected towards the Sun. At E, the direction between $\mathbf{p}$ and the tangent to the light path is $\Delta\psi$, which is

$$\Delta\psi = \frac{2\mu}{c^2E}\,\frac{\sin\psi}{1+\cos\psi} \tag{3.40}$$

where E is the distance of the Earth from the Sun, μ is the heliocentric gravitational constant, and c is the speed of light. The apparent direction of P is along the tangent to the light-path $\mathbf{p}_1$, which can be computed from

$$\mathbf{p}_1 = \mathbf{p} + \frac{g_1}{g_2}\left[(\mathbf{p}\cdot\mathbf{q})\mathbf{e} - (\mathbf{e}\cdot\mathbf{p})\mathbf{q}\right] \tag{3.41}$$

The dimensionless scalar quantities are $g_1 = \frac{2\mu}{c^2E}$ and $g_2 = 1 - \mathbf{q}\cdot\mathbf{e}$, where $\mu/c^2 = k^2\tau_A^2/86400^2$, k is the Gaussian gravitational constant, and τ_A is the light time for a unit distant (1 *astronomical unit*, AU) for the ephemeris being used. For more information, see The Explanatory Supplement (Urban and Seidelmann 2012, pp. 270–272).

In this context, the term *time dilation* refers to the observed difference of elapsed time between two observers, which are moving relative to each other, or being differently situated with respect to nearby gravitational masses.

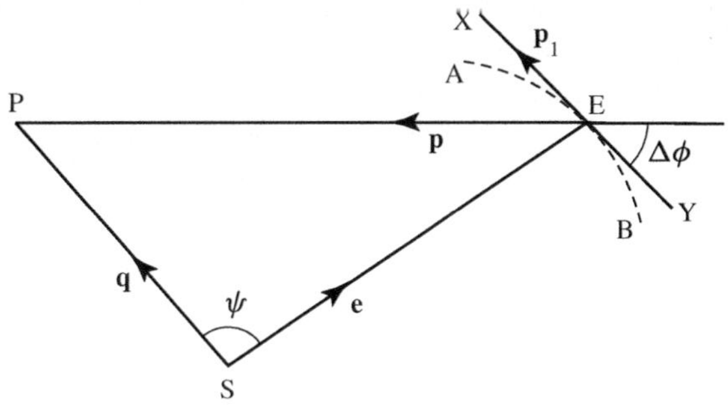

Fig. 3.15 Gravitational light deflection (reproduced from Urban and Seidelmann (2012), with permission)

3.7.7 Space Motion

The position of an object in space may be given in a form where the distance and directions are combined to form a position vector **r** at an epoch t,

$$\mathbf{r} = r \begin{bmatrix} \cos\delta\cos\alpha \\ \cos\delta\sin\alpha \\ \sin\delta \end{bmatrix} \tag{3.42}$$

where α, δ are the right ascension and declination, and r is the barycentric distance to the object in AU, which can be computed from $r = 1/\sin p$, where p is the parallax of the star. If r is known from another source, it can be used as is. If the distance, or parallax, is not known, a small positive number can be used to avoid mathematical indeterminacy. The space motion vector $\dot{\mathbf{r}}$ of the object is given then by

$$\dot{\mathbf{r}} = \begin{bmatrix} -\cos\delta\sin\alpha & -\sin\delta\cos\alpha & \cos\delta\cos\alpha \\ \cos\delta\cos\alpha & -\sin\delta\sin\alpha & \cos\delta\sin\alpha \\ 0 & \cos\delta & \sin\delta \end{bmatrix} \begin{bmatrix} 15\, srf\mu_\alpha \\ srf\mu_\delta \\ kf\dot{r} \end{bmatrix} \tag{3.43}$$

where μ_α, μ_δ are the proper motions in right ascension and declination in units of time and arc, respectively, and $\dot{r}$ is the radial velocity. The factors $15s, s$ and k convert μ_α, μ_δ and $\dot{r}$ into the required units for the space motion vector, and the 15 assumes that μ_α is in units of time. The factor f is the relativistic Doppler effect. In most cases, f may be ignored and set equal to 1. For more information, see The Explanatory Supplement (Urban and Seidelmann 2012, pp. 258–261).

3.7.8 Tidal Effects

There are periodic variations in UT1 due to tidal deformation of the solar moment of inertia and include the zonal tides of the Earth with a decoupled core, an elastic mantle and equilibrium oceans. A model uses effective *Love numbers*. Care must be used to specify any tidal model used. The IERS Conventions[8] specify the corrections to UT1, length of day, and rotation velocity in equations and tables of the terms to be calculated.

[8] Available at ftp://tai.bipm.org/iers/conv2010/.

3.8 Earth Satellite Equations of Motion in GCRS

For geocentric Earth satellites, developments may be restricted to a fictitious observer at the center of the Earth for most computations (i.e., observations are referred to the geocenter so that the GCRS is the proper reference system). A summary by Ries et al. (1991) determined the geocentric frame as being suitably accurate for near-Earth satellite motion. Equations of motion referred to the GCRS follow from that presentation, as well as the development of Huang et al. (1990). Notation is that of Ashby and Bertotti (1986); large bold letters (e.g., $\mathbf{X}, \mathbf{V}, \mathbf{A}$, and M) denote position, velocity, and acceleration vectors, and mass parameters in barycentric coordinate-reference systems, while small bold letters (e.g., $\mathbf{x}$, $\mathbf{v}$, $\mathbf{a}$, and m) denote the same in non-inertial geocentric reference systems. Subscripts A,E, and M refer to the Ath body, the Earth, and the Moon, respectively. The equations of motion for a near Earth satellite in the geocentric, kinematically non-rotating (non-inertial) system are

$$\frac{d^2\mathbf{r}}{d^2\tau} = -Gm_E\frac{\mathbf{r}}{r^3} + \mathbf{A}_{np} + \mathbf{A}_{rel-1} + \mathbf{A}_{rel-2} + \mathbf{A}_{rel-3} + \mathbf{A}_{indr} + \mathbf{A}_{ns} + \mathbf{A}_{other} \tag{3.44}$$

where $\mathbf{r}$ is the geocentric satellite location, $\mathbf{A}_{np}$, $\mathbf{A}_{ns}$, and $\mathbf{A}_{indr}$ are the Newtonian n-body perturbation, the Newtonian non-spherical geopotential perturbation, and the indirect perturbation due to the Earth's oblateness, respectively. $\mathbf{A}_{other}$ represents other Newtonian perturbations, such as solar- or Earth-radiation pressure, in the geocentric reference system. In the two-body term, $Gm_E = 3.986004418 \times 10^{14}\mathrm{m}^3/\mathrm{s}^2$ (conventional EGM96), when units of time are expressed in TCG, while $Gm_E = 3.986004415 \times 10^{14} m^3/s^2$, when units of time are expressed in TT (or equivalently, TAI), the scale difference being $(1 - L_G)$ per Eq. (3.29). The indirect perturbation term $\mathbf{A}_{indr}$ does not depend on a satellite orbit, as it is caused by the interaction between the non-spherical part of the Earth and the other bodies. To a high level of accuracy, it has the following form:

$$\begin{aligned}\mathbf{A}_{indr} = &-\frac{15}{2}\frac{GM_M}{M_E R_{ME}^7}[(\mathbf{R}_{ME}^T\, \bar{HG}^T\, \Gamma_{BF}\, \bar{HG}\, \mathbf{R}_{ME})\mathbf{R}_{ME} \\ &- \frac{2}{5}\bar{HG}^T\, \Gamma_{BF}\, \bar{HG}\, \mathbf{R}_{ME}\, R_{ME}^2]\end{aligned} \tag{3.45}$$

with

$$\Gamma_{BF} = \frac{1}{2} m_E a_{ME}^2 J_2 \begin{pmatrix} 1 & 0 & 0 \\ 0 & 1 & 0 \\ 0 & 0 & -2 \end{pmatrix} \tag{3.46}$$

where $\mathbf{R}_{ME} = \|\mathbf{X}_M - \mathbf{X}_E\|$, and a_E, J_2, and $\bar{HG}$ are the Earth's equatorial radius, dynamical form-factor for the Earth, and the coordinate-transformation matrix from the non-inertial geocentric reference system to the (central) body-fixed geocentric reference system, respectively.

The relativistic perturbations $\mathbf{A}_{rel-1}$ and $\mathbf{A}_{rel-2}$, and the Newtonian n-body perturbations $\mathbf{A}_{np}$, have the following forms:

$$\mathbf{A}_{np} = \sum_{A \neq E} \frac{GM_A}{\|\mathbf{X}(t) - \mathbf{X}_A(t)\|^3} [\mathbf{X}(t) - \mathbf{X}_A(t)] - \sum_{A \neq E} \frac{GM_A}{R_{EA}^3} \mathbf{X}_{EA} \tag{3.47}$$

The *Schwarzschild solution*, with the last term being the geodesic precession, is

$$\mathbf{A}_{rel-1} = \frac{Gm_E}{c^2 r^3} \left[\left(4 \frac{Gm_E}{r} - v^2 \right) \mathbf{r} + 4 (\mathbf{v} \cdot \mathbf{r}) \mathbf{v} \right] + 2 (\mathbf{\Omega} \times \mathbf{v}) \tag{3.48}$$

The *Lense-Thirring precession* is

$$\mathbf{A}_{rel-2} = 2 \frac{Gm_E}{c^2 r^3} \left[\frac{3}{r^2} (\mathbf{r} \times \mathbf{v}) (\mathbf{r} \cdot \mathbf{J}) + (\mathbf{v} \times \mathbf{J}) \right] \tag{3.49}$$

where

$$\mathbf{\Omega} = \frac{3}{2c^2} (\mathbf{V}_E \times \mathbf{A}_E) + 2 \sum_{A \neq E} \frac{GM_A}{c^2 R_{EA}^3} (\mathbf{V}_A \times \mathbf{X}_{EA}) \tag{3.50}$$

The first term in Eq. (3.48) is the Schwarzschild solution and the second term is the effect of geodesic precession. The average effect on the (equatorial) orbital node of a near-Earth satellite is approximately 17.6 mas/yr, sometimes known as *relativistic Coriolis acceleration*. Equation (3.49) is the relativistic effect caused by the Lense-Thirring perturbation, where $\mathbf{J}$ is the Earth's angular momentum per unit mass $\left(\|\mathbf{J}\| = 9.8 \times 10^8 \mathrm{m}^2/\mathrm{s}\right)$ (Lense and Thirring 1918). $\mathbf{A}_{rel-3}$ represents all other relativistic perturbations due to the Earth's rotation and finite size (including oblateness). Since the Sun is the only significant contributor to geodesic precession, Eq. (3.50) can be simplified to

$$\mathbf{\Omega} = \frac{3}{2} (\mathbf{V}_E - \mathbf{V}_S) \times \left[\frac{-GM_S \mathbf{X}_{ES}}{c^2 R_{ES}^3} \right] \tag{3.51}$$

This simplification introduces an error of less than 0.1 mas/yr for an Earth satellite.

The IERS Conventions (2003), in its Eq. (10.2.1), provides an alternative expression for the relativistic perturbation of an artificial satellite, comparable to Eqs. (3.48)–(3.51); however, it provides compatibility with the more general *Parameterized Post-Newtonian* (PPN) formalism. The 2000 resolutions have been discussed in the PPN context by Klioner and Soffel (2000) and Kopeikin and Vlasov (2004). As adopted, the 2000 IAU resolutions are specific to Einstein's

General Theory of Relativity, where it is assumed that the two PPN parameters are both unity (Kaplan 2005).

Barycentric equations of motion, along with their development, are given by Huang et al. (1990), and the reader is further referred to Chap. 11 of the IERS Conventions (2003) regarding the additional relativistic barycentric corrections required for laser range measurements, data timing, and station coordinates.

References

Altamini, S.P.Z., Boucher, C.: ITRF2000: A new release of the International Terrestrial Reference Frame for Earth science application. J. Geophys. Res. **107**(B10), 2214 (2002)

Ashby, N., Bertotti, B.: Relativistic effects in local inertial frames. Phys. Rev. D **34**(8), 2246 (1986)

Boucher, C., Altamimi, Z.: ITRS, PZ-90 and WGS-84: Current realizations and the related transformation parameters. J. Geodesy **75**(11), 613–619 (2001)

Boykov, V.V., Galazin, V.F., Kaplan, B.L., Maksimov, V.G., Bazlov, Y.A.: Experience in developing the PZ-90 geocentric coordinate system. Geod. Kartogr. **11**, 17–21 (1993)

Capitaine, N., Wallace, P.T.: High precision methods for locating the celestial intermediate pole and origin. Astron. Astrophys. **450**(2), 855–872 (2006)

Cutri, R., Skrutskie, M., Van Dyk, S., Beichman, C., Carpenter, J., Chester, T., Cambresy, L., Evans, T., Fowler, J., Gizis, J., et al.: 2MASS all sky catalog of point sources. NASA/IPAC Infrared Science Archive (2003)

ESA: The Hipparcos and Tycho Catalogues. European Space Agency SP-1200, Noordwijk, Netherlands (1997)

Harada, W., Fukushima, T.: Harmonic decomposition of time ephemeris TE405. Astron. J. **126**(5), 2557 (2003)

Huang, C., Ries, J.C., Tapley, B.D., Watkins, M.M.: Relativistic effects for near-Earth satellite orbit determination. Celest. Mech. Dyn. Astron. **48**(2), 167–185 (1990)

IAU-Resolutions: Transactions of the IAU, vol XXIVB. International Astronomical Union, Kluwer, Dordrecht, The Netherlands (2000)

IERS: Technical Note 35. Frankfurt am Main. Verlag des Bundsesamts fur Kattographie und Geodasie (2009)

IERS-Conventions: Technical Note 36. Frankfurt am Main. Verlag des Bundsesamts fur Kattographie und Geodasie (2010)

IERS-Conventions-2003: Technical Note 32. International Earth Rotation Service, Frankfurt am Main. Verlag des Bundsesamts fur Kattographie und Geodasie (2004)

Irwin, A.W., Fukushima, T.: A numerical time ephemeris of the Earth. Astron. Astrophys. **348**, 642–652 (1999)

Kaplan, G.: The IAU resolutions on astronomical reference systems, time scales, and Earth rotation models, explanation and implementation. US Naval Observatory Circular (179) (2005)

Klioner, S.A., Soffel, M.H.: Relativistic celestial mechanics with PPN parameters. Phys. Rev. D **62**(2), 024,019 (2000)

Kopeikin, S., Vlasov, I.: Parametrized post-Newtonian theory of reference frames, multipolar expansions and equations of motion in the n-body problem. Phys. Rep. **400**(4), 209–318 (2004)

Leick, A.: GPS Satellite Surveying, 3rd edn. Wiley, New York (2003)

Lense, J., Thirring, H.: Uber den einfluss der eigenrotation der zentralk orper auf die bewegung der planeten und monde nach der einsteinschen gravitationstheorie. Phys. Z. **19**, 156–163 (1918)

Merrigan, S.E.W.R.M., Saffel, J.: A refinement to the World Geodetic System 1984 reference frame. In: 15th International Technical Meeting of the Satellite Division of the Institute of Navigation, Institute of Navigation, Portland OR, pp. 1519–1529 (2002)

Mitrikas, V., Revnivykh, S., Bykhanov, E.: WGS-84/PZ-90 transformation parameters determination based on laser and ephemeris long-term GLONASS orbital data processing. In: Proceedings of the ION/GPS, pp. 1625–1635 (1998)
Nelson, R.: Guide to Geotemporal Sciences: Issues, Principles and References. Satellite Engineering Research Corporation, Bethesda (2000)
Ries, J.C., Huang, C., Watkins, M.M., Tapley, B.D.: Orbit determination in the relativistic geocentric reference frame. J. Astronaut. Sci. **39**, 173–181 (1991)
Soffel, M., Klioner, S.A., Petit, G., Wolf, P., Kopeikin, S.M., Bretagnon, P., Brumberg, V.A., Capitaine, N., Damour, T., Fukushima, T., et al.: The IAU 2000 resolutions for astrometry, celestial mechanics, and metrology in the relativistic framework: explanatory supplement. Astron. J. **126**(6), 2687 (2003)
Seidelmann, P.K., Seago, J.: Relativistically correct celestial reference systems. In: AIAA/AAS Astrodynamics Specialist Conference and Exhibit. Lake Tahoe, CA, Paper AAS 05-352 (2005)
Urban, S.E., Seidelmann, P.K. (eds.): Explanatory Supplement to the Astronomical Almanac. University Science Books, Mill Valley, California (2012)
Yang, T.J.Y., Han, C.: COMPASS/BeiDou Coordinate and Time Reference Systems, Global Navigation Satellite Systems. Report of a Joint Workshop of the National Academy of Engineering and the Chinese Academy of Engineering (2012)

Chapter 4
Central Force Motion

4.1 Introduction

If the force causing accelerated motion of a particle passes through a fixed point, then this yields *central force motion*. The center of force is the fixed point. Central force motion predominates in the universe:

1. Planetary motion has the force of attraction passing through the Sun.
2. One double star moves around the other due to a gravitational central force.
3. Natural and artificial satellites move around a planet.
4. Interplanetary probes have attraction forces from the Sun or a planet.

The *inverse square law* of Newton, to be discussed in Sect. 4.6, applies to astronomical and astrodynamical motions (McCuskey 1963, p. 19).

4.2 Law of Areas

We take a particle of mass m at $\mathbf{r}$ relative to O. Under a central force $\mathbf{F}$, directed toward or away from O, as shown in Fig. 4.1, the particle moves on the curve C. We denote $\mathbf{F}$ by $F\mathbf{u}_r$, where $\mathbf{u}_r$ is a unit vector along $\mathbf{r}$. According to Newton's second law, $m\dot{\mathbf{v}} = F\mathbf{u}_r$, where m is a constant mass, and

$$\mathbf{r} \times m\dot{\mathbf{v}} = \mathbf{r} \times F\mathbf{u}_r = \mathbf{0} \tag{4.1}$$

since $\mathbf{r}$ and $\mathbf{u}_r$ are collinear.

P. Gurfil, P.K. Seidelmann, *Celestial Mechanics and Astrodynamics: Theory and Practice*, Astrophysics and Space Science Library 436,
DOI 10.1007/978-3-662-50370-6_4

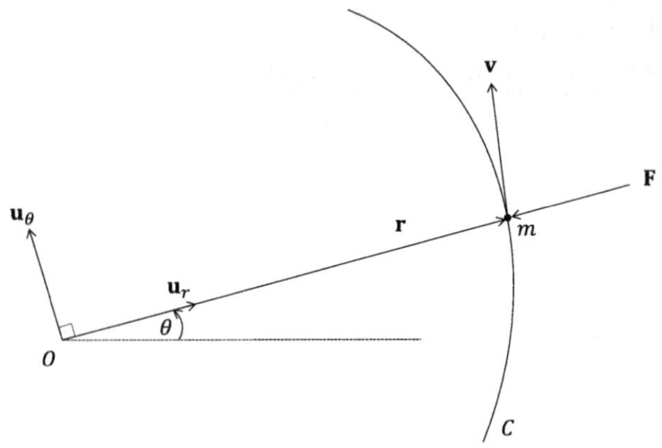

Fig. 4.1 The areal velocity due to a central force

The areal velocity (see Sect. 2.5) is

$$\dot{\mathbf{A}} = \frac{1}{2}\mathbf{r} \times \mathbf{v} \tag{4.2}$$

so

$$\frac{d\dot{\mathbf{A}}}{dt} = \frac{1}{2}[(\dot{\mathbf{r}} \times \mathbf{v}) + (\mathbf{r} \times \dot{\mathbf{v}})] = \frac{1}{2}\mathbf{r} \times \dot{\mathbf{v}} = \mathbf{0} \tag{4.3}$$

This is because $\dot{\mathbf{v}}$ is along $\mathbf{r}$ from above. Therefore, $\dot{\mathbf{A}} = constant$. The areal velocity is

$$\dot{\mathbf{A}} = \frac{1}{2}r^2\dot{\theta}\,\hat{\mathbf{u}}_A \tag{4.4}$$

where the unit vector $\hat{\mathbf{u}}_A$ is perpendicular to $\mathbf{r}$ and $\mathbf{v}$, thus perpendicular to the plane defined by $\mathbf{r}$ and $\mathbf{v}$. Since $\hat{\mathbf{u}}_A$ is a constant vector, then the mass particle, moving under a central force, moves in an orbit in a plane.

The constant areal velocity magnitude is $\frac{1}{2}h$. The area described in time t is $A = \frac{1}{2}ht + c$, where c is a constant of integration. The law of areas, mentioned in Sect. 1.2 as Kepler's second law, holds for any central force of planetary motion. The area swept out by the radius vector is directly proportional to time. Conversely, if the area swept out by the radius vector is directly proportional to time, the force is a central force. This is shown by assuming $A = pt + q$, p and q are constants. Because $\dot{A} = p$, from Eq. (4.4) $r^2\dot{\theta} = 2p$, and

$$2\frac{dp}{dt} = 0 = \frac{d}{dt}(r^2\dot{\theta}) = \frac{d\dot{A}}{dt} = r\dot{r}\dot{\theta} + r^2\ddot{\theta} = 0 \tag{4.5}$$

The expression $r\dot{r}\dot{\theta} + r\ddot{\theta}$ is the acceleration component perpendicular to the radius vector. This vanishes by Eq. (4.5), so there is no acceleration and no force perpendicular to r. Thus, the line of action of the acceleration, or force, passes through the origin. Consequently, we showed that

$$\frac{d^2\mathbf{r}}{dt^2} = (\ddot{r} - r\dot{\theta}^2)\hat{\mathbf{u}}_r + (r\ddot{\theta} + 2\dot{r}\dot{\theta})\hat{\mathbf{u}}_\theta \tag{4.6}$$

where the first term on the right is the radial component and the second the traverse component. Thus, the traverse component is zero, the acceleration is all in the radial component, and the force is a central force (McCuskey 1963, pp. 19–31).

4.3 Linear and Angular Velocities

Results follow from the fact that force, or acceleration, is central. Let p be the perpendicular distance from O, the origin, to the tangent T of the curve at a point P, as shown in Fig. 4.2. Then

$$2\dot{\mathbf{A}} = \mathbf{r} \times \mathbf{v} = h\,\hat{\mathbf{u}}_A \tag{4.7}$$

But by the cross product definition, $\mathbf{r} \times \mathbf{v} = rv \sin\alpha\, \hat{\mathbf{u}}_A$ and $r \sin\alpha = p$, so $v = h/p$. The linear speed, v, of a particle moving at point P, under a central force, is inversely proportional to the perpendicular distance from O to the instantaneous tangent to the orbit at P.

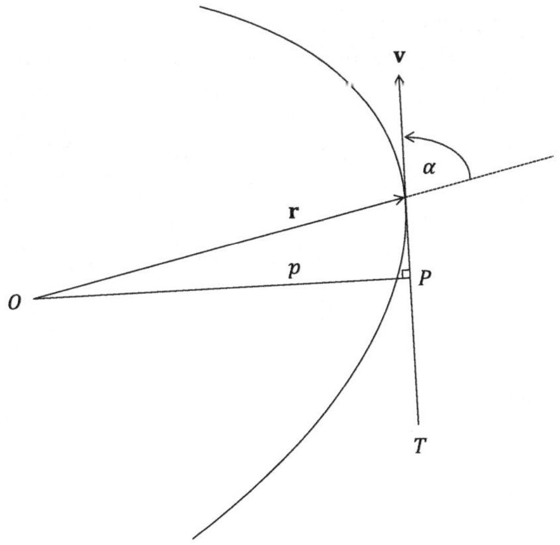

Fig. 4.2 Understanding linear and angular velocities

The angular speed of the particle at P is $\dot{\theta} = h/r^2$, since $\dot{\mathbf{A}} = \frac{1}{2}r^2\dot{\theta}\hat{\mathbf{u}}_A = \frac{1}{2}h\hat{\mathbf{u}}_A$. The angular speed of a particle moving under a central force varies inversely as the square of the distance from the origin (McCuskey 1963, pp. 20–21).

4.4 Integrals of Angular Momentum and Energy

As in the previous section, the equation of motion of a mass, m, is $m\dot{\mathbf{v}} = F\hat{\mathbf{u}}_r$, where F is the magnitude of a central force. The angular momentum is $\mathbf{L} = \mathbf{r} \times m\mathbf{v}$. So

$$\dot{\mathbf{L}} = \mathbf{r} \times m\dot{\mathbf{v}} = \mathbf{r} \times F\hat{\mathbf{u}}_r = \mathbf{0} \tag{4.8}$$

Then the angular momentum $\mathbf{L}$ of a particle moving under a central force remains constant in magnitude and direction; this is the *conservation of angular momentum*. $\mathbf{L}$ is perpendicular to the orbital plane. From Eqs. (4.4), (4.6), and (4.7),

$$\dot{\mathbf{A}} = \frac{1}{2}r^2\dot{\theta}\,\hat{\mathbf{u}}_A, \; 2\dot{\mathbf{A}} = \mathbf{r} \times \mathbf{v} = h\,\hat{\mathbf{u}}_A \tag{4.9}$$

We see that

$$\mathbf{L} = \mathbf{r} \times m\mathbf{v} = 2m\dot{\mathbf{A}} = mr^2\dot{\theta}\,\hat{\mathbf{u}}_A = mh\,\hat{\mathbf{u}}_A \tag{4.10}$$

so the magnitude of the angular momentum is

$$L = mr^2\dot{\theta} = mh \tag{4.11}$$

Then one integral of the equations of motion of the mass particle is that of constant angular momentum mh.

From the equation of motion of a mass we can write the scalar product $m\mathbf{v} \cdot \dot{\mathbf{v}} = F\mathbf{v} \cdot \hat{\mathbf{u}}_r$. But

$$\mathbf{v} \cdot \dot{\mathbf{v}} = \frac{d}{dt}\left(\frac{1}{2}\mathbf{v} \cdot \mathbf{v}\right) = \frac{d}{dt}\left(\frac{1}{2}v^2\right) \tag{4.12}$$

and $\mathbf{v} = \dot{\mathbf{r}} = \dot{r}\hat{\mathbf{u}}_r + r\dot{\theta}\hat{\mathbf{u}}_\theta$, where $\hat{\mathbf{u}}_r$ is perpendicular to $\hat{\mathbf{u}}_\theta$ (see Fig. 4.1). Since

$$\mathbf{v} \cdot \hat{\mathbf{u}}_r = \left(\dot{r}\hat{\mathbf{u}}_r + r\dot{\theta}\hat{\mathbf{u}}_\theta\right) \cdot \hat{\mathbf{u}}_r = \dot{r} \tag{4.13}$$

then

$$\frac{d}{dt}\left(\frac{1}{2}mv^2\right) = F\frac{dr}{dt} \tag{4.14}$$

If F depends only on the length of r, that is $F = F(r)$, then on integrating Eq. (4.14) we have

$$\frac{1}{2}mv^2 = \int F(r)\,dr + E \tag{4.15}$$

where E is a constant of integration depending upon the initial conditions of motion.

The integral in Eq. (4.15) is the work done by F changing the position along the orbit of the particle. From the analytical form of $F(r)$, the integral can be calculated; $\mathbf{F} = F(r)\hat{\mathbf{u}}_r$ is a *conservative force*.

A potential energy, $V(r)$, exists such that $F(r) = -dV/dr$. So, Eq. (4.15) can be rewritten in the form

$$\frac{1}{2}mv^2 + V(r) = E \tag{4.16}$$

From this equation, the sum of the kinetic energy and the potential energy of a particle moving under a central force is constant. This is the law of conservation of energy (see also Sect. 2.10), and E is a second integral of the equations of motion. Solving Eq. (4.16) for v, we have

$$v = \pm\sqrt{\frac{2}{m}(E - V)} \tag{4.17}$$

The square root depends only on r through the $V(r)$, so, at a given distance r from the center of force, the speed is the same in all orbits of the same total energy, regardless of their shapes (McCuskey 1963, pp. 21–23).

4.5 Equation of the Orbit

Denote the force by $F(r)$. From Newton's second law, the equations of motion are

$$m(\ddot{r} - r\dot{\theta}^2) = F(r), \quad mr^2\dot{\theta} = mh \tag{4.18}$$

These differential equations, in their original form, were each second order, so a complete solution would have four integration constants. Two are the angular momentum and energy integrals. Two more constants are needed from Eqs. (4.18). Two initial conditions are needed to fix the orbit.

Let $u = 1/r$ be the reciprocal radius vector. From the second of Eqs. (4.18), $\dot{\theta} = hu^2$. Also,

$$\dot{r} = -\frac{1}{u^2}\dot{u} = -h\frac{du}{d\theta} \tag{4.19}$$

and

$$\ddot{r} = -h^2u^2\frac{d^2u}{d\theta^2} \tag{4.20}$$

After simplification, the first of Eqs. (4.18) is

$$\frac{du^2}{d\theta^2} + u = -\frac{F(1/u)}{mh^2u^2} \tag{4.21}$$

This second-order differential equation for u, a function of θ, gives the polar equation of the orbit, when the proper force law in terms of u, is substituted. If the force varies as an integer power, n, of the distance, namely

$$F(r) = \alpha r^n,\ F(1/u) = \alpha u^{-n} \tag{4.22}$$

Then Eq. (4.21) becomes

$$\frac{d^2u}{d\theta^2} + u = -\frac{\alpha u^{-n-2}}{mh^2} \tag{4.23}$$

This may be integrated directly. Multiplying both sides by $2du/d\theta$ and rewriting

$$\frac{d}{d\theta}\left[\left(\frac{du}{d\theta}\right)^2 + u^2\right] = -\beta u^{-(n+2)}\frac{du}{d\theta} \tag{4.24}$$

where $\beta = 2\alpha/\left(mh^2\right)$. Integration yields

$$\left(\frac{du}{d\theta}\right)^2 + u^2 = \frac{\beta u^{-(n+1)}}{n+1} + C,\ n \neq -1 \tag{4.25}$$

where C is a constant of integration. Integrating a second time, we obtain

$$\int_{u_0}^{u} \frac{du}{\sqrt{c - u^2 + \frac{\beta u^{-(n+1)}}{n+1}}} = \theta - \theta_0,\ n \neq -1 \tag{4.26}$$

where (u_0, θ_0) define an initial starting point in the orbit.

With n, Eq. (4.26) defines u as a function of θ, which is the polar equation of the orbit. $n = -1$ is excluded, because in that case log u replaces the power of u in the right-hand side of Eq. (4.15). The integral of Eq. (4.26) is

$$\int (a + bu^2 + cu^{-n-1})^{-1/2}\, du \tag{4.27}$$

where n is an integer. Integral tables result in trigonometric functions, if the power of u does not exceed 2. So n is restricted to $n = -1, -2, -3$. Since $n = -1$ is excluded, we have $n = -2, -3$.

When $n = 1$, the integral in Eq. (4.26) is

$$\int (bu^4 + au^2 + c)^{-1/2}\, u\, du \tag{4.28}$$

A substitution of $v = u^2$ and $dv = 2udu$ produces

$$\int (bv^2 + av + c)^{-1/2}\, dv \tag{4.29}$$

which integrates into trigonometric functions. When the central force varies as r^n with $n = +1, -2, -3$, the polar equation of the orbit includes trigonometric functions. Under special circumstances, higher powers of r yield equations in circular functions. When $n = +5, +3, 0, -4, -5, -7$, the integral results in elliptic functions. We will return to this issue in Chap. 12 (McCuskey 1963, pp. 23–24).

4.6 Inverse Square Law

In most astrodynamical and astronomical applications, the orbits are derived from the inverse square law of Newton. We shall illustrate the analysis of the previous section by using

$$F(r) = -\frac{GMm}{r^2} \tag{4.30}$$

where GM is the power of the force center, and m is the mass being accelerated. The equations of motion, produced by this force law, yield for the orbit

$$\frac{d^2u}{d\theta^2} + u = \frac{GM}{h^2} \tag{4.31}$$

We shall integrate Eq. (4.31) to obtain

$$u = \frac{GM}{h^2} + A\cos(\theta - \theta_0) \tag{4.32}$$

where θ_0 and A are constants of integration. So,

$$r = \frac{h^2/GM}{1 + (Ah^2/GM)\cos(\theta - \theta_0)} \tag{4.33}$$

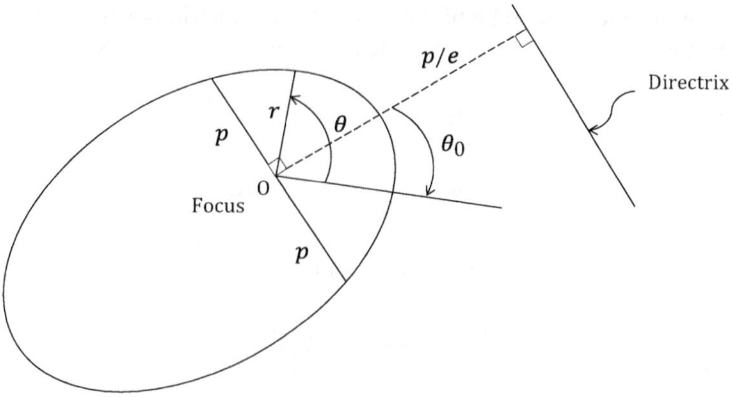

Fig. 4.3 Geometry of an ellipse

This is the polar equation of the orbit, which is the standard equation for a conic section (cf. Eq. (1.2)),

$$r = \frac{p}{1 + e\cos(\theta - \theta_0)} \tag{4.34}$$

where e is eccentricity and p/e is the distance from the focus to the directrix, as shown in Fig. 4.3.

From Eqs. (4.33) and (4.34),

$$p = \frac{h^2}{GM}, \ e = \frac{Ah^2}{GM} \tag{4.35}$$

As discussed in Sect. 1.3, the shape of the conic depends on p and e; $e < 1$ is an ellipse, $e = 1$ is a parabola, and $e > 1$ is a hyperbola. These geometrical parameters depend on the integration constant A. The physical constants of the system are h, G, and M.

Equation (4.25) shows that for the inverse square law

$$\left(\frac{du}{d\theta}\right)^2 + u^2 = \frac{2GMu}{h^2} + c \tag{4.36}$$

The velocity components in the orbit are $\dot{r} = -h du/d\theta$ and $r\dot{\theta} = hu$. So the speed v is

$$v = \left[\left(h\frac{du}{d\theta}\right)^2 + h^2u^2\right]^{1/2} \tag{4.37}$$

$-GM/r = -GMu$ is the potential energy per unit mass. Thus, Eq. (4.36) shows that the total energy in the system stays constant. So

$$\frac{1}{2}mv^2 - GMmu = \frac{1}{2}h^2cm = E \tag{4.38}$$

where E is the total energy; $c = 2E/(mh^2)$.

At the ends of the traverse axis of the conic, we have $\dot{r} = -hdu/d\theta = 0$. From Eq. (4.36),

$$u^2 - \frac{2GMu}{h^2} - \frac{2E}{mh^2} = 0 \tag{4.39}$$

and

$$u = \frac{GM}{h^2}\left[1 \pm \sqrt{1 + \frac{2Eh^2}{mG^2M^2}}\right] \tag{4.40}$$

The values of u are at the ends of the transverse axis of the conic. From Eqs. (4.32) and (4.40) we have

$$u_{\text{max}} = \frac{GM}{h^2} + A = \frac{GM}{h^2}\left[1 + \sqrt{1 + \frac{2Eh^2}{mG^2M^2}}\right] \tag{4.41}$$

$$u_{\text{min}} = \frac{GM}{h^2} - A = \frac{GM}{h^2}\left[1 - \sqrt{1 + \frac{2Eh^2}{mG^2M^2}}\right] \tag{4.42}$$

Hence,

$$A = \frac{GM}{h^2}\sqrt{1 + \frac{2Eh^2}{mG^2M^2}} \tag{4.43}$$

$e = Ah^2/(GM)$, so we have a relation between the eccentricity and the total energy of the particle, namely

$$e = \sqrt{1 + \frac{2Eh^2}{mG^2M^2}} \tag{4.44}$$

so, if

I $E = 0, e = 1$ the orbit is a parabola.
II $E < 0, e < 1$ the orbit is an ellipse.
III $E > 0, e > 1$ the orbit is a hyperbola.

Case I

Let q denote the distance from the focus to the vertex. Then

$$p = \frac{h^2}{GM} = 2q \tag{4.45}$$

The orbital equation is

$$r = \frac{2q}{1 + \cos(\theta - \theta_0)} \tag{4.46}$$

At a distance r from the center of force, the speed in the orbit is, by Eq. (4.38),

$$v_p = \sqrt{2GM/r} \tag{4.47}$$

This is the resulting speed, if a particle moved from an infinite distance to r, under the inverse square force. Or, if the particle has this speed at a distance r, it will recede indefinitely far from the center of force. This speed v_p is the escape velocity from the force center (see also Sect. 2.10.2).

Case II

Let q denote the radius vector of the vertex of the ellipse nearest to the origin, and q' denote the radius vector at the maximum distance from the origin. The origin is the focus. The major axis of the ellipse is $2a$. Then $q = p/(1 + e)$, $q' = p/(1 - e)$, and $q + q' = 2a$. So $p = a(1 - e^2)$, and the equation of the ellipse is

$$r = \frac{a(1 - e^2)}{1 + e\cos(\theta - \theta_0)} \tag{4.48}$$

$p = h^2/(GM)$, and the areal velocity constant is

$$h = \sqrt{GMa(1 - e^2)} \tag{4.49}$$

Substituting this into Eq. (4.44) and simplifying, the total energy is

$$E = -\frac{GMm}{2a} \tag{4.50}$$

Equation (4.38) yields, for the orbital speed at a distance r from the force center,

$$v^2 = GM\left(\frac{2}{r} - \frac{1}{a}\right) \tag{4.51}$$

We can write an expression for the period in elliptic motion. Let A be the area swept out by the radius vector in time t. Then, from $A = \frac{1}{2}ht + c$ given previously,

$$A = \frac{1}{2}\sqrt{GMa(1 - e^2)}\, t + c \tag{4.52}$$

where c is a constant of integration. In a period P, the radius vector sweeps out an area

$$\pi ab = \pi a^2\sqrt{1 - e^2} = \frac{1}{2}\sqrt{GMa(1 - e^2)}P \tag{4.53}$$

where $b = a\sqrt{1 - e^2}$ is the semiminor axis. So,

$$P = \frac{2\pi a^{3/2}}{\sqrt{GM}} \tag{4.54}$$

This is *Kepler's third law*, first mentioned in Sect. 1.2, which is applied to the planetary system. M is very nearly the same for each planet, and is approximately the mass of the Sun. But Kepler's third law is not quite exact, since variations of the masses of the planets cause slight variations. Accepting the approximation, if two planets have periods P_1 and P_2, and semimajor axes a_1 and a_2, then the relation

$$\left(\frac{P_1}{P_2}\right)^2 = \left(\frac{a_1}{a_2}\right)^3 \tag{4.55}$$

gives Kepler's third law.

Case III

Let $2a$ denote the transverse axis of the conic. Then the geometry of the orbit indicates $p = a(e^2 - 1)$ and

$$r = \frac{a(e^2 - 1)}{1 + e\cos(\theta - \theta_0)} \tag{4.56}$$

The areal velocity constant is

$$h = \sqrt{GMa(e^2 - 1)} \tag{4.57}$$

and the total energy will be

$$E = \frac{GMm}{2a} \tag{4.58}$$

At a distance r from the force center , the speed will be given by

$$v^2 = GM\left[\frac{2}{r} + \frac{1}{a}\right] \tag{4.59}$$

From motion under an inverse square force directed toward a fixed center, a deduction of Kepler's first law of planetary motion is implied. The differential equation defining the orbit yields, under proper conditions of total energy, an elliptical path with the force center at one focus. Kepler deduced this from observations of Mars. Kepler found Mars described an ellipse with the force center, the Sun, at one focus. The equation for such a curve is

$$r = \frac{p}{1 + e\cos\theta} \tag{4.60}$$

when the major axis of the ellipse coincides with the polar axis of the coordinates. On the assumption that the force governing the motion is a central force, we can show directly from Eq. (4.21) that the law of force must be that of an inverse square. From Eq. (4.60),

$$u = \frac{1 + e\cos\theta}{p} \tag{4.61}$$

Then

$$\frac{d^2u}{d\theta^2} + u = \frac{1}{p} \tag{4.62}$$

so we have

$$F\left(\frac{1}{u}\right) = -\frac{mh^2u^2}{p} \tag{4.63}$$

The force law is then

$$F(r) = -\frac{mh^2}{pr^2} \tag{4.64}$$

which is an inverse square law (McCuskey 1963, pp. 25–39).

4.6.1 Eccentricity Vector

We have seen in Sect. 4.4 that the angular momentum vector $\mathbf{L} = mh\hat{\mathbf{u}}_A$ is conserved in central force motion. Another important quantity in central force motion is the *Laplace–Runge–Lenz vector*, which determines the shape and orientation of orbits.

If the orbital motion is determined by an inverse square law, e.g. Newtonian gravity, the Laplace–Runge–Lenz vector is referred to as the *eccentricity vector*, and is defined as

$$\mathbf{e} = \frac{\mathbf{v} \times \mathbf{L}}{GMm} - \hat{\mathbf{u}}_r \tag{4.65}$$

By taking the time derivative of Eq. (4.65), keeping in mind that $\dot{\mathbf{L}} = \mathbf{0}$ and recalling that $\mathbf{r} \times (\mathbf{r} \times \mathbf{v}) = \mathbf{r}(\mathbf{r} \cdot \mathbf{v}) - \mathbf{v}(\mathbf{r} \cdot \mathbf{r})$, it can be shown that $\dot{\mathbf{e}} = \mathbf{0}$, so the eccentricity vector is fixed in magnitude and orientation for Newtonian gravity. Furthermore, $\mathbf{r} \cdot \mathbf{e} = re \cos f$, and the eccentricity is $e = \|\mathbf{e}\|$. The angle f is the true anomaly, which we have seen in Sect. 1.3.1. Thus, the unit vector $\hat{\mathbf{e}} = \mathbf{e}/e$ points to the *periapsis*, the point on the orbit closest to the attraction center.

4.6.2 From Orbit to Force Law

The converse of the previously discussed problem can be solved by use of Eq. (4.21).

$$\frac{d^2u}{d\theta^2} + u = -\frac{F(1/u)}{mh^2u^2} \tag{4.66}$$

Suppose a mass particle is moving on the circle $r = 2\cos\theta$ under the action of a force directed toward the origin, as shown in Fig. 4.4. Then with $\phi(u) = F(1/u)$, from Eq. (4.66) we have

$$\phi(u) = -mh^2u^2 \left[\frac{d^2u}{d\theta^2} + u \right] = -8mh^2u^5 \tag{4.67}$$

Thus, the force $F(r) = -8mh^2u^5$ is an inverse fifth power law.

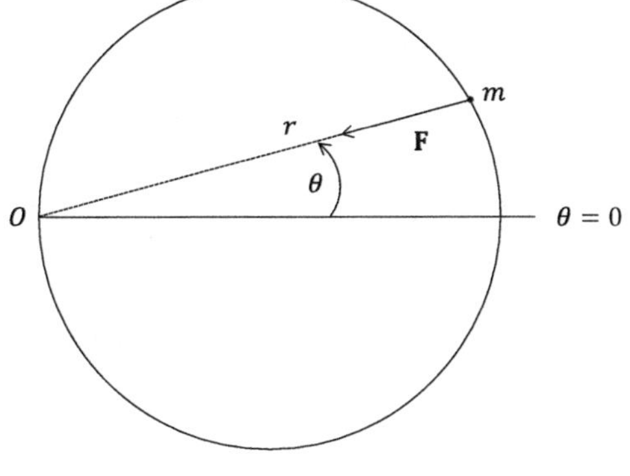

Fig. 4.4 Mass particle on a circular orbit

4.7 Einstein's Modification of the Orbit Equation

The theory of relativity requires a modification of Newton's law. The equation of the orbit becomes

$$\frac{du^2}{d\theta^2} + u = \frac{\mu}{h^2} + \alpha u^2 \tag{4.68}$$

where $\alpha = 3\mu/c^2$ and $\mu = GM$; c is the speed of light. α is a small quantity, so the square can be neglected. The ratio αu^2 to u/h^2 is three times the square of the transverse velocity in units of c.

Equation (4.68) can be solved by approximations. If we neglect α, leaving only Newtonian motion, then

$$u = \frac{\mu}{h^2}\left[1 + e\cos(\theta - \theta_0)\right] \tag{4.69}$$

For a better approximation, substitute this into the right side of Eq. (4.68),

$$\frac{du^2}{d\theta^2} + u = \frac{\mu}{h^2} + \frac{\alpha\mu^2}{h^4} + 2\alpha\frac{\mu^2}{h^4}e\cos(\theta - \theta_0) + \frac{1}{2}\alpha\frac{\mu^2}{h^4}e^2\left\{1 + \cos\left[2(\theta - \theta_0)\right]\right\} \tag{4.70}$$

If we neglect the μ/h^2 terms, we have

$$\frac{du^2}{d\theta^2} + u = \alpha\frac{\mu^2}{h^4}\left\{1 + 2e\cos(\theta - \theta_0) + \frac{1}{2}e^2\left\{1 + \cos\left[2(\theta - \theta_0)\right]\right\}\right\} \tag{4.71}$$

which has an integral

$$u = \alpha\frac{\mu^2}{h^4}\left\{1 + \frac{1}{2}e^2 + e\theta\sin(\theta - \theta_0) - \frac{1}{6}e^2\cos\left[2(\theta - \theta_0)\right]\right\} \tag{4.72}$$

Combining Eqs. (4.69) and (4.72) and letting $p = h^2/\mu$, we have a solution to Eq. (4.70),

$$u = \frac{1}{p}[1 + e\cos\,(\theta - \theta_0)] + \frac{\alpha}{p^2}\left\{\left(1 + \frac{1}{2}e^2\right) + e\theta\sin\,(\theta - \theta_0) - \frac{1}{6}e^2\cos\left[2(\theta - \theta_0)\right]\right\} \tag{4.73}$$

Consider this modification to Kepler's motion. The term

$$\frac{\alpha}{p^2}\left(1 + \frac{1}{2}e^2\right) \tag{4.74}$$

increases u by a very small constant quantity. The third term,

$$-\frac{\alpha}{6p^2}e^2\cos 2(\theta - \theta_0) \tag{4.75}$$

is very small and periodic. However, the term

$$\frac{\alpha}{p^2}e\theta\sin(\theta - \theta_0) \tag{4.76}$$

is periodic, and steadily increasing in amplitude as θ increases. So this is bound to have some effect with increased time. Considering only the observable effects,

$$u = \frac{1}{p}[1 + e\cos(\theta - \theta_0)] + \frac{\alpha e}{p^2}\theta\sin(\theta - \theta_0) \tag{4.77}$$

Let $k\theta = \alpha\theta/p$ and neglect α^2, then Eq. (4.77) can be written as

$$u = \frac{1}{p}\{1 + e[\cos(\theta - \theta_0) + k\theta\sin(\theta - \theta_0)]\} \approx \frac{1}{p}[1+e\cos(\theta-\theta_0-k\theta)] \tag{4.78}$$

$\theta_0 + k\theta$ is the angular coordinate of the perihelion. So the planet is moving in an ellipse with a moving line of apsides, which is a slowly rotating ellipse. The angular change of the line of apsides is $\Delta\omega = 2\pi\alpha/p$ per period. Substituting values for the planets we have

Mercury: $\Delta\omega = 43.03$ arcsec per century.
Venus: $\Delta\omega = 8.63$ arcsec per century.
Earth: $\Delta\omega = 3.84$ arcsec per century.

For Mercury this small effect was observable and remained unexplained until Einstein's relativity. It was fudged in Newcomb's planetary theories and called an *empirical term* (Danby 1962, pp. 66–67).

4.8 Universality of Newton's Law

Newton's law follows from Kepler's first two laws of planetary motion; therefore, any two bodies traveling around each other according to Kepler's first two laws are subject to Newton's law. When two stars are observed moving around each other, these are called *visual binaries*. In other cases of binaries, the two stars have different spectra, continuously shifting spectra, or eclipse each other. For visual binaries, when the fainter star is plotted with respect to the brighter, the orbits are ellipses, and the law of areas is followed. However, a projection of the true orbit is being plotted, so the brighter star is not at the focus. It is unlikely in any case that the orbital plane is perpendicular to the line of sight.

An ellipse always projects into another ellipse. Since the apparent orbit as plotted is an ellipse, the true orbit must be an ellipse. The law of areas of the apparent orbit will hold for the true orbit, because the law of areas depends on ratios, which are not affected by projections. For the true orbit, the theory of central forces and conservative fields applies, and an ellipse with the brighter star at the focus can be found. It can be shown that for visual binaries. elliptical motion requires the center of attraction to be at the focus of the ellipse.

So Newton's law is the only plausible law governing Keplerian motion within and exterior to the solar system. Also, this law has explained deviations from Keplerian motion, when the relativity effect is included, and has led to the correction of Kepler's third law (Danby 1962, pp. 73–76).

References

McCuskey, S.W.: Introduction to Celestial Mechanics. Addison-Wesley, Reading (1963)
Danby, J.: Fundamentals of Celestial Mechanics. The Macmillan Company, New York (1962)

Chapter 5
The Two-Body Problem

5.1 Introduction

Assume that the masses are spherically symmetrical and homogeneous in concentric layers. So they attract one another as if the mass were concentrated at spherical centers, i.e. gravitationally they act like two mass particles separated by the distance between the centers. The two masses are assumed to be isolated from other masses, so the only force acting is the inverse square force of their mutual attractions along the line joining the centers. In astronomical applications, the distance between centers is large, compared to the diameters of the spheres. This is not true for artificial satellites.

The dynamics of the motion of two masses presents two problems for celestial mechanics and astrodynamics:

1. Given the position and velocity, or three positions, of a mass as a function of time, find the elements of the orbit. This is the computation of orbits to be considered in Chap. 6.
2. Given the orbital elements, or parameters defining the orbital motion, find the position in space of the mass at a given time. This we will take up now (the first problem is by far the more difficult), with the first step being defining the *classical orbital elements*.

5.2 Classical Orbital Elements

Assume that a mass m is rotating counterclockwise in orbit when viewed from the planet's north pole. If the orbital plane and the fundamental plane of some reference frame intersect, then we define two points of interest on the line of intersection,

P. Gurfil, P.K. Seidelmann, *Celestial Mechanics and Astrodynamics: Theory and Practice*, Astrophysics and Space Science Library 436,
DOI 10.1007/978-3-662-50370-6_5

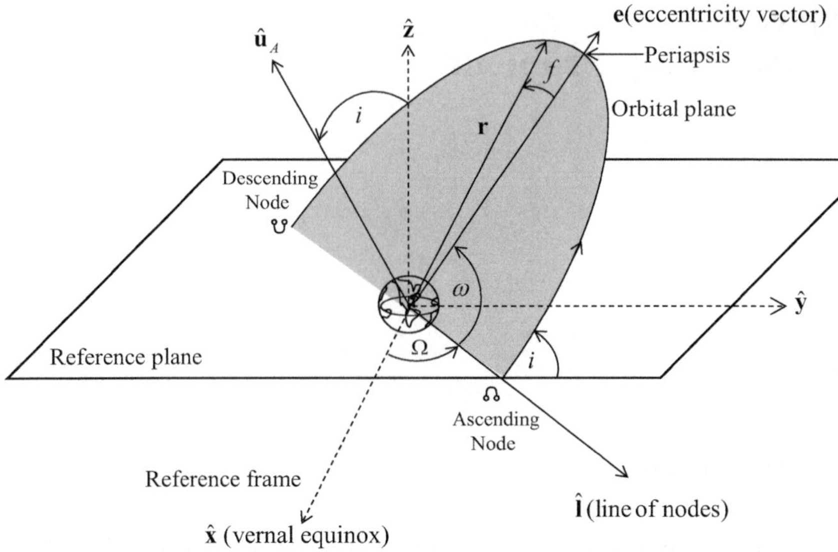

Fig. 5.1 Definitions of the right ascension of the ascending node, Ω, the argument of periapsis, ω, and the inclination, i. Also shown is the true anomaly, f

as shown in Fig. 5.1: The first is the *ascending node*, denoted by ☊. This point marks the location on the line of intersection when moving eastward; the second is the *descending node*, denoted by ☋. This point marks the location on the line of intersection when moving westward. The line connecting ☋ to ☊ is called the *line of nodes* (LON); we will use the notation $\hat{\mathbf{l}}$ to denote a unit vector that lies along the LON.

We now define three angles that determine the orientation of the orbital plane with respect to the reference frame: Ω, the *right ascension of the ascending node* (RAAN), also referred to as the *longitude of the ascending node*, an angle measured from the vernal equinox (see Sect. 3.1) to the LON; ω, the *argument of periapsis*, which is an angle measured from the LON to the eccentricity vector (see Sect. 4.6.1); and i, the *inclination*, an angle measured from the z axis of the reference frame, $\hat{\mathbf{z}}$, to the vector normal to the orbital plane, $\hat{\mathbf{u}}_A$. These angles are shown in Fig. 5.1. It is convenient to express the position vector in a *perifocal coordinate system*. This coordinate system is centered at the attraction center. The fundamental plane is the orbital plane. The unit vector $\hat{\mathbf{P}}$ is directed from the center to the periapsis (recall Sect. 4.6.1), $\hat{\mathbf{R}}$ is normal to the fundamental plane, positive in the direction of the orbital angular momentum vector, and $\hat{\mathbf{Q}}$ is pointed toward the point where the true anomaly is 90°, thus completing the right-hand Cartesian triad. Using the definition of the true anomaly (Sect. 4.6.1), we can write the position vector in the perifocal frame as

$$\mathbf{r}_p = r\,[\cos f,\ \sin f,\ 0]^T \tag{5.1}$$

where, as before, $r = p/(1 + e\cos f)$.

We can transform from the orbital frame to the reference frame using three consecutive clockwise rotations: a rotation about $\hat{\mathbf{u}}_A$ by $0 \le \omega \le 2\pi$, mapping the eccentricity vector, $\hat{\mathbf{e}}$, onto the LON, $\hat{\mathbf{l}}$; a rotation about $\hat{\mathbf{l}}$ by $0 \le i \le \pi$, mapping $\hat{\mathbf{u}}_A$ onto $\hat{\mathbf{z}}$; and a rotation about $\hat{\mathbf{z}}$ by $0 \le \Omega \le 2\pi$, mapping $\hat{\mathbf{l}}$ onto $\hat{\mathbf{x}}$.

The composite rotation, transforming any vector in the orbital frame into the inertial frame is given by

$$\mathbf{T} = \begin{bmatrix} c_\Omega c_\omega - s_\Omega s_\omega c_i & -c_\Omega s_\omega - s_\Omega c_\omega c_i & s_\Omega s_i \\ s_\Omega c_\omega + c_\Omega s_\omega c_i & -s_\Omega s_\omega + c_\Omega c_\omega c_i & -c_\Omega s_i \\ s_\omega s_i & c_\omega s_i & c_i \end{bmatrix} \tag{5.2}$$

where we used the compact notation $c_x = \cos x$, $s_x = \sin x$.

Transforming into inertial reference coordinates using Eqs. (5.1) and (5.2), we obtain the position vector

$$\mathbf{r} = \frac{p}{1 + e\cos f}\begin{bmatrix} c_{f+\omega}c_\Omega - c_i s_{f+\omega} s_\Omega \\ c_i c_\Omega s_{f+\omega} + c_{f+\omega} s_\Omega \\ s_i s_{f+\omega} \end{bmatrix} \tag{5.3}$$

The true anomaly f depends on time and on the epoch of observation, T. Thus, the inertial position and velocity depend on time t and the classical orbital elements, given by

$$\{a,\ e,\ i,\ \Omega,\ \omega,\ T\} \tag{5.4}$$

5.2.1 *Osculating Orbital Elements*

In the two-body problem, the orbital elements are constant; f is time-varying. However, in the presence of perturbations and/or thrust forces, the orbital elements may become time-varying, and are referred to as *osculating orbital elements*.

In general, elliptical motion constitutes a correct approximation to the real motion observed in the solar system. Thus, for example if, starting from an instant t_0, all the perturbing forces were neglected, the movement of a body would become exactly elliptical. It would represent the real movement quite well for a certain time, even though strictly speaking, it would not be identical with the real movement as regards position and velocity, except at the instant t_0. The elements of an ellipse that would be followed by a body after a specific time t are thus said to be *osculating*, or *instantaneous*, if starting from this instant, all the forces with the exception of the central force were to disappear. The elements of such an unperturbed orbit can be defined at any instant; they correspond to the elliptical orbit followed by a moving body, which would have at the given instant the same position and velocity as the real body. As in fact the real orbit is simply tangential to the *osculating orbit*, at an instant $t+\delta t$ the osculating orbit will be different, with different osculating elements. It follows that the osculating elements in perturbed motion are no longer constant, but are functions of time.

Osculating elements can be used to describe the *perturbed motion* of a body. They possess the advantage of having a precise and simple geometrical significance while having small variations.

The coordinates and velocity components of perturbed motion at an instant t are those which would be obtained at this instant t, assuming that the orbit is elliptical, from elements equal to the osculating elements at the same time t.

5.2.2 Nonsingular Orbital Elements

While the angles Ω, i, ω may become degenerate is some cases (for instance, ω is undefined for circular orbits; both ω and Ω are undefined for equatorial orbits), the position and velocity vectors are always well-defined. However, occasionally alternative orbital elements are used to alleviate these deficiencies. These alternative elements are collectively referred to as *nonsingular orbital elements*, see Sect. 10.12. A thorough survey of these elements was performed by Hintz (2008).

5.3 Motion of the Center of Mass

Let an origin, O, define an inertial system with Newton's laws of motion. The positions of two masses are given by vectors $\mathbf{r}_1$ and $\mathbf{r}_2$, and $\mathbf{R}$ is the vector to the center of mass of the pair, C. $\mathbf{r}$ is the position vector of m_2 relative to m_1, as shown in Fig. 5.2.

From Newton's law of gravitation, the force on m_1 due to m_2 is $\frac{1}{r^2}k^2 m_1 m_2 \hat{\mathbf{u}}_r$ and that on m_2 due to m_1 is $-\frac{1}{r^2}k^2 m_1 m_2 \hat{\mathbf{u}}_r$. $\hat{\mathbf{u}}_r$ is a unit vector in the direction of $\mathbf{r}$, and k^2 is the constant of gravitation. The reason and significance of the notation k^2 will become apparent later. The equations of motion are

$$m_1 \ddot{\mathbf{r}}_1 = \frac{k^2 m_1 m_2}{r^3}\mathbf{r} \tag{5.5}$$

$$m_2 \ddot{\mathbf{r}}_2 = -\frac{k^2 m_1 m_2}{r^3}\mathbf{r} \tag{5.6}$$

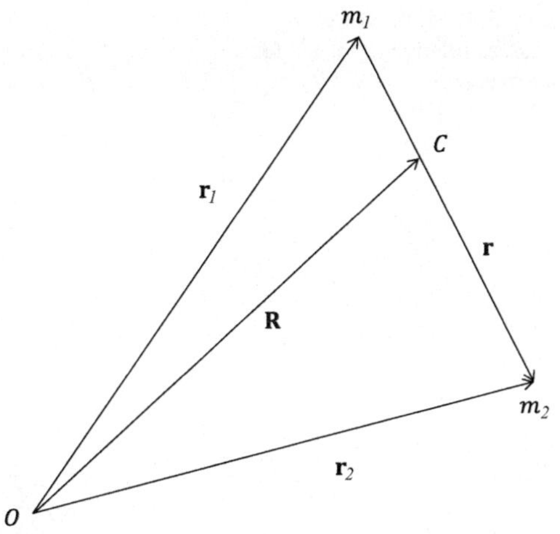

Fig. 5.2 Motion of the center of mass

Adding Eqs. (5.5) and (5.6) and integrating twice, we have

$$m_1\mathbf{r}_1 + m_2\mathbf{r}_2 = \mathbf{c}_1 t + \mathbf{c}_2 \tag{5.7}$$

where $\mathbf{c}_1$ and $\mathbf{c}_2$ are vector constants. The left side of Eq. (5.7) is $M\mathbf{R}$, by the definition of the *center of mass*, with $M = m_1 + m_2$. Thus,

$$\mathbf{R} = \frac{\mathbf{c}_1 t}{M} + \frac{\mathbf{c}_2}{M} \tag{5.8}$$

so the center of mass moves uniformly in a straight line in space.

This is in agreement with previous results and what one would expect, since there is no external force acting on this system. This result is applicable to double star observations, but of little interest otherwise (McCuskey 1963, pp. 32–33).

5.4 Relative Motion

The motions of m_1 and m_2 relative to the center of mass can be derived as follows. Let $\mathbf{r}_1 = \mathbf{R} + \mathbf{r}'_1$ and $\mathbf{r}_2 = \mathbf{R} + \mathbf{r}'_2$, where $\mathbf{r}'_1$ and $\mathbf{r}'_2$ denote position vectors to m_1 and m_2 from the center of mass, C, respectively, as shown in Fig. 5.3.

Then $\mathbf{r} = \mathbf{r}'_2 - \mathbf{r}'_1$. Since $\ddot{\mathbf{R}} = \mathbf{0}$, from definitions of $\mathbf{r}'_1$ and $\mathbf{r}'_2$, $m_1\ddot{\mathbf{r}}_1 = m_1\ddot{\mathbf{r}}'_1$ and $m_2\ddot{\mathbf{r}}_2 = m_2\ddot{\mathbf{r}}'_2$. Thus Eqs. (5.5) and (5.6) become

$$m_1\ddot{\mathbf{r}}'_1 = \frac{k^2 m_1 m_2(\mathbf{r}'_2 - \mathbf{r}'_1)}{r^3} \tag{5.9}$$

$$m_2\ddot{\mathbf{r}}'_2 = -\frac{k^2 m_1 m_2(\mathbf{r}'_2 - \mathbf{r}'_1)}{r^3} \tag{5.10}$$

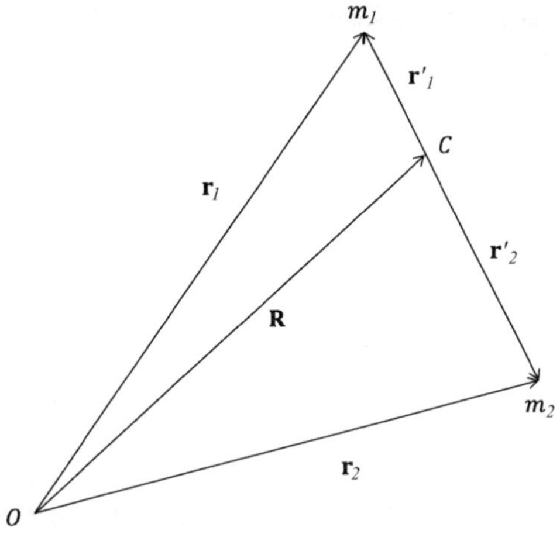

Fig. 5.3 Motion of two masses relative to the center of mass

and $m_1\mathbf{r}'_1 + m_2\mathbf{r}'_2 = 0$ due to the center of mass definition. So $\mathbf{r}'_2$ can be eliminated from Eq. (5.9) and $\mathbf{r}'_1$ can be eliminated from Eq. (5.10). Thus,

$$m_1\ddot{\mathbf{r}}'_1 = -k^2 m_1 m_2 \left[1 + \frac{m_1}{m_2}\right]\frac{\mathbf{r}'_1}{r^3} \tag{5.11}$$

$$m_2\ddot{\mathbf{r}}'_2 = -k^2 m_1 m_2 \left[1 + \frac{m_2}{m_1}\right]\frac{\mathbf{r}'_2}{r^3} \tag{5.12}$$

Since

$$r = \frac{M}{m_2}r'_1 = \frac{M}{m_1}r'_2 \tag{5.13}$$

we may write by dividing though by m_1 or m_2 and replacing the sum of the masses by M, and then substituting from above,

$$\ddot{\mathbf{r}}'_1 = -\frac{k^2 M}{r^3}\mathbf{r}'_1 = -k^2\left(\frac{m_2^3}{M^2}\right)\frac{\mathbf{r}'_1}{r_1'^3} \tag{5.14}$$

$$\ddot{\mathbf{r}}'_2 = -\frac{k^2 M}{r^3}\mathbf{r}'_2 = -k^2\left(\frac{m_1^3}{M^2}\right)\frac{\mathbf{r}'_2}{r_2'^3} \tag{5.15}$$

The accelerations of m_1 and m_2 relative to the center of mass are given. They are the same as Eqs. (5.5) and (5.6), with m_1 and m_2, respectively, replaced by the adjusted effective mass.

From Eqs. (5.14) and (5.15) and constants $\mathbf{c}_1$ and $\mathbf{c}_2$ of Eq. (5.8), the positions of m_1 and m_2 can be determined for any time. However, the constants cannot be determined, because they are with respect to an origin fixed in space. So the solution must be for one mass with respect to the other.

Consider m_1 as the origin of the two-body system. Then, from the first parts of Eqs. (5.14) and (5.15)

$$\ddot{\mathbf{r}} = -\frac{k^2 M}{r^3}\mathbf{r} \tag{5.16}$$

where $\mathbf{r}$ is the relative radius vector. This is the acceleration of m_2 around m_1. In a planetary system, m_1 is the Sun and m_2 is the planet. In the case of a satellite and a planet, m_1 is the planet and m_2 is the satellite.

For computations, the equations of relative motion can be expressed in Cartesian form, with unit vectors $\hat{\mathbf{i}}, \hat{\mathbf{j}}, \hat{\mathbf{k}}$, so that $\mathbf{r} = x\hat{\mathbf{i}} + y\hat{\mathbf{j}} + z\hat{\mathbf{k}}$. Hence, Eq. (5.16) is

$$\begin{aligned} \ddot{x} &= -k^2 M x (x^2 + y^2 + z^2)^{-3/2} \\ \ddot{y} &= -k^2 M y (x^2 + y^2 + z^2)^{-3/2} \\ \ddot{z} &= -k^2 M z (x^2 + y^2 + z^2)^{-3/2} \end{aligned} \tag{5.17}$$

The differential equations, Eqs. (5.5) and (5.6), in vector form, are three second-order equations. Each solution introduces two constants of integration, which would be the initial conditions. So there are twelve constants in the original system.

If we ignore the motion of the center of gravity, the number of constants reduces to six. So, the solution of Eqs. (5.17) will result in six constants. These six constants can be determined, if we know the 3 position coordinates and 3 velocity components at any instant. So to determine an orbit, six pieces of information are required; they can be three observations of two angles each, or two observations of two angles and a distance, each. An orbit is thus defined by six values: position and velocity or six parameters (McCuskey 1963, pp. 33–35).

5.5 The Integral of Areas

The motion of m_2 around m_1 is a central force motion, so the areal velocity is constant. That is

$$\dot{\mathbf{A}} = \frac{1}{2}(\mathbf{r} \times \mathbf{v}) = \frac{1}{2} h \hat{\mathbf{u}}_A \tag{5.18}$$

where $\hat{\mathbf{u}}_A$ is a unit vector with constant direction perpendicular to the orbital plane defined by $\mathbf{r}$ and $\mathbf{v}$. The components of areal velocity in Cartesian coordinates are

$$\begin{aligned} \frac{1}{2}(y\dot{z} - z\dot{y}) &= \frac{1}{2}c_1 \\ \frac{1}{2}(z\dot{x} - x\dot{z}) &= \frac{1}{2}c_2 \\ \frac{1}{2}(x\dot{y} - y\dot{x}) &= \frac{1}{2}c_3 \end{aligned} \tag{5.19}$$

where c_1, c_2, c_3 are constants related to h by

$$\sqrt{c_1{}^2 + c_2{}^2 + c_3^2} = h \tag{5.20}$$

From the initial coordinate and velocity components of m_2, the constants c_1, c_2, c_3 can be determined. When known, they must be related to the elements of the orbit, which were defined in Sect. 5.2 (see also Fig. 5.1). In terms of Ω, i the unit vector is

$$\hat{\mathbf{u}}_A = \sin i \sin \Omega\, \hat{\mathbf{i}} - \sin i \cos \Omega\, \hat{\mathbf{j}} + \cos i\, \hat{\mathbf{k}} \tag{5.21}$$

The areal velocity is

$$\dot{\mathbf{A}} = \frac{1}{2} h \sin i \sin \Omega\, \hat{\mathbf{i}} - \frac{1}{2} h \sin i \cos \Omega\, \hat{\mathbf{j}} + \frac{1}{2} h \cos i\, \hat{\mathbf{k}} \tag{5.22}$$

Comparing Eqs. (5.19) and (5.20) we have

$$\begin{aligned} c_1 &= h \sin i \sin \Omega \\ c_2 &= -h \sin i \cos \Omega \\ c_3 &= h \cos i \\ h &= \sqrt{c_1^2 + c_2^2 + c_3^2} \end{aligned} \tag{5.23}$$

From the initial conditions, c_1, c_2, c_3 are determined, and Eqs. (5.23) determine Ω and i. These elements orient the orbital plane with respect to a Cartesian coordinate system (McCuskey 1963, pp. 35–36).

5.6 Elements of the Orbit from Position and Velocity

The orientation of the orbital plane is established by the constants c_1, c_2, c_3. The size, shape, and orientation in the orbital plane must be determined. It will be a conic section with a central force mass, M, at the conic focus. $k^2 = G$ is the gravitational constant, m_1 is at the focus, and m_2 is the moving mass.

We seek the elements: a, the semimajor of an ellipse or semitransverse axis of a hyperbola, respectively; q, the distance from the focus to the vertex of the parabola; e, the eccentricity; ω, the argument of periapsis, which is the angle in the orbital plane between the line of nodes and the eccentricity vector, as explained in Sect. 5.2; and T, the time of periapsis passage. The *longitude of periapsis* is $\tilde{\omega} = \Omega + \omega$. As defined in Sect. 4.6.1, the true anomaly, f, is the angle in the orbital plane between the eccentricity vector and the m_2 position vector, as shown in Fig. 5.1.

The equation of the conic for the m_2 motion is

$$r = \frac{p}{1 + e \cos f} \tag{5.24}$$

The coordinates, x_0, y_0, z_0, and velocity components, $\dot{x}_0, \dot{y}_0, \dot{z}_0$, are for m_2 at time $t = 0$. Then Eqs. (5.19) and (5.23) are

$$\begin{aligned} c_1 &= y_0\dot{z}_0 - z_0\dot{y}_0 = h \sin i \sin \Omega \\ c_2 &= z_0\dot{x}_0 - x_0\dot{z}_0 = -h \sin i \cos \Omega \\ c_3 &= x_0\dot{y}_0 - y_0\dot{x}_0 = h \cos i \end{aligned} \tag{5.25}$$

which determine Ω, i, and h, as previously indicated. The initial distance and speed are $r_0 = \sqrt{x_0^2 + y_0^2 + z_0^2}$ and $v_0 = \sqrt{\dot{x}_0^2 + \dot{y}_0^2 + \dot{z}_0^2}$. Then, for an ellipse,

$$\frac{1}{a} = \frac{2}{r_0} - \frac{v_0^2}{k^2M} \tag{5.26}$$

For a hyperbola,

$$\frac{1}{a} = \frac{v_0^2}{k^2M} - \frac{2}{r_0} \tag{5.27}$$

For a parabola,

$$q = \frac{h^2}{2k^2M} \tag{5.28}$$

We can determine the eccentricity from the values of h by

$$e^2 = 1 \mp \frac{h^2}{k^2Ma} \tag{5.29}$$

where there is a minus sign for an ellipse and a plus sign for a hyperbola. The angle ω can be calculated as follows. Set the *argument of latitude* as $u = f + \omega$, then

$$\begin{aligned} r \cos u &= x \cos \Omega + y \sin \Omega \\ r \sin u &= (-x \sin \Omega + y \cos \Omega) \cos i + z \sin i \end{aligned} \tag{5.30}$$

Once Ω and i are known, and from initial conditions r_0 is known, Eq. (5.30) yields u_0 at time t_0. From Eq. (5.24),

$$e \cos f = \frac{p}{r} - 1 \tag{5.31}$$

Differentiating Eq. (5.31) with respect to time,

$$-(e\sin f)\dot{f} = -\frac{p}{r^2}\,\dot{r} \tag{5.32}$$

$r^2\dot{f} \equiv r^2\dot{\theta} = h$ is known. Furthermore,

$$\begin{aligned} r^2 &= x^2 + y^2 + z^2 \\ r\dot{r} &= x\dot{x} + y\dot{y} + z\dot{z} \end{aligned} \tag{5.33}$$

Hence, Eq. (5.32) can be written for $t = 0$,

$$e\sin f_0 = \frac{p}{h}\left[\frac{x_0\dot{x}_0 + y_o\dot{y}_0 + z_0\dot{z}_0}{r_0}\right] \tag{5.34}$$

Dividing Eq. (5.34) by Eq. (5.31) we find at $t = 0$

$$\tan f_0 = \frac{p}{h}\left[\frac{x_0\dot{x}_0 + y_o\dot{y}_0 + z_0\dot{z}_0}{p - r_0}\right] \tag{5.35}$$

We determine the argument of periapsis, $\omega = u_0 - f_0$, from u_0 computed in Eq. (5.30) and f_0 given in Eq. (5.35). In Eq. (5.35), we use $p = a(1 - e^2)$ for an ellipse, $p = a(e^2 - 1)$ for a hyperbola, and $p = 2q$ for a parabola.

The quadrant of f_0 is determined from the sign of $x_0\dot{x}_0 + y_o\dot{y}_0 + z_0\dot{z}_0$. This is equal to $r_0\dot{r}_0$ and $r_0 > 0$, so its sign will depend on $\dot{r}_0$. If $\dot{r}_0 > 0$, the radius vector is increasing, then $0 < f_0 < \pi$. If $\dot{r}_0 < 0$, then $\pi < f_0 < 2\pi$. The quadrant of ω can be calculated when the quadrant of u_0 has been established (McCuskey 1963, pp. 38–40).

5.7 Properties of Motion

At a given distance r from the force center, O, the speed in all elliptical orbits of the same major axis is the same, only the direction of the velocity vector differs. This speed is

$$v^2 = k^2M\left[\frac{2}{r} - \frac{1}{a}\right] \tag{5.36}$$

and

$$v_e = k\sqrt{\frac{2M}{r} - \frac{M}{a}} \tag{5.37}$$

where v_e is the velocity in an elliptic orbit.

At the same distance r from the force center, the speed in a parabolic orbit is

$$v_p = k\sqrt{2M/r} \tag{5.38}$$

and in a circular orbit

$$v_c = k\sqrt{M/r} \tag{5.39}$$

The circular and parabolic velocities at a given point, a distance r from the center of force, are the lower and upper limits, respectively, for the speed in elliptic orbits passing through the same point. When a mass is at **r** with a velocity **v**, between those limits, the path it follows depends on the direction of **v**.

The expression for v_p in Eq. (5.38) is identical to the escape velocity appearing in Eq. (4.47). It is the velocity required for escaping a given gravitational body at a given radius r.

For a hyperbolic orbit, utilizing the expression for the velocity in Eq. (5.36), it is seen that when $r \to \infty$, the velocity becomes

$$v_\infty = k\sqrt{-\frac{M}{a}}, \quad a < 0 \tag{5.40}$$

The velocity v_∞ is called the *hyperbolic excess velocity*. It is related to the escape velocity through the equation

$$v^2 = v_p^2 + v_\infty^2 \tag{5.41}$$

In elliptic motion Kepler's third law, given as

$$P = \frac{2\pi a^{3/2}}{k\sqrt{M}} \tag{5.42}$$

holds, where it is clear the period is independent of all orbital elements except the semimajor axis. We note again that $M = m_1 + m_2$, which is the total mass of the two-body problem.

5.8 The Constant of Gravitation

The *constant of gravitation* G as known by laboratory measurements and in the mks system is $G = 6.67428 \times 10^{-11}\text{m}^3\text{kg}^{-1}\text{s}^{-2}$. This is known to about $\pm 6.7 \times 10^{-15}$. This accuracy is not sufficient for astronomical purposes. It has been the practice since Gauss's time (ca. 1800) to adopt a value for describing the dynamical properties of the solar system. Equation (5.42) is useful to define the astronomical constant, k, the *Gaussian gravitational constant*. $k = 0.01720209895$. There are no more figures to this constant. The units for this value are: (i) The mean solar day for time; (ii) the astronomical unit (AU) for distance; (iii) the mass of the Sun for mass. In Eq. (5.42)

$$P = \frac{2\pi a^{3/2}}{k\sqrt{1+m_2}} \tag{5.43}$$

applied to the Earth-Moon system, Gauss used the values $P = 365.2563835$ mean solar days, $a = 1$ AU, and $m_2 = 1/354710$ solar masses. The value of k, given above, comes from these values. The best values today would be $P = 365.25636042 + 0.00000011T$ with T from 1900, and $m_2 = 1/332946.0487$. These values with $a = 1$ would yield a different value of k.

Until 2012, rather than change k, the value of the unit of distance has varied and k was kept constant. So the value of the semimajor axis of the Earth was $a = 1.000000031$. In 2012, it was decided to stop holding the value of k as constant, fix the real value of the AU at the determined value and use the best values of P and m_2, such that the variability of these quantities could be considered. The value of the AU was fixed at 149597870700 m exactly (Capitaine et al. 2011).

The mean daily motion of a planet is $n = 2\pi/P$, where P is the period. From Eq. (5.42)

$$n = \frac{2\pi}{P} = \frac{k\sqrt{1+m_2}}{a^{3/2}} \tag{5.44}$$

P is in mean solar days and n is in radians per day. Equation (5.44) can be written as

$$n^2 a^3 = k^2(1+m_2) \tag{5.45}$$

In some problems, it is convenient to use a system of units in which $k = 1$. We set $a = 1$, $1 + m_2 = 1$ in Eq. (5.44) and appropriately adjust the unit of time. If $a = 1$ AU, $1 + m_2 = 1$ solar mass, $k = 1$, then the unit of time is

$$365.2563835/2\pi = 58.1323589 \text{ mean solar days} \tag{5.46}$$

For the Earth-Sun two-body problem this would be a canonical system of units, where $n = 1$. Canonical units are especially useful for purely theoretical investigations (McCuskey 1963, pp. 44–45). We shall discuss this later.

5.9 Kepler's Equation

Elliptical orbital motion is the most important in the solar system and around the Earth. Consider the problem, given the elements $a, e, i, \omega, \Omega, T$ of an elliptic orbit, to find the polar coordinates—the radius, r, and the true anomaly, f—of the moving mass at any time t. To that end, the relation known as *Kepler's equation* is useful.

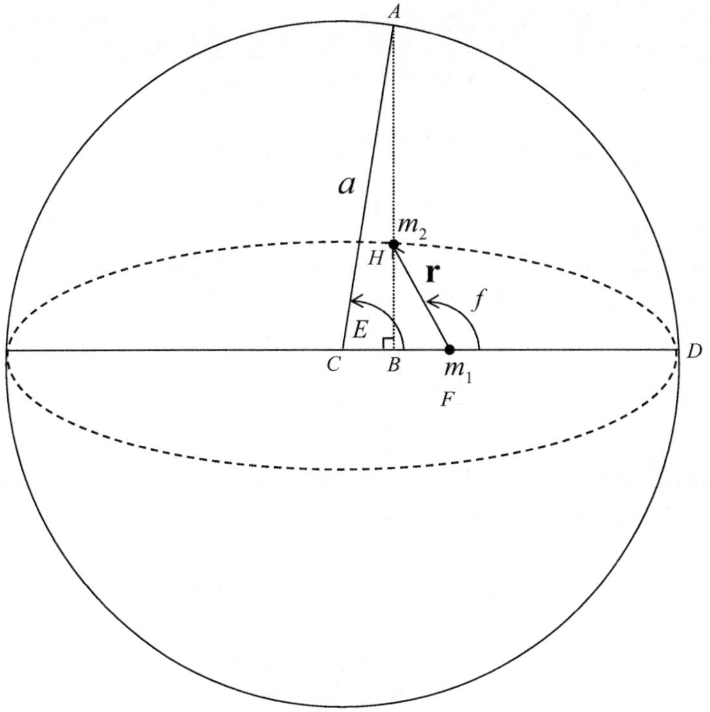

Fig. 5.4 The construction of the eccentric anomaly, E

The elliptic orbit and an auxiliary circle, whose diameter is the major axis of the ellipse, are shown in Fig. 5.4.

Mass m_1 is at the focus, F, and mass m_2 is moving in a counterclockwise direction around the ellipse. BA is perpendicular to CD through the position of m_2. The auxiliary variable, E, is the eccentric anomaly, which we mentioned in Sect. 1.3.1. As the true anomaly, f, changes by 2π radians, so does E. Then from Fig. 5.4 and the properties of an ellipse

$$r \cos f = a \cos E - ae \tag{5.47a}$$

$$r \sin f = a \sin E - HA \tag{5.47b}$$

Furthermore,

$$\frac{BH}{BA} = \frac{b}{a} = \sqrt{1 - e^2} \tag{5.48}$$

where b is the semiminor axis of the ellipse. Thus,

$$HA = BA - BH = BA(1 - \sqrt{1 - e^2}) = a\ \sin E(1 - \sqrt{1 - e^2}) \tag{5.49}$$

Hence, Eq. (5.47b) can be written as

$$r \sin f = a \sin E\sqrt{1-e^2} \tag{5.50}$$

Squaring Eq. (5.50), adding to the square Eq. (5.47a), and reducing yields

$$r = a(1 - e \cos E) \tag{5.51}$$

So once we find E as a function of time, the value of r follows from Eq. (5.51).

To find E as a function of t, let us use a geometric argument. From Fig. 5.4 it can be shown that

$$\frac{\text{area}(BDH)}{\text{area}(BDA)} = \frac{b}{a} = \sqrt{1-e^2} \tag{5.52}$$

But area(BDH) = area(DFH) − area(BFH) and area(BDA) = area(DCA) − area(BCA). By Kepler's second law,

$$\text{area}(DFH) = \frac{1}{2}na^2\sqrt{1-e^2}\,(t-T) \tag{5.53}$$

where n is the *mean motion*, T is the time of periapsis passage, and t is the time instant under consideration. Hence, Eq. (5.52) is written as

$$\sqrt{1-e^2} = \frac{\frac{1}{2}na^2\sqrt{1-e^2}\,(t-T) - \frac{1}{2}\,r^2 \sin f \cos f}{\frac{1}{2}\,a^2E - \frac{1}{2}a^2 \sin E \cos E} \tag{5.54}$$

Using Eqs. (5.49) and (5.50), this reduces to

$$n(t-T) = E - e \sin E \tag{5.55}$$

This transcendental equation in E is Kepler's equation. $n(t-T) = M$ is the *mean anomaly*, which is the angle the radius vector would describe, if it moved uniformly with the average rate $2\pi/P$. When $(t-T)$ is given, M is known, and Eq. (5.55) can be solved for E. There are a number of ways to solve Kepler's equation. We will describe two such methods.

5.9.1 Series Expansion

When e is small, E does not differ greatly from M. To the desired accuracy, E can be obtained by a series expansion. For a first approximation, $E_1 \approx M$. The second approximation is $E_2 \approx M + e \sin M$, which is approximate to first order in

e. Substituting this into Kepler's equation, we have

$$\begin{aligned} E_3 &= M + e \sin E_2 = M + e \sin(M + e \sin M) \\ &= M + e \sin M \cos(e \sin M) + e \cos M \sin(e \sin M) \end{aligned} \tag{5.56}$$

The terms in parenthesis are small, and we may write

$$\begin{aligned} \cos(e \sin M) &= 1 + \frac{e^2 \sin^2 M}{2!} + \dots \\ \sin(e \sin M) &= e \sin M - \frac{e^3 \sin^3 M}{3!} + \dots \end{aligned} \tag{5.57}$$

Substituting in Eq. (5.56), collecting like powers of e, a third-order approximation is

$$E_3 = M + e \sin\ M + \frac{e^2}{2} \sin\ 2M + \frac{e^3}{8}(\sin\ 3M - 3 \sin\ M) + \dots \tag{5.58}$$

which can be carried to a greater accuracy.

If e exceeds 0.2, the series converges too slowly. Then, an approximate value can be determined graphically and corrected differentially (McCuskey 1963, pp. 47–48).

5.9.2 *Differential Method*

Assume an approximation value, E_0, of E. Then a corresponding mean anomaly can be computed

$$M_0 = E_0 - e \sin E_0 \tag{5.59}$$

M is a continuous function of E, $M = \phi(E)$, which can be expanded into a Taylor series about the point $E = E_0$,

$$M = \phi(E_0) + \phi'(E_0)(E - E_0) + \frac{\phi''(E_0)}{2!}(E - E_0)^2 + \dots \tag{5.60}$$

where primes denote derivatives of $\phi(E)$ with respect to E. Let $E - E_0 = \Delta E_0$ and neglect powers two and greater in Eq. (5.60). $\phi(E_0) = M_0$ is given by Eq. (5.59). To this order of approximation

$$M - M_0 = \phi'(E_0)\Delta E_0 = (1 - e \cos E_0)\Delta E_0 \tag{5.61}$$

Solving for ΔE_0 we have

$$\Delta E_0 = \frac{M - M_0}{1 - e\cos E_0} \tag{5.62}$$

and we have a new value of E

$$E_1 = E_0 + \Delta E \tag{5.63}$$

By using E_1, the method is repeated with Eq. (5.59) to determine M_1, and with Eq. (5.62) for a new value of E.

Those are two of about 100 methods to solve Kepler's Equation. With high speed computers a simple iterative method can be used in the form

$$E = M + e\sin E \tag{5.64}$$

where M is the first approximation and then subsequent values of E from the left are substituted on the right until it converges to the accuracy desired. Siewert and Burniston (1972) have derived an exact analytical solution of Kepler's equation, but it probably takes more computation than any other scheme (McCuskcy 1963, pp. 48–49).

5.10 Position in the Elliptic Orbit

There is an expression for the radius vector in terms of the eccentric anomaly $r = a(1 - e\cos E)$. We need the true anomaly, f, as a function of E. From Eq. (5.47a) we have

$$a\cos E = ae + r\cos f \tag{5.65}$$

from the polar equation of the orbit (5.24),

$$r = \frac{a(1 - e^2)}{1 + e\cos f} \tag{5.66}$$

Combining this with Eq. (5.65) we have

$$\cos E = \frac{e + \cos f}{1 + e\cos f} \tag{5.67}$$

Thus,

$$1 - \cos E = \frac{(1-e)(1-\cos f)}{1+e\cos f}, \quad 1 + \cos E = \frac{(1+e)(1+\cos f)}{1+e\cos f} \tag{5.68}$$

By division

$$\tan^2\left(\frac{E}{2}\right) = \frac{1-e}{1+e}\tan^2\left(\frac{f}{2}\right) \tag{5.69}$$

Thus,

$$\tan\left(\frac{f}{2}\right) = \left(\frac{1+e}{1-e}\right)^{1/2}\tan\left(\frac{E}{2}\right) \tag{5.70}$$

The sign of the square root is not ambiguous, because $\tan(f/2)$ and $\tan(E/2)$ always have the same sign (McCuskey 1963, pp. 51).

5.11 Position in the Parabolic Orbit

When a comet is discovered near perihelion, there is no way to distinguish from the short arc of observations whether its orbit is an ellipse, a parabola, or a hyperbola. The parabolic orbit is easily calculated from observations, so it is usually a first approximation. When more observations are acquired, a definitive orbit can be calculated; most likely an eccentric ellipse. For an initial ephemeris, we must obtain polar coordinates of the object in a parabolic path.

The parabolic path equation is

$$r = \frac{2q}{1+\cos f} = q\sec^2\left(\frac{f}{2}\right) \tag{5.71}$$

which follows from Eq. (5.24). As shown in Fig. 5.5, the motion of m_2 is counterclockwise with the true anomaly f measured from the perihelion, V. The areal velocity constant for the parabolic path is

$$h = k\sqrt{2Mq} \tag{5.72}$$

The mass of a comet is usually negligible compared to the Sun's mass, so M is the solar mass. If q is in astronomical units and k is the Gaussian constant, $M = 1$. From the areal velocity $r^2\dot{f} = h$, and by combining Eqs. (5.71) and (5.72) we obtain

$$q^2\dot{f}\sec^4\left(\frac{f}{2}\right) = k\sqrt{2Mq} \tag{5.73}$$

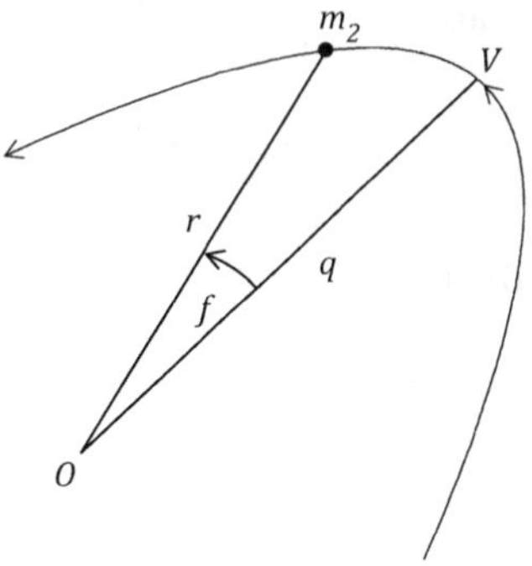

Fig. 5.5 Motion in a parabolic orbit

or

$$\sec^4\left(\frac{f}{2}\right)\, df = \frac{k\sqrt{2M}}{q^{3/2}} dt \tag{5.74}$$

and by integration

$$\tan\left(\frac{f}{2}\right) + \frac{1}{3}\tan^3\left(\frac{f}{2}\right) = k\sqrt{\frac{M}{2q^3}}\,(t-T) \tag{5.75}$$

where T is the time of perihelion passage. This is a cubic equation in $\tan(f/2)$ which can be solved for f as a function of t by graphic or tabular methods or with much computing.

Moulton (1970) derived a direct solution by a chain of calculations of auxiliary quantities:

$$\cot s = \frac{3k(t-T)}{(2q)^{3/2}} \tag{5.76}$$

$$\cot w = \left[\cot\left(\frac{s}{2}\right)\right]^{1/3} \tag{5.77}$$

$$\tan\left(\frac{f}{2}\right) = 2\cot 2w \tag{5.78}$$

where $M = 1$. Having solved Eq. (5.75) for $f/2$, the radius vector is readily found by (McCuskey 1963, pp. 52–53)

$$r = q\sec^2\left(\frac{f}{2}\right) = q\left[1 + \tan^2\left(\frac{f}{2}\right)\right] \tag{5.79}$$

5.12 Position in a Hyperbolic Orbit

A meteoroid entering the Earth's atmosphere has a hyperbolic path with the Earth's center at one focus. Finding the position of the object on the hyperbola, as shown in Fig. 5.6, is similar to that for the ellipse.
The path equation is

$$r = \frac{a(e^2 - 1)}{1 + e\cos f}, \quad e > 1 \tag{5.80}$$

where f is the true anomaly. f can only vary between values that cause the denominator to vanish, i.e.

$$\left[-180^\circ + \cos^{-1}\left(\frac{1}{e}\right)\right] < f < \left[180^\circ - \cos^{-1}\left(\frac{1}{e}\right)\right] \tag{5.81}$$

f cannot have a value in the shaded region of Fig. 5.6. Using an analytical approach to determine f and r in terms of time, from Kepler's second law for the hyperbola

$$r^2\dot{f} = \sqrt{k^2Ma(e^2 - 1)} \tag{5.82}$$

and from the energy equation

$$\dot{r}^2 + r^2\dot{f}^2 = k^2M\left[\frac{2}{r} + \frac{1}{a}\right] \tag{5.83}$$

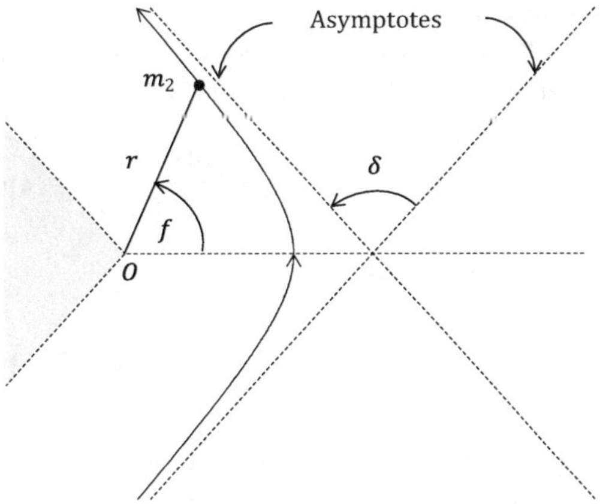

Fig. 5.6 Motion in a hyperbolic orbit

Substituting $\dot{f}$ from Eq. (5.82) into Eq. (5.83) yields the separable differential equation

$$\dot{r^2} = \frac{n^2 a^2}{r^2}\left[(a+r)^2 - a^2 e^2\right] \tag{5.84}$$

where $n^2 = k^2 M/a^3$.

The quantity n is analogous to n in the elliptic case, but the relationship between n and the period P is of course not present here.

Similar to the eccentric anomaly, E, in the elliptic orbit, we introduce an auxiliary variable, F, defined by

$$r = a[e\cosh F - 1] \tag{5.85}$$

Differentiating Eq. (5.85)

$$dr = ae\sinh F dF \tag{5.86}$$

substituting this into a separated Eq. (5.84) gives

$$n\,dt = (e\cosh F - 1)dF \tag{5.87}$$

which upon integration yields

$$n(t-T) = e\sinh F - F = M \tag{5.88}$$

This is similar to Kepler's equation. It can be solved for F, when $M = n(t-T)$ is known. This can be solved, as for Kepler's equation, by graphical, iterative, or series expansion methods.

When F is known, the radius vector can be determined from Eq. (5.85). From the relation

$$r = \frac{a(e^2-1)}{1+e\cos f} = a[e\cosh F - 1] \tag{5.89}$$

we derive the expression

$$\tan\left(\frac{f}{2}\right) = \left(\frac{e+1}{e-1}\right)^{1/2}\tanh\left(\frac{F}{2}\right) \tag{5.90}$$

which yields f when F is known (McCuskey 1963, pp. 55–57).

When $e \approx 1$, the methods discussed previously for an object in an elliptical or hyperbolic orbit, are not practicable. These very highly elliptic orbits occur among comets, artificial satellites, lunar probes and solar probes. Herget (1948) presented a special method for performing this calculation and tables of special variables.

5.13 Position on the Celestial Sphere

We have considered polar coordinates of a celestial object in its orbit from given elements. Now, consider the actual position on the sky at any time t. We seek for a planet, or a satellite, its spherical coordinates with the origin at the center of the Sun (heliocentric coordinates) or at the center of the Earth (geocentric coordinates). Here, we omit corrections due to precession, nutation, aberration, and so on.

5.13.1 Heliocentric Coordinates

The heliocentric coordinates are defined as follows. The axis x' points to the vernal equinox, $x'y'$ is the Earth's orbital plane (ecliptic plane), z' is the north pole of the ecliptic, the ascending node is N, the perihelion point P, and at any time t, m_2 is as shown in Fig. 5.7.

The heliocentric longitude, l, of m_2 is the angle, in the plane of the ecliptic, from the vernal equinox eastward to the projection of the radius vector on the ecliptic, as shown in Fig. 5.8. The heliocentric latitude, b, is the angle, in a plane perpendicular to the ecliptic, from the ecliptic to the object. We use (+) when the object is north and (−) when the object is south. So $0° \leq l < 360°$ and $-90° \leq b \leq 90°$.

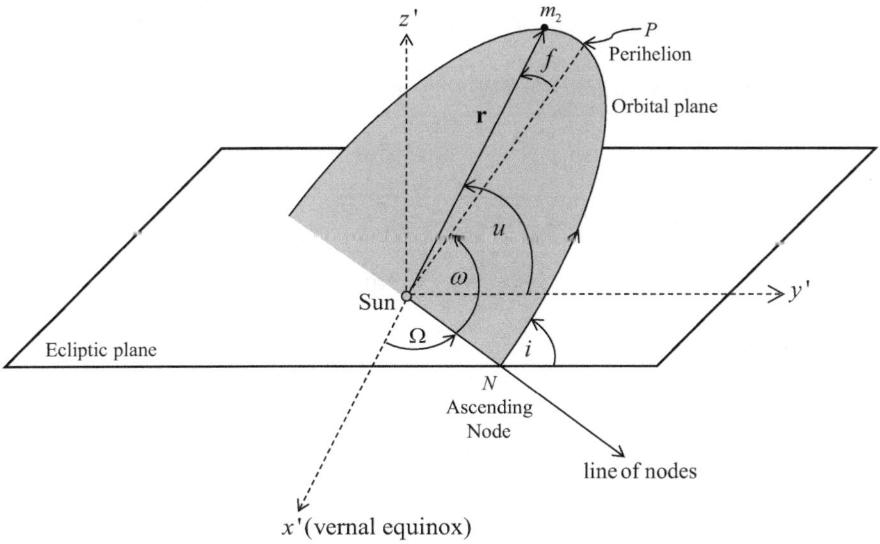

Fig. 5.7 Heliocentric coordinates

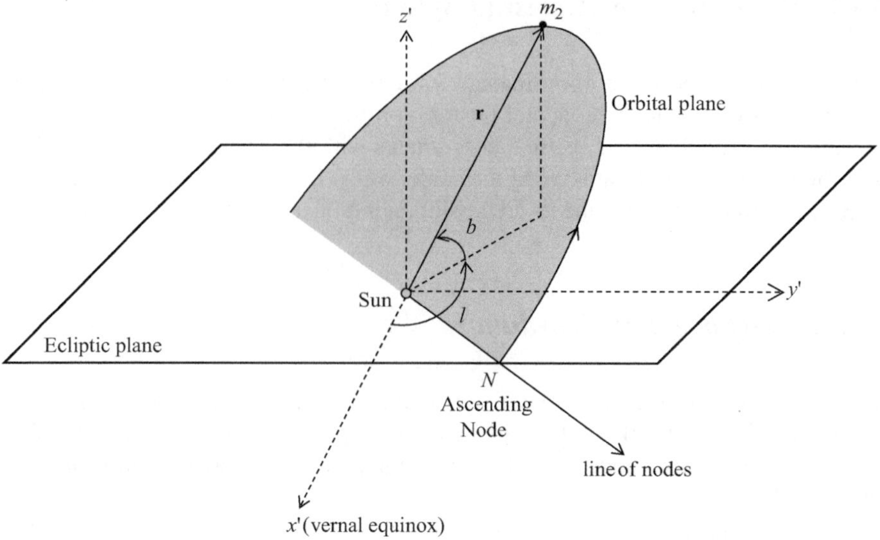

Fig. 5.8 Definition of the heliocentric longitude and latitude

Let $u = \omega + f$. The projection of the radius vector on SN is $r \cos u$. Also it is $r \cos b \cos(l - \Omega)$. Hence,

$$\cos u = \cos b \cos(l - \Omega) \tag{5.91}$$

Using projections perpendicular to SN in the ecliptic plane,

$$\sin u \cos i = \cos b \sin(l - \Omega) \tag{5.92}$$

and perpendicular to the ecliptic plane

$$\sin u \sin i = \sin b \tag{5.93}$$

From Eqs. (5.91)–(5.93) we have

$$\begin{aligned} \tan(l - \Omega) &= \cos i \tan u \\ \tan b &= \tan i \sin(l - \Omega) \end{aligned} \tag{5.94}$$

So when u, i and Ω are known, the heliocentric longitude and latitude are uniquely determined from Eqs. (5.91)–(5.94).

The rectangular ecliptic heliocentric coordinates of the moving mass m_2 are given by

$$\begin{aligned} x' &= r\cos u\cos\Omega - r\sin u\sin\Omega\cos i \\ y' &= r\cos u\sin\Omega + r\sin u\cos\Omega\cos i \\ z' &= r\sin u\sin i \end{aligned} \tag{5.95}$$

These coordinates are based on the plane of the Earth's orbit, the ecliptic. The fundamental reference plane for positions, as observed from the Earth, is the Earth's equatorial plane. This is inclined to the ecliptic by about 23° 27′. Thus, we have coordinates with the Sun at the origin. Let $x = x'$ be the vernal equinox, and y and z at angles ϵ from y' and z' as shown in Fig. 5.9. Then in this heliocentric equatorial system

$$\begin{aligned} x &= x' \\ y &= y'\cos\epsilon - z'\sin\epsilon \\ z &= y'\sin\epsilon + z'\cos\epsilon \end{aligned} \tag{5.96}$$

Substituting Eqs. (5.95) into Eqs. (5.96) yields rectangular heliocentric equatorial coordinates of the moving object. The transformation of $x, y, z \rightarrow x', y', z'$ is (McCuskey 1963, pp. 63–65).

$$\begin{aligned} x' &= x \\ y' &= y\cos\epsilon + z\sin\epsilon \\ z' &= -y\sin\epsilon + z\cos\epsilon \end{aligned} \tag{5.97}$$

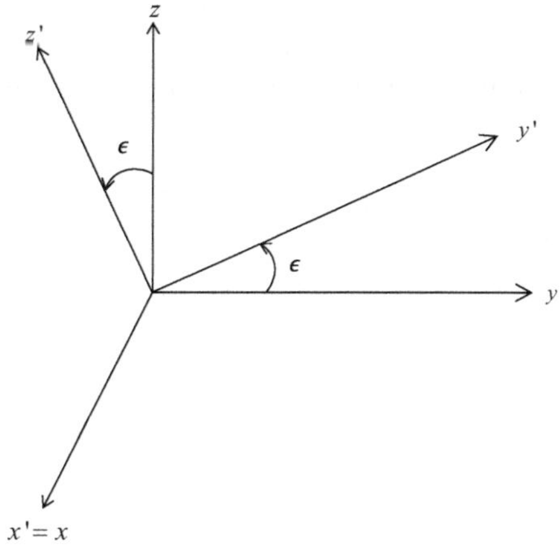

Fig. 5.9 Rotation from ecliptic to equatorial coordinates

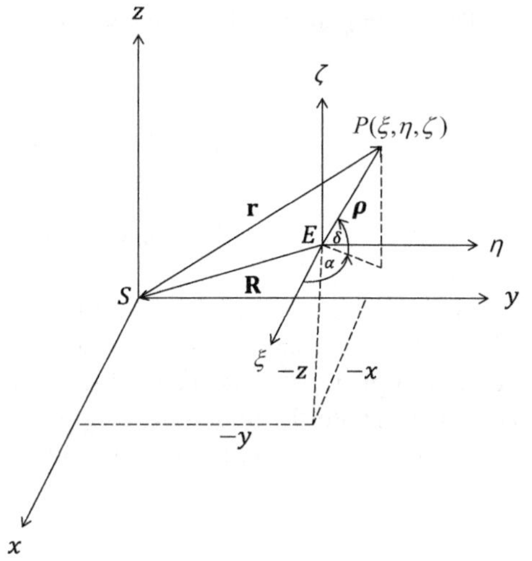

Fig. 5.10 Heliocentric and geocentric coordinates

5.13.2 Geocentric Coordinates

The heliocentric rectangular equatorial system can be related to a parallel system with the Earth at the center. The geocentric equatorial coordinates X, Y, Z of the Sun's center, given in The Astronomical Almanac,[1] are used. Let ξ, η, ζ denote geocentric equatorial coordinates of a moving object, as shown in Fig. 5.10.

Let $\boldsymbol{\rho}$ be the vector from the center of the Earth to the object P, and $\mathbf{r}$ from the center of the Sun to the object. Then

$$\begin{aligned} \xi &= x + X \\ \eta &= y + Y \\ \zeta &= z + Z \end{aligned} \tag{5.98}$$

are the geocentric coordinates of P. In vector notation,

$$\boldsymbol{\rho} = \mathbf{r} + \mathbf{R} \tag{5.99}$$

The geocentric polar coordinates are then

$$\begin{aligned} \xi &= \rho \cos\delta \cos\alpha \\ \eta &= \rho \cos\delta \sin\alpha \\ \zeta &= \rho \sin\delta \end{aligned} \tag{5.100}$$

[1]The Astronomical Almanac Online, http://asa.usno.navy.mil or http://asa.hmnao.com.

It follows that

$$
\begin{aligned}
\tan\alpha &= \eta/\xi \\
\sin\delta &= \zeta/\rho \\
\rho &= (\xi^2 + \eta^2 + \zeta^2)^{1/2}
\end{aligned} \tag{5.101}
$$

α is right ascension in the equatorial plane from the vernal equinox eastward to the projection of $\boldsymbol{\rho}$ into that plane in hours, minutes and seconds. δ is declination, the angular distance from the plane of the equator, and perpendicular to it, to the radius vector $\boldsymbol{\rho}$, (+) to the north and (−) to the south. Note these are viewed from the center of the Earth.

If we wish a geocentric system with the ecliptic as the fundamental plane, then λ is the celestial longitude, the angular distance in degrees along the ecliptic eastward from the vernal equinox to the projection of the radius vector into the ecliptic. β is the celestial latitude, the angular distance from the plane of the ecliptic and perpendicular to it to the radius vector, (+) to the North and (−) to the South. The relationships between α, δ and λ, β are (McCuskey 1963, pp. 65–67)

$$
\begin{aligned}
\cos\lambda\cos\beta &= \cos\delta\cos\alpha \\
\sin\lambda\cos\beta &= \sin\delta\sin\epsilon + \cos\delta\cos\epsilon\sin\alpha \\
\sin\beta &= \sin\delta\cos\epsilon - \cos\delta\sin\epsilon\sin\alpha
\end{aligned} \tag{5.102}
$$

References

Capitaine, N., Guinot, B., Klioner, S.: Proposal for the redefinition of the astronomical unit of length through a fixed relation to the SI metre. In: Capitaine, N. (ed.), Proceedings of the Journees 2012 Systemes de Reference Spatio-temporels, Observatoire de Paris, pp. 20–23 (2011)

Herget, P.: The Computation of Orbits. University of Cincinnati, Cincinnati (1948)

Hintz, G.: Survey of orbit element sets. J. Guid. Control Dyn. **31**(3), 785–790 (2008)

McCuskey, S.W.: Introduction to Celestial Mechanics. Addison-Wesley, Reading (1963)

Moulton, F.: An Introduction to Celestial Mechanics. Dover, New York (1970)

Siewert, C.E., Burniston, E.E.: An exact analytical solution of Kepler's equation. Celest. Mech. **6**(3), 294–304 (1972)

Chapter 6
Orbit Determination

6.1 Introduction

So far we have considered the following problem: Given the elements of the orbit, find the position of the celestial object at any time (Chap. 5). Now let us consider the other problem: Given the observations of an object, determine the elements of its orbit. We have considered determining the elements when the position and velocity are known, but these are not the usual observations. Radar observations can give the radial distance. Doppler observations can give a radial velocity, but in both cases angular observations may be lost or reduced in accuracy. In any case, six elements are to be determined, so six independent quantities must be observed, either (i) three pairs of (α, δ) from Earth, (ii) two observations of (r, α, δ), or (iii) some other combination of six quantities.

We shall assume observations have been corrected for aberration, precession, nutation, and other variations in the Earth's motion. Also, all observables are assumed in the same coordinate system. At this time, the ICRF should be used. Before considering the general problem of three pairs of angular measurements, we consider the trivial case, when the radius vector from the center of force to the moving body is known, i.e. we know its spherical coordinates at each of three times.

6.2 Known Radius Vectors

Assume an object is moving around the Sun in an ellipse with the period P. Assume a heliocentric longitude, a heliocentric latitude, and a radius vector, (l, b, r), which are known at times t_1, t_2, t_3. The radius vector at each time is

$$\mathbf{r} = x'\,\hat{\mathbf{i}} + y'\,\hat{\mathbf{j}} + z'\,\hat{\mathbf{k}} \tag{6.1}$$

P. Gurfil, P.K. Seidelmann, *Celestial Mechanics and Astrodynamics: Theory and Practice*, Astrophysics and Space Science Library 436,
DOI 10.1007/978-3-662-50370-6_6

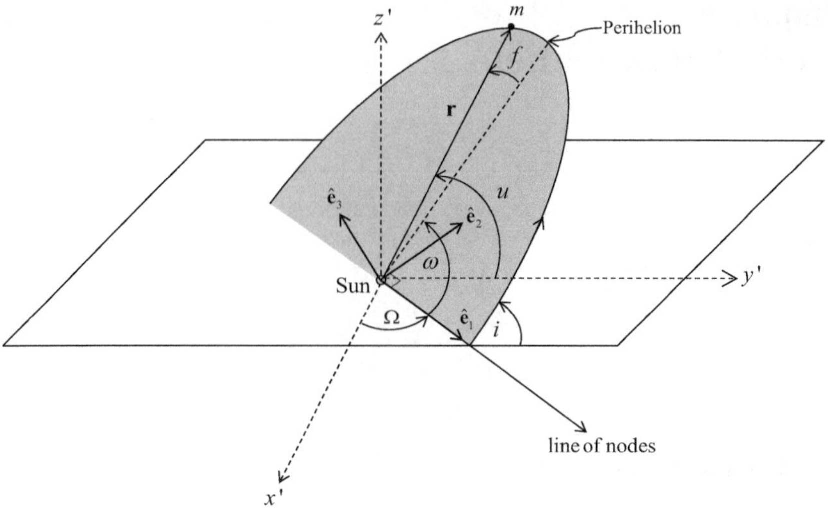

Fig. 6.1 Computation of orbits from known radius vectors

where

$$x' = r\cos b\cos l$$
$$y' = r\cos b\sin l$$
$$z' = r\sin b \tag{6.2}$$

Let the radius vectors be denoted by $\mathbf{r}_1, \mathbf{r}_2, \mathbf{r}_3$, where $t_1 < t_2 < t_3$. Let $\hat{\mathbf{e}}_1, \hat{\mathbf{e}}_2, \hat{\mathbf{e}}_3$ be a triad of unit vectors as shown in Fig. 6.1, where $\hat{\mathbf{e}}_1$ is directed along the line of nodes, $\hat{\mathbf{e}}_2$ is in the plane of the orbit and perpendicular to $\hat{\mathbf{e}}_1$, and $\hat{\mathbf{e}}_3$ is perpendicular to the orbital plane, so that $\hat{\mathbf{e}}_3 = \hat{\mathbf{e}}_1 \times \hat{\mathbf{e}}_2$.
Then

$$\hat{\mathbf{e}}_1 = \cos\Omega\,\hat{\mathbf{i}} + \sin\Omega\,\hat{\mathbf{j}}$$
$$\hat{\mathbf{e}}_2 = -\sin\Omega\cos i\,\hat{\mathbf{i}} + \cos\Omega\cos i\,\hat{\mathbf{j}} + \sin i\,\hat{\mathbf{k}}$$
$$\hat{\mathbf{e}}_3 = \sin\Omega\sin i\,\hat{\mathbf{i}} - \cos\Omega\sin i\,\hat{\mathbf{j}} + \cos i\,\hat{\mathbf{k}} \tag{6.3}$$

where $\hat{\mathbf{i}}, \hat{\mathbf{j}}, \hat{\mathbf{k}}$ are unit vectors along x', y', z', respectively. From $\mathbf{r}_1$ and $\mathbf{r}_3$, we obtain a unit vector

$$\frac{\mathbf{r}_1 \times \mathbf{r}_3}{\|\mathbf{r}_1 \times \mathbf{r}_3\|} = A_1\,\hat{\mathbf{i}} + A_2\,\hat{\mathbf{j}} + A_3\,\hat{\mathbf{k}} \tag{6.4}$$

This vector is perpendicular to the orbital plane and identical to $\hat{\mathbf{e}}_3$. From Eq. (6.3) we have

$$A_1 = \sin\Omega \sin i$$
$$A_2 = -\cos\Omega \sin i$$
$$A_3 = \cos i \tag{6.5}$$

Thus,

$$\tan\Omega = -\frac{A_1}{A_2} \tag{6.6}$$
$$\cos i = A_3 \tag{6.7}$$

and i and Ω are determined. The orbit in space is oriented. Any pair of $\mathbf{r}_1, \mathbf{r}_2, \mathbf{r}_3$ could be used in this calculation, but they must be non-collinear. The plane is best determined by their being the farthest from collinear.

Let u denote the angle in the plane of the orbit from the line of nodes (see Sect. 5.2) to the moving mass; the equation of the ellipse is written

$$r = \frac{a(1-e^2)}{1+e\cos f} = \frac{a(1-e^2)}{1+e\cos(u-\omega)} \tag{6.8}$$

where, as before, f is the true anomaly. Substituting r_1, r_2, r_3 into Eq. (6.8), rearranging terms, and differencing in pairs, we have, remembering that the elements have the same values throughout,

$$r_2 - r_1 = e\cos\omega(r_1\cos u_1 - r_2\cos u_2) + e\sin\omega(r_1\sin u_1 - r_2\sin u_2) \tag{6.9a}$$
$$r_3 - r_1 = e\cos\omega(r_1\cos u_1 - r_3\cos u_3) + e\sin\omega(r_1\sin u_1 - r_3\sin u_3) \tag{6.9b}$$

$r\cos u = \mathbf{r}\cdot\hat{\mathbf{e}}_1$ and $r\sin u = \mathbf{r}\cdot\hat{\mathbf{e}}_2$. So, Eqs. (6.9) can be written

$$r_2 - r_1 = e\cos\omega\,[(\mathbf{r}_1 - \mathbf{r}_2)\cdot\hat{\mathbf{e}}_1] + e\sin\omega\,[(\mathbf{r}_1 - \mathbf{r}_2)\cdot\hat{\mathbf{e}}_2] \tag{6.10a}$$
$$r_3 - r_1 = e\cos\omega\,[(\mathbf{r}_1 - \mathbf{r}_3)\cdot\hat{\mathbf{e}}_1] + e\sin\omega\,[(\mathbf{r}_1 - \mathbf{r}_3)\cdot\hat{\mathbf{e}}_2] \tag{6.10b}$$

We can calculate the dot products in brackets from Eqs. (6.3), (6.6), and (6.7), and the observed values of the $\mathbf{r}$'s. Equations (6.10) can be solved for $e\cos\omega$ and $e\sin\omega$. The determinant of the coefficient of $e\cos\omega$ and $e\sin\omega$ in Eqs. (6.10) does not vanish, and values of e and ω can be determined, if the three radius vectors are non-collinear.

Then, the solution of Eqs. (6.10) is $e\cos\omega = B_1$ and $e\sin\omega = B_2$, and

$$e = (B_1^2 + B_2^2)^{1/2} \tag{6.11}$$

$$\omega = \tan^{-1}\frac{B_2}{B_1} \tag{6.12}$$

The algebraic signs of $e\cos\omega$ and $e\sin\omega$ determine the quadrant of ω, because $e > 0$. The semimajor axis, a, can be determined by rearranging Eq. (6.8) into the form

$$a = \frac{r + (\mathbf{r}\cdot\hat{\mathbf{e}}_1)\, e\cos\omega + (\mathbf{r}\cdot\hat{\mathbf{e}}_2)\, e\sin\omega}{1 - e^2} \tag{6.13}$$

or equivalently

$$a = \frac{r + B_1(\mathbf{r}\cdot\hat{\mathbf{e}}_1) + B_2(\mathbf{r}\cdot\hat{\mathbf{e}}_2)}{1 - e^2} \tag{6.14}$$

Any of the observations $\mathbf{r}$ can be used, or all can be used and an average of the $a's$ taken, and it can be used as a check on the results. By Kepler's third law, the period can be found once the semimajor axis is known. The time of perihelion passage, T, can e determined by the following steps: use Kepler's equation, $M = E - e\sin E$, to find the mean anomaly, M, where E is calculated from $r = a(1 - e\cos E)$. Then, With M given for the time of one observation, the value of T follows from

$$M = \frac{2\pi}{P}(t - T) \tag{6.15}$$

Thus, the elements $a, e, i, \omega, \Omega, T$ for the orbit are determined from known space coordinates of a moving celestial object. Actually, this can be done using only the two positions and the constancy of the areal velocity, but the calculations are more complex (McCuskey 1963, pp. 70–73).

6.3 Laplace's Method

In most astronomical applications, the length of the radius vector is not known. Only the angular position in the sky as seen from the Earth is recorded. Such an observation locates an object on a line through the Earth; the distance from the Earth is unknown, so the distance and position from the Sun are unknown. In the beginning of the nineteenth century, Laplace devised a method of determining a preliminary orbit from three observations of position, referred to as *Laplace's method*.

The equation of motion of a planet around the Sun can be written

$$\ddot{\mathbf{r}} = -\frac{\mathbf{r}}{r^3} \tag{6.16}$$

if units of mass, distance, and time are chosen appropriately. The equation is in heliocentric coordinates with $\mathbf{r}(t)$, the radius vector from the Sun to the object at time t. The conditions, which determine the arbitrary constants in its solution, are expressed in terms of geocentric coordinates, because observations are made from the Earth. The solution must include the transformation from geocentric to heliocentric coordinates.

Let $\mathbf{r}_0$ be the heliocentric radius vector of the planet at the initial epoch t_0. The solution of Eq. (6.16) can be given in a Taylor series

$$\mathbf{r} = \mathbf{r}_0 + \dot{\mathbf{r}}_0\tau + \frac{1}{2!}\ddot{\mathbf{r}}_0\tau^2 + \ldots \tag{6.17}$$

where $\tau = k(t - t_0)$, and k is the Gauss gravitational constant discussed in detail in Sect. 5.8. Is assumed that $\mathbf{r}(t)$ at $t = t_0$ is differentiable. The time derivatives at $t = t_0$ are denoted by subscript 0. Using Eq. (6.16), we can express $\ddot{\mathbf{r}}$ from Eq. (6.17) in terms of $\mathbf{r}_0$ and r_0. So we have

$$\mathbf{r} = \mathbf{r}_0 + \dot{\mathbf{r}}_0\tau - \frac{1}{2!}\frac{\mathbf{r}_0}{r_0^3}\tau^2 + \ldots \tag{6.18}$$

Writing Eq. (6.16) for time $t = t_0$

$$\ddot{\mathbf{r}}_0 = -r_0^{-3}\mathbf{r}_0 \tag{6.19}$$

and differentiating

$$\frac{d^3\mathbf{r}_0}{dt^3} = 3r_0^{-4}\mathbf{r}_0\dot{r}_0 - r_0^{-3}\dot{\mathbf{r}}_0 \tag{6.20}$$

and again in order

$$\begin{aligned}\frac{d^4\mathbf{r}_0}{dt^4} &= -12r_0^{-5}\mathbf{r}_0\dot{r}_0^2 + 3r_0^{-4}\dot{\mathbf{r}}_0\dot{r}_0 \\ &+ 3r_0^{-4}\mathbf{r}_0\ddot{r}_0 + 3r_0^{-4}\dot{\mathbf{r}}_0\dot{r}_0 - r_0^{-3}\ddot{\mathbf{r}}_0 \end{aligned} \tag{6.21}$$

Substituting $\ddot{\mathbf{r}}_0 = -r_0^{-3}\mathbf{r}_0$ and $\ddot{r}_0 = -r^{-2}$ and collecting terms

$$\frac{d^4\mathbf{r}_0}{dt^4} = (-2r_0^{-6} - 12r_0^{-5}\dot{r}_0{}^2)\mathbf{r}_0 + 6r_0^{-4}\dot{r}_0\dot{\mathbf{r}}_0 \tag{6.22}$$

In like manner higher derivatives can be written in terms of $\mathbf{r}_0$ and $\dot{\mathbf{r}}_0$.

Substituting Eqs. (6.20) and (6.22) into Eq. (6.17) reduces Eq. (6.17) to a linear form,

$$\mathbf{r} = f\mathbf{r}_0 + g\dot{\mathbf{r}}_0 \tag{6.23}$$

where

$$f = 1 - \frac{1}{2}\tau^2 r_0^{-3} + \frac{1}{2}\tau^3 r_0^{-4}\dot{r}_0 - \frac{1}{12}\tau^4(r_0^{-6} + 6r_0^{-5}\dot{r}_0^2) + \ldots \tag{6.24}$$

$$g = \tau - \frac{1}{6}\tau^3 r_0^{-3} + \frac{1}{24}\tau^4 r_0^{-4}\dot{r}_0 - \ldots \tag{6.25}$$

where $\tau = k(t - t_0)$.

Thus, in principle, if the f and g series in Eqs. (6.24) and (6.25) converge, given the position and velocity vectors $\mathbf{r}_0$ and $\dot{\mathbf{r}}_0$ at the initial epoch t_0, we can find $\mathbf{r}(t)$. So the problem is to find these vectors from the observations.

Consider Fig. 6.2. Let x, y, z denote a heliocentric equatorial coordinate system. S denotes the Sun, ♈ indicates the x-axis toward the vernal equinox. Let ξ, η, ζ denote a parallel geocentric coordinate system centered at the Earth, E. The unit vector, $\hat{\mathbf{u}}$, defines the position of the mass, m,

$$\hat{\mathbf{u}} = \cos\alpha\cos\delta\,\hat{\mathbf{i}} + \sin\alpha\cos\delta\,\hat{\mathbf{j}} + \sin\delta\,\hat{\mathbf{k}} \tag{6.26}$$

where $\hat{\mathbf{i}}, \hat{\mathbf{j}}, \hat{\mathbf{k}}$ are unit vectors along ξ, η, ζ, respectively. The observations are (α_1, δ_1) at t_1, (α_0, δ_0) at t_0, (α_3, δ_3) at t_3, where $t_1 < t_0 < t_3$. By Eq. (6.26), these can be

Fig. 6.2 Heliocentric and parallel geocentric coordinates

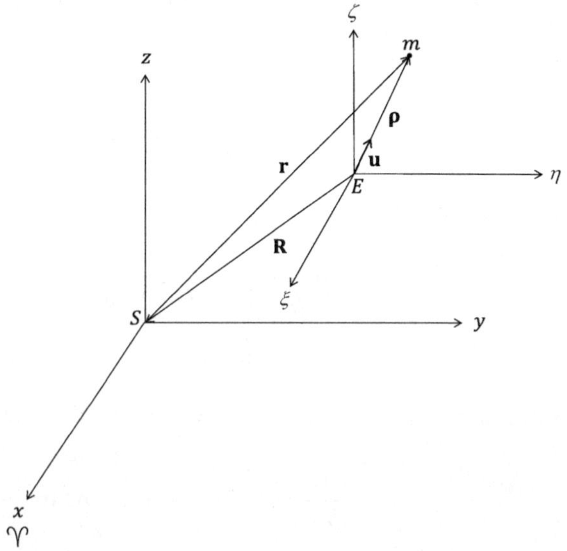

expressed in unit vectors $\hat{\mathbf{u}}_1, \hat{\mathbf{u}}_0, \hat{\mathbf{u}}_3$. From Fig. 6.2 it is seen that

$$\mathbf{r} = \boldsymbol{\rho} - \mathbf{R} = \rho\,\hat{\mathbf{u}} - \mathbf{R} \tag{6.27}$$

Hence,

$$\dot{\mathbf{r}} = \dot{\rho}\,\hat{\mathbf{u}} + \rho\,\dot{\hat{\mathbf{u}}} - \dot{\mathbf{R}} \tag{6.28}$$

and

$$\ddot{\mathbf{r}} = \ddot{\rho}\,\hat{\mathbf{u}} + 2\dot{\rho}\,\dot{\hat{\mathbf{u}}} + \rho\,\ddot{\hat{\mathbf{u}}} - \ddot{\mathbf{R}} \tag{6.29}$$

From Eq. (6.16), $\ddot{\mathbf{r}} = -r^{-3}\mathbf{r}$. So Eq. (6.29) becomes

$$\ddot{\rho}\,\hat{\mathbf{u}} + 2\dot{\rho}\,\dot{\hat{\mathbf{u}}} + \rho\,\ddot{\hat{\mathbf{u}}} = \ddot{\mathbf{R}} - r^{-3}\,\mathbf{r} \tag{6.30}$$

Take the triple scalar product, $\left[\hat{\mathbf{u}} \times \dot{\hat{\mathbf{u}}}\cdot\right]$, of both sides of Eq. (6.30), remembering that such a product, containing two identical elements, vanishes. So,

$$\rho\left[\hat{\mathbf{u}} \times \dot{\hat{\mathbf{u}}} \cdot \ddot{\hat{\mathbf{u}}}\right] = \left[\hat{\mathbf{u}} \times \dot{\hat{\mathbf{u}}} \cdot \ddot{\mathbf{R}}\right] - r^{-3}\left[\hat{\mathbf{u}} \times \dot{\hat{\mathbf{u}}} \cdot \mathbf{r}\right] \tag{6.31}$$

If we substitute in the last bracket $\mathbf{r} = \rho\hat{\mathbf{u}} - \mathbf{R}$, we have

$$\rho\left[\hat{\mathbf{u}} \times \dot{\hat{\mathbf{u}}} \cdot \ddot{\hat{\mathbf{u}}}\right] = \left[\hat{\mathbf{u}} \times \dot{\hat{\mathbf{u}}} \cdot \ddot{\mathbf{R}}\right] + \frac{\left[\hat{\mathbf{u}} \times \dot{\hat{\mathbf{u}}} \cdot \mathbf{R}\right]}{r^3} \tag{6.32}$$

This equation is of the form

$$\rho = A + Br^{-3} \tag{6.33}$$

where

$$A = \frac{\hat{\mathbf{u}} \times \dot{\hat{\mathbf{u}}} \cdot \ddot{\mathbf{R}}}{\hat{\mathbf{u}} \times \dot{\hat{\mathbf{u}}} \cdot \ddot{\hat{\mathbf{u}}}}, \quad B = \frac{\hat{\mathbf{u}} \times \dot{\hat{\mathbf{u}}} \cdot \mathbf{R}}{\hat{\mathbf{u}} \times \dot{\hat{\mathbf{u}}} \cdot \ddot{\hat{\mathbf{u}}}} \tag{6.34}$$

Another relation between ρ and r is evident from Fig. 6.2. From the cosine law

$$r^2 = R^2 + \rho^2 - 2\rho\,(\hat{\mathbf{u}} \cdot \mathbf{R}) \tag{6.35}$$

So we want to solve Eqs. (6.33) and (6.35) simultaneously for ρ and r at time t_0; i.e., the problem is to simultaneously solve Eqs. (6.33) and (6.35) for ρ and r at time t_0. The scalars A and B, which depend on the unit vectors and the position vector of the Sun, must be known. For each day, The Astronomical Almanac gives the Cartesian coordinates of the Sun, from which $R = X\hat{\mathbf{i}} + Y\hat{\mathbf{j}} + Z\hat{\mathbf{k}}$ can be computed, and it gives the magnitude of this vector and the geocentric equatorial coordinates of the Sun. From $\ddot{\mathbf{R}} = -R^{-3}\mathbf{R}$, $\ddot{\mathbf{R}}$ is known. So substitution can be made for A in Eq. (6.34).

Let $\tau_1 = k(t_1 - t_0)$ and $\tau_3 = k(t_3 - t_0)$ be the time intervals between the middle and first and the middle and last observations. By a Taylor series, we write

$$\hat{\mathbf{u}}_1 = \hat{\mathbf{u}}_0 + \dot{\hat{\mathbf{u}}}_0\,\tau_1 + \frac{1}{2}\ddot{\hat{\mathbf{u}}}_0\,\tau_1^2 + \ldots \tag{6.36a}$$

$$\hat{\mathbf{u}}_3 = \hat{\mathbf{u}}_0 + \dot{\hat{\mathbf{u}}}_0\,\tau_3 + \frac{1}{2}\ddot{\hat{\mathbf{u}}}_0\,\tau_3^2 + \ldots \tag{6.36b}$$

Note $\tau_1 = k(t_1 - t_0)$, $k = 0.017202\ldots$, so $\tau = 1$, if $(t_1 - t_0) = 58.3$ days.

If the observations are not too far apart in time, these expressions can be truncated to include only second degree terms in τ. They may be written as

$$\hat{\mathbf{u}}_1 - \hat{\mathbf{u}}_0 = \dot{\hat{\mathbf{u}}}_0\tau_1 + \frac{1}{2}\ddot{\hat{\mathbf{u}}}_0\tau_1^2 \tag{6.37a}$$

$$\hat{\mathbf{u}}_3 - \hat{\mathbf{u}}_0 = \dot{\hat{\mathbf{u}}}_0\tau_3 + \frac{1}{2}\ddot{\hat{\mathbf{u}}}_0\tau_3^2 \tag{6.37b}$$

$\hat{\mathbf{u}}_1, \hat{\mathbf{u}}_0, \hat{\mathbf{u}}_3, \tau_1, \tau_3$ are known, so we can solve for $\dot{\hat{\mathbf{u}}}_0$ and $\ddot{\hat{\mathbf{u}}}_0$, which will be used in Eq. (6.34), to yield A and B at $t = t_0$. To solve Eqs. (6.37), we introduce the notation

$$\mathbf{u}_{(1,0)} \triangleq \frac{\hat{\mathbf{u}}_1 - \hat{\mathbf{u}}_0}{\tau_1}, \quad \mathbf{u}_{(3,0)} \triangleq \frac{\hat{\mathbf{u}}_3 - \hat{\mathbf{u}}_0}{\tau_3} \tag{6.38}$$

Then Eqs. (6.37) can be written as

$$\mathbf{u}_{(1,0)} = \dot{\hat{\mathbf{u}}}_0 + \frac{1}{2}\tau_1\ddot{\hat{\mathbf{u}}}_0 \tag{6.39a}$$

$$\mathbf{u}_{(3,0)} = \dot{\hat{\mathbf{u}}}_0 + \frac{1}{2}\tau_3\ddot{\hat{\mathbf{u}}}_0 \tag{6.39b}$$

and we find

$$\dot{\hat{\mathbf{u}}}_0 = \frac{\tau_3\mathbf{u}_{(1,0)} - \tau_1\mathbf{u}_{(3,0)}}{\tau_3 - \tau_1} \tag{6.40a}$$

$$\ddot{\hat{\mathbf{u}}}_0 = \frac{2[\mathbf{u}_{(3,0)} - \mathbf{u}_{(1,0)}]}{\tau_3 - \tau_1} \tag{6.40b}$$

If there are more than three observations, the series expansion in Eqs. (6.36) can include higher order terms. The higher order terms in $\hat{\mathbf{u}}$ and its derivatives can be eliminated, resulting in higher accuracy for $\dot{\hat{\mathbf{u}}}_0$ and $\ddot{\hat{\mathbf{u}}}_0$.

Equations (6.33) and (6.35) can be solved by an iterative process as follows. Assume a value for r_1 in Eq. (6.33), and calculate a corresponding ρ_1. Then

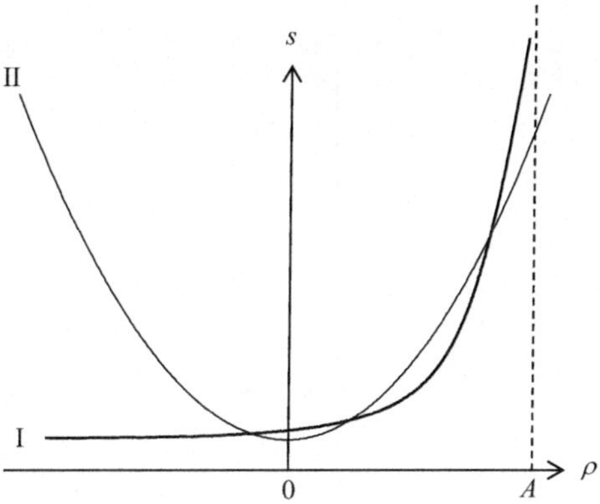

Fig. 6.3 Illustration of solutions based on Laplace's method

substitute this in Eq. (6.35) to determine a new value of r_2. Repeat the calculation of Eq. (6.33) for a new value of ρ_2. The speed of convergence depends on the wisdom in guessing an initial r value. If mean daily motion, or the period of the object can be estimated, Kepler's third law can be used to estimate r.

The solution process can be visualized, as shown in Fig. 6.3. Let $r^2 = s$, then Eq. (6.33) is

$$s = \left(\frac{B}{\rho - A}\right)^{2/3} \tag{6.41}$$

The graph of s against ρ is shown by curve I in Fig. 6.3, with a vertical asymptote at $\rho = A$.

Likewise Eq. (6.35) can be written

$$s = \rho^2 - 2(\hat{\mathbf{u}} \cdot \mathbf{R})\rho + R^2 \tag{6.42}$$

This is parabola II in Fig. 6.3. There may be three intersections of curves I and II. One of these values can be used to begin the iterative process for determining the exact numerical solution described above. Some of the values of $r^2 = s$ can be eliminated as unreasonable, such as r less than the Earth's radius. The slopes of curves I and II in the neighborhood of the true values are indicative of how rapidly the process converges.

We assume r_0 and ρ_0 are known from the above calculations and they satisfy Eq. (6.27). To determine $\dot{\mathbf{r}}_0$, we need to evaluate $\dot{\rho}_0$. Operating on Eq. (6.30) with the product $[\cdot(\hat{\mathbf{u}} \times \ddot{\hat{\mathbf{u}}}]$ and eliminating terms that vanish

$$2\dot{\rho}[\dot{\hat{\mathbf{u}}} \cdot \hat{\mathbf{u}} \times \ddot{\hat{\mathbf{u}}}] = [\ddot{\mathbf{R}} \cdot \hat{\mathbf{u}} \times \ddot{\hat{\mathbf{u}}}] + \frac{1}{r^3}[\mathbf{r} \cdot \hat{\mathbf{u}} \times \ddot{\hat{\mathbf{u}}}] \tag{6.43}$$

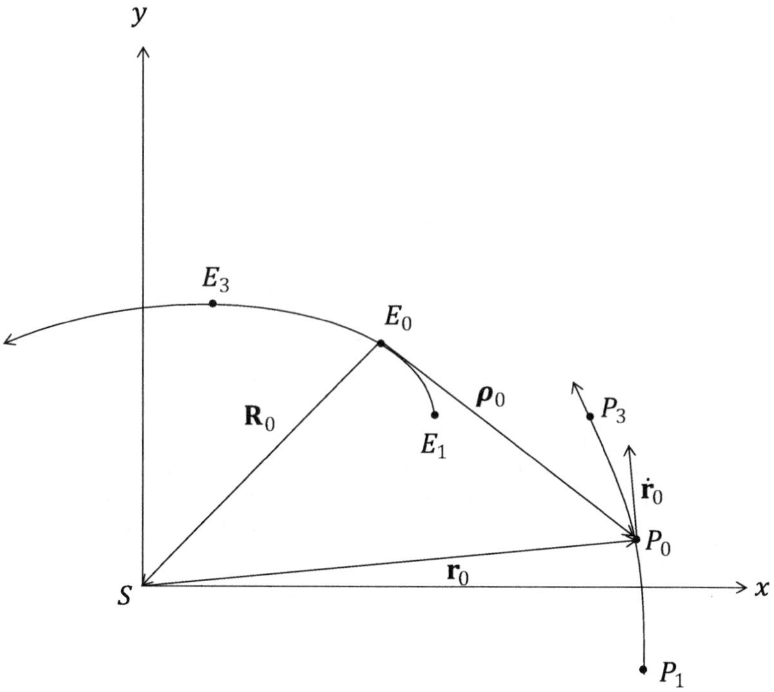

Fig. 6.4 Successive positions of the Earth and the object, projected on the equatorial plane

The quantities in brackets and r_0 at $t = t_0$ are known and $\dot{\rho}_0$ can be calculated. From Eq. (6.28) we have $\dot{r}_0$ from

$$\dot{\mathbf{r}}_0 = \dot{\rho}\,\hat{\mathbf{u}}_0 + \rho_0\,\dot{\hat{\mathbf{u}}}_0 - \dot{\mathbf{R}}_0 \tag{6.44}$$

With $\mathbf{r}_0$ and $\dot{\mathbf{r}}_0$ we can find $\mathbf{r}$ at any other time t from Eq. (6.23). Thus, Laplace's method yields the values of $\mathbf{r}_1, \mathbf{r}_0, \mathbf{r}_3$, so the orbital elements can be calculated by the method previously described. Alternatively, from the initial the position and velocity vectors, $\mathbf{r}_0$ and $\dot{\mathbf{r}}_0$, the elements can be calculated, as described in an earlier section. The reader is referred elsewhere for an example (Herget 1948, p. 44).

A sketch of successive positions of the Earth and the object would appear as shown in Fig. 6.4, projected on the equatorial plane.

The following checks can be applied to calculations:

1. Visual check of consistency between values.
2. From the calculated $\mathbf{r}_1$ and $\mathbf{r}_3$, calculate values of α_c and δ_c to compare with the observed values of α and δ. This indicates how well the preliminary orbit fits the observations. The middle observation is fit exactly in this method. The other two

observations only enter through approximations to the derivatives $\dot{\hat{\mathbf{u}}}_0, \ddot{\hat{\mathbf{u}}}_0, \dot{\mathbf{R}}_0, \ddot{\mathbf{R}}_0$, and so forth. Thus, exact agreement would not be expected.

This Laplace method represents the dynamical motion of the object, assuming purely Keplerian motion. The first and third positions are dynamically correct. To improve the orbital calculations and better fit the observations, the times of the observations should be corrected for the light time, or the time for the light to travel from the object to the Earth. Light travels 1 AU in 0.00577 days; the planetary aberration correction is given by

$$t(\text{true}) = t(\text{obs}) - 0.00577\rho \quad \text{days} \tag{6.45}$$

There are indeterminate cases with the Laplace method:

1. The inclination is zero. Then, i and Ω cannot be found. Four elements, a, e, ω, T, can be determined. When we convert to heliocentric coordinates, we have only three longitudes and all latitudes are zero. Hence, a fourth observation is required.
2. If (a) the three observations are on a great circle and (b) the Sun crosses the great circle during the time interval of the observations; then the situation is similar to an object moving in the ecliptic plane. So more observations are required (McCuskey 1963, pp. 76–84), (Herget 1948, pp. 40–51).

6.4 Gauss's Method

Laplace's method yields an orbit which fits the middle observation exactly. For times t_1 and t_3 the fit depends on the truncated series f and g. In *Gauss's method*, the observed positions at t_1 and t_3 enter directly, thus strengthening the solution immediately.

$\mathbf{r}_1, \mathbf{r}_2, \mathbf{r}_3$ are unknown heliocentric equatorial position vectors at times t_1, t_2, t_3. Since the motion takes place in a plane, one of these can be a linear combination of the other two. So, we write

$$\mathbf{r}_2 = c_1\,\mathbf{r}_1 + c_3\,\mathbf{r}_3 \tag{6.46}$$

where it is assumed $\mathbf{r}_1, \mathbf{r}_2, \mathbf{r}_3$ are not collinear. Set

$$\mathbf{r} = \rho\,\hat{\mathbf{u}} - \mathbf{R} \tag{6.47}$$

where $\hat{\mathbf{u}}$ is a unit vector for the geocentric position of the celestial object. Then Eq. (6.46) can be

$$c_1\rho_1\,\hat{\mathbf{u}}_1 - \rho_2\,\hat{\mathbf{u}}_2 + c_3\rho_3\,\hat{\mathbf{u}}_3 = c_1\,\mathbf{R}_1 - \mathbf{R}_2 + c_3\,\mathbf{R}_3 \tag{6.48}$$

If c_1 and c_3 are known, this equation provides three simultaneous equations in ρ_1, ρ_2, ρ_3, which are geocentric distances of the object at the times of the observations.

So far we have discussed a geometrical relationship. So, c_1 and c_3 must be determined such that as the object moves in its orbit, the dynamical conditions are satisfied, that is

$$\ddot{\mathbf{r}}_i = -r_i^{-3}\mathbf{r}_i, \ i = 1, 2, 3 \tag{6.49}$$

There are interesting geometrical interpretations of c_1 and c_3. Vector products of Eq. (6.46) by $\mathbf{r}_1$ and by $\mathbf{r}_3$ yield

$$\mathbf{r}_1 \times \mathbf{r}_2 = c_3\mathbf{r}_1 \times \mathbf{r}_3 \tag{6.50a}$$

$$\mathbf{r}_2 \times \mathbf{r}_3 = c_1\mathbf{r}_1 \times \mathbf{r}_3 \tag{6.50b}$$

Consider Fig. 6.5, in which $\|\mathbf{r}_1 \times \mathbf{r}_2\|$ is twice the area of triangle SP_1P_2, $\|\mathbf{r}_2 \times \mathbf{r}_3\|$ is twice the area of triangle SP_2P_3, and $\|\mathbf{r}_1 \times \mathbf{r}_3\|$ is twice the area of triangle SP_1P_3. Since $\mathbf{r}_1, \mathbf{r}_2, \mathbf{r}_3$ are coplanar, the cross products are collinear. From the areas

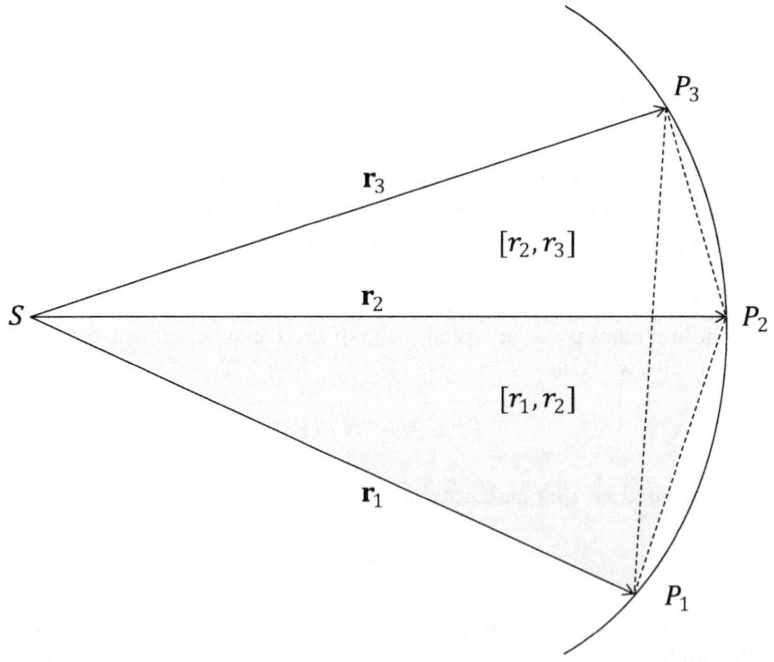

Fig. 6.5 Geometry for orbit determination using the Gauss method

involved by $[r_1, r_2]$, $[r_2, r_3]$, and $[r_1, r_3]$, we have

$$c_1 = \frac{[r_2, r_3]}{[r_1, r_3]} \tag{6.51a}$$

$$c_3 = \frac{[r_1, r_2]}{[r_1, r_3]} \tag{6.51b}$$

The solutions devised by Gauss for c_1 and c_3 use *sector-triangle ratios*. This is the ratio between the sectors of the ellipse bounded by the radius vectors to the corresponding triangles. Let (r_1, r_3), (r_1, r_2), and (r_2, r_3) denote the bounded sectors of the ellipse. The sector-triangle ratios are

$$y_1 = \frac{(r_2, r_3)}{[r_2, r_3]} \tag{6.52a}$$

$$y_2 = \frac{(r_1, r_3)}{[r_1, r_3]} \tag{6.52b}$$

$$y_3 = \frac{(r_1, r_2)}{[r_1, r_2]} \tag{6.52c}$$

Then we can write

$$c_1 = \frac{(r_2, r_3)}{y_1} \frac{y_2}{(r_1, r_3)} = \frac{(t_3 - t_2)}{(t_3 - t_1)} \frac{y_2}{y_1} \tag{6.53a}$$

$$c_3 = \frac{(r_1, r_2)}{y_3} \frac{y_2}{(r_1, r_3)} = \frac{(t_2 - t_1)}{(t_3 - t_1)} \frac{y_2}{y_3} \tag{6.53b}$$

The last terms in Eqs. (6.53) are based on Kepler's second law, specifically that the areas of the sectors, swept out by the radius vector, are proportional to the time intervals.

The evaluation of c_1 and c_3 depends on a method of calculating the ratios y_2/y_1 and y_2/y_3. Methods are given in books on orbit computation, such as Williams (1934) and Herget (1948). We can obtain series expansions for c_1 and c_3, which satisfy the dynamical conditions of the motion. Let

$$T_1 = k(t_3 - t_2), \; T_2 = k(t_3 - t_1), \; T_3 = k(t_2 - t_1) \tag{6.54}$$

We have seen previously

$$\mathbf{r}_1 = f_1 \, \mathbf{r}_2 + g_1 \, \dot{\mathbf{r}}_2 \tag{6.55a}$$

$$\mathbf{r}_3 = f_3 \, \mathbf{r}_2 + g_3 \, \dot{\mathbf{r}}_2 \tag{6.55b}$$

where approximately

$$f_1 = 1 - \frac{1}{2}\sigma\, T_3^2,\ g_1 = -T_3(1 - \frac{1}{6}\sigma\, T_3^2) \tag{6.56a}$$

$$f_3 = 1 - \frac{1}{2}\sigma\, T_1^2,\ g_3 = -T_1(1 - \frac{1}{6}\sigma\, T_1^2) \tag{6.56b}$$

and $\sigma = 1/r_2^3$. Taking the f and g series further requires $\dot{r}_2$, which would be too many unknowns. From Eqs. (6.55), substituting for $\mathbf{r}_1$ and $\mathbf{r}_3$

$$\mathbf{r}_1 \times \mathbf{r}_2 = -g_1\, \mathbf{r}_2 \times \dot{\mathbf{r}}_2 = 2[r_1, r_2] \tag{6.57a}$$

$$\mathbf{r}_2 \times \mathbf{r}_3 = g_3\, \mathbf{r}_2 \times \dot{\mathbf{r}}_2 = 2[r_2, r_3] \tag{6.57b}$$

$$\mathbf{r}_1 \times \mathbf{r}_3 = g_2\, \mathbf{r}_2 \times \dot{\mathbf{r}}_2 = 2[r_1, r_3] \tag{6.57c}$$

where $g_2 = f_1 g_3 - f_3 g_1$ based on $\mathbf{r}_1$ and $\mathbf{r}_3$ in terms of $\mathbf{r}_2$ and $\dot{\mathbf{r}}_2$. From Eqs. (6.51) then

$$c_1 = \frac{g_3}{g_2},\ c_3 = -\frac{g_1}{g_2} \tag{6.58}$$

Substituting Eqs. (6.56) into g_2 we have

$$g_2 = f_1 g_3 - f_3 g_1 = T_2 - \frac{1}{6}\sigma\, T_2^3 - \frac{1}{4}\sigma T_2^3 (T_3 - T_1) + \ldots \tag{6.59}$$

Substituting from Eqs. (6.56) and (6.59) into Eqs. (6.58) and expanding in powers of T

$$c_1 = \frac{T_1}{T_2}\left[1 + \frac{1}{6}\sigma(T_2^2 - T_1^2)\right] \tag{6.60a}$$

$$c_3 = \frac{T_3}{T_2}\left[1 + \frac{1}{6}\sigma(T_2^2 - T_3^2)\right] \tag{6.60b}$$

where the derivation of the $c's$ should not be of concern. Gibbs (1889) gave a more accurate expression for the $c's$. The expressions given are of the form

$$c_1 = a_1 + \frac{b_1}{r_2^3},\ c_3 = a_3 + \frac{b_3}{r_2^3} \tag{6.61}$$

where a_1, b_1, a_3, b_3 can be calculated from the times of observations. From Eq. (6.48)

$$c_1\rho_1\, \hat{\mathbf{u}}_1 - \rho_2\, \hat{\mathbf{u}}_2 + c_3\rho_3\, \hat{\mathbf{u}}_3 = c_1\, \mathbf{R}_1 - \mathbf{R}_2 + c_3\, \mathbf{R}_3 \tag{6.62}$$

We isolate the $\rho_2\hat{\mathbf{u}}_2$ term by taking the triple scalar product of both sides by $[\cdot\hat{\mathbf{u}}_1 \times \hat{\mathbf{u}}_3]$

$$-\rho_2[\hat{\mathbf{u}}_2 \cdot \hat{\mathbf{u}}_1 \times \hat{\mathbf{u}}_3] = c_1[\mathbf{R}_1 \cdot \hat{\mathbf{u}}_1 \times \hat{\mathbf{u}}_3] - [\mathbf{R}_2 \cdot \hat{\mathbf{u}}_1 \times \hat{\mathbf{u}}_3] + c_3[\mathbf{R}_3 \cdot \hat{\mathbf{u}}_1 \times \hat{\mathbf{u}}_3] \tag{6.63}$$

Substituting values of the $c's$ from Eq. (6.61), there results (reversing the order of vectors on the left side)

$$-\rho_2[\hat{\mathbf{u}}_2 \cdot \hat{\mathbf{u}}_1 \times \hat{\mathbf{u}}_3] = \left(a_1 + \frac{b_1}{r_2^3}\right)[\mathbf{R}_1 \cdot \hat{\mathbf{u}}_1 \times \hat{\mathbf{u}}_3] - [\mathbf{R}_2 \cdot \hat{\mathbf{u}}_1 \times \hat{\mathbf{u}}_3] + \left(a_3 + \frac{b_3}{r_2^3}\right)[\mathbf{R}_3 \cdot \hat{\mathbf{u}}_1 \times \hat{\mathbf{u}}_3] \tag{6.64}$$

which yields

$$\rho_2 = A + \frac{B}{r_2^3} \tag{6.65}$$

where

$$A = \frac{a_1[\mathbf{R}_1 \cdot \hat{\mathbf{u}}_1 \times \hat{\mathbf{u}}_3] - [\mathbf{R}_2 \cdot \hat{\mathbf{u}}_1 \times \hat{\mathbf{u}}_3] + a_3[\mathbf{R}_3 \cdot \hat{\mathbf{u}}_1 \times \hat{\mathbf{u}}_3]}{[\hat{\mathbf{u}}_1 \cdot \hat{\mathbf{u}}_2 \times \hat{\mathbf{u}}_3]} \tag{6.66a}$$

$$B = \frac{b_1[\mathbf{R}_1 \cdot \hat{\mathbf{u}}_1 \times \hat{\mathbf{u}}_3] - b_3[\mathbf{R}_3 \cdot \hat{\mathbf{u}}_1 \times \hat{\mathbf{u}}_3]}{[\hat{\mathbf{u}}_1 \cdot \hat{\mathbf{u}}_2 \times \hat{\mathbf{u}}_3]} \tag{6.66b}$$

Also we have

$$r_2^2 = \rho_2^2 + R_2^2 - 2(\hat{\mathbf{u}}_2 \cdot \mathbf{R}_2)\rho_2 \tag{6.67}$$

So Eqs. (6.65) and (6.67) are solved simultaneously for ρ_2 and r_2, similarly to the procedure in the Laplace method. Operating on Eq. (6.48) with $[\cdot\hat{\mathbf{u}}_2 \times \hat{\mathbf{u}}_3]$, we obtain

$$c_1\rho_1[\hat{\mathbf{u}}_1 \cdot \hat{\mathbf{u}}_2 \times \hat{\mathbf{u}}_3] = c_1[\mathbf{R}_1 \cdot \hat{\mathbf{u}}_2 \times \hat{\mathbf{u}}_3] - [\mathbf{R}_2 \cdot \hat{\mathbf{u}}_2 \times \hat{\mathbf{u}}_3] + c_3[\mathbf{R}_3 \cdot \hat{\mathbf{u}}_2 \times \hat{\mathbf{u}}_3] \tag{6.68}$$

and using $[\cdot\hat{\mathbf{u}}_1 \times \hat{\mathbf{u}}_2]$ on Eq. (6.48) we have

$$c_3\rho_3[\hat{\mathbf{u}}_3 \cdot \hat{\mathbf{u}}_1 \times \hat{\mathbf{u}}_2] = c_1[\mathbf{R}_1 \cdot \hat{\mathbf{u}}_1 \times \hat{\mathbf{u}}_2] - [\mathbf{R}_2 \cdot \hat{\mathbf{u}}_1 \times \hat{\mathbf{u}}_2] + c_3[\mathbf{R}_3 \cdot \hat{\mathbf{u}}_1 \times \hat{\mathbf{u}}_2] \tag{6.69}$$

Since we know r_2, we can determine c_1 and c_3 from Eq. (6.61) and then ρ_1 and ρ_3 from Eqs. (6.68) and (6.69). The geometrical conditions of the observations at all three times are satisfied by the values of ρ found. The values of c_1 and c_3 were determined from truncated series expansions, so the dynamical conditions imposed upon the solution are only approximately satisfied. In Eq. (6.60) the truncation error enters the values of b_1 and b_3. There are more accurate formulas for c_1 and c_3, when r_1, r_2, r_3 are approximately known. Using more accurate values, new values of b_1 and b_3 can be determined from

$$b_1 = (c_1' - a_1)r_2^3, \; b_3 = (c_3' - a_3)r_2^3 \tag{6.70}$$

where c_1' and c_3' are improved values of c_1 and c_3. Gibbs (1889) gives the formulas for c_1 and c_3

$$c_1 = \frac{T_1}{T_2}\left[\frac{1 + B_1 r_1^{-3}}{1 - B_2 r_2^{-3}}\right], \; c_3 = \frac{T_3}{T_2}\left[\frac{1 + B_3 r_3^{-3}}{1 - B_2 r_2^{-3}}\right] \tag{6.71}$$

where

$$B_1 = \frac{1}{12}(mn + n - m)T_2^2 \tag{6.72a}$$

$$B_2 = \frac{1}{12}(mn + 1)T_2^2 \tag{6.72b}$$

$$B_3 = \frac{1}{12}(mn - n + m)T_2^2 \tag{6.72c}$$

$$m = \frac{T_1}{T_2}, \; n = \frac{T_3}{T_2} \tag{6.72d}$$

These values can improve ρ_1, ρ_2, ρ_3 and, hence, r_1, r_2, r_3. The times used in this improvement should be corrected for light time by using the expression $t_{\text{corr}} = t_{\text{obs}} - 0.00577\rho$ in days, where ρ is the distance computed in the first approximation.

Now the calculation is again determining the orbital elements from the values of $\mathbf{r}_1, \mathbf{r}_2, \mathbf{r}_3$ (Herget 1948, pp. 62–65), (McCuskey 1963, pp. 86–91), (Danby 1962, 175–177).

6.5 Lambert's Theorem

Suppose E_1 and E_2, $E_2 > E_1$, are eccentric anomalies for points P_1 and P_2 in an elliptic orbit. Let

$$2G = E_2 + E_1 \tag{6.73}$$

and

$$2g = E_2 - E_1 > 0 \tag{6.74}$$

From the ellipse equation

$$r_1 = a(1 - e\cos E_1),\ r_2 = a(1 - e\cos E_2) \tag{6.75}$$

and

$$r_1 + r_2 = a[2 - e(\cos E_1 + \cos E_2)] = 2a[1 - e\cos G\cos g] \tag{6.76}$$

c is the length of the chord P_1P_2; using the equations

$$x = a\cos E \tag{6.77}$$

$$y = b\sin E = a\sqrt{1 - e^2}\sin E \tag{6.78}$$

$$\begin{aligned} c^2 &= a^2(\cos E_2 - \cos E_1)^2 + a^2(1 - e^2)(\sin E_2 - \sin E_1)^2 \\ &= 4a^2\sin^2 G\sin^2 g + 4a^2(1 - e^2)\cos^2 G\sin^2 g \\ &= 4a^2\sin^2 g(\sin^2 G + \cos^2 G) - 4a^2e^2\cos^2 G\sin^2 g \end{aligned} \tag{6.79}$$

as well as writing $e\cos G = \cos j$,

$$c^2 = 4a^2\sin^2 g(1 - \cos^2 j) \tag{6.80}$$

So

$$c = 2a\sin g\sin j \tag{6.81}$$

From Eq. (6.76)

$$r_1 + r_2 = 2a(1 - \cos g\cos j) \tag{6.82}$$

$$\epsilon = j + g,\ \delta = j - g \tag{6.83}$$

From Eqs. (6.81) and (6.82)

$$r_1 + r_2 + c = 2a[1 - \cos(g + j)] = 4a\sin^2\frac{1}{2}\epsilon \tag{6.84a}$$

$$r_1 + r_2 - c = 2a[1 - \cos(g - j)] = 4a\sin^2\frac{1}{2}\delta \tag{6.84b}$$

Let t be the time interval in the orbit between the two positions. From Kepler's equation

$$nt = E_2 - E_1 - e(\sin E_2 - \sin E_1) \tag{6.85a}$$

$$nt = (\epsilon - \delta) - (\sin \epsilon - \sin \delta) \tag{6.85b}$$

The variables ϵ and δ in terms of $(r_1 + r_2), c$ and a are given by Eqs. (6.84). These equations are *Lambert's theorem* for elliptic motion. The ϵ and δ values are not unique, but, for a small arc, ϵ and δ can be taken as the smallest values satisfying Eqs. (6.84). A geometric discussion is given by Plummer (1918).

A limiting case of Lambert's theorem exists for parabolic motion, when a tends to infinity. When a is large, ϵ and δ are small, so approximately

$$a\epsilon^2 = r_1 + r_2 + c \tag{6.86a}$$

$$a\delta^2 = r_1 + r_2 - c \tag{6.86b}$$

Replacing n by $\sqrt{\mu/a^3}$ and expanding the sine function as a series expansion ($\sin x = x + \frac{1}{6}x^3 + \ldots$) Eqs. (6.85) become

$$\sqrt{\mu}t = \frac{1}{6}a^{3/2}(\epsilon^3 - \delta^3) \tag{6.87}$$

and substituting from Eqs. (6.86)

$$\sqrt{\mu}t = \frac{1}{6}(r_1 + r_2 + c)^{3/2} \pm \frac{1}{6}(r_1 + r_2 - c)^{3/2} \tag{6.88}$$

If this is to be applied to comets, it can be written

$$6kt = (r_1 + r_2 + c)^{3/2} \pm (r_1 + r_2 - c)^{3/2} \tag{6.89}$$

This is *Euler's theorem*. Concerning the sign ambiguity, for small arcs use the minus sign. When the difference between the true anomalies is 180°, the second term is zero. For larger differences, use the plus sign. So the plus, or minus, sign are used based on whether the arc contains, or does not contain, the focus (Danby 1962, p. 141–3).

6.6 Parabolic Orbits, Olber's Method

Olber's method for parabolic orbit determination can be described as follows. Using Euler's equation for parabolic motion (6.89)

$$(r_1 + r_3 + c)^{3/2} - (r_1 + r_3 - c)^{3/2} = 6k(t_3 - t_1) = 6T_2 \tag{6.90}$$

where c is the length of the chord $(\mathbf{r}_3 - \mathbf{r}_1)$. Divide through Eq. (6.90) by $(r_1 + r_3)^{3/2}$ and put

$$\frac{2T_2}{(r_1 + r_3)^{3/2}} = \eta \tag{6.91}$$

Then Eq. (6.90) becomes

$$\left(1 + \frac{c}{r_1 + r_3}\right)^{3/2} - \left(1 - \frac{c}{r_1 + r_3}\right)^{3/2} = 3\eta \tag{6.92}$$

Solve this equation for $c/(r_1 + r_3)$ as a function of η. Call this function $f(\eta)$, which can be tabulated. Consider the equation from Gauss's method

$$c_1\,\boldsymbol{\rho}_1 - \boldsymbol{\rho}_2 + c_3\,\boldsymbol{\rho}_3 = c_1\,\mathbf{R}_1 - \mathbf{R}_2 + c_3\,\mathbf{R}_3 = \mathbf{V} \tag{6.93}$$

Let $\mathbf{U}$ be a vector coplanar with $\mathbf{V}$ and $\boldsymbol{\rho}_2$, which we will specify later. Multiply Eq. (6.93) scalarly by $(\boldsymbol{\rho}_2 \times \mathbf{U})$; then only the terms in ρ_1 and ρ_3 remain, so we have

$$\rho_3 = -\frac{c_1[\hat{\boldsymbol{\rho}}_1, \hat{\boldsymbol{\rho}}_3, \mathbf{U}]}{c_3[\hat{\boldsymbol{\rho}}_3, \hat{\boldsymbol{\rho}}_2, \mathbf{U}]}\rho_1 = M\rho_1 \tag{6.94}$$

Now

$$r_1^2 = (\mathbf{R}_1 - \boldsymbol{\rho}_1)^2 = R_1^2 - 2(\mathbf{R}_1 \cdot \hat{\boldsymbol{\rho}}_1)\rho_1 + \rho_1^2 = \alpha_1 + \beta_1\rho_1 + \gamma_1\rho_1^2 \tag{6.95}$$

and

$$\begin{aligned} r_3^2 &= (\mathbf{R}_3 - \boldsymbol{\rho}_3)^2 = (\mathbf{R}_3 - M\rho_1\hat{\boldsymbol{\rho}}_3)^2 \\ &= R_3^2 - 2M(\mathbf{R}_3 \cdot \hat{\boldsymbol{\rho}}_3)\rho_1 + M^2\rho_1^2 \\ &= \alpha_3 + \beta_3\rho_1 + \gamma_3\rho_1^2 \end{aligned} \tag{6.96}$$

$$\begin{aligned} c^2 &= (\mathbf{r}_3 - \mathbf{r}_1)^2 = [(M\hat{\boldsymbol{\rho}}_3 - \hat{\boldsymbol{\rho}}_1)\rho_1 - (\mathbf{R}_3 - \mathbf{R}_1)]^2 \\ &= (M\hat{\boldsymbol{\rho}}_3 - \hat{\boldsymbol{\rho}}_1) \cdot (M\hat{\boldsymbol{\rho}}_3 - \hat{\boldsymbol{\rho}}_1)\rho_1^2 - 2(M\hat{\boldsymbol{\rho}}_3 - \hat{\boldsymbol{\rho}}_1) \cdot (\mathbf{R}_3 - \mathbf{R}_1)\rho_1 \\ &+ (\mathbf{R}_3 - \mathbf{R}_1) \cdot (\mathbf{R}_3 - \mathbf{R}_1) \\ &= \alpha_3 + \beta_3\rho_1 + \gamma_3\rho_1^2 \end{aligned} \tag{6.97}$$

If $\mathbf{U}$ is known, then M is known and also all the coefficients. If we take a trial value of ρ_1, we can calculate c from Eq. (6.97). Also, we can calculate r_1 and r_3 from Eqs. (6.95) and (6.96), so we can calculate η and $f(\eta)$ and $c = (r_1 + r_3)f(\eta)$. The two values of c will not agree, the difference being due to the trial value of ρ_1, so we call this difference $\Delta(\rho_1)$. By varying ρ_1 values, a table of $\Delta(\rho_1)$ as a function of ρ_1 can be prepared. By inverse interpolation, a value of ρ_1, when $\Delta(\rho_1)$ is zero, can be found. Once ρ_1 is found, ρ_3, r_1, and r_3 can be found and, hence, the elements. q is found from η by

$$q = \frac{\|\mathbf{r}_1 \times \mathbf{r}_3\|^2}{2T_2^2} \left[\frac{1}{3} \left(1 + \frac{2}{\sqrt{1 - f(\eta)^2}} \right) \right]^2 \tag{6.98}$$

and v, the true anomaly, from

$$\tan^2 \frac{v}{2} = \frac{r - q}{q} \tag{6.99}$$

In this method, the approximation is the preliminary value of $\mathbf{U}$. We have

$$\mathbf{R}_2 = C_1\mathbf{R}_1 + C_3\mathbf{R}_3 \tag{6.100}$$

where C_1 and C_3 are triangle ratios for the motion of the Earth about the Sun. To a good approximation,

$$\frac{C_1}{C_3} = \frac{T_1}{T_3} \tag{6.101}$$

If the time intervals are short, T_1/T_3 is a first approximation to the $c's$ ratios, so to this order

$$c_1 = aC_1, \; c_3 = aC_3 \tag{6.102}$$

for some unspecified a. From Eq. (6.93)

$$\mathbf{V} = c_1\mathbf{R}_1 - \mathbf{R}_2 + c_3\mathbf{R}_3 = aC_1\mathbf{R}_1 - \mathbf{R}_2 + aC_3\mathbf{R}_3 = \mathbf{R}_2(a-1) \tag{6.103}$$

For a first approximation, we take

$$\mathbf{U} = \mathbf{R}_2, \; \frac{c_1}{c_3} = \frac{T_1}{T_3} \tag{6.104}$$

When $\mathbf{r}_1$ and $\mathbf{r}_3$ are known, $\mathbf{r}_2$ is found from

$$c_1\boldsymbol{\rho}_1 - \boldsymbol{\rho}_2 + c_3\boldsymbol{\rho}_3 = c_1\mathbf{R}_1 - \mathbf{R}_2 + c_3\mathbf{R}_3 \tag{6.105}$$

and $\mathbf{r} = \boldsymbol{\rho} - \mathbf{R}$.

With a rough value of $\mathbf{r}_2$, better values of the $c's$ can be found, and an improved value of $\mathbf{r}_2$. From a final value we can determine values of α and δ of the second observation, but these are not likely to agree with the observed values. To eliminate the discrepancies, the improved values of the $c's$ can be used to find $\mathbf{V}$. Then take

$\mathbf{U} = \mathbf{V}$ and repeat the solution, or repeat the solution with different values of M. The best result is found by inverse interpolation.

Another approach, due to Strömgren (1914), is the *method of false position*. If $\mathbf{U}$ is fixed, then M can be a function of $\hat{\rho}_2$. Since $\hat{\rho}_2$ (observed) produces a solution that yields the middle position at $\hat{\rho}_2$ (computed), then if we use a fictitious

$$\hat{\rho}_2 = 2\hat{\rho}_2\,(\text{obs}) - \hat{\rho}_2\,(\text{comp}) \tag{6.106}$$

in M, we expect to get a solution yielding the middle position $\hat{\rho}_2$ (obs). Thus, M is recomputed as before with the fictitious $\hat{\rho}_2$, and the solution is repeated.

It may be that the residuals refuse to disappear; it could be that the orbit is not parabolic. Three observations give too many quantities for the 5 unknowns of the parabolic orbit. When a comet is discovered, it is usually moving in a nearly parabolic orbit. A search ephemeris for a newly discovered comet is usually calculated by this method. For a minor planet, a parabolic orbit normally is unsatisfactory. When Uranus was discovered, it was thought to be a comet; it was only recognized as a planet, when its orbit could not be fit by a parabolic orbit (Danby 1962, pp. 180–182).

6.7 Circular Orbits

We include this mainly for completeness. Only four elements are required for a circular orbit, so two observations are adequate. An orbit based on two observations is unlikely to be of much value, unless the eccentricity is small. Thus, it might be sufficient for a minor planet, but useless for a comet. The times of observations are t_1 and t_2, and $T = k(t_2 - t_1)$. During interval t_1 to t_2, the object revolved through an angle M around the Sun, where

$$\frac{M}{2\pi} = \frac{T}{period} \tag{6.107}$$

The period from Kepler's third law is $2\pi a^{3/2}$, so

$$M = Ta^{-3/2} \tag{6.108}$$

Since it is a circular orbit,

$$r_2 = r_1 = a \tag{6.109}$$

So,

$$\|\mathbf{r}_1 \times \mathbf{r}_2\| = a^2 \sin M = a^2 \left(M - \frac{1}{6} M^3 + \frac{1}{120} M^5 - \ldots \right) \tag{6.110}$$

$$\|\mathbf{r}_1 \times \mathbf{r}_2\| = \sqrt{a} T \left(1 - \frac{1}{6} \frac{T^2}{a^3} + \frac{1}{120} \frac{T^4}{a^6} - \ldots \right) \tag{6.111}$$

Since $r_1^2 = a^2 = r_2^2$,

$$\rho_1^2 - 2R_1\rho_1 \cos\psi_1 + R_1^2 = a^2 = \rho_2^2 - 2R_2\rho_2 \cos\psi_2 + R_2^2 \tag{6.112}$$

where ψ_1 and ψ_2 are the angles between $\boldsymbol{\rho}_1$ and $\mathbf{R}_1$ and $\boldsymbol{\rho}_2$ and $\mathbf{R}_2$. If a value of ρ_1 is assumed, then $a, \rho_2, \mathbf{r}_1$, and $\mathbf{r}_2$ can be found. Then each side of Eq. (6.111) can be calculated independently. They are not likely to agree, but, if their difference is tabulated for different values of ρ_1, the value of ρ_1, which makes the difference zero, can be found by inverse interpolation. This gives the solution (Danby 1962, p. 182).

References

Danby, J.: Fundamentals of Celestial Mechanics. The Macmillan Company, New York (1962)

Gibbs, J.W.: On the determination of elliptic orbits from three complete observations, vol. 4. National Academy of Sciences, Washington DC (1889)

Herget, P.: The Computation of Orbits. University of Cincinnati, Cincinnati (1948)

McCuskey, S.W.: Introduction to Celestial Mechanics. Addison-Wesley, Reading (1963)

Plummer, H.C.K.: An Introductory Treatise on Dynamical Astronomy. University Press, Cambridge (1918)

Strömgren, E.: Ueber den ursprung der kometen. Publications of the Kobenhavns Observatory **19**, 1–62 (1914)

Williams, K.P.: The Calculation of the Orbits of Asteroids and Comets. The Principia Press, Bloomington (1934)

Chapter 7
The n-Body Problem

7.1 Introduction

When going from two bodies to three, or more, bodies, the complexity increases significantly, due to their mutual attractions. The two-body problem can be mathematically formulated so a closed-form solution is possible. With more than two bodies, it is impossible to formulate such a solution. There are some special cases, however, that can be handled.

The solar system planets are an *n-body problem*. Here, the Sun is the center of force, and the individual planetary motions approximate the two-body problem. The deviations from the two-body problem are due to the other planets, and are called *perturbations*. The minor planet motions approach the three-body problem (see also Sect. 1.5), because the effects of the Sun and Jupiter dominate the motions. The *Trojan minor planets* are a special case of the three-body problem.

When n is a limited number, less than ten such as for the solar system, the computations of the motions of the n bodies can be done by general theories, where the expressions for the bodies are generally in Fourier series or Chebyshev series, or they can be done by numerical integrations. These will be discussed in subsequent chapters. When n becomes a large number, as in the case of galaxies or star clusters, special numerical techniques need to be applied for making the computer computations of reasonable times.

We will start our discussion with analyzing the dynamics of the three-body problem, and then generalize to the n-body case.

P. Gurfil, P.K. Seidelmann, *Celestial Mechanics and Astrodynamics: Theory and Practice*, Astrophysics and Space Science Library 436,
DOI 10.1007/978-3-662-50370-6_7

7.2 Equations of Motion

As usual, we assume the masses are spherically symmetrical in homogeneous layers, so they attract one another like point masses. Let us consider three masses m_1, m_2, m_3 located in a primary-fixed reference system by position vectors $\mathbf{r}_1, \mathbf{r}_2, \mathbf{r}_3$, as shown in Fig. 7.1.

Only the mutual Newtonian attractions of the bodies on each other are the forces. The equations of motion are

$$m_1\ddot{\mathbf{r}}_1 = k^2\left[\frac{m_1 m_2}{\rho_{12}^3}\boldsymbol{\rho}_{12} + \frac{m_1 m_3}{\rho_{13}^3}\boldsymbol{\rho}_{13}\right] \tag{7.1}$$

$$m_2\ddot{\mathbf{r}}_2 = k^2\left[\frac{m_2 m_3}{\rho_{23}^3}\boldsymbol{\rho}_{23} - \frac{m_2 m_1}{\rho_{12}^3}\boldsymbol{\rho}_{12}\right] \tag{7.2}$$

$$m_3\ddot{\mathbf{r}}_3 = k^2\left[-\frac{m_3 m_2}{\rho_{23}^3}\boldsymbol{\rho}_{23} - \frac{m_3 m_1}{\rho_{13}^3}\boldsymbol{\rho}_{13}\right] \tag{7.3}$$

where the $\boldsymbol{\rho}_{ij}$ vectors join the masses with the orientation as in Fig. 7.1. Adding Eqs. (7.1), (7.2), and (7.3) and integrating twice, we obtain

$$m_1\mathbf{r}_1 + m_2\mathbf{r}_2 + m_3\mathbf{r}_3 = \mathbf{c}_1 t + \mathbf{c}_2 \tag{7.4}$$

where $\mathbf{c}_1$ and $\mathbf{c}_2$ are vector constants of integration. By definition, the center of mass $\mathbf{R}$ of the three mass points is given by

$$\mathbf{R} = \frac{m_1\mathbf{r}_1 + m_2\mathbf{r}_2 + m_3\mathbf{r}_3}{M} \tag{7.5}$$

where $M = m_1 + m_2 + m_3$. Therefore, from Eq. (7.4),

$$\mathbf{R} = \frac{\mathbf{c}_1 t}{M} + \frac{\mathbf{c}_2}{M} \tag{7.6}$$

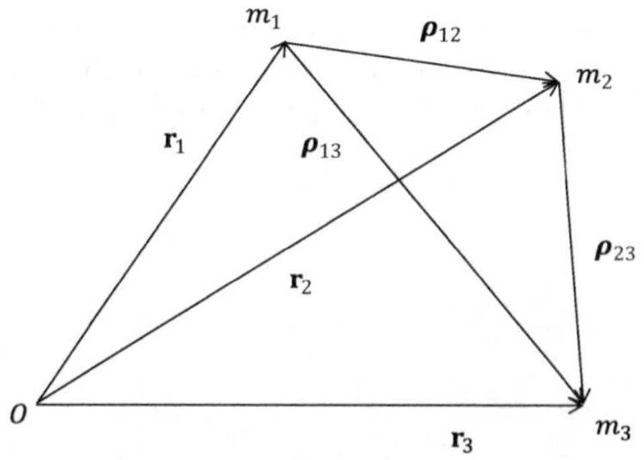

Fig. 7.1 Setup for the three-body problem

The center of mass remains at rest, or moves uniformly in a straight line, which agrees with the same observation we made in the two-body problem (see Sect. 5.3). This is the first integral of the equations of motion, and has six arbitrary constants, the three components of each of the vectors $\mathbf{c}_1$ and $\mathbf{c}_2$.

With n bodies in the system, the ith member differential equation is

$$m_i\ddot{\mathbf{r}}_i = k^2 \sum_{j=1}^{n-1} \frac{m_i m_j}{\rho_{ij}^3} \boldsymbol{\rho}_{ij}, \; i,j = 1,2,3,\ldots .n, \; i \neq j \tag{7.7}$$

where $\boldsymbol{\rho}_{ij}$ is from m_i to m_j. Remembering $\boldsymbol{\rho}_{ij} = -\boldsymbol{\rho}_{ji}$, summing Eq. (7.7) over i we have

$$\sum_{i=1}^{n} m_i\ddot{\mathbf{r}}_i = 0 \tag{7.8}$$

Two integrations of this equation give

$$\sum_{i=1}^{n} m_i\mathbf{r}_i = \mathbf{c}_1 t + \mathbf{c}_2 \tag{7.9}$$

and since

$$\mathbf{R} = \frac{1}{M}\sum_{i=1}^{n} m_i\mathbf{r}_i \tag{7.10}$$

then

$$\mathbf{R} = \frac{\mathbf{c}_1}{M}t + \frac{\mathbf{c}_2}{M} \tag{7.11}$$

where

$$M = \sum_{i=1}^{n} m_i \tag{7.12}$$

So from the two-body problem, and any number of body problem, the center of mass remains at rest, or moves uniformly in space on a straight line (McCuskey 1963, pp. 92–94).

7.3 Angular Momentum, or Areal Velocity, Integral

$\hat{\mathbf{u}}_{12}, \hat{\mathbf{u}}_{13}, \hat{\mathbf{u}}_{23}$ are unit vectors in the directions of $\hat{\boldsymbol{\rho}}_{12}, \hat{\boldsymbol{\rho}}_{13}, \hat{\boldsymbol{\rho}}_{23}$, respectively. Equations (7.1), (7.2), (7.3) are written

$$m_1\dot{\mathbf{v}}_1 = k^2\left[\frac{m_1m_2}{\rho_{12}^2}\hat{\mathbf{u}}_{12} + \frac{m_1m_3}{\rho_{13}^2}\hat{\mathbf{u}}_{13}\right] \tag{7.13a}$$

$$m_2\dot{\mathbf{v}}_2 = k^2\left[\frac{m_2m_3}{\rho_{23}^2}\hat{\mathbf{u}}_{23} - \frac{m_2m_1}{\rho_{12}^2}\hat{\mathbf{u}}_{12}\right] \tag{7.13b}$$

$$m_3\dot{\mathbf{v}}_3 = k^2\left[-\frac{m_3m_2}{\rho_{23}^2}\hat{\mathbf{u}}_{23} - \frac{m_3m_1}{\rho_{13}^2}\hat{\mathbf{u}}_{13}\right] \tag{7.13c}$$

Taking the vector products of Eqs. (7.13) by $\mathbf{r}_1, \mathbf{r}_2, \mathbf{r}_3$, respectively, and adding, we have

$$\begin{aligned}\sum_{i=1}^{3}\mathbf{r}_i \times m_i\dot{\mathbf{v}}_i = k^2[&\frac{m_1m_2}{\rho_{12}^2}(\mathbf{r}_1 - \mathbf{r}_2) \times \hat{\mathbf{u}}_{12} \\ &+ \frac{m_2m_3}{\rho_{23}^2}(\mathbf{r}_2 - \mathbf{r}_3) \times \hat{\mathbf{u}}_{23} + \frac{m_3m_1}{\rho_{12}^2}(\mathbf{r}_1 - \mathbf{r}_3) \times \hat{\mathbf{u}}_{13}]\end{aligned} \tag{7.14}$$

$\mathbf{r}_1 - \mathbf{r}_2$ is collinear with $\hat{\mathbf{u}}_{12}$; $\mathbf{r}_2 - \mathbf{r}_3$ is collinear with $\hat{\mathbf{u}}_{23}$; and $\mathbf{r}_1 - \mathbf{r}_3$ is collinear with $\hat{\mathbf{u}}_{13}$. Thus, each vector product is zero, and the right-hand side of Eq. (7.14) is zero. The left-hand side can be written as

$$\frac{d}{dt}\left[\sum_{i=1}^{3}\mathbf{r}_i \times m\,\mathbf{v}_i\right] = \mathbf{0} \tag{7.15}$$

The total angular momentum of the three masses about the origin O appears in the brackets; we denote it by $\mathbf{L}$. Equation (7.15) states that "$\mathbf{L}$ = constant vector". This is the second integral of the equations of motion. $\mathbf{L}$ has three arbitrary scalar constants. Equation (7.15) expresses the conservation of angular momentum for the system of three masses.

We represent the angular momentum vector in a Cartesian coordinate system, with the origin at O, as shown in Fig. 7.2, by

$$\mathbf{L} = a_1\,\hat{\mathbf{i}} + a_2\,\hat{\mathbf{j}} + a_3\,\hat{\mathbf{k}} \tag{7.16}$$

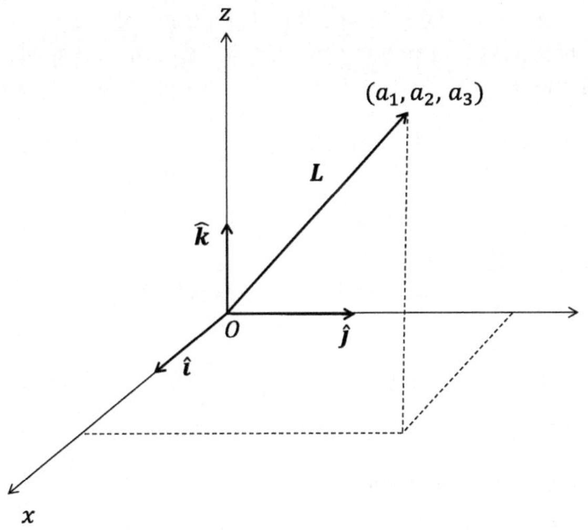

Fig. 7.2 The angular momentum vector shown in a Cartesian coordinate system

The Cartesian equations for the angular momentum are

$$\sum_{i=1}^{3} m_i \left[y_i \dot{z}_i - z_i \dot{y}_i\right] = a_1 \tag{7.17a}$$

$$\sum_{i=1}^{3} m_i \left[z_i \dot{x}_i - x_i \dot{z}_i\right] = a_2 \tag{7.17b}$$

$$\sum_{i=1}^{3} m_i \left[x_i \dot{y}_i - y_i \dot{x}_i\right] = a_3 \tag{7.17c}$$

The brackets in these equations are the projections of the areal velocities of the different bodies upon the three coordinate planes. They are the *integrals of area*,

$$\frac{1}{2}\left[y\dot{z} - z\dot{y}\right] = \frac{1}{2}c_1 \tag{7.18a}$$

$$\frac{1}{2}\left[z\dot{x} - x\dot{z}\right] = \frac{1}{2}c_2 \tag{7.18b}$$

$$\frac{1}{2}\left[x\dot{y} - y\dot{x}\right] = \frac{1}{2}c_3 \tag{7.18c}$$

and $h = \sqrt{c_1^2 + c_2^2 + c_3^2}$. These expressions have a counterpart in two-body motion (see Sect. 5.5).

Let A_{i1}, A_{i2}, A_{i3} denote the projections of the areas swept out by the radius vector of the mass m_i upon the yz, xz, and xy planes, respectively. Equations (7.17) may be written as

$$\sum_{i=1}^{3} m_i \dot{A}_{i1} = a_1 \tag{7.19a}$$

$$\sum_{i=1}^{3} m_i \dot{A}_{i2} = a_2 \tag{7.19b}$$

$$\sum_{i=1}^{3} m_i \dot{A}_{i3} = a_3 \tag{7.19c}$$

Integrated, these yield

$$\sum_{i=1}^{3} m_i A_{i1} = a_1 t + b_1 \tag{7.20a}$$

$$\sum_{i=1}^{3} m_i A_{i2} = a_2 t + b_2 \tag{7.20b}$$

$$\sum_{i=1}^{3} m_i A_{i3} = a_3 t + b_3 \tag{7.20c}$$

Consequently, the sums of the products of the masses and the projections of the areas swept out by the corresponding radius vectors increase uniformly with time. As the three bodies move, their position and velocity vectors are oriented so the vector **L** has a fixed direction in space and a constant magnitude, $(a_1^2 + a_2^2 + a_3^2)^{1/2}$. **L** is directed along a line that is the *invariable line*. A plane, which is perpendicular to invariable line, and passes through the center of mass of the system, is the invariable plane (see also Sect. 3.5.1). This term was introduced by Laplace; the Laplacian plane is used for natural satellite systems, i.e. around Jupiter and Saturn. The invariable plane has the following properties:

(a) The angular momentum about any line in the plane is zero.
(b) The angular momentum about a line normal to the plane is a maximum.

In our notation, the direction numbers of the normal to the invariable plane are a_1, a_2, a_3.

The orbital angular momentum of Jupiter and Saturn is nearly 87 % of the whole solar system. As a consequence, and because the two planets' orbital planes are nearly in the ecliptic, the vector **L** is directed only about $1°35'$ from the pole of the ecliptic. The invariable plane lies between the orbita l planes of these two planets. Since the masses, positions, and velocities of planets of the solar system are well known, the constants a_1, a_2, a_3 can be determined with considerable accuracy. So, the solar system's invariable plane is relatively well determined.

What is stated here about the three-body system can be generalized to the n-body problem, and will be shown later (McCuskey 1963, pp. 94–97).

7.4 Integral of Energy

The energy integral of the equations of motion in the three-body problem is analogous to $\frac{1}{2}mv^2 + V(r) = E$ under central force motion (see Sect. 4.4).

Take the scalar product of Eqs. (7.13) by $\dot{\mathbf{r}}_1, \dot{\mathbf{r}}_2, \dot{\mathbf{r}}_3$, respectively, and add the results. Then

$$\sum_{i=1}^{3} m_i \dot{\mathbf{r}}_i \cdot \dot{\mathbf{v}}_i = k^2 \left[\frac{m_1 m_2}{\rho_{12}^2} \hat{\mathbf{u}}_{12} \cdot (\dot{\mathbf{r}}_1 - \dot{\mathbf{r}}_2) + \frac{m_1 m_3}{\rho_{13}^2} \hat{\mathbf{u}}_{13} \cdot (\dot{\mathbf{r}}_1 - \dot{\mathbf{r}}_3) + \frac{m_2 m_3}{\rho_{23}^2} \hat{\mathbf{u}}_{23} \cdot (\dot{\mathbf{r}}_2 - \dot{\mathbf{r}}_3) \right] \tag{7.21}$$

with $\dot{\mathbf{r}}_1 - \dot{\mathbf{r}}_2 = -\dot{\boldsymbol{\rho}}_{12}$, $\dot{\mathbf{r}}_1 - \dot{\mathbf{r}}_3 = -\dot{\boldsymbol{\rho}}_{13}$, $\dot{\mathbf{r}}_2 - \dot{\mathbf{r}}_3 = -\dot{\boldsymbol{\rho}}_{23}$ and

$$\dot{\boldsymbol{\rho}}_{ij} = \frac{d}{dt}(\rho_{ij}\, \hat{\mathbf{u}}_{ij}) = \dot{\rho}_{ij}\, \hat{\mathbf{u}}_{ij} + \rho_{ij}\, \dot{\hat{\mathbf{u}}}_{ij}, \ i, j = 1, 2, 3, \ i \neq j \tag{7.22}$$

Since $\hat{\mathbf{u}}_{ij}$ is a unit vector, $\dot{\hat{\mathbf{u}}}_{ij} \cdot \hat{\mathbf{u}}_{ij} = 0$ and $\hat{\mathbf{u}}_{ij} \cdot \hat{\mathbf{u}}_{ij} = 1$. Then, using this in Eq. (7.21)

$$\sum_{i=1}^{3} m_i \dot{\mathbf{r}}_i \cdot \dot{\mathbf{v}}_i = -k^2 \left[\frac{m_1 m_2}{\rho_{12}^2} \dot{\rho}_{12} + \frac{m_1 m_3}{\rho_{13}^2} \dot{\rho}_{13} + \frac{m_2 m_3}{\rho_{23}^2} \dot{\rho}_{23} \right] \tag{7.23}$$

which can be written

$$\frac{d}{dt} \left[\frac{1}{2} \sum_{i=1}^{3} m_i \dot{r}_i^2 \right] = k^2 \frac{d}{dt} \left[\frac{m_1 m_2}{\rho_{12}} + \frac{m_1 m_3}{\rho_{13}} + \frac{m_2 m_3}{\rho_{23}} \right] \tag{7.24}$$

By definition, the kinetic energy of the system is

$$T = \frac{1}{2} \sum_{i=1}^{3} m_i \dot{\mathbf{r}}_i^2 = \frac{1}{2} \sum_{i=1}^{3} m_i v_i^2 \tag{7.25}$$

where v_i is the speed of the ith mass. The potential energy of the system is defined as

$$V = -k^2 \left[\frac{m_1 m_2}{\rho_{12}} + \frac{m_1 m_3}{\rho_{13}} + \frac{m_2 m_3}{\rho_{23}} \right] \tag{7.26}$$

Integration of Eq. (7.24) yields

$$T + V = E = \text{constant} \tag{7.27}$$

Equation (7.27), the energy integral, shows the conservation of energy for the system of three masses. This result can be generalized to n bodies.

7.5 Stationary Solutions of the Three-Body Problem

Two special solutions of the three-body problem, designated *stationary solutions*, were discovered in 1772 by Lagrange. Assume three mass points, m_1, m_2, m_3, are projected in one plane. A stationary solution is one where the geometric configuration of the three masses is invariant with time. If the masses move such that their mutual distances remain unchanged, the configuration simply rotates around the center of mass in a plane. Alternatively, a contraction, which does not alter the pattern of the points, may take place.

First, consider the special case where the three masses revolve with constant angular speed around the center of mass in coplanar circular orbits. Their position vectors relative to the center of mass C are $\mathbf{r}_1 = r_1\hat{\mathbf{u}}_1$, $\mathbf{r}_2 = r_2\hat{\mathbf{u}}_2$, $\mathbf{r}_3 = r_3\hat{\mathbf{u}}_3$. $\hat{\mathbf{u}}_1, \hat{\mathbf{u}}_2, \hat{\mathbf{u}}_3$ are the unit vectors in the directions $\mathbf{r}_1, \mathbf{r}_2, \mathbf{r}_3$, respectively, as shown in Fig. 7.3. For circular motion, r_1, r_2, r_3 are constant.

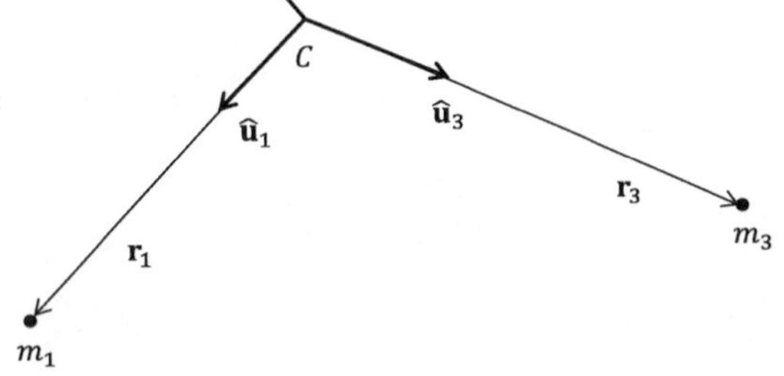

Fig. 7.3 Coplanar orbits in the three-body problem

The total planar acceleration of each mass point is given by (recall Eq. (4.6))

$$\mathbf{a} = (\ddot{r} - r\dot{\theta}^2)\,\hat{\mathbf{u}}_r + (2\dot{r}\dot{\theta} + r\ddot{\theta})\,\hat{\mathbf{u}}_\theta \tag{7.28}$$

where $\hat{\mathbf{u}}_r$ and $\hat{\mathbf{u}}_\theta$ are unit vectors in the radial and transverse directions, respectively. For the special case of the three-body problem considered here, for each mass $\dot{r} = \ddot{r} = 0,\ \ddot{\theta} = 0,\ \dot{\theta} = n$, the constant speed of the revolution about C. Thus, the accelerations become

$$\mathbf{a}_i = -r_i n^2\,\hat{\mathbf{u}}_r,\ i = 1, 2, 3 \tag{7.29}$$

With these values in Eqs. (7.1)–(7.3), we have

$$-n^2 r_1\hat{\mathbf{u}}_1 = k^2\left[\frac{m_2}{\rho_{12}^3}(r_2\hat{\mathbf{u}}_2 - r_1\hat{\mathbf{u}}_1) + \frac{m_3}{\rho_{13}^3}(r_3\hat{\mathbf{u}}_3 - r_1\hat{\mathbf{u}}_1)\right] \tag{7.30a}$$

$$-n^2 r_2\hat{\mathbf{u}}_2 = k^2\left[\frac{m_3}{\rho_{23}^3}(r_3\hat{\mathbf{u}}_3 - r_2\hat{\mathbf{u}}_2) - \frac{m_1}{\rho_{12}^3}(r_2\hat{\mathbf{u}}_2 - r_1\hat{\mathbf{u}}_1)\right] \tag{7.30b}$$

$$-n^2 r_3\hat{\mathbf{u}}_3 = k^2\left[-\frac{m_2}{\rho_{23}^3}(r_3\hat{\mathbf{u}}_3 - r_2\hat{\mathbf{u}}_2) - \frac{m_1}{\rho_{13}^3}(r_3\hat{\mathbf{u}}_3 - r_1\hat{\mathbf{u}}_1)\right] \tag{7.30c}$$

However, the origin is selected so

$$m_1 r_1\hat{\mathbf{u}}_1 + m_2 r_2\hat{\mathbf{u}}_2 + m_3 r_3\hat{\mathbf{u}}_3 = 0 \tag{7.31}$$

If we multiply Eq. (7.30a), by m_1 and (7.30b) by m_2 and add, we derive an equation, which upon substitution from Eq. (7.31) yields Eq. (7.30c). Thus, we can use Eq. (7.31) in place of Eq. (7.30c) in the unit vectors $\hat{\mathbf{u}}_1, \hat{\mathbf{u}}_2, \hat{\mathbf{u}}_3$. Rewrite the system of equations in the form

$$\left(-n^2 + \frac{k^2 m_2}{\rho_{12}^3} + \frac{k^2 m_3}{\rho_{13}^3}\right) r_1\hat{\mathbf{u}}_1 - \frac{k^2 m_2}{\rho_{12}^3} r_2\hat{\mathbf{u}}_2 - \frac{k^2 m_3}{\rho_{13}^3} r_3\hat{\mathbf{u}}_3 = 0 \tag{7.32a}$$

$$-\frac{k^2 m_1}{\rho_{12}^3} r_1\hat{\mathbf{u}}_1 + \left(-n^2 + \frac{k^2 m_3}{\rho_{23}^3} + \frac{k^2 m_1}{\rho_{12}^3}\right) r_2\hat{\mathbf{u}}_2 - \frac{k^2 m_3}{\rho_{23}^3} r_3\hat{\mathbf{u}}_3 = 0 \tag{7.32b}$$

$$m_1 r_1\hat{\mathbf{u}}_1 + m_2 r_2\hat{\mathbf{u}}_2 + m_3 r_3\hat{\mathbf{u}}_3 = 0 \tag{7.32c}$$

These conditions must be fulfilled for the three mass points to move with uniform angular speed in circular orbits around the center of mass in a plane.

If we have a rectangular coordinate system, origin at C, rotating in the counter-clockwise direction with a constant angular speed, then the unit vectors $\hat{\mathbf{u}}_1, \hat{\mathbf{u}}_2, \hat{\mathbf{u}}_3$ are fixed in position. The angles in the equations

$$u_i = \cos\theta_i\,\hat{\mathbf{i}} + \sin\theta_i\,\hat{\mathbf{j}},\ i = 1, 2, 3 \tag{7.33}$$

denote the orientations relative to these Cartesian coordinates. Taking the scalar products of $\hat{\mathbf{i}}$ and Eqs. (7.32), after simplification

$$-n^2x_1+\frac{k^2m_2}{\rho_{12}^3}(x_1-x_2)+\frac{k^2m_3}{\rho_{13}^3}(x_1-x_3)=0 \tag{7.34}$$

$$-n^2x_2+\frac{k^2m_1}{\rho_{12}^3}(x_2-x_1)+\frac{k^2m_3}{\rho_{23}^3}(x_2-x_3)=0 \tag{7.35}$$

$$m_1x_1+m_2x_2+m_3x_3=0 \tag{7.36}$$

where

$$x_i=r_i\cos\theta_i,\ \ i=1,2,3 \tag{7.37}$$

Taking the scalar products of $\hat{\mathbf{j}}$ and the same Eqs. (7.32), we have

$$-n^2y_1+\frac{k^2m_2}{\rho_{12}^3}(y_1-y_2)+\frac{k^2m_3}{\rho_{13}^3}(y_1-y_3)=0 \tag{7.38}$$

$$-n^2y_2+\frac{k^2m_1}{\rho_{12}^3}(y_2-y_1)+\frac{k^2m_3}{\rho_{23}^3}(y_2-y_3)=0 \tag{7.39}$$

$$m_1y_1+m_2y_2+m_3y_3=0 \tag{7.40}$$

where

$$y_i=r_i\sin\theta_i,\ \ i=1,2,3 \tag{7.41}$$

Equations (7.34)–(7.36) and (7.38)–(7.40) are a system of six simultaneous equations in the unknowns $(x_i, y_i),\ i = 1, 2, 3$. Consider a case where the masses are at the vertices of an equilateral triangle. Then, $\rho_{12} = \rho_{23} = \rho_{13} = \rho$ at all times. If the scale of the length is adjusted so that ρ is unity at all times, the equations may be written with $k^2 = 1$ by proper choice of time units,

$$(-n^2+m_2+m_3)x_1-m_2x_2-m_3x_3=0 \tag{7.42a}$$

$$-m_1x_1+(-n^2+m_1+m_3)x_2-m_3x_3=0 \tag{7.42b}$$

$$m_1x_1+m_2x_2+m_3x_3=0 \tag{7.42c}$$

$$(-n^2+m_2+m_3)y_1-m_2y_2-m_3y_3=0 \tag{7.43a}$$

$$-m_1y_1+(-n^2+m_1+m_3)y_2-m_3y_3=0 \tag{7.43b}$$

$$m_1y_1+m_2y_2+m_3y_3=0 \tag{7.43c}$$

These have a nontrivial solution if

$$\det \begin{bmatrix} M' & -m_2 & -m_3 & 0 & 0 & 0 \\ -m_1 & M' & -m_3 & 0 & 0 & 0 \\ m_1 & m_2 & m_3 & 0 & 0 & 0 \\ 0 & 0 & 0 & M' & -m_2 & -m_3 \\ 0 & 0 & 0 & -m_1 & M' & m_3 \\ 0 & 0 & 0 & m_1 & m_2 & m_3 \end{bmatrix} = 0 \tag{7.44}$$

where $M' = (-n^2 + m_2 + m_3)$.

The solution $x_i = 0,\ y_i = 0,\ i = 1, 2, 3$ is inconsistent with the requirement that $\rho = 1$. Let $M = m_1 + m_2 + m_3$ denote the total mass of the system. By the usual reduction rules for determinants, we have Eq. (7.44)

$$m_3^2(M - n^2)^4 = 0 \tag{7.45}$$

This is satisfied when $n^2 = M$. If we left k^2 in the equations and the equal values of $\rho = \rho_{ij},\ i, j = 1, 2, 3; i \neq j$, this condition would be

$$n^2 = \frac{k^2 M}{\rho^3} \tag{7.46}$$

Equation (7.46) is dimensionally consistent and specifies that n is in radians per unit of time.

With $n^2 = M$ in Eqs. (7.42) and (7.43), any two pairs of coordinates (x_i, y_i) can arbitrarily be chosen, the scale adjusted, and the third pair determined such that $\rho = 1$. Therefore, there is a solution to the equations in this special problem of three bodies.

There is a second solution for Eqs. (7.34)–(7.36) and (7.38)–(7.40). If $y_1 = y_2 = y_3 = 0$, then all three mass points lie on the x-axis, and Eqs. (7.38)–(7.40) are satisfied. Arrange the masses on the x-axis as shown in Fig. 7.4.

Denote the distance between x_3 and x_2 by ρ and adjust the scale so the distance between x_2 and x_1 is unity. Due to the inequalities, $x_1 < x_2 < x_3$, Eqs. (7.34)–(7.36) become

$$n^2 x_1 + k^2 m_2 + k^2 m_3 (1 + \rho)^{-2} = 0 \tag{7.47a}$$

$$n^2(1 + x_1) - k^2 m_1 + k^2 m_3 \rho^{-2} = 0 \tag{7.47b}$$

$$m_1 x_1 + m_2(1 + x_1) + m_3(1 + x_1 + \rho) = 0 \tag{7.47c}$$

We eliminate n^2 and x_1 from these to obtain

$$(m_1 + m_2)\rho^5 + (3m_1 + 2m_2)\rho^4 + (3m_1 + m_2)\rho^3 - (m_2 + m_3)\rho^2 - (2m_2 + 3m_3)\rho - (m_2 + m_3) = 0 \tag{7.48}$$

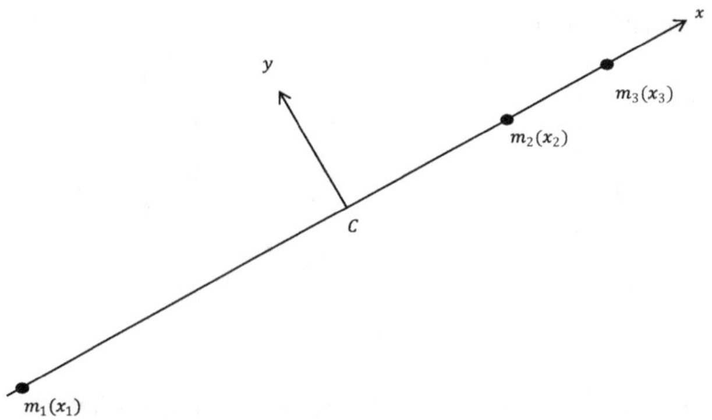

Fig. 7.4 Three masses arranged on the x-axis

This 5th degree equation in ρ, originally due to Lagrange, has only one real positive root, because there is only one sign change in the coefficients. With this arrangement of the masses along the x axis, only one positive solution for ρ exists. When ρ is found, m_3 is uniquely located, because m_1 and m_2 are arbitrarily located to set the distance scale, $x_2 - x_1 = 1$. If the location of the masses is interchanged, a similar unique solution is obtained for each arrangement. The solution for Eq. (7.48) can be found by numerical, or iterative, means.

Lagrange found two distinct solutions to the three-body problem:

(i) The equilateral triangle solution
(ii) The straight-line solution

These solutions are valid for any masses moving with uniform angular speed in coplanar circular orbits around their center of mass. Let us consider under what dynamical conditions such motions take place.

Since each mass moves uniformly in a circular path around the center of mass, C, the areal velocity in each orbit is constant. So, the resultant force, acting on each mass, must pass through C.

From Eq. (7.29) the acceleration of each mass is $-r_i n^2 \hat{\mathbf{u}}_i,\ i = 1, 2, 3$. So, the force is $-m_i r_i n^2 \hat{\mathbf{u}}_i$. The resultant force on each mass is directly proportional to the distance from C, and is directed toward that point.

The initial velocity vectors, required for circular motion of the system, are perpendicular to the initial position vectors of the respective masses. For each mass, m_i, the velocity, $\mathbf{v}_i = r_i \dot{\hat{\mathbf{u}}}_i$, so the velocity is proportional to the distance of m_i from C and is perpendicular to $\hat{\mathbf{u}}_i$. The motion in these cases is a rotation of the system around the center of mass, C.

If the initial velocity vectors are not perpendicular to the position vectors of the respective mass points, the configuration will change with time. If each vector, $\mathbf{v}_i,\ i = 1, 2, 3$, makes the same angle with respect to its corresponding $\mathbf{r}_i$, then the configuration will expand or contract such that the problem is still solvable. Each mass will follow a conic section, the distances between the bodies varying with time,

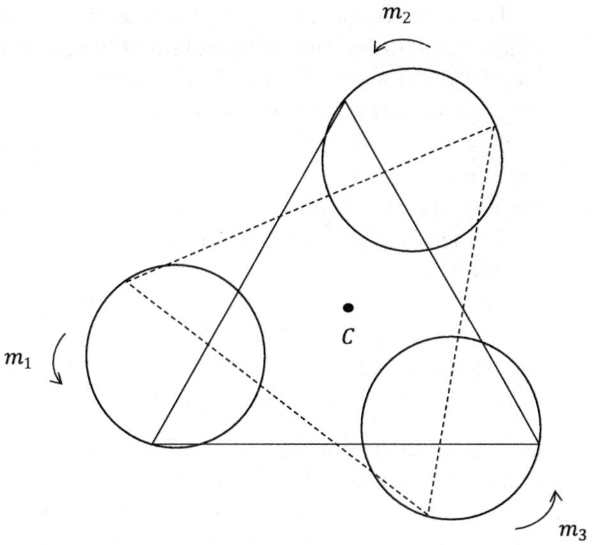

Fig. 7.5 A particular solution of the three-body problem, wherein each mass follows a conic section, resulting in a rotating equilateral triangle

but such that their mutual distances remain in the same ratio. Thus, if $\rho_{12}, \rho_{23}, \rho_{13}$ are initial distances at time t, then the distances become $\alpha\rho_{12}, \alpha\rho_{23}, \alpha\rho_{13}$, where α is the proportional factor. The distances of m_1, m_2, m_3 from the center of mass at any time are $r_1 = \alpha r_{01}$, $r_2 = \alpha r_{02}$, $r_3 = \alpha r_{03}$, where r_{0i}, $i = 1, 2, 3$ are the initial position vector lengths. The shape of the configuration remains invariant with time. α may be a function of time. Each mass traverses a conic section with a resultant pattern as shown in Fig. 7.5, with each triangle being equilateral.

The vertices of the equilateral triangle solution and the points on the straight line solution of the three-body problem are the Lagrangian points (see also Chap. 1). These are fixed in a planar coordinate system rotating with constant angular speed around the center of mass. A body, situated and initially at rest at one of the Lagrangian points, will remain so unless disturbed by external forces. The gravitational and centrifugal forces balance each other at these points (McCuskey 1963, pp. 102–108).

7.6 Generalization to n Bodies

Previously, important characteristics of three body motions, which can be generalized to the n-body problem, have been developed. These characteristics are summarized as follows:

(a) The system potential energy is

$$V = -k^2 \sum_{i,j=1}^{n} \frac{m_i m_j}{\rho_{ij}}, \; i \neq j \tag{7.49}$$

This depends on the relative distances between the mass points. This value is independent of location of the origin of the coordinate system and any arbitrary angular rotation of the axes of the system.

(b) Because of this independence, the integrals yielding the motion of the center of mass and areas, follow. There are nine constants of integration, six define the center of mass position at any instant, and three define the orientation of the angular momentum vector. A fixed origin and orientation of axes cannot be defined in space, so these nine constants cannot be determined from the observations.

(c) There is a tenth constant, E, the energy integral. These are ten constants of integration for the n-body problem. From the equations of motion, $6n$ constants are needed to solve the problem completely. For the three-body problem we have only 10 of 18 constants needed.

Other integrals of the equations of motion, other constants, have been sought by many. These efforts have not been successful. These integrals, such as the energy integral, are relations between coordinates and velocities. They are called the *classical integrals* of the three-body problem.

Bruns and Poincaré have shown that these 10 integrals are the only independent integrals, and all others are combinations of these 10.

Thus, the coordinates and velocity components of the three mass points as functions of time cannot be solved for, when given their values at some initial epoch. This is also true for the n-body problem. A closed form solution of the n-body problem is impossible. Specifically, given the coordinates and velocities of n mass points, which are moving under mutual gravitational attractions, the motion for any arbitrary succeeding time interval can not be predicted. Whether masses will collide or escape in an arbitrary finite time interval cannot be predicted. Only some special cases of the three-body problem can be addressed (McCuskey 1963, pp. 98–99).

7.7 Equations of Relative Motion

Since the motions of celestial bodies cannot be referred to a fixed coordinate system in space, the relative motions of $n - 1$ bodies with respect to the nth body can be established. In the solar system, the Sun is the origin of coordinates, and the plane of the ecliptic or equator is the fundamental reference plane. The other bodies motions, planets, comets, and asteroids, are given relative to the Sun. Now, we generalize the three-body problem to the n body problem.

Let m_1, shown in Fig. 7.6, be chosen as the origin. Taking the first three equations from the three-body problem

$$m_1\ddot{\mathbf{r}}_1 = k^2\left[\frac{m_1m_2}{\rho_{12}^3}\boldsymbol{\rho}_{12} + \frac{m_1m_2}{\rho_{13}^3}\boldsymbol{\rho}_{13}\right] \tag{7.50}$$

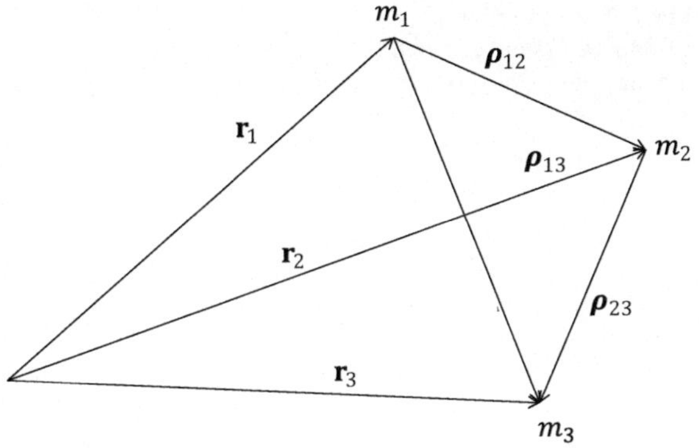

Fig. 7.6 Relative motion in the three-body problem

$$m_2\ddot{\mathbf{r}}_2 = k^2\left[\frac{m_2m_3}{\rho_{23}^3}\boldsymbol{\rho}_{23} - \frac{m_2m_1}{\rho_{12}^3}\boldsymbol{\rho}_{12}\right] \tag{7.51}$$

$$m_3\ddot{\mathbf{r}}_3 = k^2\left[-\frac{m_3m_2}{\rho_{23}^3}\boldsymbol{\rho}_{23} - \frac{m_3m_1}{\rho_{13}^3}\boldsymbol{\rho}_{13}\right] \tag{7.52}$$

So we will subtract Eq. (7.50) from Eqs. (7.51) and (7.52), and after some simplification

$$\ddot{\boldsymbol{\rho}}_{12} = -\frac{k^2(m_1+m_2)}{\rho_{12}^2}\hat{\mathbf{u}}_{12} + k^2m_3\left[\frac{1}{\rho_{23}^2}\hat{\mathbf{u}}_{23} - \frac{1}{\rho_{13}^2}\hat{\mathbf{u}}_{13}\right] \tag{7.53}$$

$$\ddot{\boldsymbol{\rho}}_{13} = -\frac{k^2(m_1+m_3)}{\rho_{13}^2}\hat{\mathbf{u}}_{13} - k^2m_2\left[\frac{1}{\rho_{23}^2}\hat{\mathbf{u}}_{23} + \frac{1}{\rho_{12}^2}\hat{\mathbf{u}}_{12}\right] \tag{7.54}$$

These are the equations of motion of m_2 relative to m_1 and m_3 relative to m_1. The right hand side of the first equation has three terms which represent, respectively: (i) the acceleration of m_2 due to m_1; (ii) the acceleration of m_2 due to m_3; and (iii) the reactive, or negative, acceleration of m_1 due to m_3. A similar explanation can be given for the terms of Eq. (7.51).

If $m_3 = 0$, Eq. (7.53) describes the two-body motion of m_2 around m_1. Likewise, if $m_2 = 0$, Eq. (7.54) gives the two-body motion of m_3 with respect to m_1.

Let m_1 be the dominant mass of the three masses and at the origin of a Cartesian coordinate system, as shown in Fig. 7.7.

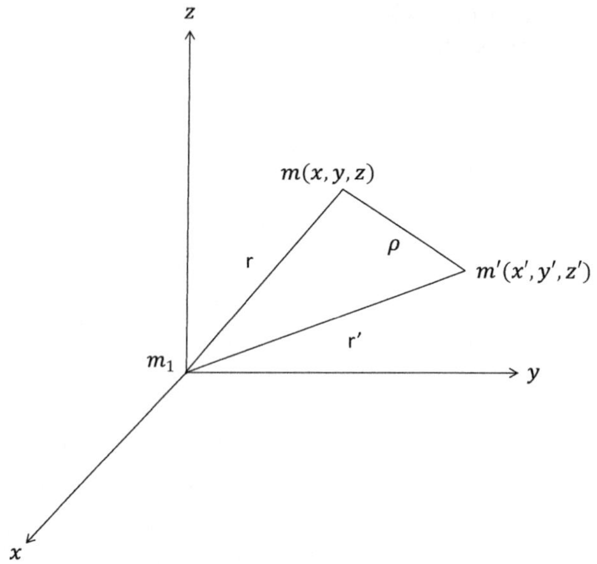

Fig. 7.7 A Cartesian coordinate system centered at the mass m_1

Let m be the body to be studied, and m' a mass disturbing the motion of m around m_1. The equations of motion for m are

$$\ddot{x} = -\frac{k^2 M x}{r^3} + k^2 m' \left[\frac{x' - x}{\rho^3} - \frac{x'}{r'^3} \right] \tag{7.55a}$$

$$\ddot{y} = -\frac{k^2 M y}{r^3} + k^2 m' \left[\frac{y' - y}{\rho^3} - \frac{y'}{r'^3} \right] \tag{7.55b}$$

$$\ddot{z} = -\frac{k^2 M z}{r^3} + k^2 m' \left[\frac{z' - z}{\rho^3} - \frac{z'}{r'^3} \right] \tag{7.55c}$$

where $M = m_1 + m$. Since $\rho^2 = (x' - x)^2 + (y' - y)^2 + (z' - z)^2$, then

$$\frac{x' - x}{\rho^3} = \frac{\partial}{\partial x} \rho^{-1} \tag{7.56}$$

$$\frac{y' - y}{\rho^3} = \frac{\partial}{\partial y} \rho^{-1} \tag{7.57}$$

$$\frac{z' - z}{\rho^3} = \frac{\partial}{\partial z} \rho^{-1} \tag{7.58}$$

Since x, y, z are independent of x', y', z', we can write

$$\frac{\partial}{\partial x} \left[\frac{x x' + y y' + z z'}{r'^3} \right] = \frac{x'}{r'^3} \tag{7.59}$$

$$\frac{\partial}{\partial y}\left[\frac{xx' + yy' + zz'}{r'^3}\right] = \frac{y'}{r'^3} \tag{7.60}$$

$$\frac{\partial}{\partial z}\left[\frac{xx' + yy' + zz'}{r'^3}\right] = \frac{z'}{r'^3} \tag{7.61}$$

Substituting these expressions into Eqs. (7.55) gives the symmetric forms

$$\ddot{x} = -\frac{k^2 M x}{r^3} + \frac{\partial R}{\partial x} \tag{7.62}$$

$$\ddot{y} = -\frac{k^2 M y}{r^3} + \frac{\partial R}{\partial y} \tag{7.63}$$

$$\ddot{z} = -\frac{k^2 M z}{r^3} + \frac{\partial R}{\partial z} \tag{7.64}$$

where

$$R = k^2 m' \left[\frac{1}{\rho} - \frac{xx' + yy' + zz'}{r'^3}\right] \tag{7.65}$$

R is called the *disturbing function* or *perturbing function*. $R = 0$ leads to the simple two-body problem.

We can now generalize to the n-body problem by noting that, if more than one disturbing mass is present, the disturbing function due to all masses is the sum of the disturbing functions for the individual masses. Let

$$R_i = k^2 m_i' \left[\frac{1}{\rho_i} - \frac{xx_i' + yy_i' + zz_i'}{r_i'^3}\right] \tag{7.66}$$

be the disturbing function for the ith mass acting on m. The equations of motion of m become

$$\ddot{x} = -\frac{k^2 M x}{r^3} + \sum_{i=1}^{n-2} \frac{\partial R}{\partial x} \tag{7.67}$$

$$\ddot{y} = -\frac{k^2 M y}{r^3} + \sum_{i=1}^{n-2} \frac{\partial R}{\partial y} \tag{7.68}$$

$$\ddot{z} = -\frac{k^2 M z}{r^3} + \sum_{i=1}^{n-2} \frac{\partial R}{\partial z} \tag{7.69}$$

where there are n masses in the system (McCuskey 1963, pp. 100–102).

7.8 Energy Integral and Force Function

Define the force function, U, of a system as

$$U = k^2 \sum_{i<j} \sum_{j=1}^{n} \frac{m_i m_j}{\rho_{ij}} \tag{7.70}$$

Then

$$\frac{\partial U}{\partial x_i} = k^2 m_i \frac{\partial}{\partial x_i} \sum_{j=1}^{n} \frac{m_j}{\rho_{ij}} \tag{7.71}$$

$$\frac{\partial U}{\partial x_i} = k^2 m_i \sum_{j=1}^{n} m_j \frac{x_i - x_j}{\rho_{ij}^3} \tag{7.72}$$

The equation of motion of m_i,

$$m_i \ddot{\mathbf{r}}_i = -k^2 m_i \sum_{j=1}^{n} m_j \frac{\mathbf{r}_i - \mathbf{r}_j}{\rho_{ij}^3} \tag{7.73}$$

can be written

$$m_i \ddot{\mathbf{r}}_i = \nabla_i U \tag{7.74}$$

where

$$\nabla_i = \hat{\mathbf{i}} \frac{\partial}{\partial x_i} + \hat{\mathbf{j}} \frac{\partial}{\partial y_i} + \hat{\mathbf{k}} \frac{\partial}{\partial z_i}$$

If we assemble a system from an infinite diffusion state, we start with m_i at $\mathbf{r}_i$. In moving m_2 from infinite to $\mathbf{r}_2$ $-k^2 m_1 m_2/\rho_{12}$ work is performed. Moving m_3 to $\mathbf{r}_3$, $-k^2 m_1 m_3/\rho_{13} - k^2 m_2 m_3/\rho_{23}$ additional work is performed. As particles are assembled, the function U is built up, so $-U$ is the total system potential energy.

Multiply Eq. (7.74) by $\dot{\mathbf{r}}_i$ and add all n equations, then

$$\sum_{i=1}^{n} m_i \dot{\mathbf{r}} \cdot \ddot{\mathbf{r}}_i = \sum_{i=1}^{n} \dot{\mathbf{r}}_i \cdot \nabla_i U = \frac{dU}{dt} \tag{7.75}$$

by definition of the total derivative. Integrating this we find

$$\frac{1}{2} \sum_{i=1}^{n} m_i \dot{\mathbf{r}}_i^2 = U + C \tag{7.76}$$

which is the energy integral. If T is the system kinetic energy, then

$$T = \frac{1}{2}\sum_{i=1}^{n} m_i \dot{\mathbf{r}}_i^2 \tag{7.77}$$

and the energy integral can be written

$$T = U + C \tag{7.78}$$

(Danby 1962, pp. 206–207).

References

Danby, J.: Fundamentals of Celestial Mechanics. The Macmillan Company, New York (1962)
McCuskey, S.W.: Introduction to Celestial Mechanics. Addison-Wesley, Reading (1963)

Chapter 8
The Restricted Three-Body Problem

8.1 Introduction

An important particular solution of the three-body problem results when one of the three masses is so small, in comparison to the other two, that its gravitational effects can be neglected. This may be called an *infinitesimal body* compared with the two finite bodies. This is the restricted three-body problem (Szebehely 1967), as mentioned in Sect. 1.5.

The restricted three-body problem has attracted the attention of numerous mathematicians and astronomers since it was first considered by Euler and Jacobi. The most obvious reason for this continued interest is that the model of the restricted problem can serve as a good approximation in a number of real situations in astronomy and astronautics.

A general solution to the problem has not been found, in spite of the apparent simplicity of the problem. In 1913, Sundman obtained a solution to the problem in the form of a convergent series (Rauschenbakh et al. 2003). However, its rate of convergence is so slow that it is not possible to use it in practical calculations. Estimates showed that to carry out calculations with an acceptable precision, numerous terms would have to be employed.

In this chapter, we will consider some aspects of the restricted three-body problem, including modelling, regions of motion, equilibria and their stability, as well as some other models related to the restricted three-body problem.

8.2 Equations of Motion

Take two, spherically symmetric, massive bodies moving in circular orbits around their center of mass. A third infinitesimal mass moves under the combined gravitational attraction of the two, but it does not affect their motion. This is the circular

P. Gurfil, P.K. Seidelmann, *Celestial Mechanics and Astrodynamics: Theory and Practice*, Astrophysics and Space Science Library 436,
DOI 10.1007/978-3-662-50370-6_8

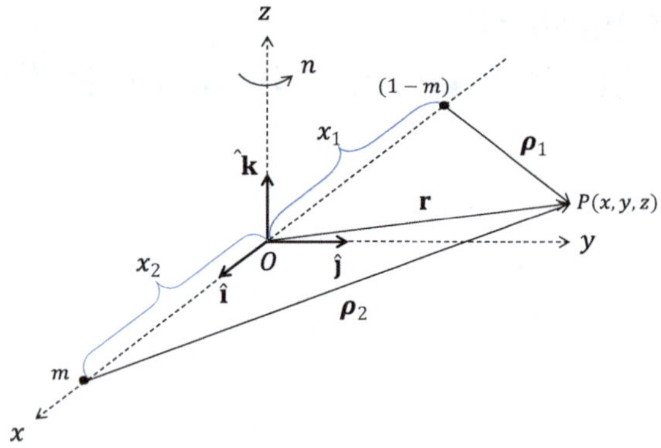

Fig. 8.1 Coordinate system for the circular restricted three-body problem

restricted three-body problem, which we mentioned in Sect. 1.5. For example, if we neglect (i) the presence of the Sun, (ii) non-sphericity of the Earth, (iii) the eccentricity of the Moon's orbit, then the Earth-Moon system with an artificial satellite is such a system of masses.

Denote the smaller of the two finite masses, m, the *secondary*, and the larger mass, $1 - m$, the *primary*. The unit of mass, then, is such that the sum of the masses is 1. Let the unit of time be such that the gravitational constant $k^2 = 1$. P denotes the position of the infinitesimal mass; $\mathbf{r}$ is its radius vector from the origin, O, which is the center of mass of m and $1 - m$. Denote the distances from $1 - m$ and m as ρ_1, ρ_2, respectively. The orbital plane of the finite masses is the xy plane. This is shown in Fig. 8.1.

Consider the rotating coordinate system shown in Fig. 8.1, whose origin is located at the center of mass. This coordinate system is also known as a *barycentric coordinate system*. Let $\mathbf{a} = \ddot{\mathbf{r}}$ be the acceleration of P, and $\mathbf{v} = \dot{\mathbf{r}}$ be the velocity of P in the rotating barycentric coordinate system, in which the finite masses revolve around the z axis with angular speed n radians per second. In this rotating system, the equation of motion of the infinitesimal mass is

$$\mathbf{a} + 2\boldsymbol{\omega} \times \mathbf{v} + \boldsymbol{\omega} \times (\boldsymbol{\omega} \times \mathbf{r}) = -\frac{(1-m)\,\boldsymbol{\rho}_1}{\rho_1^3} - \frac{m\,\boldsymbol{\rho}_2}{\rho_2^3} \tag{8.1}$$

where $\boldsymbol{\omega} = n\hat{\mathbf{k}}$. The vectors in Eq. (8.1) can be given in their Cartesian components as

$$\mathbf{a} = \ddot{x}\hat{\mathbf{i}} + \ddot{y}\hat{\mathbf{j}} + \ddot{z}\hat{\mathbf{k}} \tag{8.2}$$

$$\mathbf{v} = \dot{x}\hat{\mathbf{i}} + \dot{y}\hat{\mathbf{j}} + \dot{z}\hat{\mathbf{k}} \tag{8.3}$$

$$\boldsymbol{\omega} \times \mathbf{v} = -n\left[\dot{y}\hat{\mathbf{i}} - \dot{x}\hat{\mathbf{j}}\right] \tag{8.4}$$

$$\boldsymbol{\omega} \times (\boldsymbol{\omega} \times \mathbf{r}) = -n^2\left[x\hat{\mathbf{i}} + y\hat{\mathbf{j}}\right] \tag{8.5}$$

Then the Cartesian equations of motion are

$$\ddot{x} - 2n\dot{y} = n^2 x - \frac{(1-m)}{\rho_1^3}(x - x_1) - \frac{m}{\rho_2^3}(x - x_2) \tag{8.6a}$$

$$\ddot{y} + 2n\dot{x} = n^2 y - \frac{(1-m)}{\rho_1^3}y - \frac{m}{\rho_2^3}y \tag{8.6b}$$

$$\ddot{z} = -\frac{(1-m)}{\rho_1^3}z - \frac{m}{\rho_2^3}z \tag{8.6c}$$

where $(x_1, 0, 0)$ and $(x_2, 0, 0)$ are the coordinates of $1 - m$ and m, respectively, and (see Fig. 8.1)

$$\rho_1 = \sqrt{(x - x_1)^2 + y^2 + z^2}, \ \rho_2 = \sqrt{(x - x_2)^2 + y^2 + z^2} \tag{8.7}$$

With these equations, the positions of the finite masses do not change with time in this rotating coordinate system. From Kepler's third law, the orbital angular speed, n, is

$$n = \frac{2\pi}{P} = \frac{k\sqrt{m_1 + m_2}}{a^{3/2}} \tag{8.8}$$

If we choose the distance scale so $x_2 - x_1 = 1$, then, with the assumptions about k and the masses, we have $n = 1$. Equations (8.6) become (where we have eliminated n since it is 1)

$$\ddot{x} - 2\dot{y} = x - \frac{(1-m)}{\rho_1^3}(x - x_1) - \frac{m}{\rho_2^3}(x - x_2) \tag{8.9a}$$

$$\ddot{y} + 2\dot{x} = y - \frac{(1-m)}{\rho_1^3}y - \frac{m}{\rho_2^3}y \tag{8.9b}$$

$$\ddot{z} = -\frac{(1-m)}{\rho_1^3}z - \frac{m}{\rho_2^3}z \tag{8.9c}$$

Thus, as usual, we need six constants to determine the motion of the infinitesimal mass (McCuskey 1963, pp. 109–110).

8.3 The Jacobi Constant

These equations of motion yield one integral, which is similar to the three-body problem energy integral. Define a function

$$U = \frac{1}{2}(x^2 + y^2) + \frac{1-m}{\rho_1} + \frac{m}{\rho_2} \tag{8.10}$$

Then

$$U_x \triangleq \frac{\partial U}{\partial x} = x - \frac{(1-m)}{\rho_1^3}(x - x_1) - \frac{m}{\rho_2^3}(x - x_2) \tag{8.11a}$$

$$U_y \triangleq \frac{\partial U}{\partial y} = y - \frac{(1-m)}{\rho_1^3}y - \frac{m}{\rho_2^3}y \tag{8.11b}$$

$$U_z \triangleq \frac{\partial U}{\partial z} = -\frac{(1-m)}{\rho_1^3}z - \frac{m}{\rho_2^3}z \tag{8.11c}$$

Substituting Eqs. (8.11) into Eq. (8.9) we have

$$\ddot{x} - 2\dot{y} = \frac{\partial U}{\partial x} \tag{8.12a}$$

$$\ddot{y} + 2\dot{x} = \frac{\partial U}{\partial y} \tag{8.12b}$$

$$\ddot{z} = \frac{\partial U}{\partial z} \tag{8.12c}$$

Multiplying these by $2\dot{x}, 2\dot{y}$, and $2\dot{z}$, respectively, and adding, we have

$$2\dot{x}\ddot{x} + 2\dot{y}\ddot{y} + 2\dot{z}\ddot{z} = 2\dot{x}\frac{\partial U}{\partial x} + 2\dot{y}\frac{\partial U}{\partial y} + 2\dot{z}\frac{\partial U}{\partial z} \tag{8.13}$$

or

$$\frac{d}{dt}\left(\dot{x}^2 + \dot{y}^2 + \dot{z}^2\right) = 2\frac{dU}{dt} \tag{8.14}$$

Integrating we have

$$v^2 = 2U - C \tag{8.15}$$

where the speed of the infinitesimal mass is v.

U, originally introduced by Jacobi, looks like a potential energy, when the energy from the rotation of the coordinate system, specifically $(x^2 + y^2)/2$, is included. Equation (8.15) involves U and the constant of integration, C, which is one constant of motion. Then, five constants remain to be found. The constant C is named the *Jacobi constant*.

If we further restrict the motion of the infinitesimal mass to the xy plane, the number of constants required can be reduced to three. Jacobi (1884) has shown that two of these are related to the third. For a complete solution, one new integral must be found, but Bruns (1887) demonstrated that no new algebraic integrals exist in rectangular coordinates (McCuskey 1963, pp. 110–111).

8.4 Zero Velocity Curves

The equations of motion in the rotating coordinate system, and particularly Eq. (8.15), are useful in discussing qualitatively the behavior of the infinitesimal mass. For simplification, assume that the infinitesimal mass moves in the xy plane. Equation (8.15) shows that its speed is a function of its position in the plane. The Jacobi constant, C, depends on its initial position and velocity. There will be *zero velocity curves* given by $2U - C = 0$, and in Cartesian coordinates by

$$x^2 + y^2 + \frac{2(1-m)}{\sqrt{(x-x_1)^2 + y^2}} = \frac{2m}{\sqrt{(x-x_2)^2 + y^2}} = C \tag{8.16}$$

The particle motion can only occur in those xy plane regions where $2U - C > 0$ (i.e., we cannot have a negative velocity squared). The contour curves from Eq. (8.16) specify the boundaries of the regions within which motion can take place. The contour curves for different values of C are shown in Fig. 8.2 where $C_1 > C_2 > C_3$.

Let us now consider areas of possible motion:

Case I
C is very large if x and y are very large, or if ρ_1 or ρ_2 are very small. As C becomes larger, the oval outer contour asymptotically approaches the boundary circle. Then the terms involving $\frac{2(1-m)}{\sqrt{(x-x_1)^2+y^2}}$ and $\frac{2m}{\sqrt{(x-x_1)^2+y^2}}$ become very small in Eq. (8.16). Motion can take place outside the curve C_1. Similarly, for small ρ_1 and ρ_2 and large C, the x^2 and y^2 terms in Eq. (8.16) become insignificant compared with the third and fourth terms. Thus, we have a pair of ovals surrounding $1-m$ and m. With a large value of C, motion cannot take place in the region between the ovals and the outer contour. Motion can only take place within the C_1 ovals, or outside the nearly circular contour, C_1.

Case II
If C decreases, the ovals around $(1-m)$ and m expand and merge into a single closed contour, and the outer contour moves toward the center of the figure. Motion can take place inside the merged figure marked C_2, or outside the larger contour marked C_2.

Case III
If C decreases further, the regions where motion can occur become larger. The enlarged oval pattern around the finite masses merge into that outside the exterior oval. We have only a small region, enclosed by C_3, where motion is impossible (Moulton 1970, pp. 112–113).

L_1, L_2, L_3, L_4, L_5 are the Lagrangian points of the system, which are equilibrium points for system (8.9). The stability of these points is analyzed in Sect. 8.6. L_1, L_3 are the points where the inner and outer systems of ovals merge to a common tangent. L_2 is where the ovals around the individual masses merge.

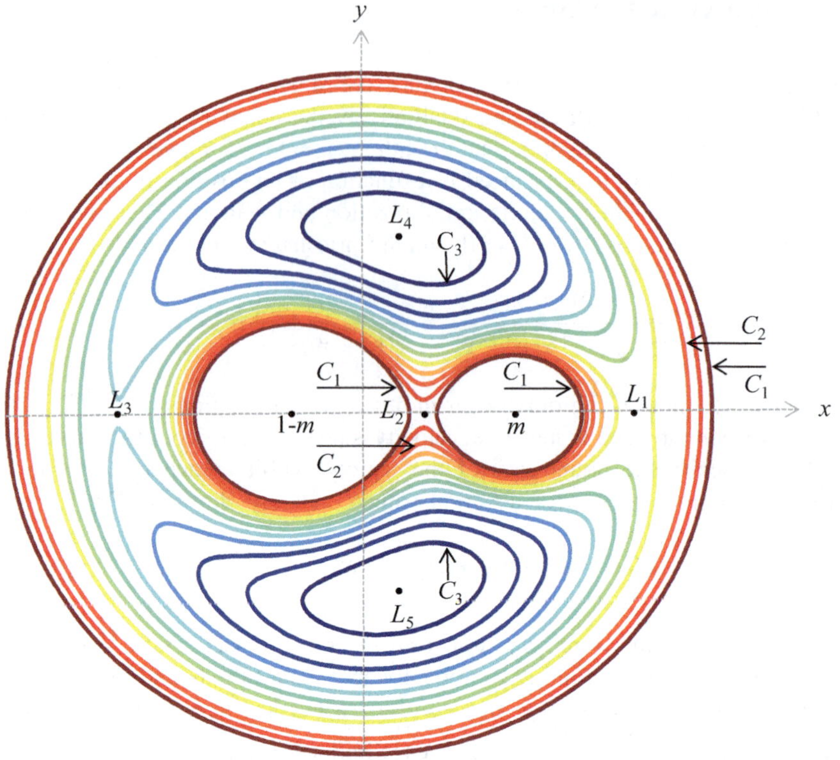

Fig. 8.2 xy projection of the zero velocity curves, with $C_1 > C_2 > C_3$. Also shown are the Lagrangian equilibrium points L_1-L_5

We note that, in recent years, a different convention for numbering the Lagrangian points has been widespread. Under this convention, L_2 is the exterior equilibrium point and L_1 is the interior equilibrium point. The notation for the remaining points is the same as used here.

In the discussion above, the infinitesimal mass's motion has been restricted to the xy plane. In the more general case, where the particle moves in space, analysis of zero velocity curves in xz and yz planes, similar to the xy described, is possible. In the xz plane, there exist ovals surrounding $(1 - m)$ and m, together with exterior contours approaching the straight lines $x = \pm C$, as shown in Fig. 8.3.

In the yz plane, the outer contours asymptotically approach the lines $y = \pm C$ for large values of C. Also, there are closed oval contours surrounding the origin of coordinates, as shown in Fig. 8.4 (Moulton 1970, pp 283–285).

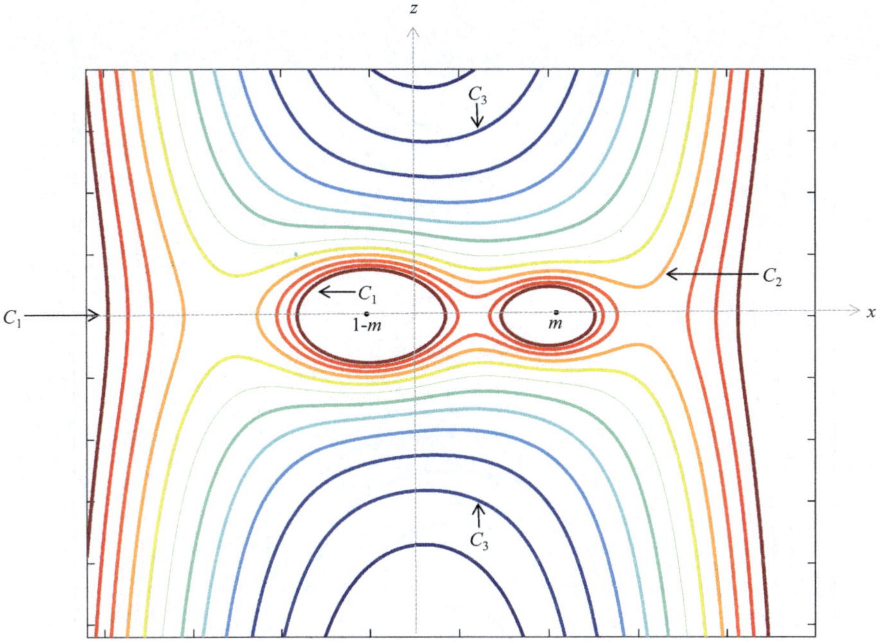

Fig. 8.3 xz projection of the zero velocity curves, with $C_1 > C_2 > C_3$

8.5 The Lagrangian Points

The Lagrangian points, which we also mentioned in Sects. 1.5.1 and 7.5, should be discussed next. The equations of the contour curves for motion in the xy plane may be written as

$$f(x, y) = x^2 + y^2 + \frac{2(1-m)}{\rho_1} + \frac{2m}{\rho_2} = C \tag{8.17}$$

As shown in Fig. 8.2, at a Lagrangian point L_1, L_2, or L_3, double tangents appear on these curves. In the geometry of algebraic curves, it is shown that a singular point, or a point having two tangents to the curve, will occur where $\partial f/\partial x = 0$ and $\partial f/\partial y = 0$. Applying these criteria to Eq. (8.17), we have

$$x - \frac{(1-m)(x-x_1)}{\rho_1^3} - \frac{m(x-x_2)}{\rho_2^3} = 0 \tag{8.18a}$$

$$y - \frac{(1-m)y}{\rho_1^3} - \frac{my}{\rho_2^3} = 0 \tag{8.18b}$$

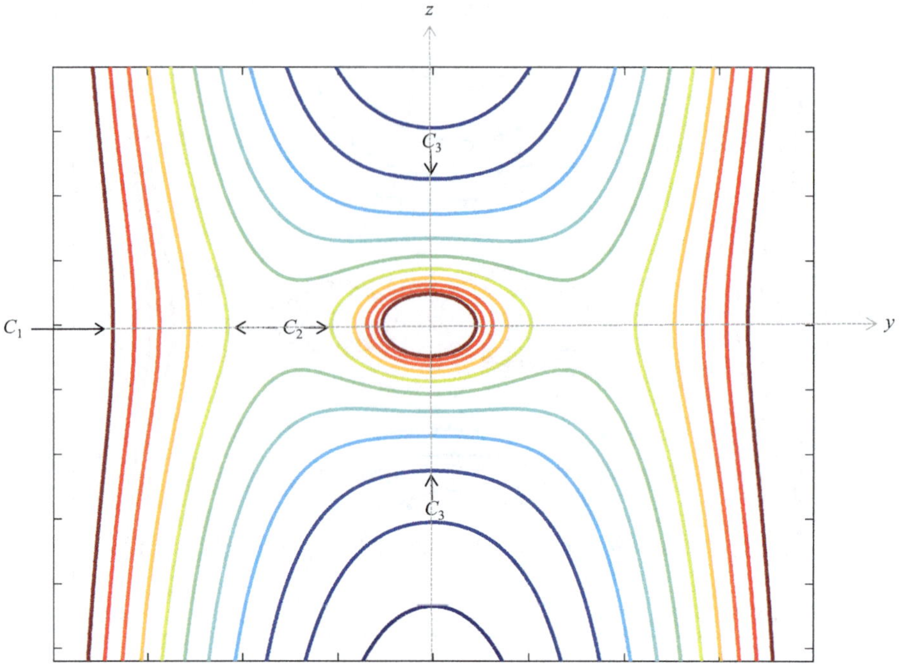

Fig. 8.4 *yz* projection of the zero velocity curves, with $C_1 > C_2 > C_3$

These are the values of $\partial U/\partial x$ and $\partial U/\partial y$ on the right-hand sides of Eqs. (8.9). Since such a double point lies on the curve $f(x, y) = C$, then the velocity components $\dot{x} = \dot{y} = 0$. Examination of Eqs. (8.9) shows that at a double point $\ddot{x} = \ddot{y} = 0$. Since $z = 0$, then $\ddot{z} = 0$.

So, there is no acceleration of a particle located at one of these points. If a particle is placed there at rest, it remains at rest, unless acted upon by a force extraneous to the system.

The Lagrangian points on the x axis, often called the *collinear Lagrangian points* or *libration points*, are determined by setting $y = 0$ in Eqs. (8.18) and solving for values of x, the roots of Eq. (8.18a). This equation has one root $x > x_2$, one root $x_1 < x < x_2$, and a root $x < x_1$. In addition, there are also *triangular Lagrangian points*, denoted by L_4 and L_5, which solve Eqs. (8.18) for $y \neq 0$. We will discuss the collinear solutions first.

Case I, Collinear Points, $x > x_2$

Let $x - x_2 = \rho_2 = \rho$, $x - x_1 = \rho_1 = 1 + \rho$, $(1 - m)x_1 + mx_2 = 0$. From the last of these $x = 1 - m + \rho$. Substituting in Eq. (8.18)

$$\rho^5 + (3 - m)\rho^4 + (3 - 2m)\rho^3 - m\rho^2 - 2m\rho - m = 0 \tag{8.19}$$

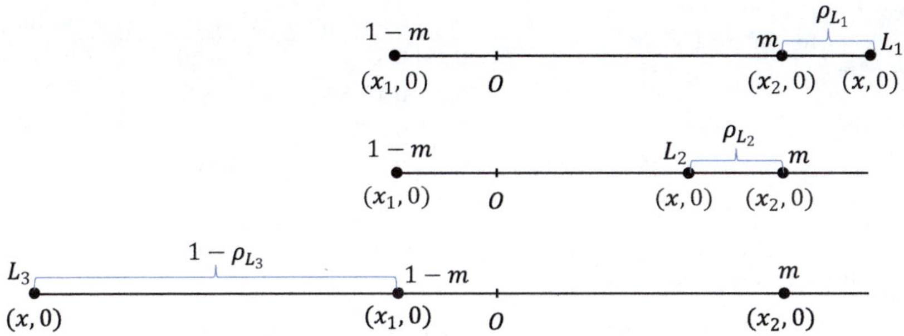

Fig. 8.5 From top to bottom, location of the collinear Lagrangian points L_1, L_2 and L_3

This equation has only one real positive root. This root can be found by modern numerical methods such as Graeffe's root squaring process (Wylie 1975), or, since by definition $m < 1/2$, Eq. (8.19) can be solved with $m = 0$, and then the solution for $m \neq 0$ as a power series in $m^{1/3}$ is

$$\rho_{L_1} = \left(\frac{m}{3}\right)^{1/3} + \frac{1}{3}\left(\frac{m}{3}\right)^{2/3} - \frac{1}{9}\left(\frac{m}{3}\right)^{3/3} + \ldots \tag{8.20}$$

With ρ, $\rho_1 = 1 + \rho$ is the distance from the mass $(1 - m)$ to L_1, the Lagrangian point exterior to m, as shown in Fig. 8.5. We note that in modern literature, the exterior point is sometimes referred to as L_2, and the interior point, discussed next, is referred to as L_1.

Case II, Collinear Points, $x_1 < x < x_2$
Let $\rho_2 = x_2 - x = \rho$, $\rho_1 = x - x_1 = 1 - \rho$. From the definition of the center of mass, $x = (1 - m - \rho)$. Then the 5th degree equation in ρ is

$$\rho^5 - (3 - m)\rho^4 + (3 - 2m)\rho^3 - m\rho^2 + 2m\rho - m = 0 \tag{8.21}$$

From the same analysis as used in Case I,

$$\rho_{L_2} = \left(\frac{m}{3}\right)^{1/3} - \frac{1}{3}\left(\frac{m}{3}\right)^{2/3} - \frac{1}{9}\left(\frac{m}{3}\right)^{3/3} - \ldots \tag{8.22}$$

and hence $\rho_1 = 1 - \rho$. This is the location of the Lagrangian collinear point L_2, lying between the primaries, as shown in Fig. 8.5.

Case III, Collinear Points, $x < x_1$
Take $1 - \rho$ as the distance from the larger mass $(1 - m)$ to the collinear point L_3. Then $x_1 - x = 1 - \rho$, $x_2 - x = 2 - \rho$ and, from the center of mass equation, $x = -(m + 1) + \rho$. The polynomial equation in ρ is

$$\rho^5 - (7 + m)\rho^4 + (19 + 6m)\rho^3 - (24 + 13m)\rho^2 + (12 + 14m)\rho - 7m = 0 \tag{8.23}$$

Table 8.1 Values of ρ_{L_1}, ρ_{L_2}, ρ_{L_3} for different values of m

m	Case I (ρ_{L_1})	Case II (ρ_{L_2})	Case III (ρ_{L_3})
0.1	0.35264	0.28360	0.05839
0.2	0.45289	0.34327	0.11710
0.3	0.52486	0.38124	0.17647
0.4	0.58306	0.40906	0.23681

The solution of this equation is

$$\rho_{L_3} = \frac{7}{12}m + \frac{1127}{20736}m^3 + \dots \tag{8.24}$$

From this we find $\rho_1 = 1 - \rho$, $\rho_2 = 2 - \rho$. The location of the Lagrangian point L_3 is shown in Fig. 8.5. To give some idea of the magnitudes of ρ, values are given in Table 8.1 for Cases I–III as a function of m.

Case IV, Triangular Points

The Lagrangian points away from the x axis, denoted as L_4 and L_5, as mentioned before, follow from Eqs. (8.18),

$$x - \frac{(1-m)(x-x_1)}{\rho_1^3} - \frac{m(x-x_2)}{\rho_2^3} = 0 \tag{8.25a}$$

$$y - \frac{(1-m)y}{\rho_1^3} - \frac{my}{\rho_2^3} = 0 \tag{8.25b}$$

Since $y \neq 0$, we can divide Eq. (8.25b) by y to obtain

$$1 - \frac{(1-m)}{\rho_1^3} - \frac{m}{\rho_2^3} = 0 \tag{8.26}$$

Multiplying this by $(x - x_1)$ gives

$$(x-x_1) - \frac{(1-m)}{\rho_1^3}(x-x_1) - \frac{m}{\rho_2^3}(x-x_1) = 0 \tag{8.27}$$

Multiplying Eq. (8.26) by $(x - x_2)$, we have

$$(x-x_2) - \frac{(1-m)}{\rho_1^3}(x-x_2) - \frac{m}{\rho_2^3}(x-x_2) = 0 \tag{8.28}$$

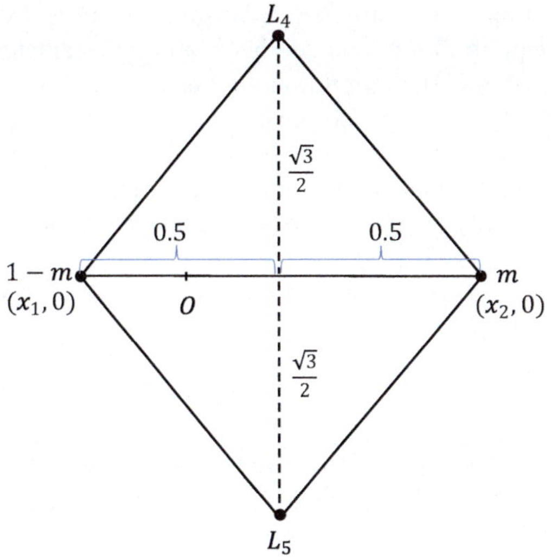

Fig. 8.6 Location of the triangular Lagrangian points L_4 and L_5

Eliminating x from Eqs. (8.25), (8.27) and (8.28) gives

$$x_2 - \frac{(1-m)}{\rho_1^3}(x_2 - x_1) = 0 \tag{8.29a}$$

$$x_1 + \frac{m}{\rho_2^3}(x_2 - x_1) = 0 \tag{8.29b}$$

However, $x_2 - x_1 = 1$ and $(1-m)x_1 + mx_2 = 0$, so $x_1 = -m$ and $x_2 = 1 - m$. From Eqs. (8.29) it follows that

$$1 - \frac{1}{\rho_1^3} = 0, \ \frac{1}{\rho_2^3} - 1 = 0 \tag{8.30}$$

Only if $\rho_1 = 1, \rho_2 = 1$ are these satisfied. Thus, L_4 and L_5 are at the vertices of equilateral triangles, as shown in Fig. 8.6, whose common base is the segment (x_1, x_2). The equilateral triangle solutions are independent of the relative sizes of the two finite masses. This result is consistent with the stationary solution of the three-body problem discussed in Chap. 7 (McCuskey 1963, pp. 112–117).

8.6 Stability of Motion Near the Lagrangian Points

Stability of an infinitesimal mass near one of the Lagrangian points is claimed when, after a small displacement and small velocity, the particle oscillates for some time around the point. It is unstable if it departs from the point with time. The stability of motion means bounded displacements and velocities as a functions of time in the equilibrium points neighborhoods. A mathematical investigation of such stability follows. The infinitesimal mass equations of motion are given by Eqs. (8.9).

Denote the coordinates of any Lagrangian point by (x_0, y_0, z_0). These points occur in the xy plane, so $z_0 = 0$. Let (α, β, γ) denote small displacements of the infinitesimal mass from the Lagrangian point as functions of time, so there are velocity components $\dot{\alpha}, \dot{\beta}, \dot{\gamma}$ and acceleration components $\ddot{\alpha}, \ddot{\beta}, \ddot{\gamma}$. We may write

$$x = x_0 + \alpha,\ \dot{x} = \dot{x_0} + \dot{\alpha},\ \ddot{x} = \ddot{x_0} + \ddot{\alpha} \tag{8.31}$$

$$y = y_0 + \beta,\ \dot{y} = \dot{y_0} + \dot{\beta},\ \ddot{y} = \ddot{y_0} + \ddot{\beta} \tag{8.32}$$

$$z = z_0 + \gamma,\ \dot{z} = \dot{z_0} + \dot{\gamma},\ \ddot{z} = \ddot{z_0} + \ddot{\gamma} \tag{8.33}$$

The displacements are assumed to be sufficiently small, so Taylor expansions of U_x, U_y, U_z in Eqs. (8.11) about the Lagrangian point can be written as

$$U_x = (U_x)_0 + \alpha(U_{xx})_0 + \beta(U_{xy})_0 + \gamma(U_{xz})_0 \tag{8.34}$$

$$U_y = (U_y)_0 + \alpha(U_{yx})_0 + \beta(U_{yy})_0 + \gamma(U_{yz})_0 \tag{8.35}$$

$$U_z = (U_z)_0 + \alpha(U_{zx})_0 + \beta(U_{zy})_0 + \gamma(U_{zz})_0 \tag{8.36}$$

where the partial derivatives are evaluated at the Lagrangian point. This assumption means that the forces, acting on the infinitesimal mass, when it is displaced from the equilibrium position, are proportional to the displacement only to the first power.

Equations (8.25) show that at a Lagrangian point

$$(U_x)_0 = (U_y)_0 = (U_z)_0 = 0 \tag{8.37}$$

Moreover, $\ddot{x_0} = \ddot{y_0} = \ddot{z_0} = 0$ and $\dot{x_0} = \dot{y_0} = \dot{z_0} = 0$ at one of these points. Thus, from Eqs. (8.12), (8.31) and (8.34) we have

$$\ddot{\alpha} - 2\dot{\beta} = \alpha(U_{xx})_0 + \beta(U_{xy})_0 + \gamma(U_{xz})_0 \tag{8.38a}$$

$$\ddot{\beta} + 2\dot{\alpha} = \alpha(U_{yx})_0 + \beta(U_{yy})_0 + \gamma(U_{yz})_0 \tag{8.38b}$$

$$\ddot{\gamma} = \alpha(U_{zx})_0 + \beta(U_{zy})_0 + \gamma(U_{zz})_0 \tag{8.38c}$$

When the displacements are small, these are the equations of motion of the infinitesimal mass particle in the neighborhood of a Lagrangian point. From Eqs. (8.11) we have

$$U_{xx} = 1 - \frac{1-m}{\rho_1^3} - \frac{m}{\rho_2^3} + \frac{3(1-m)(x-x_1)^2}{\rho_1^5} + \frac{3m(x-x_2)^2}{\rho_2^5} \tag{8.39a}$$

$$U_{xy} = \frac{3(1-m)(x-x_1)y}{\rho_1^5} + \frac{3m(x-x_2)y}{\rho_2^5} \tag{8.39b}$$

$$U_{xz} = \frac{3(1-m)(x-x_1)z}{\rho_1^5} + \frac{3m(x-x_2)z}{\rho_2^5} \tag{8.39c}$$

$$U_{yy} = 1 - \frac{(1-m)}{\rho_1^3} - \frac{m}{\rho_2^3} + \frac{3(1-m)y^2}{\rho_1^5} + \frac{3my^2}{\rho_2^5} \tag{8.39d}$$

$$U_{yz} = \frac{3(1-m)yz}{\rho_1^5} + \frac{3myz}{\rho_2^5} \tag{8.39e}$$

$$U_{zz} = -\frac{(1-m)}{\rho_1^3} - \frac{m}{\rho_2^3} + \frac{3(1-m)z^2}{\rho_1^5} + \frac{3mz^2}{\rho_2^5} \tag{8.39f}$$

Equations (8.38) and (8.39) will be applied to study the stability of motion near the triangular point L_4 and the collinear point L_1.

Case I, Motion Around L_4

We have seen that all the Lagrangian points lie in the xy plane, so $z_0 = 0$ in each case. For $L_4, x_0 - x_1 = 1/2, z_0 = 0, x_0 - x_2 = -1/2, \rho_1 = \rho_2 = 1, y_0 = \sqrt{3}/2$ (see Fig. 8.6). From Eqs. (8.39), the partial derivatives are

$$U_{xx} = \frac{3}{4}, \; U_{yz} = U_{zy} = 0 \tag{8.40}$$

$$U_{xy} = U_{yx} = \frac{3\sqrt{3}}{2}\left(\frac{1}{2} - m\right), \; U_{yy} = 9/4 \tag{8.41}$$

$$U_{xz} = U_{zx} = 0, \; U_{zz} = -1 \tag{8.42}$$

The equations of motion (8.38) are

$$\ddot{\alpha} - 2\dot{\beta} = \frac{3}{4}\alpha + \frac{3\sqrt{3}}{2}\left(\frac{1}{2} - m\right)\beta \tag{8.43a}$$

$$\ddot{\beta} + 2\dot{\alpha} = \frac{3\sqrt{3}}{2}\left(\frac{1}{2} - m\right)\alpha + \frac{9}{4}\beta \tag{8.43b}$$

$$\ddot{\gamma} = -\gamma \tag{8.43c}$$

Equation (8.43c) has the solution $\gamma = c_1 \cos t + c_2 \sin t$, where c_1 and c_2 are constants of integration. The z direction displacement is periodic with the period 2π, which is the period of revolution of the two larger masses around their center of mass. The disturbed motion of the infinitesimal mass is bounded and stable perpendicular to the xy plane.

To solve Eqs. (8.43a) and (8.43b), let $\alpha = Ae^{\lambda t}$, $\beta = Be^{\lambda t}$, where A, B, λ are parameters to be determined. Substitution in the equations of motion and rearrangement yields

$$A\left(\lambda^2 - \frac{3}{4}\right) + B\left[-2\lambda - \frac{3\sqrt{3}}{2}\left(\frac{1}{2} - m\right)\right] = 0 \tag{8.44}$$

$$A\left[2\lambda - \frac{3\sqrt{3}}{2}\left(\frac{1}{2} - m\right)\right] + B\left[\lambda^2 - \frac{9}{4}\right] = 0 \tag{8.45}$$

These have a non trivial solution for A and B, if

$$\begin{bmatrix} \lambda^2 - \dfrac{3}{4} & -2\lambda - \dfrac{3\sqrt{3}}{2}\left(\dfrac{1}{2} - m\right) \\ -2\lambda - \dfrac{3\sqrt{3}}{2}\left(\dfrac{1}{2} - m\right) & \lambda^2 - \dfrac{9}{4} \end{bmatrix} = 0 \tag{8.46}$$

which simplifies to

$$4\lambda^4 + 4\lambda^2 + 27m(1 - m) = 0 \tag{8.47}$$

The solution as a quadratic equation in λ^2 is

$$\lambda^2 = -\frac{1}{2} \pm \frac{1}{2}\sqrt{1 - 27m(1 - m)} \tag{8.48}$$

From the solutions $\alpha = Ae^{\lambda t}$ and $\beta = Be^{\lambda t}$ it is recognized that these are periodic and bounded, only if λ is imaginary. Thus, we must choose m in Eq. (8.48) so that $\lambda^2 < 0$. Hence, if stable motion is to be ensured

$$1 - 27m(1 - m) \geq 0 \tag{8.49}$$

Otherwise λ^2 would have an imaginary part, which would lead to a real part of λ. Solving Eq. (8.49) we find that $m < 0.0385$, if λ^2 is to be negative. The other root of Eq. (8.49) is greater than $1/2$, and is excluded due to our hypothesis that $m < 1/2$. Thus, stable motion around the L_4 point takes place if the mass m does not exceed 0.0385.

The Trojan group of asteroids are an example of the motion discussed here. These bodies are located near the points L_4 and L_5 of the Sun-Jupiter system. They have closed orbits around equilibrium points, while moving as a group with the same period as Jupiter around the Sun. Jupiter's mass is about 0.001 the solar mass, so the stability condition of Eq. (8.49) is met. The initial displacement of position and velocity determines the motion around the Lagrangian point.

Case II, Motion Around L_1
In this case $x_0 - x = \rho_1 = 1 + \rho,\ x_0 - x_2 = \rho_2 = \rho,\ y_0 = z_0 = 0$ (see Fig. 8.5). The partial derivatives of Eqs. (8.39) become

$$U_{xx} = 1 - \frac{(1-m)}{(1+\rho)^3} - \frac{m}{\rho^3} + \frac{3(1-m)}{(1+\rho)^3} + \frac{3m}{\rho^3} \tag{8.50a}$$

$$= 1 + \frac{2(1-m)}{(1+\rho)^3} + \frac{2m}{\rho^3} \tag{8.50b}$$

$$U_{xy} = U_{yx} = U_{xz} = U_{zx} = U_{yz} = U_{zy} = 0 \tag{8.50c}$$

$$U_{yy} = 1 - \frac{(1-m)}{(1+\rho)^3} - \frac{m}{\rho^3} \tag{8.50d}$$

$$U_{zz} = -\frac{(1-m)}{(1+\rho)^3} - \frac{m}{\rho^3} \tag{8.50e}$$

Substituting these expressions into Eqs. (8.38), the motion around L_1 is

$$\ddot{\alpha} - 2\dot{\beta} = \alpha[1 + 2f] \tag{8.51a}$$

$$\ddot{\beta} + 2\dot{\alpha} = \beta[1 - f] \tag{8.51b}$$

$$\ddot{\gamma} = -f\gamma \tag{8.51c}$$

where

$$f = \frac{1-m}{(1+\rho)^3} + \frac{m}{\rho^3} \tag{8.52}$$

Equation (8.51c) indicates that the motion perpendicular to the xy plane is periodic with frequency $\omega = \sqrt{f}$ and, therefore, bounded with

$$\gamma = C_3 \cos\left(\sqrt{f}t\right) + C_4 \sin\left(\sqrt{f}t\right) \tag{8.53}$$

where C_3 and C_4 are constants of integration.

For motion in the xy plane, let $\alpha = Ae^{\lambda t}$ and $\beta = Be^{\lambda t}$. Using Eqs. (8.51) and combining terms

$$A[\lambda^2 - (1 + 2f)] + -2B\lambda = 0 \tag{8.54a}$$

$$2A\lambda + B[\lambda^2 - (1 - f)] = 0 \tag{8.54b}$$

These have a nontrivial solution if

$$\begin{bmatrix} \lambda^2 - (1+2f) & -2\lambda \\ 2\lambda & \lambda^2 - (1-f) \end{bmatrix} = 0 \tag{8.55}$$

Simplified, this is

$$\lambda^4 + (2 - f)\lambda^2 + (1 + 2f)(1 - f) = 0 \tag{8.56}$$

From the definition of f, the fact that $m < 1/2$, and the value of ρ given by Eq. (8.20)

$$\rho = \left(\frac{m}{3}\right)^{1/3} + \frac{1}{3}\left(\frac{m}{3}\right)^{2/3} - \frac{1}{9}\left(\frac{m}{3}\right)^{3/3} + \dots \tag{8.57}$$

Then we realize that

$$(1 + 2f)(1 - f) = 1 + f - 2f^2 < 0 \tag{8.58}$$

This is because with $m < 1/2, \rho \approx 1/2$, so the second term of f in Eq. (8.52) is greater than 3, and the first term of f in Eq. (8.52) is positive. Thus, $f > 1$ and Eq. (8.58) is evident. Therefore, the solution of Eq. (8.56), which is

$$\lambda^2 = \frac{-(2 - f) \pm \sqrt{(2 - f)^2 - 4(1 + 2f)(1 - f)}}{2} \tag{8.59}$$

is positive when the plus sign is taken, and negative when the minus sign is taken.

This is explained as follows. Since $-4(1 + 2f)(1 - f) < 0$, then

$$\sqrt{(2 - f)^2 - 4(1 + 2f)(1 - f)} > 2 - f$$

so when the plus sign is taken, $\lambda^2 > 0$, and when the minus sign is taken, $\lambda^2 < 0$.

The two real roots of Eq. (8.59) are equal numerically, but opposite in sign. The two remaining roots of the characteristic equation (8.56) can be denoted by

$$\lambda_1 = a,\ \lambda_2 = -a,\ \lambda_3 = b_i,\ \lambda_4 = -b_i \tag{8.60}$$

where

$$a = \left[\frac{-(2 - f) + \sqrt{(2 - f)^2 - 4(1 + f - 2f^2)}}{2}\right]^{1/2} \tag{8.61}$$

$$b = \left[\frac{(2 - f) + \sqrt{(2 - f)^2 - 4(1 + f - 2f^2)}}{2}\right]^{1/2} \tag{8.62}$$

Equations (8.51) then become upon substitution

$$\alpha = A_1 e^{at} + A_2 e^{-at} + A_3 e^{ibt} + A_4 e^{-ibt} \tag{8.63a}$$

$$\beta = B_1 e^{at} + B_2 e^{-at} + B_3 e^{ibt} + B_4 e^{-ibt} \tag{8.63b}$$

where B_j are related to A_j ($j = 1, 2, 3, 4$) through Eq. (8.54a); that is

$$B_j = \left[\frac{\lambda_j^2 - (1 + 2f)}{2\lambda_j} \right] A_j, \; j = 1, 2, 3, 4 \tag{8.64}$$

The exponential factors e^{at} and e^{-at} in Eqs. (8.63) show there is unbounded motion in the xy plane. Thus, the L_1 point is an unstable equilibrium point. Similar analysis shows that the behavior of an infinitesimal mass near L_2 and L_3 is similar. In general, these Lagrangian points are unstable. Assigning appropriate initial conditions for the disturbed motion around L_1, this point can be a point of stability.

Let

$$B_1 = cA_1, \; B_2 = -cA_2 \tag{8.65}$$

$$B_3 = idA_3, \; B_4 = -idA_4 \tag{8.66}$$

where

$$c = \frac{a^2 - (1 + 2f)}{2a}, \; d = \frac{b^2 + (1 + 2f)}{2b} \tag{8.67}$$

Then Eqs. (8.63), with the corresponding velocities, are

$$\alpha = A_1 e^{at} + A_2 e^{-at} + A_3 e^{ibt} + A_4 e^{-ibt} \tag{8.68a}$$

$$\beta = cA_1 e^{at} - cA_2 e^{-at} + idA_3 e^{ibt} - idA_4 e^{-ibt} \tag{8.68b}$$

$$\dot{\alpha} = aA_1 e^{at} - aA_2 e^{-at} + ibA_3 e^{ibt} - idA_4 e^{-ibt} \tag{8.68c}$$

$$\dot{\beta} = acA_1 e^{at} + acA_2 e^{-at} - bdA_3 e^{ibt} - bdA_4 e^{-ibt} \tag{8.68d}$$

Set the initial displacements and velocities to $\alpha_0, \beta_0, \dot{\alpha}_0, \dot{\beta}_0$. For $t = 0$, from Eqs. (8.68) we have

$$\alpha_0 = A_1 + A_2 + A_3 + A_4 \tag{8.69a}$$

$$\beta_0 = c(A_1 - A_2) + id(A_3 - A_4) \tag{8.69b}$$

$$\dot{\alpha}_0 = a(A_1 - A_2) + ib(A_3 - A_4) \tag{8.69c}$$

$$\dot{\beta}_0 = ac(A_1 + A_2) - bd(A_3 - A_4) \tag{8.69d}$$

Assuming the motion is bounded and periodic, then $A_1 + A_2 = 0$ and $A_1 - A_2 = 0$, so only the imaginary exponents remain. Hence, $A_1 = A_2 = 0$. From that constraint, the relationship among the initial conditions to guarantee periodicity can be worked out. With the restriction $A_1 = A_2 = 0$, from Eqs. (8.69a)–(8.69b) we find

$$A_3 = \frac{\alpha_0}{2} - \frac{i\beta_0}{2d}, \; A_4 = \frac{\alpha_0}{2} + \frac{i\beta}{2d} \tag{8.70}$$

Equations (8.68a)–(8.68b) then become

$$\alpha = \left(\frac{\alpha_0}{2} - i\frac{\beta_0}{2d}\right) e^{ibt} + \left(\frac{\alpha_0}{2} + i\frac{\beta_0}{2d}\right) e^{-ibt} \tag{8.71}$$

$$\beta = id\left(\frac{\alpha_0}{2} - i\frac{\beta_0}{2d}\right) e^{ibt} - id\left(\frac{\alpha_0}{2} + i\frac{\beta_0}{2d}\right) e^{-ibt} \tag{8.72}$$

which, by means of the Euler relations between trigonometric functions and exponents, can be simplified to

$$\alpha = \alpha_0 \cos(bt) + \frac{\beta_0}{d} \sin(bt) \tag{8.73}$$

$$\beta = \beta_0 \cos(bt) - d\alpha_0 \sin(bt) \tag{8.74}$$

These are the equations of the motion of the infinitesimal mass around L_1. t may be eliminated by solving for $\cos(bt)$ and $\sin(bt)$, then by squaring the results and adding,

$$\frac{\alpha^2}{(d^2\alpha_0^2 + \beta_0^2)/d^2} + \frac{\beta^2}{d^2\alpha_0^2 + \beta_0^2} = 1 \tag{8.75}$$

This is the equation of an ellipse, whose center is L_1 and with axes parallel to the x and y axes of the rotating coordinate system. It can be shown that $d^2 > 1$, so the ellipse's major axis is parallel to the y-axis. Also, the eccentricity of the ellipse is given by $e^2 = 1 - (1/d^2)$. Thus, the ellipse's shape depends only on the relative mass distribution through the constant d, and not on the disturbed motions initial conditions.

There is a hypothesis that the counterglow, or *gegenschein*, a hazy patch of light 180° from the Sun and near the ecliptic, is many small dust particles near the L_1 point of the Sun-Earth system. Even if A_1 and A_2 in Eqs. (8.68) are not zero, but are small in comparison with A_3 and A_4, particles could remain for a long time near L_1, before departing from it. The initial conditions for meteoric particles to be near the L_1 point are critical for this explanation of the gegenschein (McCuskey 1963, pp. 118–126).

8.7 Hill's Restricted Three-Body Problem

The discussion in previous sections dealt with a special case of the three-body problem, in which one of the three masses, m_3, was infinitesimally small compared to the other two masses, m_1 and m_2, i.e. $m_3 \ll m_2, m_1$. Another special case of importance is when two masses are much smaller than the remaining mass, i.e., $m_3, m_2 \ll m_1$. This problem is usually referred to as *Hill's restricted three-body problem* (Hill 1878). This is a problem of practical importance both in celestial mechanics and astrodynamics. The most important feature of Hill's restricted three-body problem compared to the original restricted three-body problem is that the dynamics of the former are not dependant on the masses, and can hence be used to any celestial or astrodynamical system in which two masses are much smaller than the remaining one. We will present some interesting periodic orbits around the secondary mass that emerge in Hill's restricted problem.

8.7.1 Equations of Motion

Let m denote the mass of the secondary and M be the mass of the primary. In previous sections, a barycentric rotating coordinate system, the origin of which was located at the center of mass of m and M, was used (Fig. 8.1). Here we will use a similar coordinate system; the only difference compared to the previous one is that the origin is shifted to the center of m, as shown in Fig. 8.7. This coordinate system was used by Hill in his lunar theory (1878).

Let $\mathbf{r} = [x, y, z]^T$ be the position vector of point $P(x, y, z)$ relative to m, $\mathbf{R}$ be the position of m relative to M, and $\boldsymbol{\rho}$ be the position of $P(x, y, z)$ relative to M, as shown in Fig. 8.7. Also we assume that m moves on a circular orbit about M. The equations of motion of the point P are

$$\frac{d^2\boldsymbol{\rho}}{dt^2} = -\frac{k^2 m\mathbf{r}}{\|\mathbf{r}\|^3} - \frac{k^2 M\boldsymbol{\rho}}{\|\boldsymbol{\rho}\|^3} \tag{8.76}$$

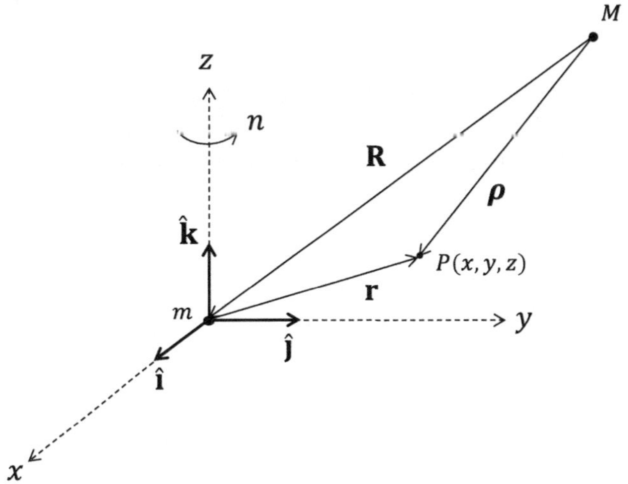

Fig. 8.7 Coordinate system for Hill's restricted three-body problem

Substituting $\boldsymbol{\rho} = \mathbf{R} + \mathbf{r}$ into Eq. (8.76) yields

$$\frac{d^2\mathbf{r}}{dt^2} = -\frac{k^2 m\mathbf{r}}{\|\mathbf{r}\|^3} - \frac{k^2 M(\mathbf{R}+\mathbf{r})}{\|\mathbf{R}+\mathbf{r}\|^3} - \frac{d^2\mathbf{R}}{dt^2} \tag{8.77}$$

The gravitational acceleration of m relative to M is given by

$$\frac{d^2\mathbf{R}}{dt^2} = -\frac{k^2 M\mathbf{R}}{\|\mathbf{R}\|^3} \tag{8.78}$$

We now substitute Eq. (8.78) into Eq. (8.77), and write $\mathbf{R} = [R, 0, 0]^T$ where R is the constant orbital radius of m. The angular velocity is $\boldsymbol{\omega} = [0, 0, n]^T$, where $n^2 = k^2(M + m)/R^3$. Using Eq. (8.1) and substituting for the gravitational terms yields

$$\ddot{x} - 2\dot{y} - n^2 x = -\frac{k^2 m x}{[x^2+y^2+z^2]^{\frac{3}{2}}} - \frac{k^2 M(x+R)}{[(x+R)^2+y^2+z^2]^{\frac{3}{2}}} + \frac{k^2 M}{R^2} \tag{8.79a}$$

$$\ddot{y} + n\dot{x} - n^2 y = -\frac{k^2 m y}{[x^2+y^2+z^2]^{\frac{3}{2}}} - \frac{k^2 M y}{[(x+R)^2+y^2+z^2]^{\frac{3}{2}}} \tag{8.79b}$$

$$\ddot{z} = -\frac{k^2 m z}{[x^2+y^2+z^2]^{\frac{3}{2}}} - \frac{k^2 M z}{[(x+R)^2+y^2+z^2]^{\frac{3}{2}}} \tag{8.79c}$$

To obtain a normalized set of equations, we divide the position components by R, so that

$$\xi = \frac{x}{R},\ \eta = \frac{y}{R}, \zeta = \frac{z}{R} \tag{8.80}$$

and time is normalized by n. A new dimensionless gravitational parameter μ is defined as

$$\mu = \frac{m}{m+M} \tag{8.81}$$

Thus, the velocity and the acceleration are normalized as

$$\dot{\xi} = \frac{\dot{x}}{nR},\ \ddot{\xi} = \frac{\ddot{x}}{n^2 R} \tag{8.82}$$

We now rewrite Eq. (8.79a) into

$$\ddot{\xi}n^2R - 2\dot{\eta}n^2R - \xi n^2R = -\frac{\mu\xi}{(\xi^2+\eta^2+\zeta^2)^{\frac{3}{2}}}\frac{k^2m+k^2M}{R^3}R - \frac{(1-\mu)(\xi+1)}{\left[(\xi-1)^2+\eta^2+\zeta^2\right]^{\frac{3}{2}}}\frac{k^2m+k^2M}{R^3}R + \frac{1-\mu}{R^2}(k^2m+k^2M) \tag{8.83}$$

If we substitute $n^2 = k^2(m+M)/R^3$ into Eq. (8.83), we obtain the normalized equation. In the same manner, we can manipulate the other two equations and get a set of normalized equations,

$$\ddot{\xi} = 2\dot{\eta} + \xi - \frac{\mu\xi}{[\xi^2+\eta^2+\zeta^2]^{\frac{3}{2}}} - \frac{(1-\mu)(\xi+1)}{[(\xi+1)^2+\eta^2+\zeta^2]^{\frac{3}{2}}} + 1 - \mu \tag{8.84a}$$

$$\ddot{\eta} = -2\dot{\xi} + \eta - \frac{\mu\eta}{[\xi^2+\eta^2+\zeta^2]^{\frac{3}{2}}} - \frac{(1-\mu)\eta}{[(\xi+1)^2+\eta^2+\zeta^2]^{\frac{3}{2}}} \tag{8.84b}$$

$$\ddot{\zeta} = -\frac{\mu\zeta}{[\xi^2+\eta^2+\zeta^2]^{\frac{3}{2}}} - \frac{(1-\mu)\zeta}{[(\xi+1)^2+\eta^2+\zeta^2]^{\frac{3}{2}}} \tag{8.84c}$$

8.7.2 Hill's Equations of Motion

As noted previously, $m \ll M$ and hence $\mu \ll 1$ and moreover, $\mu \to 0$. To use this fact, apply the following transformation to Eqs. (8.84),

$$\xi = 1 - \mu + \mu^{\frac{1}{3}}x \tag{8.85a}$$

$$\eta = \mu^{\frac{1}{3}}y \tag{8.85b}$$

$$\zeta = \mu^{\frac{1}{3}}z \tag{8.85c}$$

When we substitute Eqs. (8.85) into Eqs. (8.84) and let $\mu \to 0$, we obtain *Hill's equations* (Hénon 1969),

$$\ddot{x} - 2\dot{y} - 3x = \frac{\partial W}{\partial x} \tag{8.86a}$$

$$\ddot{y} + 2\dot{x} = \frac{\partial W}{\partial y} \tag{8.86b}$$

$$\ddot{z} + z = \frac{\partial W}{\partial z} \tag{8.86c}$$

where

$$W \equiv \frac{1}{r} = \frac{1}{\sqrt{x^2 + y^2 + z^2}} \tag{8.87}$$

Obviously, Eqs. (8.86) are independent of μ. Another interesting observation is the location of the Lagrangian collinear points; In Hill's model $L_3 \rightarrow -\infty$ and $(L_1, L_2) = \pm(\frac{1}{3})^{\frac{1}{3}} = \pm 0.69336$. The Jacobi constant is given by

$$\Gamma = 3x^2 + z^2 + \frac{2}{r} - (\dot{x}^2 + \dot{y}^2 + \dot{z}^2) \tag{8.88}$$

8.7.3 Families of Periodic Orbits

Periodic orbits about the smaller primary m can be found using a numerical search. The question is which initial conditions to choose in the six-dimensional state space. To simplify the problem, we can eliminate some of the parameters that need to be found.

One possibility is to search for orbits only in the xy plane. This reduces the problem into a four-dimensional search. We can simplify the search further if we choose to find only symmetrical orbits with respect to the x-axis. In this case we know that the orbit must intersect the x-axis, so we can define the initial condition in the intersection point, x_0. Also, symmetry implies that the velocity at the intersection point will only be in the y direction. Thus, looking for symmetric orbits simplifies the problem to finding x_0 and $\dot{y}_0$. It is generally more convenient to use the Jacobi integral instead of $\dot{y}_0$.

In order to find orbits in the planar case, we need to know the value of the Jacobi constant, Γ, and the initial condition x_0, where the orbit crosses the x axis with $\dot{y} > 0$. For that point we have $y_0 = 0$; $\dot{x}_0$, and $\dot{y}_0$ are found from Eq. (8.88).

It is convenient to represent an orbit by a point in the (Γ, x_0) plane. In Fig. 8.8, we depict the characteristic families of periodic orbits, which are denoted by $a, c, f, g, g', g'_2, g'_3$ (Hénon 1969, 1970).

The hatched areas in Fig. 8.8 are "forbidden", i.e. areas in which $\dot{y}_0 < 0$. All the orbits found are symmetric with respect to the x axis because $\dot{x}_0 = 0$.

We can use the Jacobi constant as an energy measure, where near m we have $\Gamma \rightarrow +\infty$. From Fig. 8.8, it is evident that at the energy level of the L_1, L_2 points, there are periodic orbits are closer to the secondary than L_1. To find larger orbits, we

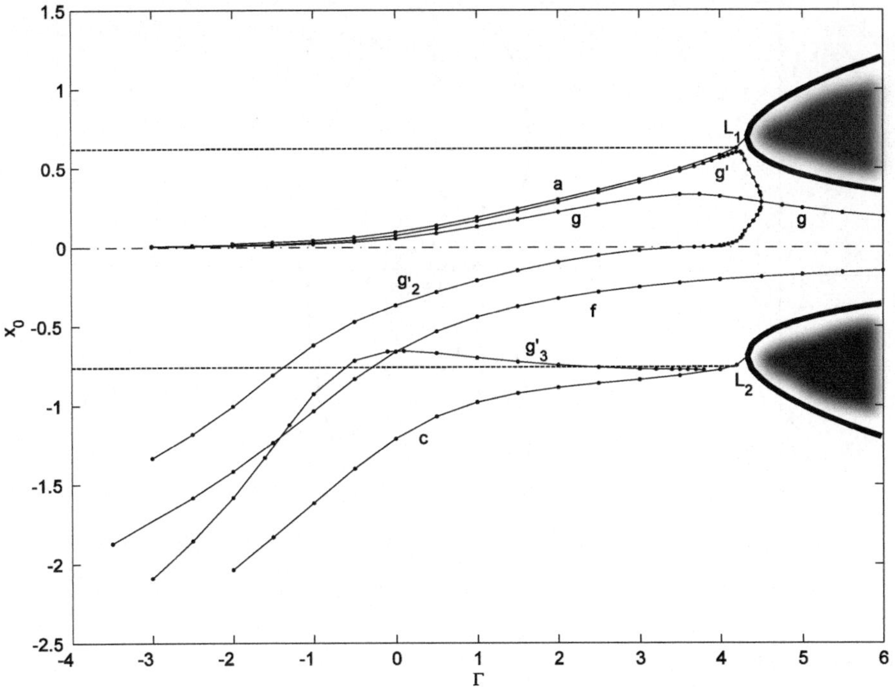

Fig. 8.8 General map of periodic orbits in Hill's problem

need to change the velocity of P, and thereby change the value of Γ to the value of the desired orbit.

Families *a* and *c*

Families a and c include periodic orbits about the collinear points L_1 and L_2. Some orbits of Family a can be seen in Fig. 8.9. Table 8.2 displays the Jacobi constant Γ, the initial condition x_0 and the orbital period. The orbits of Family c are symmetric with respect to Family a as can be seen in Fig. 8.10 and in Table 8.3.

Family *f*

Family f includes *distant retrograde orbits* about the secondary. These orbits are stable and symmetric with respect to the x axis (Fig. 8.11). From Fig. 8.8 it is seen that these orbits can be found for any $\Gamma < 0$. For $\Gamma \to -\infty$ the orbits become ellipses centered at the secondary with a major to minor axis ratio of 2 as can be seen in Fig. 8.12. Orbits that are very close to the secondary become circles due to the decreasing gravitational effect of the primary. Figure 8.11 shows some of the characteristic orbits in this family. The orbit parameters are given in Table 8.4.

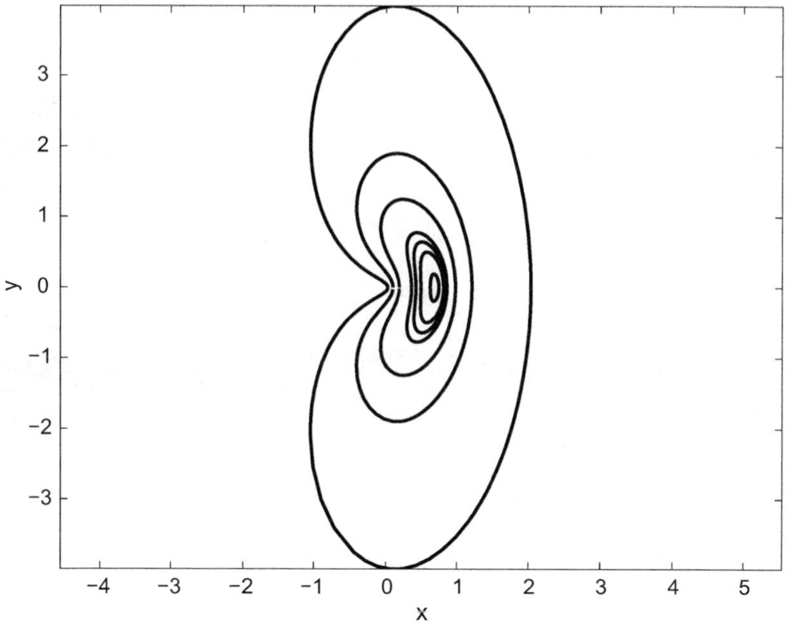

Fig. 8.9 Family *a* in Hill's problem

Table 8.2 Family *a*: Jacobi constant and initial conditions

Γ	x_0	T	Γ	x_0	T
4.327	0.69336	3.0513	1	0.18797	4.88
4.2	0.62698	3.084	0.5	0.13756	5.6
4	0.5802	3.172	0	0.09515	6.3504
3.5	0.4958	3.288	−0.5	0.06402	6.95
3	0.42585	3.44	−1	0.04383	6.95
2.5	0.36181	3.464	−1.5	0.03121	7.37
2	0.30114	3.928	−2	0.02314	7.656
1.5	0.24307	4.32			

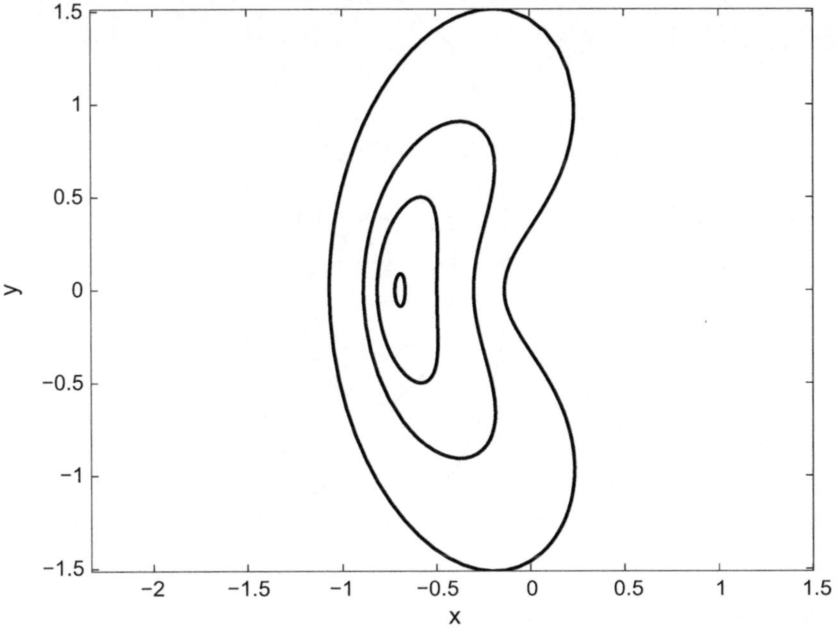

Fig. 8.10 Family c in Hill's problem

Table 8.3 Family c: Jacobi constant and initial conditions

Γ	x_0	T	Γ	x_0	T
4.327	−0.69336	3.0513	1	−0.9778	4.88
4.2	−0.74757	3.084	0.5	−1.0677	6.3504
4	−0.77522	3.172	0	−1.2082	6.3504
3.5	−0.81245	3.288	−0.5	−1.3992	6.95
3	−0.83714	3.44	−0.5	−1.3992	6.95
2.5	−0.8597	3.646	−1	−1.615	6.95
2	−0.88586	4.32	−1.5	−1.8304	7.37
1.5	−0.92204	4.32	−2	−2.0352	7.656

Family *g*

Family g can be divided into several groups according to the value of Γ. For $\Gamma \geq 4.5$, we get a group of stable *distant prograde orbits*. They can be seen in Fig. 8.13. These orbits are almost circular in form and resemble the orbits of the two-body problem.

For $2 \leq \Gamma < 4.5$ we obtain unstable distant prograde orbits. Figure 8.14 shows that the orbits become ellipses, and in the critical value $\Gamma = 2$ the orbit resembles a parabolic orbit, in which the velocity reduces to zero at infinity.

When $\Gamma < 2$, we obtain unstable orbits as seen in Fig. 8.15. These orbits can be used as transfer orbits. In the WIND mission, these orbits were used to get to

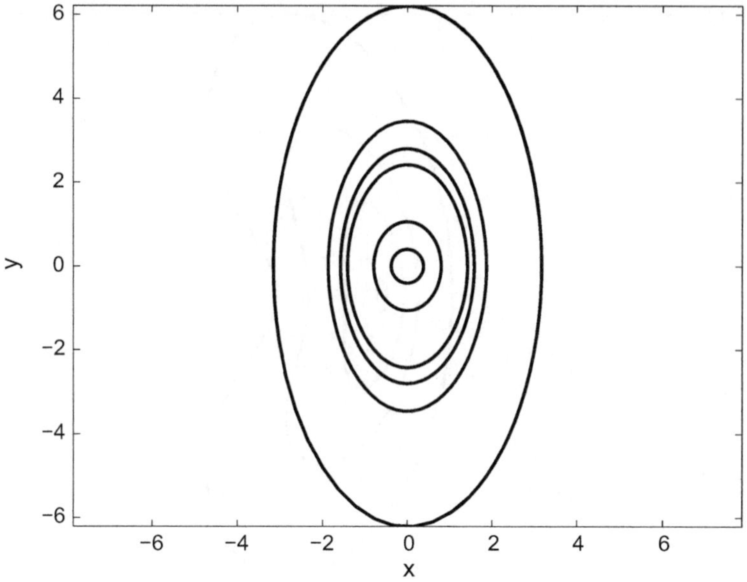

Fig. 8.11 Family f in Hill's problem

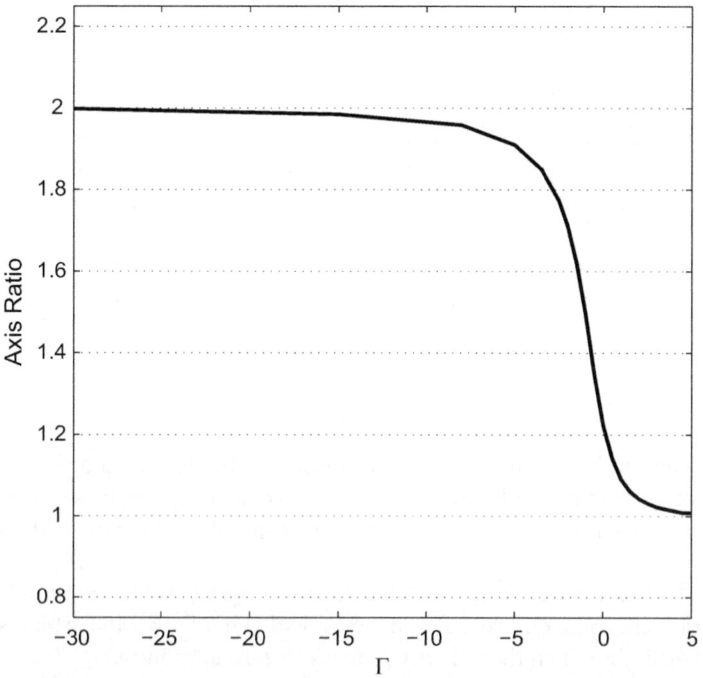

Fig. 8.12 Family f major to minor axis ratio

Table 8.4 Family f: Jacobi constant and initial conditions

Γ	x_0	T	Γ	x_0	T
6	−0.14779	0.3394	1	−0.43991	1.52
5.5	−0.15888	0.3794	0.5	−0.53182	1.938
5	−0.17169	0.422	0	−0.65966	2.526
4.5	−0.18661	0.474	−0.5	−0.83185	3.292
4	−0.20421	0.537	−1	−1.034	4.08
3.5	−0.22523	0.616	−1.5	−1.2341	4.7
3	−0.25071	0.714	−2	−1.4168	5.12
2.5	−0.28212	0.84	−2.5	−1.5817	5.38
2	−0.32163	1.02	−3.5	−1.8705	5.694
1.5	−0.37252	1.222			

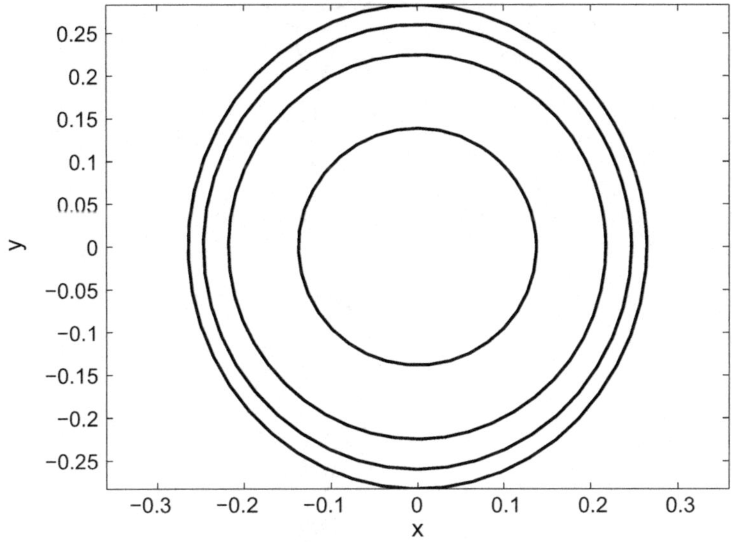

Fig. 8.13 Family g with $\Gamma \geq 4.5$: Stable distant prograde orbits

and from a distant retrograde orbit.[1] As mentioned previously, in Hill's problem the Lagrangian points L_1 and L_2 are located at $x = \pm 0.69336$, so the orbits of family g may reach to about five times this value. The initial conditions of these orbits are displayed in Table 8.5.

Family g'

The orbits in family g' pass close to the primary; they are unstable. The orbits are shown in Figs. 8.16 and 8.17. There are two branches of the Family that split from family g at the critical point $\Gamma = 4.5$.

[1] See http://wind.nasa.gov/.

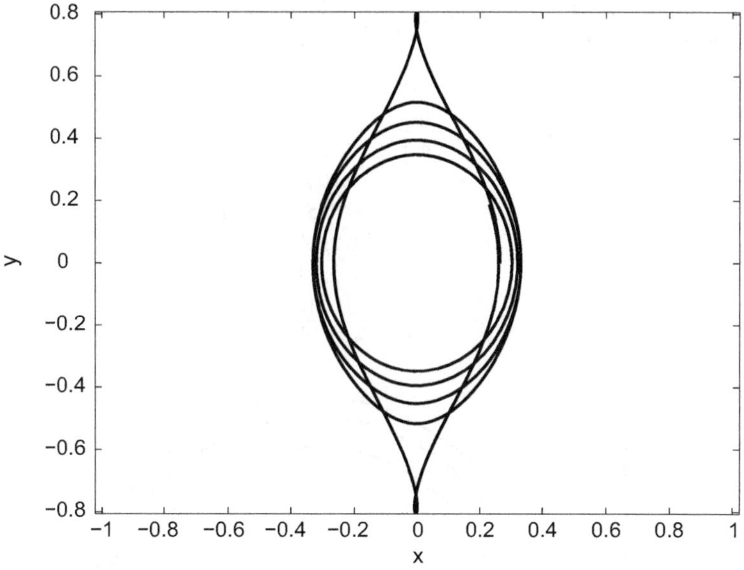

Fig. 8.14 Family g with $2 \leq \Gamma < 4.5$: Unstable prograde orbits

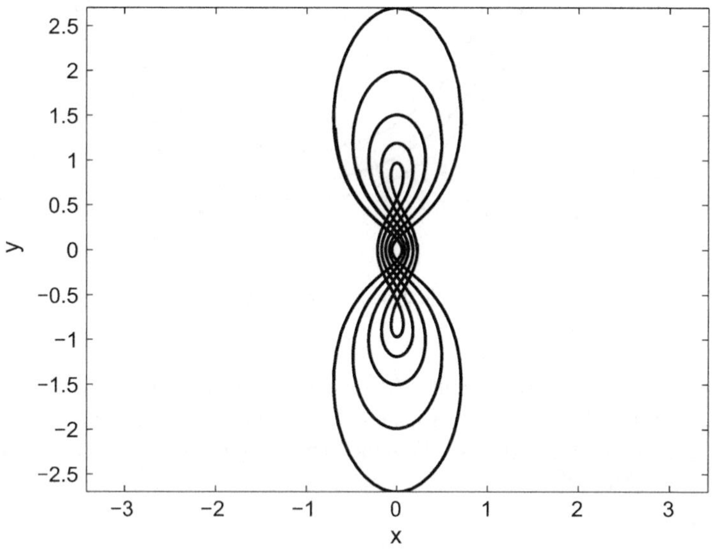

Fig. 8.15 Family g with $\Gamma < 2$: Unstable prograde orbits

Table 8.5 Family g: Jacobi constant and initial conditions

Γ	x_0	T	Γ	x_0	T
6	0.19489	0.62	1.5	0.17545	4.95
5.5	0.21788	0.734	1	0.13032	5.9
5	0.247	0.94	0.5	0.08902	7.094
4.75	0.26435	1.06	0	0.05587	8.44
4.5	0.2835	1.226	−0.5	0.03392	9.52
4.25	0.30343	1.728	−1	0.02104	10.26
3.75	0.33178	2.054	−1.5	0.01356	10.72
3.5	0.33173	2.4	−2	0.00909	11.02
3	0.3069	3.02	−2.5	0.006313	11.26
2.5	0.26679	3.6	−3	0.004523	11.42
2	0.22168	4.22			

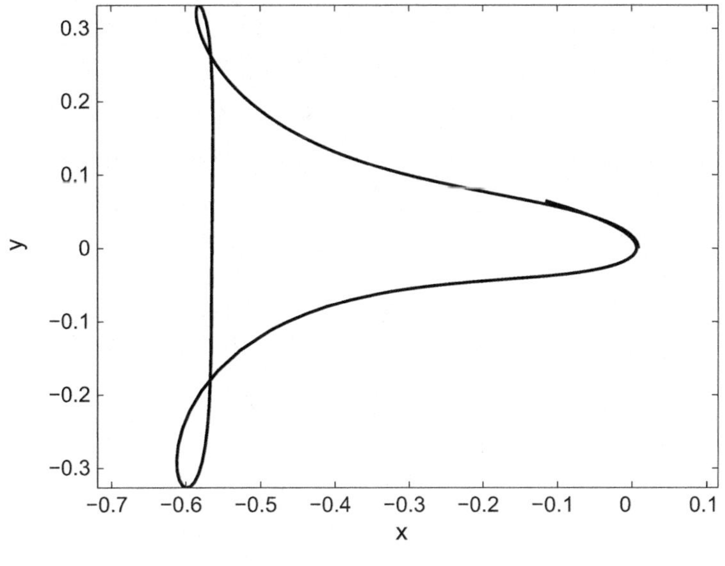

Fig. 8.16 Family g': Unstable orbits

The trajectory close to the secondary is equivalent to a hyperbolic flyby as seen in the two-body problem. Figure 8.17 demonstrates a "slingshot" effect. A particle, e.g. a spacecraft, will be diverted into an escape orbit with respect to a planet due to this effect.

The orbits in this family display a new kind of motion. The orbits revolve around the secondary only once, and can be characterized by two points x_{01} and x_{02} along the orbit as noted in Table 8.6. The two branches of this family can be seen in Fig. 8.8, where the branch of Family g' splits from Family g at the critical point $\Gamma = 4.5$.

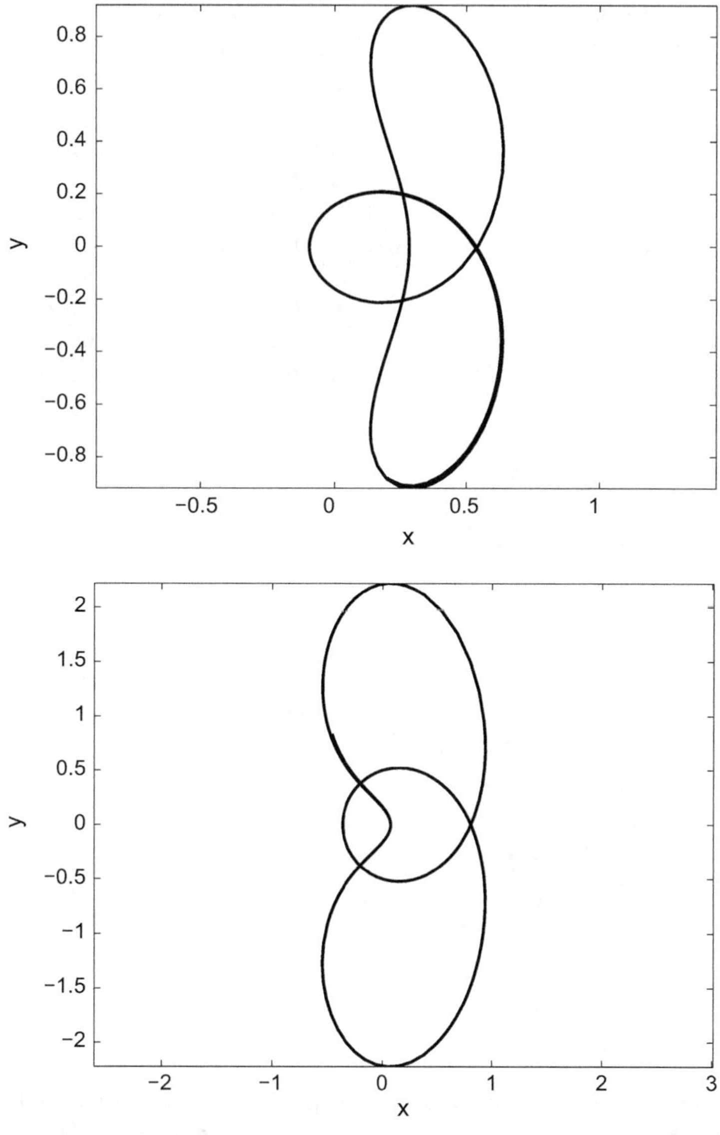

Fig. 8.17 Family g': Delayed escape

Family g_3

An additional family of orbits is g_3. This family can be seen in Figs. 8.18 and 8.19 and its parameters are given in Table 8.7. The orbits in this family can be used for libration point missions. In Fig. 8.18 the orbit passes close to the libration points and, therefore, can be used as a transfer orbit between libration points. The orbit in Fig. 8.19 moves from the vicinity of the secondary to a large orbit around

Table 8.6 Family g': Jacobi constant and initial conditions

Γ	x_{01}	x_{02}	T	Γ	x_{01}	x_{02}	T
4.5	0.2835	0.2839	1.226	3.6845	0.50908	0	3.484
4.49	0.32486	0.24509	1.242	3.5	0.4808	−0.00184	3.56
4.48	0.34334	0.22965	1.26	3	0.41052	−0.02001	3.84
4.45	0.34684	0.19924	1.43	2.5	0.34555	−0.0515	4.12
4.4	0.43684	0.16385	1.43	2	0.28365	−0.0938	4.51
4.35	0.48918	0.13329	1.62	1.5	0.22422	−0.14642	5.06
4.3	0.55029	0.09921	1.94	1	0.16778	−0.20938	5.82
4.2714	0.58769	0.07366	2.302	0.5	0.11637	−0.28226	6.92
4.25	0.6009	0.057	2.6	0	0.07422	−0.36577	8.36
4.2	0.6004	0.03613	2.92	−0.5	0.0457	−0.47125	9.84
4.15	0.59171	0.025	3.06	−1	0.02928	−0.6191	11.08
4.1	0.58201	0.01766	3.16	−1.5	0.02011	−0.80868	12.2
4.05	0.57234	0.01241	3.22	−2	0.01464	−1.0044	12.8
4	0.56291	0.008533	3.28	−2.5	0.01107	−1.1795	13.34
3.9	0.54488	0.003494	3.28	−3	0.008613	−1.3322	13.72
3.8	0.52782	0.0009045	3.44				

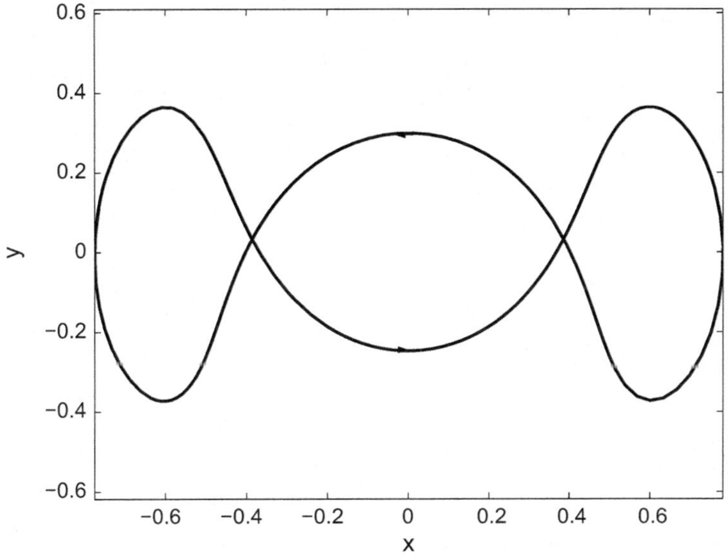

Fig. 8.18 Family g_3: Libration point transfer

the libration points; this kind of orbit can be used as an orbiter or a carrier that always travels between a planet and the collinear points, thus reducing the velocity requirements for transfer maneuvers.

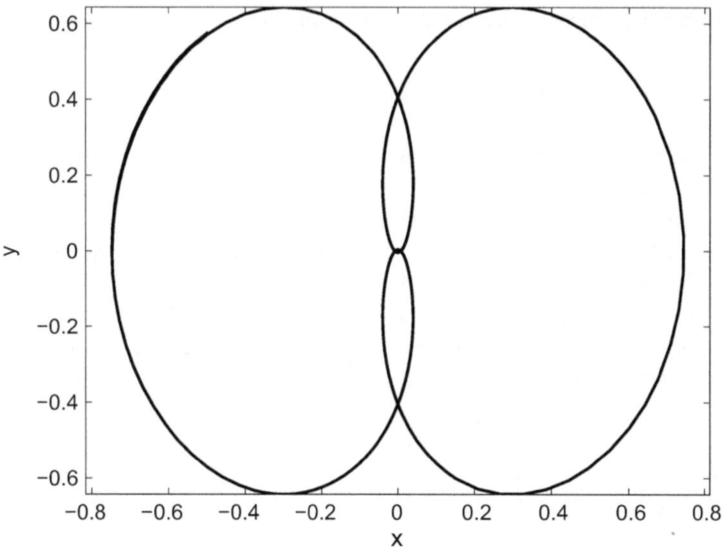

Fig. 8.19 Family g_3: Libration point and planet orbiter

Table 8.7 Family g_3: Jacobi constant and initial conditions

Γ	x_0	T	Γ	x_0	T
3.5	−0.77365	5.16	0.1	−0.65482	7.18
3.4	−0.77323	5.02	−0.1	−0.65866	8.02
3.2	−0.77176	4.82	−0.5	−0.7161	9.92
3	−0.76969	4.68	−1	−0.92602	12.22
2.5	−0.75961	4.52	−1.3	−1.1221	13.42
2	−0.74453	4.54	−1.6	−1.3295	14.36
1.5	−0.72413	4.72	−2	−1.5823	15.82
1	−0.69838	5.14	−2.5	−1.8537	15.82
0.5	−0.66969	5.98	−3	−2.089	16.204

References

Bruns, H.: Ueberdie integrale des vierkörper problems. Acta Math. **11**, 25–96 (1887)

Hénon, M.: Numerical exploration of the restricted problem, V. Astron. Astrophys. **1**, 223–238 (1969)

Hénon, M.: Numerical exploration of the restricted problem, VI. Astron. Astrophys. **9**, 24–36 (1970)

Hill, G.W.: Researches in the lunar theory. Am. J. Math. **1**(1), 5–26 (1878)

Jacobi, C.G.J.: Vorlesungen Über Dynamik. Collected Works Supplement (1884)

McCuskey, S.W.: Introduction to Celestial Mechanics. Addison-Wesley, Reading (1963)

Moulton, F.: An Introduction to Celestial Mechanics. Dover, New York (1970)

Rauschenbakh, B.V., Ovchinnkov, Y.U., McKenna-Lewlor, S.: Essential Spaceflight Dynamics and Magnetospherics. Springer, New York (2003)

Szebehely, V.: Theory of Orbits: The Restricted Problem of Three Bodies. Academic Press, New York (1967)

Wylie, C.R.: Advanced Engineering Mathematics. McGraw-Hill, New York (1975)

Chapter 9
Numerical Procedures

9.1 Differences and Sums

This section is meant as an introduction to notation and procedures involved in numerical work that occurs in celestial mechanics and astrodynamics. No theoretical developments will be presented; for theory the reader might consult a work on finite differences. The notation to be used and reference to more information is available from Nautical Almanac Office (1956).

Assume a function $f(x)$ of an independent variable, x, is tabulated at equal intervals, h. If x_0 is a tabular point, then we can make a table of $f(x_0+ph)$ values for integer values of p, which could be written $f(x_p)$, or f_p. The argument is $(x_0 + ph)$, but, without ambiguity, this could be designated as p. The first difference, $\delta_{p+1/2}$ is determined by subtracting f_p from f_{p+1}. This is shown in Table 9.1. Note that more values of the function are required to generate the differences shown.

If we form a column of first differences, they can be differenced to form second differences,

$$\delta_{p+1/2} - \delta_{p-1/2} - \delta_p^2 \tag{9.1}$$

The superscript denotes the difference number, the subscript denotes the position with respect to the argument. In principle, the differencing can be carried out indefinitely.

Table 9.1 can be changed by adding arithmetical means of the entries, immediately above and below a space. These "half differences" are preceded by a μ. Thus,

$$\mu\delta f_p = \frac{1}{2}\left(\delta_{p+1/2} + \delta_{p-1/2}\right) \tag{9.2}$$

The half differences are listed in Table 9.2.

P. Gurfil, P.K. Seidelmann, *Celestial Mechanics and Astrodynamics: Theory and Practice*, Astrophysics and Space Science Library 436,
DOI 10.1007/978-3-662-50370-6_9

Table 9.1 Finite differences

Argument	Function	Differences 1st	2nd	3rd	4th
-2	f_{-2}		δ^2_{-2}		δ^4_{-2}
		$\delta_{-1\ 1/2}$		$\delta^3_{-1\ 1/2}$	
-1	f_{-1}		δ^2_{-1}		δ^4_{-1}
		$\delta_{-1/2}$		$\delta^3_{-1/2}$	
0	f_0		δ^2_0		δ^4_0
		$\delta_{1/2}$		$\delta^3_{1/2}$	
$+1$	f_{+1}		δ^2_{+1}		δ^4_{+1}
		$\delta_{+1\ 1/2}$		$\delta^3_{+1\ 1/2}$	
$+2$	f_{+2}		δ^2_{+2}		δ^4_{+2}

Table 9.2 Half differences

Argument	Function	Differences 1st	2nd	3rd	4th
-1	f_{-1}	$[\mu\delta_{-1}]$	δ^2_{-1}	$[\mu\delta^3_{-1}]$	δ^4_{-1}
	$[\mu f_{-1/2}]$	$\delta_{-1/2}$	$[\mu\delta^2_{-1/2}]$	$\delta^3_{-1/2}$	$[\mu\delta^4_{-1/2}]$
0	f_0	$[\mu\delta_0]$	δ^2_0	$[\mu\delta^3_0]$	δ^4_0
	$[\mu f_{1/2}]$	$\delta_{1/2}$	$[\mu\delta^2_{1/2}]$	$\delta^3_{1/2}$	$[\mu\delta^4_{1/2}]$
1	f_1	$[\mu\delta_1]$	δ^2_1	$[\mu\delta^3_1]$	δ^4_1

Table 9.3 Sums

Argument	Sums 2nd	1st	Function
-2	δ^{-2}_{-2}		f_{-2}
		$\delta^{-1}_{-1\ 1/2}$	
-1	δ^{-2}_{-1}		f_{-1}
		$\delta^{-1}_{1/2}$	
0	δ^{-2}_0		f_0
		$\delta^{-1}_{1/2}$	
1	δ^{-2}_1		f_1
		$\delta^{-1}_{1\ 1/2}$	
2	δ^{-2}_2		f_2

Differencing is a means of checking some calculations and tabulations of numbers. The f_p values can be considered as first differences of another function, which we can call 1st sums of f_p. The 1st sums can be viewed as differences of 2nd sums, and so on. The sums have an uncertainty of an additive constant. See Table 9.3 for sums.

There is an analog between summing and integrating and between differencing and differentiating. The δ's can be considered operators that obey the laws of algebra (Danby 1962, pp. 214–215).

9.2 Interpolation

The tables discussed referred to values of $f(x_0 + ph)$ for integral values of p; once the table is constructed, with some constraints, a value of f can be determined for any value of p, which is within the range of the table. This process is *interpolation*.

Let us now assume that $0 \le p < 1$. If the second differences are negligible or zero,

$$f_p = f_0 + p\delta_{1/2} \tag{9.3}$$

This is equivalent to linear interpolation. In general, f_p should be given by a formula involving successive differences in a way depending on p. We will give two relations where this is the case. The first is *Bessel's formula*,

$$f_p = f_0 + p\,\delta_{1/2} + B_2\,(\delta_0^2 + \delta_1^2) + B_3\,\delta_{1/2}^3 + B_4\,(\delta_0^4 + \delta_1^4) + \ldots \tag{9.4}$$

where the B's are *Bessel interpolation coefficients* and functions of p. For example,

$$B_2 = \frac{1}{4}p(p-1) \tag{9.5}$$

$$B_3 = \frac{1}{12}p(p-1)(2p-1) \tag{9.6}$$

$$B_4 = \frac{1}{48}p(p^2-1)(p-2) \tag{9.7}$$

These functions are tabulated in many references. The second relation is *Everett's formula*,

$$f_p = (1-p)\,f_0 + p\,f_1 + E_2\,\delta_0^2 + F_2\,\delta_1^2 + E_4\,\delta_0^4 + F_4\,\delta_1^4 + \ldots \tag{9.8}$$

The *Everett interpolation coefficients* are related to the Bessel coefficients by

$$E + F = 2B \tag{9.9}$$

and also $E_{(p)} = F_{(1-p)}$ and $E_{(1-p)} = F_{(p)}$.

Differences must be taken out to the point where they are small and irregular. This is not a condition simply detected on computers. To reduce the interpolation formula and the differences required, a *throw back scheme* may be used. For instance, if the 4th differences are small (less than 1000), then the B_4 term can be neglected, provided that the 2nd difference is modified by

$$\delta_{mod}^2 = \delta^2 - 0.184\,\delta^4 \tag{9.10}$$

Then, the 4th differences have been thrown back. A similar process can be used in Everett's formula, and more complicated throw backs can be used with higher differences. When differences are smooth, it is possible with a minimum of error to *extrapolate* the table of differences beyond the functions available. Likewise, it is possible to determine the argument for a given value of the function by inverse interpolation. This is usually done by some scheme of successive approximation (Danby 1962, pp. 215–218).

9.3 Lagrangian Methods

It is frequently desirable to have a formula expressed explicitly in terms of functional values, rather than in terms of the differences. Such formulae permit a direct consideration of the effect on the value due to a change, or error, in the ordinates, and the use does not require the calculation and tabulation of differences. Apparently, the basic formula is due to Waring, but it is given the name Lagrange.

Lagrange's form of the polynomial

$$y(x) \equiv y_0, y_1, \ldots, y_n(x) \quad \text{of degree n} \tag{9.11}$$

takes on the same values as a given function $f(x)$ for the $n+1$ distinct abscissas $x_0, x_1, x_2, \ldots, x_n$. It differs from the Newtonian form

$$\begin{aligned} f(x) &= f[x_0] + (x - x_0)f[x_0, x_1] \\ &\quad + (x - x_0)(x - x_1)f[x_0, x_1, x_2] + \ldots + (x - x_0)\ldots(x - x_{n-1})f[x_0, \ldots, x_n] \\ &\quad + E(x) \end{aligned} \tag{9.12}$$

where $E(x) \triangleq \frac{f(x_1)-f(x_0)}{(x_1-x_0)}$ is a first divided difference.

The ordinates are displayed explicitly in the Lagrangian form, while the Newtonian form explicitly involves divided differences of those ordinates. While Lagrange's form may be derived from Newton's form, its importance justifies a separate consideration.

We can write $y(x)$ in the form

$$y(x) = A_0 + A_1x + A_2x^2 + \ldots + A_nx^n = \sum_{k=0}^{n} A_kx^k \tag{9.13}$$

where the $A's$ are to be determined such that

$$y(x_i) = f(x_i),\ i = 0, 1, 2, \ldots, n \tag{9.14}$$

These requirements can be given by $n + 1$ linear equations

$$\begin{aligned} A_0 + A_1x_0 + A_2x_0^2 + \cdots + A_nx_0^n &= f(x_0) \\ &\vdots \\ A_0 + A_1x_n + A_2x_n^2 + \cdots + A_nx_n^n &= f(x_n) \end{aligned} \tag{9.15}$$

These equations can be solved by use of determinants, where the special properties of the determinants can lead to simple expressions for the $A's$ in terms of the ordinates. The requirement that the $A's$ satisfy Eqs. (9.13) and (9.15) can be expressed by

$$\begin{bmatrix} y & | & x & x^2 & \dots & x^n \\ f(x_0) & | & x_0 & x_0^2 & \dots & x_0^n \\ \vdots & \vdots & \vdots & \vdots & \vdots & \vdots \\ f(x_n) & | & x_n & x_n^2 & \dots & x_n^n \end{bmatrix} = 0 \tag{9.16}$$

whose the expanded form gives the equation of the interpolation polynomial $y = y_0, \dots, y_n(x)$. Alternatively, we could write $y(x)$ as

$$y(x) = l_0(x)f(x_0) + l_1(x)f(x_1) + \ldots + l_n(x)f(x_n) \triangleq \sum_{k=0}^{n} l_k(x)f(x_k) \tag{9.17}$$

where $l_0(x), \dots, l_n(x)$ are polynomials of degree n or less, which are determined based on the requirement that replacing $y(x)$ by $f(x)$ be an identity, when $f(x)$ is an arbitrary polynomial of degree n or less. This situation will be true, if and only if, replacing $y(x)$ by $f(x)$ is an identity when $f(x) = 1, x, x^2, \dots, x^n$. These requirements are represented by $n + 1$ equations,

$$l_0(x) + l_1(x) + \ldots + l_n(x) = 1 \tag{9.18a}$$

$$x_0l_0(x) + x_1l_1(x) + \ldots + x_nl_n(x) = x \tag{9.18b}$$

$$\vdots \tag{9.18c}$$

$$x_0^nl_0(x) + x_1^nl_1(x) + \ldots + x_n^nl_n(x) = x^n \tag{9.18d}$$

where the coefficient functions can be determined as ratios of determinants, which can be expanded in simple forms.

Rather than pursuing either of these approaches, we can avoid a lengthy calculation by noticing that Eq. (9.17) will take on the value of $f(x_i)$, when $x = x_i$, if $l_i(x_i) = 1$, and if $l_i(x_j) = 0$, when $j \neq i$. Using the notation called the *Kronecker delta*,

$$\delta_{ij} = 0 \text{ if } i \neq j, \qquad \delta_{ij} = 1 \text{ if } i = j \tag{9.19}$$

the requirement becomes

$$l_i(x_j) = \delta_{ij},\ i = 0, \ldots, n,\ j = 0, \ldots, n \tag{9.20}$$

Since $l_i(x)$ is to be a polynomial of degree n, which vanishes when $x = x_0, x_1, \ldots, x_{i-1}, x_{i+1}, \ldots, x_n$, it follows

$$l_i(x) = C_i[(x - x_0) \ldots (x - x_{i-1})(x - x_{i+1}) \ldots (x - x_n)] \tag{9.21}$$

where C_i is a constant. C_i is determined from $l_i(x_i) = 1$ in the form

$$C_i = \frac{1}{(x_i - x_0) \ldots (x_i - x_{i-1})(x_i - x_{i+1}) \ldots (x_i - x_n)} \tag{9.22}$$

and the Lagrangian coefficient functions, $l_i(x)$, are determined by introducing Eqs. (9.21) and (9.22). Putting this in a more compact form, let

$$\pi(x) = (x - x_0)(x - x_1) \ldots (x - x_n) \tag{9.23}$$

The derivative of $\pi(x)$ can be expressed as the sum of $n + 1$ terms, in each of which one of the factors of $\pi(x)$ is deleted. If we set $x = x_i$ in this expression, we obtain

$$\pi'(x_i) = (x_i - x_0) \ldots (x_i - x_n) = \frac{1}{C_i'} \tag{9.24}$$

where $(x_i - x_i)$ is to be omitted in the product. After introducing Eqs. (9.17), (9.21), (9.22) the Lagrangian interpolation polynomial of degree n is obtained

$$y(x) = \sum_{k=0}^{n} \frac{\pi(x)}{(x - x_k)\pi'(x_k)} f(x_k) \triangleq \sum_{k=o}^{n} l_k(x) f(x_k) \tag{9.25}$$

where

$$\begin{aligned} l_i(x) &= \frac{\pi(x)}{(x - x_i)\pi'(x_i)} \\ &= \frac{(x - x_0) \ldots (x - x_{i-1})(x - x_{i+1}) \ldots (x - x_n)}{(x_i - x_0) \ldots (x_i - x_{i-1})(x_i - x_{i+1}) \ldots (x_i - x_n)} \end{aligned} \tag{9.26}$$

The first expression for $l_i(x)$ is useful for theoretical considerations, the second expression is useful for the actual calculation. The error committed by replacing $f(x)$ by $y(x)$ is

$$f(x) = \sum_{k=0}^{n} l_k(x) f(x_k) + E(x) \tag{9.27}$$

where

$$E(x) = \pi(x)f[x_0, \ldots, x_n, x] = \pi(x)\frac{f^{n+1}(\xi)}{(n+1)!} \tag{9.28}$$

where ξ is some number in the interval limited by the largest and smallest of the numbers $x_0, x_1, \ldots, x_n$, and x.

The use of the Lagrangian formula is shown by taking the interpolation polynomial of degree three related to the data

x	-1	0	1	2
$f(x)$	1	1	1	-5

in the form

$$\begin{aligned} y &= 1 \cdot \frac{(x-0)(x-1)(x-2)}{(-1-0)(-1-1)(-1-2)} + 1 \cdot \frac{(x+1)(x-1)(x-2)}{(0+1)(0-1)(0-2)} \\ &+ 1 \cdot \frac{(x+1)(x-0)(x-2)}{(1+1)(1-0)(1-2)} - 5 \cdot \frac{(x+1)(x-0)(x-1)}{(2+1)(2-0)(2-1)} \\ y &= -\frac{1}{6}x(x-1)(x-2) + \frac{1}{2}(x+1)(x-1)(x-2) \\ &- \frac{1}{2}(x+1)x(x-2) - \frac{5}{6}(x+1)x(x-1) \\ y &= -x^3 + x + 1 \end{aligned} \tag{9.29}$$

The Lagrangian form of the interpolation formula has the advantage that it does not require differencing of data. It has the disadvantage that, if $f(x)$ is not given analytically, the truncation error of the result afforded by interpolation from a given number of ordinates, or the number of ordinates required to reduce the truncation error below prescribed limits, are difficult to estimate (Hildebrand 1956, pp. 60–64).

9.4 Differentiation

If the 2nd differences in a table were all zero, the function would be an arithmetical series, and its differential coefficient would be exactly

$$h\frac{df}{dx} = hf' = \frac{df}{dp} = \delta_{1/2} \tag{9.30}$$

This is the linear case, so, in general, one would expect f' to be given by a function involving higher differences. The appearance of h in the equation must not be forgotten. The Bessel interpolation coefficients are expressed in terms of p in Eq. (9.4), so it can be differentiated any number of times to give formulas for the successive differential coefficients. For instance, differentiating Eq. (9.4) with the expressions of different $B's$

$$hf'_p = \delta_{1/2} + \frac{1}{2}(2p - 1)(\delta_0^2 + \delta_1^2) + B'_3\,\delta_{1/2}^3 + B'_4\,(\delta_0^4 + \delta_1^4) + \ldots \tag{9.31}$$

The $B's$ are tabulated in references. Similar to interpolation, there are alternative formulas. Usually, the derivatives at tabular, or half-tabular, points are only required. The coefficients can be given definite values. For tabular points we have

$$hf'_0 = \mu\,\delta_0 - \frac{1}{6}\mu\,\delta_0^3 + \frac{1}{30}\mu\,\delta_0^5 - \frac{1}{140}\mu\,\delta_0^7 + \ldots \tag{9.32}$$

and

$$h^2 f''_0 = \delta_0^2 - \frac{1}{12}\delta_0^4 + \frac{1}{90}\delta_0^6 + \ldots \tag{9.33}$$

At half tabular points,

$$hf'_{1/2} = \delta_{1/2} - \frac{1}{24}\delta_{1/2}^3 + \frac{3}{640}\delta_{1/2}^5 - \ldots \tag{9.34}$$

and

$$h^2 f''_{1/2} = \mu\,\delta_{1/2}^2 - \frac{5}{24}\mu\,\delta_{1/2}^4 + \frac{259}{5760}\mu\,\delta_{1/2}^6 - \ldots \tag{9.35}$$

(Danby 1962, pp. 219–220).

9.5 Integration

From the analogy between sums and integrals, or direct integration of Bessel's interpolation formula, we expect a formula like

$$\int^p f_p dp = \delta_{1/2}^{-1} + A'_0(f_0 + f_1) + A'_1\delta_{1/2} + A'_2(\delta_0^2 + \delta_1^2) + \ldots \tag{9.36}$$

for the integral of f_p. Such is the case, and the $A's$ are tabulated in references. There is a similar formula for double integrals. Then

$$\int f(x)dx = h\int f_p dp \tag{9.37}$$

If we are only interested in evaluating integrals at tabular points, where $p = 0$, then

$$\frac{1}{h}\int^{0} f(x)dx = \int^{0} f_p dp = \mu\,\delta_0^{-1} - \frac{1}{12}\mu\,\delta_0 + \frac{11}{720}\mu\,\delta_0^3 - \frac{191}{60480}\mu\,\delta_0^5 + \ldots \quad (9.38)$$

and

$$\frac{1}{h^2}\int\int^{0} f(x)dx^2 = \int\int^{0} f_p dp^2 = \delta_0^{-2} + \frac{1}{12}f_0 - \frac{1}{240}\delta_0^2 + \frac{31}{60480}\delta_0^4 - \ldots \quad (9.39)$$

These expressions will be needed for the numerical solution of differential equations. The notation $\int^p$ causes uncertainty, because a definite value requires two limits. When only the upper limit is specified, the value of an added constant is unknown. In Eq. (9.36), this is the uncertainty in forming first sums. A definite integral can be expressed as

$$\int_q^p fdp = \int_C^p fdp - \int_C^q fdp \quad (9.40)$$

where C is arbitrary. If we omit C and apply Eq. (9.36) to each integral, the result for the definite integral will be independent of the arbitrary constant, when forming the first sums, as the constant disappears in subtraction.

When solving a differential equation, the sums must be known definitely and they can be found from initial conditions. Definite integrals can be determined by using tabular point values of the function to be integrated, instead of using sums and differences. Many formulas are available for this; for example, the repeated *Simpson rule*: Integrate $y = f(x)$ from $x_0 = a$ to $x_n = b$, and let y be tabulated for all x_i, so

$$y_i = f(x_i), \; i = 1, 2, 3, \ldots, n \quad (9.41)$$

Assuming n is even, then

$$\int_a^b ydx = \frac{b-a}{3n}(y_0 + 4y_1 + 2y_2 + 4y_3 + 2y_4 + \ldots + 2y_{n-2} + 4y_{n-1} + y_n) \quad (9.42)$$

Only if the third differences of the y_i are approximately constant should this formula be used (Danby 1962, pp. 220–221).

9.6 Differential Equations

The numerical solution of ordinary differential equations can be performed in many ways. The purpose here is to prepare for some standard methods for calculation of special perturbations. The situation faced, when solving a differential equation numerically, has been described by Herget (1948) as the problem of filling in quantities on an initially blank sheet of paper, where only one quantity (the starting value) may be known. For other values, we have to make tentative entries, and proceed by successive approximation. Once the integration has been started, we proceed step by step to build up the solution, using methods to be described. To start the integration, we could solve the equation

$$h\frac{dx}{dt} = f(x, t) \tag{9.43}$$

in a series that rapidly converges over a limited range of t; the series is used to find values for a numerical solution. A series expansion, however, is not practical in celestial mechanics.

If we seek the perturbations of a cometary orbit caused by the planets, and if they are not too large at the time of the start of the integration, the ephemeris of the unperturbed orbit can be used as the start. This establishes a starting table, from which the integration proceeds. The arbitrary constants in the sums are determined from initial conditions. For Eq. (9.43), Table 9.4 shows the initial values schematically. x_0 is the starting value and x^a are approximate, and to be modified as the solution proceeds.

Here we have assumed that the interval is so chosen that differences beyond the fourth are negligible. Normally, higher differences are required. x_1 can be found from Eq. (9.38) by

$$x_1 = \mu\,\delta_1^{-1} - \frac{1}{12}\mu\,\delta_1 + \frac{11}{720}\mu\,\delta_1^3 \tag{9.44}$$

Here, $\delta^3_{1\,1/2}$ must be extrapolated, but x_1 will not be affected much, unless this extrapolation is very wrong. Suppose, for example, that $\delta^3_{1\,1/2}$ is wrong by 20, then $\mu\,\delta_1^3$ will be wrong by 10, and $\frac{11}{720}\mu\,\delta_1^3$ will be wrong by 0.1, which does not affect x_1. This illustrates the normal strength of convergence of this process.

Table 9.4 Initial and propagated values

t	x	x^a	δ^{-1}	f	δ	δ^2	δ^3	δ^4
0	x_0			f_0		δ_0^2		δ_0^4
			$\delta_{1/2}^{-1}$		$\delta_{1/2}$		$\delta_{1/2}^3$	
1		x_1^a		f_1		δ_1^2		
			$\delta_{1\,1/2}^{-1}$		$\delta_{1\,1/2}$			
2		x_2^a		f_2				
			$\delta_{2\,1/2}^{-1}$					

Table 9.5 Extrapolated values

t	x	x^a	δ^{-1}	f	δ	δ^2	δ^3	δ^4
0	x_0			f_0		δ_0^2		δ_0^4
			$\delta_{1/2}^{-1}$		$\delta_{1/2}$		$\delta_{1/2}^3$	
1	x_1	x_1^a		f_1		δ_1^2		δ_1^{4a}
			$\delta_{1\,1/2}^{-1}$		$\delta_{1\,1/2}$		$\delta_{1\,1/2}^{3a}$	
2		x_2^a		f_2		δ_2^{2a}		
			$\delta_{2\,1/2}^{-1}$		$\delta_{2\,1/2}^a$			
3		x_3^a		f_3^a				
			$\delta_{3\,1/2}^{-1}$					

We can extrapolate forward by assuming $\delta_1^{4a} = \delta_0^4$ and we can build up extra values as shown in Table 9.5. Then we can approximate x_3 from

$$x_3^a = \mu\,\delta_3^{-1} - \frac{1}{12}\mu\,\delta_3 + \frac{11}{720}\mu\,\delta_3^3 \tag{9.45}$$

where more judicious extrapolation would be needed; then a better value of f_3 can be found and the approximate differences improved (we may find it necessary to improve the value of x_1).

We assume here that f will not be too sensitive to small errors in x, so this value may be final. But any, or all, of these steps may have to be revised as the solution is built up. Table 9.5 then is similar to Table 9.4, but one step further on. We are ready to find x_2 and so on. When the x value differs from the x^a to affect f, the table must be revised. This normally happens in the early stages of the integration, so it is normal to iterate on the *starting table*, as it is called. After the starting table, the deviations should not be that great, unless the interval of tabulation is too great. An alternative method to find x_3^a, instead of using Eq. (9.45), is to use an extrapolation formula such as

$$x_3^a = x_1 + 2f_2 + \frac{1}{3}\delta_2^2 - \frac{1}{90}\delta_2^4 \tag{9.46}$$

Now, if we consider a second-order differential equation

$$h^2\frac{d^2x}{dt^2} = f(x,t) \tag{9.47}$$

we tabulate second sums, so the starting table will be of the form of Table 9.6. Now, we have from Eq. (9.39),

$$x_1 = \delta_1^{-2} + \frac{1}{12}f_1 + \frac{1}{240}\delta_1^2 + \frac{31}{60480}\delta_1^4 \tag{9.48}$$

Table 9.6 Starting values with second sums

t	x	x^a	δ^{-2}	δ^{-1}	f	δ	δ^2	δ^3	δ^4
0	x_0		δ_0^{-2}		f_0		δ_0^2		δ_0^4
				$\delta_{1/2}^{-1}$		$\delta_{1/2}$		$\delta_{1/2}^3$	
1		x_1^a	δ_1^{-2}		f_1		δ_1^2		
				$\delta_{1\ 1/2}^{-1}$		$\delta_{1\ 1/2}$			
2		x_2^a	δ_2^{-2}		f_2				
				$\delta_{2\ 1/2}^{-1}$					
3			δ_3^{-2}						

As before, if the 4th differences are small and steady (or higher differences, but the table should extend to even differences), we can develop tentative values for $\delta_1^4, \delta_{1\ 1/2}^3, \delta_2^2, \delta_{2\ 1/2}$, and f_3. These provide tentative values of δ_3^2 and, hence, of x_3^a, using

$$x_3^a = \delta_3^{-2} + \frac{1}{12} f_3 - \frac{1}{240} \delta_3^2 \tag{9.49}$$

x_3^a is used to find f_3 more accurately, and we can proceed to the next step.

In perturbation problems, seen in previous chapters (e.g. Eqs. (7.67)–(7.69)), we have to solve simultaneously three equations of the form

$$\ddot{x} = X(x, y, z, t) \quad \ddot{y} = Y(x, y, z, t) \quad \ddot{z} = Z(x, y, z, t) \tag{9.50}$$

The three tables must proceed together with the same method. Tentative values, x^a, y^a, z^a, are used to correct extrapolated values of X, Y, Z (Danby 1962, pp. 221–224).

9.7 Errors

We will primarily be concerned with unavoidable errors, i.e. we are differentiating between errors and blunders. All effort possible should be made to avoid blunders, and there are some common checks which should be applied. For example,

1. In orbit calculations it is common to use a perifocal frame, as defined in Sect. 5.2. Using the eccentric anomaly and the relations defined in Eqs. (5.47)–(5.50), we have for the position in the perifocal frame

$$\mathbf{r} = a\,\hat{\mathbf{P}}(\cos E - e) + b\,\hat{\mathbf{Q}} \sin E \tag{9.51}$$

where $\hat{\mathbf{P}} = [P_x, P_y, P_z]^T$ and $\hat{\mathbf{Q}} = [Q_x, Q_y, Q_z]^T$. When using these unit vectors, it is advisable to apply the checks that

$$P_x^2 + P_y^2 + P_z^2 = 1 = Q_x^2 + Q_y^2 + Q_z^2$$
$$P_x Q_x + P_y Q_y + P_z Q_z = 0 \tag{9.52}$$

2. When the integration of equations is being performed and the energy integral is known, but not being used, it should be calculated at intervals and used as a check during the course of the integration. This is also a check on the accumulation of accidental errors.
3. When a table is constructed, such as a calculation of an ephemeris, the figures should be differenced. These differences should be carried until they become small. If they do not become small, there is probably a mistake and blunders usually show up clearly as a triangle of unusual values.
4. When laying out the computation of a complex problem, it is advisable to build in check computations at various stages, when this is possible.
5. It is very helpful to have some idea of the answer to be expected. Drawings of orbital positions are quite helpful in this regard. Quadrant errors are quite common and easily picked up by rough sketches.
6. Critical examination of results is helpful; a minor planet on a hyperbolic orbit should be reviewed suspiciously. An artificial satellite with a radius vector shorter than the Earth's equatorial radius is highly questionable. Experience is a big help in this regard.
7. Be careful of the location of decimal points.
8. Be careful of plus and minus signs. Particularly when dealing with latitude and declination where $\pm$ degrees, minutes, and seconds are involved.
9. Hand checks of computer programs are usually well worth the effort, both in clarifying the process before programming and in checking sneaky errors that will crop up.
10. Signs and quadrants from trigonometric functions bear special checking.
11. Choose the trigonometric function that is best behaved in the area to be used. Cotangent near 90° is better than a tangent.
12. Interpolate on a smooth function, not one changing radically. Azimuth and altitude interpolation should be avoided.
13. Limit significant figures to what is justifiable in the result.
14. Remember, few numbers are completely accurate, usually just so many significant figures of an accurate value. It is advisable to carry extra significant figures during a calculation.
15. It is usually best to have a set rule for rounding when the next figure is 5, i.e. how to round 0.20015, or 0.20025 to 4 significant figures. The American rule is usually to make the last figure even, so both are rounded to 0.2002.
16. In solving a differential equation, an early round-off error can have a considerable effect. To see the effect of rounding errors, consider the differences of the sequence $+\frac{1}{2}, -\frac{1}{2}, +\frac{1}{2}, -\frac{1}{2}, +\frac{1}{2}$. This is shown in Table 9.7.

Table 9.7 Effect of rounding errors

	δ	δ^2	δ^3	δ^4	δ^5	δ^6
+1/2		−2		+8		−32
	−1		+4		−16	
−1/2		+2		−8		+32
	+1		−4		+16	
+1/2		−2		+8		−32
	−1		+4		−16	
−1/2		+2		−8		
	+1		−4			
+1/2		−2				
	−1					
−1/2						

Table 9.8 Effect of an error in differencing a sequence

	δ	δ^2	δ^3	δ^4	δ^5	δ^6
0		0		0		+1
	0		0		+1	
0		0		+1		−6
	0		+1		−5	
0		+1		−4		+15
	+1		−3		+10	
1		−2		+6		−20
	−1		+3		−10	
0		+1		−4		+15
	0		−1		+5	
0		0		+1		−6
	0		0		−1	
0		0		0		+1

The magnitude of δ^n can be 2^{n-1}. So if we calculated a table and found that the 6th difference fluctuated between ± 32, this would not signify any avoidable error. A larger fluctuation would signify an avoidable error.

17. Precision versus accuracy. Precision is the number of digits given, accuracy is the number of digits that have meaning.
18. What is the effect of a mistake on the differences? Suppose one tabular value is in error by one, so we have $0, 0, 0, 1, 0, 0, 0$ from differencing the sequence. The effect is shown in Table 9.8.

The numbers occurring in δ^n are the binomial coefficients in the expansion of $(1 - x)^n$. If we difference a table that, aside from errors, should produce negligible 6th differences, the presence and location of the error would be apparent when the 6th differences are written down. The location of the error is opposite the largest value in the even differences (Danby 1962, pp. 219–228).

9.8 Numerical Integration

Let us illustrate the process of *numerical integration* of a hypothetical planet with negligible mass, which is moving from perihelion to aphelion under the attraction of only the Sun. The orbit is an ellipse with a semimajor axis of 2 AU and an eccentricity of 0.2. The plane of the orbit is the reference plane, with the x axis directed toward perihelion. The equations of motion are

$$\frac{d^2x}{dt^2} = -k^2\frac{x}{r^3} \tag{9.53}$$

$$\frac{d^2y}{dt^2} = -k^2\frac{y}{r^3} \tag{9.54}$$

where $r^2 = x^2 + y^2$ and k is the Gaussian constant, 0.01720209895. The units for these equations are based on the value of k, with x, y and r given in astronomical units, and t in ephemeris days. Based on experience, an orbit of this size and shape can be integrated with a tabular interval of 20 days to acquire the desired degree of convergence in differences. For simplicity, we will use a 10 day interval. So, denoting the interval by w, we have $w^2 = 100$ for converting the second summations of $\ddot{x}$ and $\ddot{y}$ into x and y. It is inconvenient to multiply $-x/r^3$ and $-y/r^3$ by the small factor k^2, and then the second summations by 100. It is desirable for the second summations to be the same magnitude as x and y, which can be accomplished by integrating the equations

$$f_x = -w^2k^2\frac{x}{r^3},\ f_y = -w^2k^2\frac{y}{r^3} \tag{9.55}$$

where $w^2k^2 = 0.02959122$ for a 10 day interval. We take the time origin to be the instant of perihelion passage. From the dimensions and orientation of the orbit, two constants of integration are

$$x_0 = r_0 = a(1-e) = 1.6\ \text{AU},\ y_0 = 0.0 \tag{9.56}$$

$w\dot{x}_0$ and $w\dot{y}_0$ are two more constants. At perihelion $\dot{x}_0 = 0$, so $wk\dot{x}_0 = 0$ and $wk\dot{y}_0 = wk\ a^{1/2}r^{-1}(1-e^2)^{1/2} = 0.1489745$, since the velocity is all in the y direction. To start the integration, we calculate seven consecutive values of f_x and f_y. The mean motion is $57.2957795°ka^{-3/2} = 0.348464933°$, degrees per day, from which the values of the mean anomaly are determined. The eccentric anomaly is obtained by Kepler's equation (see Sect. 5.9) and x and y by

$$x = a(\cos u - e),\ y = a(1-e^2)^{1/2}\sin u \tag{9.57}$$

Table 9.9 Values related to x

t	δ^{-2}	δ^{-1}	f_x	δ	δ^2	δ^3	δ^4	δ^5
−30			−109729					
				−3220				
−20			−112949		1245			
				−1975		+63		
−10			−114924		1308		−37	
				−667		+26		−15
0	1.60093638^x		−115591		1334		−52	
		57796^x		+667		−26		+15
10	1.5951842		−114924		1308		−37	
		−172720		+1975		−63		(−7)
20	15779122		−112949		1245		(−44)	
		−285669		+3220		(−107)		+15
30	1.5493453		−109729		(1138)		−29	
		−395398		(4358)		−136		
40	1.5098055		(−105371)		+1002			
		(−500769)		5360				
50	(1.4597286)		100011					
		−600780						
..	..	..	..	..	..	..	..	..
500	−2.3933823		52314		−45		+1	
		−59327		+51		+1		−3
510	−2.3993150		+51365		−44		−2	
		−7962		+7		−1		
520	−2.4001112		+51372		−45			
		+43410		−38				
530	−2.3957702		51334					
		+94744						
540	−2.3862958		+51253					

From x and y, we have r, $1/r^3$, w^2k^2/r^3; f_x and f_y are determined from Eqs. (9.55). From these, the first seven lines of Tables 9.9 and 9.10 are constructed.

The starting values of δ_0^{-2} and $\delta_{1/2}^{-1}$ are obtained by Eqs. (9.38) and (9.39), which are written as

$$\delta_0^{-2} = x_0 - \frac{1}{12}f_0 + \frac{1}{240}\delta_0^2 - \frac{31}{60480}\delta_0^4 \tag{9.58}$$

$$\delta_{1/2}^{-1} = \omega\dot{x}_0 + \frac{1}{2}f_0 + \frac{1}{12}\mu\,\delta_0 - \frac{11}{720}\mu\delta_0^3 + \frac{191}{60480}\mu\delta_0^5 \tag{9.59}$$

Table 9.10 Values related to y

t	δ^{-2}	δ^{-1}	f_y	δ	δ^2	δ^3	δ^4	δ^5
−30			+31331					
				−10093				
−20			+21238		−419			
				−19512		205		
−10			+10726		−214		9	
				−10726		214		9
0	0.0000000^x		0		0		0	
		1488848^x		10726		214		−9
10	0.11488848		−10726		214		−9	
		1478122		−10512		205		(−15)
20	0.2966970		−21238		419		(−24)	
		1456884		−10093		(+181)		−3
30	0.4423854		−31331		(+600)		−27	
		1425553		(−9493)		+154		
40	0.5849407		(−40824)		+754			
		(1384729)		−8739				
50	(0.7234136)		−49563					
		1335166						
..	..	..	..	..	..	..	..	..
500	+0.1642454		−3521		−7		−4	
		−991836		+2128		+5		−4
510	+0.0650618		−1393		−2		−1	
		−993229		+2126		+4		
520	−0.0342611		+733		+2			
		−992496		+2128				
530	−0.1335107		+2861					
		−989635						
540	−0.2324742		+4993					

using $\mu\delta_0 = \frac{1}{2}(\delta_{-1/2} + \delta_{1/2})$, and having two similar equations for y and $\dot{y}$. So the two values marked with superscript x's in Tables 9.9 and 9.10 may be inserted. Now, the table can be extended down and to the left through $t = 40$ using

$$\delta^{-1}_{n+1/2} - \delta^{-1}_{n-1/2} = f_n \qquad (9.60)$$

$$\delta^{-2}_{n} - \delta^{-2}_{n-1} = \delta^{-1}_{n-1/2} \qquad (9.61)$$

and rearrangements of these equations. So the table can be extended by forming differences.

Next, we calculate f_x and f_y for $t = 40$, which requires the values of x and y. Provisional values, as indicated by parentheses, may be obtained by means of Eq. (9.39),

$$\frac{1}{h^2}\int\int^0 f(x)dx^2 = \int\int^0 f_p dp^2 = \delta_0^{-2} + \frac{1}{12}f_0 - \frac{1}{240}\delta_0^2 + \frac{31}{60480}\delta_0^4 \quad (9.62)$$

if the necessary differences were extrapolated.

The values of x and y, thus obtained, are generally sufficiently accurate, but not the differences written in by inspection. Instead of using an equation such as (9.39), it is better to use a formula that will give x and y in terms of the diagonal differences already known. Newton formulas of diagonal differences are an example;

$$\begin{aligned} f_n = f_0 + n\delta_{1/2} + \frac{n(n-1)}{2!}\delta_1^2 + \frac{n(n-1)(n-2)}{3!}\delta_{3/2}^3 \\ + \frac{n(n-1)(n-2)(n-3)}{4!}\delta_2^4 + \ldots \\ + \frac{n(n-1)(n-2)\ldots(n-m+1)}{m!}\delta_{m/2}^m \end{aligned} \quad (9.63)$$

which is the forward difference formula and

$$\begin{aligned} f_n = f_0 + n\delta_{-1/2} + \frac{n(n+1)}{2!}\delta_{-1}^2 + \frac{n(n+1)(n+2)}{3!}\delta_{-3/2}^3 \\ + \frac{n(n+1)(n+2)(n+3)}{4!}\delta_{-2}^4 + \ldots \\ + \frac{n(n+1)(n+2)\ldots(n+m-1)}{m!}\delta_{-m/2}^m \end{aligned} \quad (9.64)$$

which is the backward difference formula.

Integrating Newton's backward interpolation twice, and manipulating a little, we have with sufficient accuracy for most applications

$$\begin{aligned} x_0 = \delta_0^{-2} + \frac{1}{12}f_{-1} + \frac{1}{12}\delta_{-3/2} + 0.0791667\,\delta_{-2}^2 + 0.075\,\delta_{-5/2}^3 \\ + 0.07135\delta_{-3}^4 + 0.0682\delta_{-7/2}^5 + 0.065\delta_{-4}^6 + 0.06\delta_{-9/2}^7 \end{aligned} \quad (9.65)$$

The forward formula for backward integration can be obtained by changing the signs of all the subscripts and of the coefficients of the odd-ordered terms. With Eq. (9.65) a third table can be started with values of x, y, r^2 and $1/r^3$. The differences in the integration tables become a check on the work; care must be taken to obtain accurate differences. Due to the accumulation of rounding errors, the last decimal of x and y is uncertain after a few integration steps. With f_x and f_y calculated by Eq. (9.55) after determining x and y from Eq. (9.65), the diagonal differences above and to the right may be entered. Before filling in the sums, the provisional values of x and y should be tested by the more accurate formula

$$x_0 = \delta_0^{-2} + \frac{1}{12} f_0 - \frac{1}{240}\,\delta_{-1}^2 - \frac{1}{240}\delta_{-3/2}^3 - 0.00365\,\delta_{-2}^4 \\ - 0.0031\delta_{-5/2}^5 - 0.003\delta_{-3}^6 \tag{9.66}$$

If new values differ from the provisional values by half a unit or more, they need to be corrected, and f_x and f_y values examined to determine if they need correction, or not. If they do, corrected values of f_x and f_y must be entered in the integration table, the differences corrected, and Eq. (9.66) reapplied. Generally, the tabular interval should be chosen, so the provisional values of x and y might require correction by a few units in the last decimal, and small enough that the provisional values of f_x and f_y seldom need correction. The general rule is a step size about 1/100 of the period of the orbit. When the final values of f_x and f_y are obtained, the tables may be extended to the left giving δ_x^{-2} and δ_y^{-2} for $t = 50$, and the first integration step is completed. Now, the values in parenthesis can be entered. The second step is like the first. The rules for each step are the following:

1. Having completed a diagonal of the integration tables, determine provisional values of x and y by Eq. (9.65).
2. Calculate f_x and f_y from the differential equations (9.55), enter them in the tables, and fill in the differences above and to the right.
3. Test x and y by Eq. (9.66); if necessary, correct them, and test f_x and f_y. Then, complete the diagonal of the tables.

In this example, the tabular interval is so small that Step 3 is hardly necessary. The calculation of a single step is simple, but more steps are required, there is no savings in effort, and the accumulation of end-figure errors is greater than necessary. The table can be given after progressing as shown in subsequent values (Brouwer and Clemence 1961, pp. 146, 153–157).

9.9 Numerical Integration by Runge-Kutta Methods

Just as we can interpolate using a difference table, or directly from function values, similarly, numerical integration can be performed directly from functional values without the difference table. There are a wide variety of methods available; we will just give a few examples. *Runge-Kutta* methods are single-step methods, so each step is like starting an initial value problem, independent of what went before. Consider the equation

$$\frac{dx}{dt} = f(t, x), \;\; x(t_0) = x_0 \tag{9.67}$$

The Runge-Kutta method has for a stepsize h, the operations

$$f_0 = f(t_0, x_0) \tag{9.68}$$

$$f_1 = f(t_0 + \frac{1}{2}h, x_0 + \frac{1}{2}hf_0) \tag{9.69}$$

$$f_2 = f(t_0 + \frac{1}{2}h, x_0 + \frac{1}{2}hf_1) \tag{9.70}$$

$$f_3 = f(t_0 + \frac{1}{2}h, x_0 + \frac{1}{2}hf_2) \tag{9.71}$$

$$x(t_0 + h) \simeq x_1 = x_0 + \frac{1}{6}(f_0 + 2f_1 + 2f_2 + f_3) \tag{9.72}$$

This is a fourth-order method. Then, for h small enough, the error of approximation can be estimated by Ah^5, with A independent of h, or

$$x(t_0 + h) - x_1 \simeq Ah^5 \tag{9.73}$$

This algorithm is quick and easy to program, but for most orbital calculations it is neither accurate nor fast enough to be satisfactory. One modification of this method allows for automatic control of the stepsize from one step to the next, based on an error bound set by the operation. This is one of the methods by Fehlberg (1968). The method described will give fifth-order accuracy. The method runs two algorithms in tandem; at the end of a step there are two separate approximations for x; x_1 has fourth-order and $\hat{x}_1$ has fifth-order accuracy,

$$x(t_0 + h) \simeq x_1 + Ah^5 \tag{9.74}$$

$$x(t_0 + h) \approx \hat{x}_1 + Bh^6 \tag{9.75}$$

The difference $(\hat{x}_1 - x_1)$ is an approximation of the local truncation error of the approximation x_1. The stepsize is calculated, or adjusted, so this difference will be less than a set amount. If the error of x_1 is less than the amount, so will the error of $\hat{x}_1$. $\hat{x}_1$ is used as the starting value of the next step. This procedure is most valuable in cases of close approaches, or where the optimum step size changes. Let

$$TE = |\hat{x}_1 - x_1| \tag{9.76}$$

Suppose we have just calculated a step in the integration. If TE is too large, or if $TE > TL$, where TL is the greatest tolerable local truncation error, then the step must be repeated with a smaller step size. If $TE < TL$, then the step is good and we need the stepsize for the next step. When a step is calculated, a new value of h must be found in each case. Approximately,

$$TE \simeq Ah^5 \rightarrow A \simeq \frac{TE}{h^5} \tag{9.77}$$

If the local truncation error resulting from a new stepsize, h_1, were to equal *TL*, then

$$Ah_1^5 \simeq TL \tag{9.78}$$

So from Eq. (9.78)

$$h_1 \simeq \left(\frac{TL}{A}\right)^{1/5} \simeq h\left(\frac{TL}{TE}\right)^{1/5} \tag{9.79}$$

It is not correct to replace $\simeq$ by $=$ in a formula for h_1, due to all the approximations made. So, we can set

$$h_1 = CH \cdot h\left(\frac{TL}{TE}\right)^{1/5} \tag{9.80}$$

where *CH* is the "chicken factor", some number less than 1. A value of $CH = 0.9$ can usually ensure that *TE* will be less than *TL*. The basic formulae from Fehlberg (1968) follow,

$$\frac{dx}{dt} = f(t,x), \;\; x(t_0) = x_0 \tag{9.81}$$

and

$$f_1 = f(t_0, x_0) \tag{9.82}$$

$$f_2 = f(t_0 + a_2 h, x_0 + h(b_{21}, f_1)) \tag{9.83}$$

$$f_3 = f(t_0 + a_3 h, x_0 + h(b_{31} f_1 + b_{32} f_2)) \tag{9.84}$$

$$f_4 = f(t_0 + a_4 h, x_0 + h(b_{41} + b_{42} f_2 + b_{43} f_3)) \tag{9.85}$$

$$f_5 = f(t_0 + a_5 h, x_0 + h(b_{51} f_1 + b_{52} f_2 + b_{53} f_3 + b_{54} f_4)) \tag{9.86}$$

$$f_6 = f(t_0 + a_6 h, x_0 + h(b_{61} f_1 + b_{62} f_2 + b_{63} f_3 + b_{64} f_4 + b_{65} f_5)) \tag{9.87}$$

$$x_1 = x_0 + h(c_1 f_1 + c_2 f_2 + c_3 f_3 + c_4 f_4 + c_5 f_5) \tag{9.88}$$

$$\hat{x}_1 = x_0 + h(\hat{c}_1 f_1 + \hat{c}_2 f_2 + \hat{c}_3 f_3 + \hat{c}_4 f_4 + \hat{c}_5 f_5 + \hat{c}_6 f_6) \tag{9.89}$$

The constants are given in the Table 9.11. Then,

$$\begin{aligned} TE &= \hat{x}_1 - x_1 \\ &= h\left(-\frac{1}{150} f_1 + \frac{3}{100} f_3 - \frac{16}{75} f_4 - \frac{1}{20} f_5 + \frac{6}{25} f_6\right) \end{aligned} \tag{9.90}$$

Table 9.11 Constants for Runge-Kutta-Fehlberg integration

k	a_k	$l = 1$	$l = 2$	$l = 3$	$l = 4$	$l = 5$	c_k	$\hat{c}_k$
1	0						1/9	47/450
2	2/9	2/9					0	0
3	1/3	1/12	1/4				9/20	12/25
4	3/4	69/128	−243/128	135/64			16/45	32/225
5	1	−17/12	27/4	−27/5	16/15		1/12	1/30
6	5/6	65/432	−5/16	13/16	4/27	5/144		6/25

This method can be generalized to apply to a system of equations. In that case, the greatest absolute value for a *TE* must be used to determine the step size (Danby 1988, pp. 296–298).

9.10 Accumulation of Errors in Numerical Integration

The error accumulated in an integration cannot be determined in general, but the probable error to be expected after any number of steps can be estimated by the theory of errors. The theory of elliptic motion permits the determination of the error for the example in Sect. 9.8. The real coordinates and velocities at aphelion are $x = -2.4$ AU, $y = 0, \dot{x} = 0, \dot{y} = -0.5773503$ AU/day. The time of aphelion passage is $t = 516.551259$ days. Interpolating the table for this time, we have $x = 2.3999888, wk\dot{x} = +0.0000007, \dot{x} = +0.0000041, y = -0.0000047, wk\dot{y} = -0.0993166$, and $\dot{y} = -0.5773516$, where w, k appear in Eq. (9.55). The error in x and y is greater than the error in $\dot{x}$ and $\dot{y}$.

The theory of the accumulation of errors in numerical integration indicates that after n steps, the probable error of a double integral is $0.1124\ n^{3/2}$ in units of the last decimal. This means in many examples, that half will have errors greater than this and half smaller. In this case $n = 52$, so the probable error is 42, and both cases have given larger errors. The above error expression permits one to decide the number of decimals required before starting an integration, when the length and required accuracy is known. If an asteroid is to be integrated for 10 years at a 10 day interval, that is 365 steps, or 183 each way, when the integration is started at the middle of the arc. The probable error after 183 steps is about 280. If the largest error to be tolerated is 0.1 s of arc, or 0.0000005 radians, then nine decimals are necessary in the calculation.

The theory of the accumulation of errors is applicable to the errors of osculating orbital elements (see Sect. 5.2.1) of an orbit determined by numerical integration. While the mean error of the mean orbital longitude is proportional the number of steps to the three halves power, the mean errors of the other five elements are proportional to the square root of the number of steps. This is true if the elements are determined directly by integration, or obtained from transforming rectangular coordinates and velocities. The perihelion can have the same size error as the orbital longitude (Brouwer and Clemence 1961, pp. 158–159).

9.11 Numerical Integration of Orbits

Numerical integration is undoubtedly the best method for calculating the motion of any body in the solar system for a few revolutions about its primary, with full precision of observations. Analytic methods of general theories were preferred until the availability of computers made numerical integration accessible and more accurate. A general theory amounts to a large number of terms of a Fourier series, or a Chebyshev series, which can be evaluated and summed for any particular time to give the position of the body at that time. Both general theories and numerical integration can be performed on computers with the integration being easier to understand, program, and perform.

We will discuss the two most common methods of numerical integration of orbits, *Cowell's method* and *Encke's method*. These methods are popular due to the availability of computers and software to implement them.

Cowell's method does not use the conic section as a first approximation of the orbit explicitly. The equations of motion in rectangular coordinates are integrated giving rectangular coordinates of the disturbed body. The process is like the example in Sect. 9.8, except:

1. Three coordinates are necessary instead of two.
2. The attractions of the disturbing planets are added to those of the Sun.

Usually, the origin is at the primary, but this restriction is not necessary. The center of mass of the system or any disturbing mass can be used. The motions of all bodies exerting appreciable effects can be taken from previous computations, or integrated simultaneously. Since the conic section is not used in the method, the system need not be dominated by a single mass. So, the method could be applied to the satellite of a binary star system, or to an artificial satellite going from one planet to another in the solar system. The disadvantage of the method is that the integrals contain many significant figures and they change rapidly with time. So, the integration tables are slowly convergent, which compels the use of a small tabular interval.

In Encke's method, the coordinates are not determined directly, but the integration calculates the difference between the actual coordinates and an osculating orbit coordinates, which is the position the body would have had, if it had continued to move in the conic section corresponding to the orbital parameters at a particular instant, called the *epoch of osculation*. The differences from the osculating orbit are perturbations, and are zero at the epoch of osculation. The advantage of the method is that the perturbations are small for times near the epoch; they only require a few significant figures, which permits a larger tabular interval than Cowell's method. The disadvantage is that the perturbations increase to a large size with time, which requires *orbit rectification*. The rectification requires the coordinates and velocities to be determined at a new epoch, and the integration restarted. This rectification introduces an error. The difficulty could be avoided, if the choice of the first approximation conic section more closely matched the true motion over a longer period of time, than the osculating orbit does. Generally, there is no information to determine such a first approximation. Another method of integration is the *variation of parameters*, the equations of which will be discussed in Chap. 10. In this method, the quantities obtained by the integration are the six osculating elements. They change comparatively slowly, which means that a larger tabular interval can be used. The disadvantages are that the differential equations are more complicated in form than the equations in rectangular coordinates, and it is necessary at each step to determine the rectangular coordinates to evaluate the perturbations. There are other more complicated and more elegant methods, which have not seen much use due to the disadvantages of complications.

9.11.1 Equations for Cowell's Method

The equations of motion of two point-masses, m_a and m_b, under their mutual action for the ξ component are

$$m_a\ddot{\xi}_a = k^2 m_a m_b \frac{\xi_b - \xi_a}{r^3} \tag{9.91}$$

$$m_b\ddot{\xi}_b = k^2 m_a m_b \frac{\xi_a - \xi_b}{r^3} \tag{9.92}$$

with similar equations for η and ζ, and $r^2 = (\xi_a - \xi_b)^2 + (\eta_a - \eta_b)^2 + (\zeta_a - \zeta_b)^2$. If we introduce additional point masses $m_1, m_2, m_3, \ldots$ and denote any of these by m_j, the attractions of m_j on m_a and m_b are given by summing all these attractions,

$$m_a\ddot{\xi}_a = k^2 m_a m_b \frac{\xi_b - \xi_a}{r^3} + \sum_j k^2 m_a m_j \frac{\xi_j - \xi_a}{\rho_{ja}^3} \tag{9.93}$$

$$m_b\ddot{\xi}_b = k^2 m_a m_b \frac{\xi_a - \xi_b}{r^3} + \sum_j k^2 m_b m_j \frac{\xi_j - \xi_b}{\rho_{jb}^3} \tag{9.94}$$

and similar equations for η and ζ, where

$$\rho_{ja}^2 = (\xi_a - \xi_j)^2 + (\eta_a - \eta_j)^2 + (\zeta_a - \zeta_j)^2 \tag{9.95}$$

$$\rho_{jb}^2 = (\xi_b - \xi_j)^2 + (\eta_b - \eta_j)^2 + (\zeta_b - \zeta_j)^2 \tag{9.96}$$

We take the origin of the coordinates at m_a, which is a linear transformation

$$\xi_b - \xi_a = x,\ \xi_j - \xi_a = x_j \tag{9.97}$$

$$\eta_b - \eta_a = y,\ \eta_j - \eta_a = y_j \tag{9.98}$$

$$\zeta_b - \zeta_a = z,\ \zeta_j - \zeta_a = z_j \tag{9.99}$$

and let

$$r_j^2 = x_j^2 + y_j^2 + z_j^2 \tag{9.100}$$

$$\rho_j^2 = (x_j - x)^2 + (y_j - y)^2 + (z_j - z)^2 \tag{9.101}$$

Then divide Eq. (9.93) by m_a and Eq. (9.94) by m_b, and subtract the first from the second. Then the equation of motion of m_b relative to m_a is,

$$\ddot{x} = -k^2(m_a + m_b)\frac{x}{r^3} - \sum_j k^2 m_j \frac{x_j}{r_j^3} + \sum_j k^2 m_j \frac{x_j - x}{\rho_j^3} \tag{9.102}$$

with similar equations for y and z.

Assume that m_a is the unit of mass, and all the other masses are measured in this unit. Then put $m_a = 1$, and drop the subscript from m_b; Eq. (9.102) is now

$$\ddot{x} = -k^2(1 + m)\frac{x}{r^3} - \sum_j k^2 m_j \left(\frac{x_j - x}{\rho_j^3} - \frac{x_j}{r_j^3}\right) \tag{9.103}$$

This, with similar equations for y and z, are Cowell's method's fundamental equations. If m_a represents the Sun, then x, y, z are the heliocentric coordinates of the body to be integrated, and x_j, y_j, z_j are the heliocentric coordinates of the bodies affecting m. From the derivation of the equations, it should be evident that in Eq. (9.103):

(a) The first term represents the action of the Sun on m;
(b) the first term in the parenthesis represents the action of m_j on m (the direct term);
(c) and the second term in the parenthesis represents the action of m_j on the Sun (the indirect term).

The equations can be used for satellite motion, taking the origin at the primary and one m_j for the Sun. For an asteroid or comet, m is equal to zero, the m_j are small compared to unity, and the x_j, y_j, z_j can be regarded as known. The solution may be obtained by successive approximations. At each step of the integration, approximate coordinates are obtained by extrapolation and used to calculate $w^2\ddot{x}, w^2\ddot{y}, w^2\ddot{z}$, as described earlier, except that the disturbing planets' portions must be calculated separately for each planet, and added to the portion contributed by the Sun.

The body will not depart much from the elliptical orbit at the osculation epoch over a few revolutions, and unless the body comes close to a major planet. In this case, the planetary attractions may be calculated in advance, using the position of the body in the osculating orbit instead of in its actual orbit. When the integration extends over many revolutions, such an approximation is not satisfactory (Brouwer and Clemence 1961, pp. 169–170).

9.11.2 Equations for Encke's Method

Take the heliocentric rectangular coordinates, x_0, y_0, z_0, of a point-mass, m, moving under the attraction of the Sun alone. The equations of the orbit are

$$\ddot{x}_0 = -k^2(1+m)\frac{x_0}{r_0^3} \tag{9.104}$$

$$\ddot{y}_0 = -k^2(1+m)\frac{y_0}{r_0^3} \tag{9.105}$$

$$\ddot{z}_0 = -k^2(1+m)\frac{z_0}{r_0^3} \tag{9.106}$$

where $r_0^2 = x_0^2 + y_0^2 + z_0^2$. Take ξ, η, ζ as increments of x_0, y_0, z_0, caused by the attraction of the planets. The actual coordinates, x, y, z, of m at any time are

$$x = x_0 + \xi, \; y = y_0 + \eta, \; z = z_0 + \zeta \tag{9.107}$$

and the actual equations of motion are as in Eq. (9.103), with similar equations for y and z. Subtracting Eq. (9.104) from Eq. (9.103) gives

$$\ddot{x} - \ddot{x}_0 = \ddot{\xi} = k^2(1+m)\left(\frac{x_0}{r_0^3} - \frac{x}{r^3}\right) + \sum_j k^2 m_j \left(\frac{x_j - x}{\rho_j^3} - \frac{x_j}{r_j^3}\right) \tag{9.108}$$

with similar equations for $\ddot{\eta}$ and $\ddot{\zeta}$.

The perturbations ξ, η, ζ could be calculated by direct integration of Eq. (9.108). From the laws of elliptic motion, x_0/r_0^3 could be calculated for every integration step, and x/r^3 determined at each step by extrapolating ξ and adding it to x_0 to get x, and so on. This procedure is not convenient in practice. ξ is a small quantity, x_0/r_0^3 is almost equal to x/r^3, so these terms would have to be calculated to many more significant figures than their differences require. Encke sought a transformation to avoid this difficulty. Taking the equation for $\ddot{\xi}$, knowing that $\ddot{\eta}$ and $\ddot{\zeta}$ are similar, we have

$$\frac{x_0}{r_0^3} - \frac{x}{r^3} = \frac{1}{r_0^3}\left(x_0 - \frac{r_0^3}{r^3}x\right) = \frac{1}{r_0^3}\left(x - \xi - \frac{r_0^3}{r^3}x\right)$$
$$= \frac{1}{r_0^3}\left[\left(1 - \frac{r_0^3}{r^3}\right)x - \xi\right] \tag{9.109}$$

But

$$r^2 = x^2 + y^2 + z^2$$
$$= (x_0 + \xi)^2 + (y_0 + \eta)^2 + (z_0 + \zeta)^2$$
$$= r_0^2 + 2x_0\,\xi + 2y_0\,\eta + 2z_0\,\zeta + \xi^2 + \eta^2 + \zeta^2 \tag{9.110}$$

and

$$\frac{r^2}{r_0^2} = 1 + 2\frac{(x_0 + \frac{1}{2}\xi)\xi + (y_0 + \frac{1}{2}\eta)\eta + (z_0 + \frac{1}{2}\zeta)\zeta}{r_0^2} \tag{9.111}$$

Put

$$q = \frac{(x_0 + \frac{1}{2}\xi)\xi + (y_0 + \frac{1}{2}\eta)\eta + (z_0 + \frac{1}{2}\zeta)\zeta}{r_0^2} \tag{9.112}$$

Then

$$\frac{r^2}{r_0^2} = 1 + 2q \tag{9.113}$$

$$\frac{r_0^3}{r^3} = (1 + 2q)^{-3/2} \tag{9.114}$$

$$1 - \frac{r_0^3}{r^3} = 1 - (1 + 2q)^{-3/2} \tag{9.115}$$

Assume ξ, η, ζ are so small compared with x_0, y_0, z_0 that their squares may be neglected, then

$$q = \frac{x_0\xi + y_0\eta + z_0\zeta}{r_0^2} \tag{9.116}$$

which is easily calculated, while $1 - (1 + 2q)^{-3/2}$ is not. Tables for this function could be calculated, but they would be extensive and inconvenient to interpolate. If

q is small compared with unity, then an approximate value of $1 - (1 + 2q)^{-3/2}$ can be determined from the first few terms of the binomial expansion. Approximately,

$$1 - (1 + 2q)^{-3/2} \approx 3q - \frac{15}{2}q^2 \tag{9.117}$$

Define a function, f (not to be confused with the previous uses of f) as

$$f = \frac{1 - (1 + 2q)^{-3/2}}{q} \tag{9.118}$$

When q is small, we can see from Eqs. (9.117) and (9.118) that f will be close to 3, and, since f changes much less rapidly than q, a table of f as a function of q can be interpolated. Through multiplication by w^2, Eq. (9.108) becomes

$$w^2\ddot{\xi} = w^2k^2(1 + m)\frac{1}{r_0^3}(fqx - \xi) + \sum_j w^2k^2m_j\left(\frac{x_j - x}{\rho_j^3} - \frac{x_j}{r_j^3}\right) \tag{9.119}$$

$$w^2\ddot{\eta} = w^2k^2(1 + m)\frac{1}{r_0^3}(fqy - \eta) + \sum_j w^2k^2m_j\left(\frac{y_j - y}{\rho_j^3} - \frac{y_j}{r_j^3}\right) \tag{9.120}$$

$$w^2\ddot{\zeta} = w^2k^2(1 + m)\frac{1}{r_0^3}(fqz - \zeta) + \sum_j w^2k^2m_j\left(\frac{z_j - z}{\rho_j^3} - \frac{z_j}{r_j^3}\right) \tag{9.121}$$

These are Encke's method equations to be integrated. The solution requires six constants of integration, selected so that the coordinates and velocity components in the undisturbed orbit, $(x_0, y_0, z_0, \dot{x}_0, \dot{y}_0, \dot{z}_0)$, are the same as those in the actual orbit at a selected time; so $\xi, \eta, \zeta, \dot{\xi}, \dot{\eta}, \dot{\zeta}$ are all zero at that time. The selected time is the epoch of osculation, as mentioned before. Equations (9.119)–(9.121) are rigorous, if q is calculated by Eq. (9.112). In practice, Eq. (9.116) will be used for starting the integration. The perturbations, ξ, η, ζ, will gradually increase in size, and when their squares become significant, Eq. (9.112) must be used.

Since x, y, z differ from x_0, y_0, z_0 by the small quantities ξ, η, ζ, and since m_j is small, we may substitute x_0, y_0, z_0 for x, y, z in Eqs. (9.119)–(9.121). The resulting errors will be of the order of m_j^2. This substitution permits the attractions to be calculated in advance for a large number of steps. These perturbations are said to be accurate to the first order of the disturbing forces. However, if the integration covers a long time interval, these perturbations will become so large that the error committed becomes appreciable, and then the rigorous formulae must be used. This may be avoided by orbit rectification, mentioned earlier. For rectification at a new time, the coordinates of the body of interest are obtained by calculating accurate values of x_0, y_0, z_0 and adding them to ξ, η, ζ; the velocity components are obtained by adding $w\dot{x}_0, w\dot{y}_0, w\dot{z}_0$ to values of $w\dot{\xi}, w\dot{\eta}, w\dot{\zeta}$, which are derived from ξ, η, ζ by numerical differentiation. The values of a, e, n and the vectorial constants of the ellipse, corresponding to the new coordinates and velocities at the epoch of osculation, are calculated, and the integration is restarted at the new epoch of osculation. Usually, it is preferable to rectify the orbit rather than use the rigorous formulae for the planetary attractions (Brouwer and Clemence 1961, pp. 176–178).

9.11.3 Comparison of Cowell's and Encke's Methods

1. Neither is clearly superior to the other.
2. Encke's method permits larger tabular intervals, but each step requires more calculations than Cowell's method.
3. For comets, it is recommended to use Encke's method, when the comet is near the Sun, and Cowell's method, when it is far from the Sun.
4. When close approaches occur, Encke's perturbations increase rapidly in size and it loses all advantages.
5. It is easy to switch from one to the other by calculating coordinates and velocities at a new epoch of osculation.
6. Cowell's method is certainly simpler, since no judgements are required; such as the formula to be used for q, and when to rectify to a new epoch of osculation. These problems are especially true when using a computer, where the decisions must be programmed (Brouwer and Clemence 1961, pp. 185–186).

9.12 Equations with Origin at the Center of Mass

The central force equations of motion may be extended to any number of disturbing point masses simply by

$$\ddot{x} = \sum_j k^2 m_j \frac{x_j - x}{\rho_j^3} \tag{9.122}$$

$$\ddot{y} = \sum_j k^2 m_j \frac{y_j - y}{\rho_j^3} \tag{9.123}$$

$$\ddot{z} = \sum_j k^2 m_j \frac{z_j - z}{\rho_j^3} \tag{9.124}$$

where $\rho_j^2 = (x_j - x)^2 + (y_j - y)^2 + (z_j - z)^2$, in which the origin of coordinates is the whole system's center of mass. These equations are simpler than those of heliocentric motion, because they avoid terms for the action of the disturbing planets on the Sun. The coordinates of the Sun and disturbing planets relative to the center of mass of the solar system are now available, because they are needed for millisecond pulsar observations reductions.

There are cases where the use of the center of mass is an advantage. In fact, the *Heidelberg n-body program* integrates with respect to the center of mass, and separately integrates the position of the Sun. When the Sun is used as the origin and a body is very distant from the Sun, the indirect term x_j/r_j^3 may become greater than the direct term $(x_j - x)/\rho_j^3$. For example, the action of Jupiter on Pluto, the ratio of these two terms averages to about $40^2/5^2$ or 64. So if the center of mass is used as the origin, a longer tabular interval can be used than when the Sun is taken as the origin. Reduction of observational data from millisecond pulsars requires reduction to the center of mass of the solar system.

Separating the solar term from the others and denoting the barycentric coordinates of the Sun by x_s, y_s, z_s, Eqs. (9.122)–(9.124) for $\ddot{x}$ become

$$\ddot{x} = k^2 \frac{x_s - x}{\rho_s^3} + \sum_j k^2 m_j \frac{x_j - x}{\rho_j^3} \tag{9.125}$$

and similar equations for $\ddot{y}$ and $\ddot{z}$.

These are Cowell's method equations referred to the barycenter. These require the barycentric coordinates,$x_s, y_s, z_s, x_j, y_j, z_j$, rather than heliocentric coordinates. For now, we denote the heliocentric coordinates by x_{jh}, y_{jh}, z_{jh}, so

$$x_j = \frac{1}{1 + m_j} x_{jh}, \; x_s = -\sum_j \frac{m_j}{1 + m_j} x_{jh} \tag{9.126}$$

with similar equations for the other coordinates. If the coordinates and velocity components at the epoch of osculation have been obtained from heliocentric elements, the following transformations are necessary before calculating the starting values for an integration. The coordinates are

$$x = \frac{1}{1 + m} x_h - \sum_j \frac{m_j}{1 + m} x_{jh} \tag{9.127}$$

with similar equations for y and z. The barycentric velocity components can be obtained from

$$\dot{x} = \frac{1}{1 + m} \dot{x}_h - \frac{d}{dt} \sum_j \frac{m_j}{1 + m_j} x_{jh} \tag{9.128}$$

with similar equations for $\dot{y}$ and $\dot{z}$. The second part of Eq. (9.128) may be calculated by numerical differentiation or the velocity may be known. Upon completion of the integration, the coordinates may be referred to the Sun by

$$\frac{1}{1 + m} x_h = x + \sum_j \frac{m_j}{1 + m_j} x_{jh} \tag{9.129}$$

with similar equations for y_h and z_h. For Encke's method the equations referred to the barycenter are

$$\ddot{\xi} = k^2 \frac{1}{r_0^3} [fq(x - x_s) - \xi] + \sum_j k^2 m_j \frac{x_j - x}{\rho_j^3} \tag{9.130}$$

and similar equations for $\ddot{\eta}$ and $\ddot{\zeta}$, where q and r_0^3 are determined from

$$x_0 = \frac{1}{1+m} x_{0h} - \sum_j \frac{m_j}{1+m_j} x_{jh} \tag{9.131}$$

and x_j and x_s derived from Eq. (9.126).

9.13 Integration with Augmented Mass of the Sun

When the disturbed body is very far from the Sun, then one or more of the ρ_j is nearly equal to ρ_s. Then, in Eq. (9.128), we can denote such a ρ_j by ρ_p, and the corresponding disturbing mass by m_p. If m_p is sufficiently small, then

$$k^2 \frac{x_s - x}{\rho_s^3} + k^2 m_p \frac{x_p - x}{\rho_p^3} = k^2 (1 + m_p) \frac{x_s - x}{\rho_p^3} \tag{9.132}$$

to the same number of significant figures as are needed in the attractions. This can be the case with the action of Mercury on an asteroid, or the four inner planets on Pluto or Kuiper Belt objects. Instead of Eq. (9.128), we write

$$\ddot{x} = k^2 (1 + m_p) \frac{x_s - x}{\rho_s^3} + \sum_j k^2 m_j \frac{x_j - x}{\rho_j^3} \tag{9.133}$$

This is equivalent to assuming that the body moves in an elliptic orbit about the center of mass of the Sun and m_p. Rigorously, the coordinates of m_p should be included in Eqs. (9.126) and (9.127), but this is normally not done; the planet's mass is just added to the Sun's mass.

The center of mass of Mercury and the Sun is about 0.000000007 AU from the center of the Sun, and for most asteroid orbits such a small correction to the disturbing body coordinates may be neglected. When an augmented mass of the Sun is used for integration, that same augmented mass must be used in calculating the integration starting values. The method of doing this depends on the method of integration. For Cowell's method and Encke's method the constants of integration come from the coordinates and velocities at the epoch of osculation; these are not the same for the augmented mass as for the un-augmented mass. The only change involves the relation between the mean motion, n, and the semimajor axis in applying Kepler's third law. The term $k^2(1 + m + m_p)$ should be substituted for $k^2(1 + m)$. It is not normal to give the value of a or n based on the augmented mass. So when the values are given, unless there is a specific note to the contrary, the elements can be assumed referred to the Sun only.

References

Brouwer, D., Clemence, G.M.: Methods of Celestial Mechanics. Academic Press, New York (1961)
Danby, J.: Fundamentals of Celestial Mechanics. The Macmillan Company, New York (1962)
Danby, J.: Fundamentals of Celestial Mechanics, 2nd edn. Willmann-Bell, Richmond (1988)
Fehlberg, E.: Classical fifth-, sixth-, seventh- and eighth-order Runge-Kutta formulas with stepsize control. Tech. Rep. 287, NASA, Huntsville, Alabama (1968)
Herget, P.: The Computation of Orbits. University of Cincinnati, Cincinnati (1948)
Hildebrand, F.B.: Introduction to Numerical Analysis. McGraw-Hill, New York (1956)
Nautical Almanac Office: Interpolation and Allied Tables. H.M. Stationery Office, London (1956)

Chapter 10
Canonical Equations

10.1 Introduction

Now we want to look at a completely different approach to solving the equations of motion, and a different type of variables. We must first establish our notation and the equations of motion which underlie this approach. We again assume a relative coordinate system with one body at the origin, P_1. The coordinates of bodies P_j are

$$X_j = x_j - x_1,\ Y_j = y_j - y_1,\ Z_j = z_j - z_1 \tag{10.1}$$

where the little letters represent coordinates with respect to some fixed reference system, and the capital letters represent coordinates with respect to P_1. Also, let Δ_{1j} be the distance between P_1 and P_j. The equations of motion of P_j with respect to P_1 are

$$\begin{aligned} \frac{d^2X_j}{dt^2} &= -k^2(m_1 + m_j)\frac{X_j}{\Delta_{J1}^3} \\ &+ \sum_{i=2,i\neq j}^{n} k^2 m_i \left(\frac{X_i - X_j}{\Delta_{ij}^3} - \frac{X_i}{\Delta_{1i}^3}\right),\ i,j = 2,3,4,\dots n \end{aligned} \tag{10.2}$$

with similar equations for Y and Z. By calculating the partial derivatives of the function

$$V_j = \frac{k^2(m_1 + m_j)}{\Delta_{j1}} + \sum_{i=2,i\neq j}^{n} k^2 m_i \left(\frac{1}{\Delta_{ij}} - \frac{X_iX_j + Y_iY_j + Z_iZ_j}{\Delta_{1i}^3}\right) \tag{10.3}$$

P. Gurfil, P.K. Seidelmann, *Celestial Mechanics and Astrodynamics: Theory and Practice*, Astrophysics and Space Science Library 436,
DOI 10.1007/978-3-662-50370-6_10

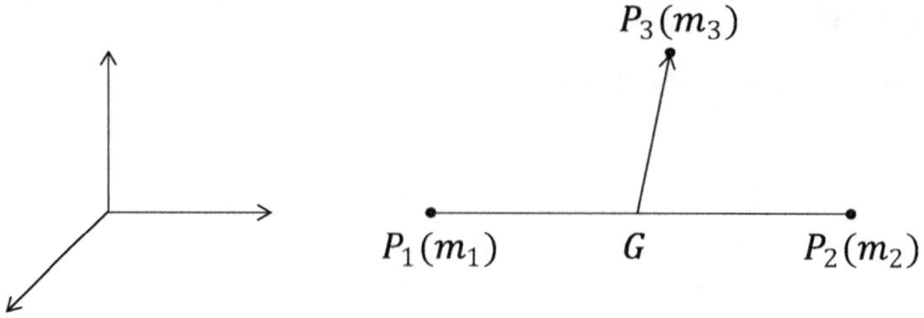

Fig. 10.1 Coordinates and mass locations

Equations (10.2) become

$$\frac{d^2X_j}{dt^2} = \frac{\partial V_j}{\partial X_j}, \; \frac{d^2Y_j}{dt^2} = \frac{\partial V_j}{\partial Y_j}, \; \frac{d^2Z_j}{dt^2} = \frac{\partial V_j}{\partial Z_j} \tag{10.4}$$

There are as many functions V_j as there are bodies minus one.

Let us now review various notations and coordinate systems. Denote the coordinates of P_2 with respect to P_1 by x, y, z. Denote the coordinates of P_3 with respect to the center of gravity G of P_1 and P_2 (see Fig. 10.1) by x', y', z'. The two sets of axes are parallel.

Then

$$x' = X_3 - \frac{m_2}{m_1 + m_2} X_2 \tag{10.5}$$

$$y' = Y_3 - \frac{m_2}{m_1 + m_2} Y_2 \tag{10.6}$$

$$z' = Z_3 - \frac{m_2}{m_1 + m_2} Z_2 \tag{10.7}$$

and

$$\Delta_{13}^2 = \sum \left(x' + \frac{m_2}{m_1 + m_2} x \right)^2 \tag{10.8}$$

$$\Delta_{23}^2 = \sum \left(x' - \frac{m_1}{m_1 + m_2} x \right)^2 \tag{10.9}$$

The equations of motion of P_2 relative to P_1 are then

$$\frac{d^2x}{dt^2} = -\frac{k^2(m_1 + m_2)x}{\Delta_{12}^3} + k^2 m_3 \left[\frac{x' - \frac{m_1}{m_1+m_2} x}{\Delta_{23}^3} - \frac{x' + \frac{m_2}{m_1+m_2} x}{\Delta_{13}^3} \right] \tag{10.10}$$

and analogous equations for y and z. There are similar equations in x', y', z', from those in X_3, Y_3, Z_3,

$$\frac{d^2x'}{dt^2} + \frac{m_2}{m_1+m_2}\frac{d^2x}{dt^2} = -\frac{k^2(m_1+m_3)(x'+\frac{m_2}{m_1+m_2}x)}{\Delta_{13}^3} \tag{10.11}$$

$$-k^2m_2\left[\frac{x'-\frac{m_1}{m_1+m_2}x}{\Delta_{23}^3}+\frac{x}{\Delta_{12}^3}\right] \tag{10.12}$$

and similarly for y' and z'. Subtracting Eqs. (10.11) and (10.10) term by term, d^2x/dt^2 is eliminated from Eq. (10.11). If we put

$$V = \frac{k^2m_1m_2}{\Delta_{12}} + \frac{k^2m_1m_3}{\Delta_{13}} + \frac{k^2m_2m_3}{\Delta 23} \tag{10.13}$$

the three-body problem equations become

$$\frac{m_1m_2}{m_1+m_2}\frac{d^2x}{dt^2} = \frac{\partial V}{\partial x}, \quad \frac{(m_1+m_2)m_3}{m_1+m_2+m_3}\frac{d^2x'}{dt^2} = \frac{\partial V}{\partial x'} \tag{10.14}$$

and similar equations for y, z and y', z' (Kovalevsky 1963, pp. 20–22).

10.2 Canonical Form of the Equations

The system of equations is

$$m_j\frac{d^2x_j}{dt^2} = \frac{\partial V}{\partial x_j}, \; j = 1, 2, \ldots, n \tag{10.15}$$

We introduce n new variables $y_j = m_j(dx_j/dt)$, where y_j is not the y coordinate, and putting

$$T = \frac{1}{2}\sum_{j=1}^{n}\frac{y_j^2}{m_j} \tag{10.16}$$

Then

$$\frac{\partial T}{\partial x_j} = 0, \; \frac{\partial T}{\partial y_j} = \frac{y_j}{m_j}, \; \frac{\partial V}{\partial y_j} = 0 \tag{10.17}$$

Equation (10.15) can be written as a system of linear equations of order $2n$,

$$\frac{dx_j}{dt} = \frac{\partial F}{\partial y_j}, \frac{dy_j}{dt} = -\frac{\partial F}{\partial x_j}, j = 1, 2, \ldots, n \tag{10.18}$$

with $F = T - V$.

This is a *canonical form* of the equations of motion. The common function F is called the *Hamiltonian*, or *characteristic function*, and x_j and y_j are *conjugate variables*. Problems of celestial and quantum mechanics use these equations. The function F may, or may not, be a function of time. In the three-body case, the canonical system is of sixth order; from the positions of P_3, F is a function of time (Kovalevsky 1963, pp. 22–23).

10.3 Eliminating the Time Dependency

In using canonical systems, eliminating the explicit presence of t in F is frequently desirable. So with $2n$ canonical equations

$$\frac{dq_j}{dt} = \frac{\partial F}{\partial p_j}(q_i, p_i, t), \frac{dp_j}{dt} = -\frac{\partial F}{\partial q_j}(q_i, p_i, t), \ 1 \leq i \leq n, \ 1 \leq j \leq n \tag{10.19}$$

we have that $\partial F/\partial t \neq 0$ if F explicitly depends on t. So two new variables, q_{n+1} and p_{n+1}, are introduced. The explicit variable t in F is replaced by q_{n+1}. p_{n+1} is the variable conjugate to q_{n+1}, and t remains the independent variable. p_{n+1} is not present in the Hamiltonian F, but a function of p_{n-1} could be added without affecting Eqs. (10.19). The new system with this function is

$$\frac{dq_j}{dt} = \frac{\partial F^*}{\partial p_j}(q_i, p_i, q_{n+1}, p_{n+1}), \frac{dp_j}{dt} = -\frac{\partial F^*}{\partial q_j}(q_i, p_i, q_{n+1}, p_{n+1}) \tag{10.20}$$

with $j = 1, 2, \ldots, n, n+1$, and the solution for $q_{n+1} = t$. We take $F^* = F + p_{n+1}$. The last two equations become

$$\frac{dq_{n+1}}{dt} = \frac{\partial F^*}{\partial p_{n+1}} = 1 \tag{10.21}$$

$$\frac{dp_{n+1}}{dt} = -\frac{\partial F^*}{\partial q_{n+1}} = -\left(\frac{\partial F*}{\partial t}\right)_{t=q_{n+1}} \tag{10.22}$$

The initial system has been reduced to a more general canonical system of order $2n+2$, and includes the solution of the first system. The characteristic function F does not contain t (Kovalevsky 1963, pp. 23–24).

10.4 Integral of a System of Canonical Equations

Let us take a system of $2n$ canonical equations, whose characteristic function does not depend on t,

$$\frac{dq_j}{dt} = \frac{\partial F(q_i, p_i)}{\partial p_j}, \quad \frac{dp_j}{dt} = -\frac{\partial F(q_i, p_i)}{\partial q_j}, \quad 1 \leq i \leq n, \ 1 \leq j \leq n \tag{10.23}$$

Since F is not an explicit function of t, its total derivative is

$$\frac{dF}{dt} = \sum_{j=1}^{n} \left(\frac{\partial F}{\partial q_j} \frac{dq_j}{dt} + \frac{\partial F}{\partial p_j} \frac{dp_j}{dt} \right) \tag{10.24}$$

We substitute for the $2n$ functions, q_j and p_j, from Eqs. (10.23), which are a solution of the system; then

$$\frac{dF}{dt} = \sum_{j=1}^{n} \left(\frac{\partial F}{\partial q_j} \frac{\partial F}{\partial p_j} - \frac{\partial F}{\partial p_j} \frac{\partial F}{\partial q_j} \right) = 0 \tag{10.25}$$

Integrating this equation, the equality $F(q_i, p_i) = C$ is satisfied for any solution $q_i(t), p_i(t)$, and it is an integral of Eqs. (10.23). The partial derivatives of F are also independent of time; t only occurs as the differential dt, and by choosing q_n as a new independent variable, we write the system (10.23) in the form

$$\frac{dq_j}{dq_n} = \frac{\partial F / \partial p_j}{\partial F / \partial p_n}, \quad 1 \leq j \leq n - 1 \tag{10.26}$$

$$\frac{dp_j}{dp_n} = -\frac{\partial F / \partial q_j}{\partial F / \partial q_n}, \quad 1 \leq j \leq n \tag{10.27}$$

This system is of order $2n - 1$, with $F = C$ as its integral. If we have a function of q_n as a solution of the system, then we obtain t from the integration

$$t - t_0 = \int dq_n / (\partial F / \partial p_n) \tag{10.28}$$

since the right-hand side is a function of q_n alone. The result is a canonical system of order $2n - 2$, with a Hamiltonian dependent on an independent variable and an integration. This process on a $2n + 2$ order system, with a characteristic function F^*, produces $F^* = C$ as an integral, and Eq. (10.28) reduces to $\partial F^* / \partial p_{n+1} = 1$. Now, the integral $F^* = C$ is a function of p_{n+1}, an additional variable. In celestial mechanics, usually the elimination of t from a characteristic function is sought (Kovalevsky 1963, pp. 24–25).

10.5 Canonical Transformation of Variables

In celestial mechanics and astrodynamics, a frequent method of solving equations of motion is by transformation of variables. With equations in canonical form, it is very effective. The variables $p_i, q_i, (1 \leq i \leq n)$ are transformed into $P_i, Q_i, (1 \leq i \leq n)$, resulting in simpler equations. This is a canonical transformation, if the new equations are canonical. This process can be continued as long as such transformations can be found to reach an easily integrated system of equations.

We now seek the necessary and sufficient conditions under which a transformation of variables is canonical.

10.5.1 Necessary Condition

Consider the differential system (10.23), rewritten here for convenience,

$$\frac{dq_j}{dt} = \frac{\partial F(q_i, p_i)}{\partial p_j}, \quad \frac{dp_j}{dt} = -\frac{\partial F(q_i, p_i)}{\partial q_j}, \quad 1 \leq i, \ j \leq n \tag{10.29}$$

where F does not depend explicitly on t, and let $P_j, Q_j, \ 1 \leq j \leq N$ be $2N$ new variables, which are canonical. Consider

$$d\Theta = \sum_j p_j dq_j - Fdt \tag{10.30}$$

We have $dp_j/dt = -\partial F/\partial q_j$ for all values of j and p_j depends only on t, since it is a solution; we can write

$$\frac{\partial p_j}{\partial t} = -\frac{\partial F}{\partial q_j} \tag{10.31}$$

A necessary and sufficient condition for $\sum X_i dx_i$ to be a total differential is that the quantities $(\partial X_i/\partial x_k - \partial X_k/\partial x_i)$ are zero. Applying this condition to the right-hand side of Eq. (10.30), and remembering that only F depends on the q_j, and p_j depends on t alone, we obtain (10.31)

$$\frac{\partial p_j}{\partial t} + \frac{\partial F}{\partial q_j} = 0 \tag{10.32}$$

The quantity $\sum_j p_j dq_j - Fdt$ is a total differential. Applying the same reasoning to the new canonical system

$$\frac{dQ_j}{dt} = \frac{\partial F^*(P_i, Q_i)}{\partial P_j}, \quad \frac{dP_j}{dt} = -\frac{\partial F^*(P_i, Q_i)}{\partial Q_j} \tag{10.33}$$

The new Hamiltonian F^* is not necessarily identical with F. $\sum P_j dQ_j - F^* dt = d\Theta^*$, is a total differential. Subtracting term by term, a necessary condition for the transformation of the variables to be canonical is

$$\sum_j P_j dQ_j - \sum_j p_j dq_j = d(\Theta^* - \Theta) + (F^* - F)dt \tag{10.34}$$

Putting $K = F^* - F$, this becomes

$$\sum_j P_j dQ_j - \sum_j p_j dq_j - K dt = dW \tag{10.35}$$

where K is a function of the variables and dW is a total differential.

10.5.2 Sufficient Condition

We shall show that condition (10.35) is sufficient. Since the initial system is canonical,

$$\sum_j p_j dq_j - F dt = d\Theta \tag{10.36}$$

Substituting into Eq. (10.35) we obtain

$$\sum_j P_j dQ_j - (F + K)dt = d(W + \Theta) \tag{10.37}$$

The right-hand side is a total differential, so the left-hand side is one. The condition for this is

$$\frac{\partial P_j}{\partial t} = -\frac{\partial (F + K)}{\partial Q_j} \tag{10.38}$$

Since we suppose that P_j is a variable, it depends only on t, which gives

$$\frac{dP_j}{dt} = -\frac{\partial (F + K)}{\partial Q_j} \tag{10.39}$$

We also know that the total differential of $\sum_j P_j Q_j$ is

$$d\left(\sum_j P_j Q_j\right) = \sum_j P_j dQ_j + \sum_j Q_j dP_j \tag{10.40}$$

Equation (10.37) can be written

$$d\left(\sum_j P_j Q_j\right) - \sum_j Q_j dP_j - (F+K)dt = d(W+\Theta) \tag{10.41}$$

or

$$\sum_j Q_j dP_j + (F+K)dt = d\left(\sum_j P_j Q_j - W - \Theta\right) \tag{10.42}$$

The left-hand side must be a total differential. This implies

$$\frac{dQ_j}{dt} = \frac{\partial(F+K)}{\partial P_j} \tag{10.43}$$

Equations (10.38) and (10.43) show that the system of equations in P_j and Q_j, with a Hamiltonian $(F+K)$, is canonical. So we have shown that the condition

$$\sum_j P_j dQ_j - \sum_j p_j dq_j - Kdt = dW \tag{10.44}$$

is a necessary and sufficient condition for the change of variables from p_j, q_j to P_j, Q_j to be canonical. The new characteristic function is $(F+K)$, regardless of the explicit presence of t in the Hamiltonian.

The transformation $(p_i, q_i) \to (P_i, Q_i)$ of Eq. (10.44) is called a *contact transformation*, which is important in the theory of partial differential equations (Kovalevsky 1963, pp. 25–27).

10.6 Examples of Canonical Transformations

10.6.1 Change of Variables by Means of a Generating Function

Consider a completely general set of canonical equations

$$\frac{dq_i}{dt} = \frac{\partial F}{\partial p_i}, \quad \frac{dp_i}{dt} = -\frac{\partial F}{\partial q_i}, \quad 1 \le i \le n \tag{10.45}$$

and a change of variables $(q_j, p_j) \to (Q_j, P_j)$. We have an arbitrary function S of $2n$ variables, called a *generating function* and written as a function of the new variables Q_j and the old variables p_j; $S(Q_j, p_j)$, $1 \le j \le n$. The change of variables is defined by $2n$ implicit equations

$$q_j = \frac{\partial S}{\partial p_j}, \; P_j = \frac{\partial S}{\partial Q_j} \tag{10.46}$$

For any S, the change of variables is canonical and does not change the characteristic function. We evaluate the quantity

$$E = \sum_j (P_j dQ_j - p_j dq_j) \tag{10.47}$$

Differentiation of $S(Q_i, p_i)$ gives the identity

$$dS = \sum_j \frac{\partial S}{\partial Q_j} dQ_j + \sum_j \frac{\partial S}{\partial p_j} dp_j \tag{10.48}$$

From the definitions of P_j and q_j in Eq. (10.46)

$$dS = \sum_j P_j dQ_j + \sum q_j dp_j \tag{10.49}$$

Then

$$E = dS - \sum_j q_j dp_j - \sum_j p_j dq_j = d\left(S - \sum_j p_j q_j\right) \tag{10.50}$$

This is a total differential. Condition (10.35) is satisfied by $K = 0$ (the invariance of the Hamiltonian) and $W = S - \sum_j p_j q_j$ (Kovalevsky 1963, pp. 27–28).

10.6.2 Conjugate Variables to Q_j

Consider a system with canonical variables x_j and y_j. We wish to make a canonical transformation, which leaves the Hamiltonian unchanged and such that the q_j are given functions of the x_j. The relation (10.35) gives

$$\sum_j y_j dx_j - \sum_j p_j dq_j = dW \tag{10.51}$$

Since the x_j are functions of the q_j,

$$dx_j = \sum_i \frac{\partial x_j}{\partial q_i} dq_i, \ i, j = 1, 2, \ldots .n \tag{10.52}$$

If we put $dW = 0$, Eq. (10.51) is identically satisfied, if

$$p_i = \sum_j y_j \frac{\partial x_j}{\partial q_i}, \ i, j = 1, 2, \ldots .n \tag{10.53}$$

If T has the previous form with respect to the x_j, from Eq. (10.52)

$$T = \frac{1}{2} \sum_j m_j \left(\frac{dx_j}{dt} \right)^2 = \frac{1}{2} \sum_j m_j \left(\sum_i \frac{\partial x_j}{\partial q_i} \frac{dq_i}{dt} \right)^2 \tag{10.54}$$

with $q_i' = (dq_i/dt)$,

$$\frac{\partial T}{\partial q_i'} - \sum_j m_j \frac{\partial x_j}{\partial q_i} \frac{dx_j}{dt} = \sum_j y_j \frac{\partial x_j}{\partial q_i} \tag{10.55}$$

This is from the definition of y_j given after Eq. (10.15), i.e. $y_j = m_j(dx_j/dt)$. Then, from Eqs. (10.53) and (10.55), we have

$$p_i = \frac{\partial T}{\partial q_i'} \tag{10.56}$$

where T is a function of q_i and q_i' (Kovalevsky 1963, pp. 28–29).

10.7 Jacobi's Theorem

We want to establish a theorem to define an important system of canonical variables for the two-body problem. We want a canonical transformation that makes the new Hamiltonian zero. For the canonical system

$$\frac{dq_j}{dt} = \frac{\partial F}{\partial p_j}, \ \frac{dp_j}{dt} = -\frac{\partial F}{\partial q_j}, \ 1 \le j \le n \tag{10.57}$$

we make the canonical transformation $(p_j, q_j) \rightarrow (P_j, Q_j)$ such that

$$\sum_j P_j dQ_j - \sum_j p_j dq_j + F dt = -dW \tag{10.58}$$

We write $-dW$ to simplify future notation. W is a function of t, since F is nonzero. We equate term by term the coefficients of the $(n+1)$ differentials,

$$P_j = -\frac{\partial W}{\partial Q_j},\ p_j = \frac{\partial W}{\partial q_j},\ F(q_j, p_j, t) = -\frac{\partial W}{\partial t} \tag{10.59}$$

Replacing p_j by $\partial W/\partial q_j$ in the last equation, we obtain *Jacobi's equation*

$$F\left(q_j, \frac{\partial W}{\partial q_j}, t\right) + \frac{\partial W}{\partial t} = 0 \tag{10.60}$$

If the change in variables has been made, the new Hamiltonian is $F - F = 0$. The equations become

$$\frac{dP_j}{dt} = 0,\ \frac{dQ_j}{dt} = 0 \tag{10.61}$$

or $P_j = b_j, Q_j = a_j$, where a_j and b_j are constants. If we can find this change in variables, the problem is thus solved. If we have a solution of Eq. (10.60), which depends on n linearly independent arbitrary constants a_j, the n variables q_j, and t, then $W(q_j, a_j, t) = 0$.

The canonical transformation is such that the numbers a_j are solutions to the new variables Q_j. Equations (10.59) define the conjugate variables P_j.

$$P_j = b_j = -\frac{\partial W(q_j, a_j, t)}{\partial a_j} \tag{10.62}$$

The constant values b_j are the solution to P_j. The n relations (10.62) determine n variables q_j, which are functions of $2n$ integration constants a_j and b_j, and t; this requires each q_j to be in each of the n equations. This condition prevents one a_j as an additive constant, so W cannot be of the form $W(q_j, a_j, \ldots, a_{n-1}, t) + a_n$. We substitute the values of q_j in the second of the series of Eqs. (10.59), $p_j = \partial W(q_j, a_j, t)/\partial q_j$, obtaining the n variables p_j as a function of a_j, b_j, t. *Jacobi's theorem* can be summarized as follows:
To integrate a system of $2n$ canonical equations

$$\frac{dq_j}{dt} = \frac{\partial F}{\partial p_j},\ \frac{dp_j}{dt} = -\frac{\partial F}{\partial q_j} \tag{10.63}$$

find a complete solution of the Jacobi equation

$$F\left(q_j, \frac{\partial W}{\partial q_j}, t\right) + \frac{\partial W}{\partial t} = 0 \tag{10.64}$$

This solution depends on n linearly independent arbitrary constants a_j. Then solve for q_j, p_j in the system of equations

$$b_j = -\frac{\partial W(q_j, a_j, t)}{\partial a_j}, \; p_j = \frac{\partial W(q_j, a_j, t)}{\partial q_j} \tag{10.65}$$

where the n quantities b_j are the missing n integration constants.

We can introduce one of the integration constants, say a_n, into Jacobi's equation by writing the equation as

$$F\left(q_j, \frac{\partial W}{\partial q_j}, t\right) + \frac{\partial W}{\partial t} = a_n \tag{10.66}$$

This is adding $a_n dt$ to both sides of Eq. (10.58), which does not change the condition, except that the Hamiltonian is no longer zero; it is equal to a_n, i.e. Q_n (Kovalevsky 1963, pp. 29–30).

10.8 Canonical Equations for the Two-Body Problem

Jacobi's theorem can be applied to the elliptical case of the two-body problem. The orbital elements for a system of canonical conjugate variables can be shown. After division by m_j in Eqs. (10.17), the equations of motion of one of the bodies are

$$\frac{dx_j}{dt} = \frac{\partial F}{\partial y_j}, \; \frac{dy_j}{dt} = -\frac{\partial F}{\partial x_j}, \; j = 1, 2, 3 \tag{10.67}$$

where x_1, x_2, x_3 are the Cartesian coordinates of the body and

$$F = T - V = \frac{1}{2}(y_1^2 + y_2^2 + y_3^2) - \frac{\mu}{r} \tag{10.68}$$

retaining only the first term of V. From Fig. 10.2, the equations in spherical coordinates are

$$x = r\cos\varphi\cos\theta, \; y = r\cos\varphi\sin\theta, \; z = r\sin\varphi \tag{10.69}$$

The Hamiltonian F in these coordinates is

$$\left(\frac{dx}{dt}\right)^2 + \left(\frac{dy}{dt}\right)^2 + \left(\frac{dz}{dt}\right)^2 = \left(\frac{dr}{dt}\right)^2 + r^2\left(\frac{d\varphi}{dt}\right)^2 + r^2\cos^2\varphi\left(\frac{d\theta}{dt}\right)^2 \tag{10.70}$$

Fig. 10.2 Spherical coordinates

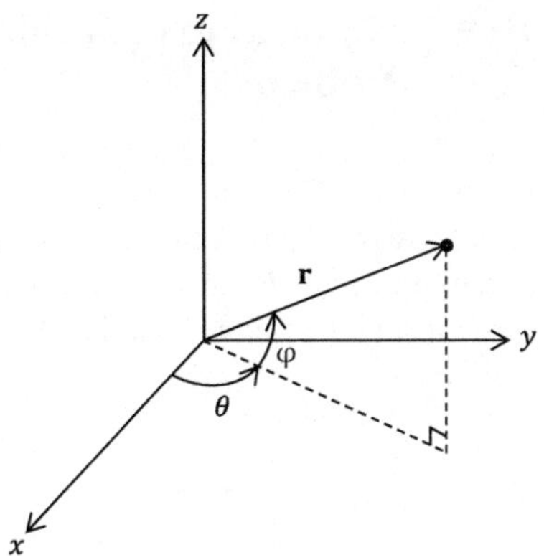

Take the variables

$$q_1 = r,\ q_2 = \varphi,\ q_3 = \theta \tag{10.71}$$

and let q_1', q_2', q_3' be their derivatives $dr/dt, d\varphi/dt$, and $d\theta/dt$. In this notation, from $T = \frac{1}{2}\sum_{j=1}^{n}\frac{y_j^2}{m_j}$

$$T = \frac{1}{2}q_1'^2 + \frac{1}{2}q_1^2q_2'^2 + \frac{1}{2}q_1^2q_3'^2\cos^2 q_2 \tag{10.72}$$

According to Eq. (10.56), the variables conjugate to q_i are

$$p_1 = \frac{\partial T}{\partial q_1'} = q_1' \tag{10.73}$$

$$p_2 = \frac{\partial T}{\partial q_2'} = q_1^2 q_2' \tag{10.74}$$

$$p_3 = \frac{\partial T}{\partial q_3'} = q_1^2 q_3' \cos^2 q_2 \tag{10.75}$$

In these new variables, the characteristic function $F = T - V$ is

$$F = \frac{1}{2}p_1^2 + \frac{1}{2q_1^2}p_2^2 + \frac{1}{2q_1^2\cos^2 q_2}p_3^2 - \frac{\mu}{q_1} \tag{10.76}$$

and the equations are

$$\frac{dq_i}{dt} = \frac{\partial F}{\partial p_i},\ \frac{dp_i}{dt} = -\frac{\partial F}{\partial q_i},\ i = 1, 2, 3 \tag{10.77}$$

(Kovalevsky 1963, pp. 31–32).

10.9 Application of Jacobi's Theorem to the Two-body Problem

From Eq. (10.76), the characteristic function, F, does not depend on t, so $F = h$ is an integral of the system of equations, and h is a constant. This integral is the kinetic energy integral. F can be replaced everywhere by $F - h$. We can substitute $F - h$ for F in Jacobi's equation, thus introducing an arbitrary constant. Jacobi's equation is then

$$\frac{1}{2}\left(\frac{\partial W}{\partial q_1}\right)^2 + \frac{1}{2q_1^2}\left(\frac{\partial W}{\partial q_2}\right)^2 + \frac{1}{2q_1^2\cos^2 q_2}\left(\frac{\partial W}{\partial q_3}\right)^2 - \frac{\mu}{q_1} - h = 0 \quad (10.78)$$

We want a solution depending on three arbitrary constants, not a complete solution.

Since the derivatives are separable, we want a solution W, whose variables are separable. We seek W in the form

$$W = W_1(q_1) + W_2(q_2) + W_3(q_3) \quad (10.79)$$

Jacobi's equation is then

$$\frac{1}{2}\left(\frac{\partial W_1}{\partial q_1}\right)^2 + \frac{1}{2q_1^2}\left(\frac{\partial W_2}{\partial q_2}\right)^2 + \frac{1}{2q_1^2\cos^2 q_2}\left(\frac{\partial W_3}{\partial q_3}\right)^2 - \frac{\mu}{q_1} - h = 0 \quad (10.80)$$

Since the three bracketed quantities are independent, the equation is satisfied by the three equations

$$\frac{dW_3}{dq_3} = a_3 \quad (10.81)$$

$$\frac{1}{2}\left(\frac{dW_2}{dq_2}\right)^2 + \frac{a_3^2}{2\cos^2 q_2} = \frac{a_2^2}{2} \quad (10.82)$$

$$\frac{1}{2}\left(\frac{dW_1}{dq_1}\right)^2 + \frac{a_2^2}{2q_1^2} - \frac{\mu}{q_1} - h = 0 \quad (10.83)$$

Direct substitution makes the left-hand side of Jacobi's equation zero. From Eq. (10.79) we have

$$W = \int \left(2h + \frac{2\mu}{q_1} - \frac{a_2^2}{q_1^2}\right)^{1/2} dq_1 + \int \left(a_2^2 - \frac{a_3^2}{\cos^2 q_2}\right)^{1/2} dq_2 + \int a_3 dq_3 \quad (10.84)$$

which is a solution of Eq. (10.78) up to an additive constant from the indefinite integrals. There are three arbitrary constants a_3, a_2, and $h \equiv a_1$. The signs of the square root are not required at this stage (Kovalevsky 1963, pp. 31–32).

10.9.1 Meaning of the Constants a

The integration constants, a_1, a_2, a_3, appearing in W, are the values of the new variables Q_1, Q_2, Q_3 of a system of canonical equations equivalent to the initial system, whose characteristic function is zero. The solution is of the form

$$Q_1 = a_1,\ Q_2 = a_2,\ Q_3 = a_3 \tag{10.85}$$

What is the meaning of these three canonical variables in elliptic motion? a_1 is the energy constant, $h = -\mu/(2a)$, where a is the semimajor axis.

The basic equation for a canonical transformation is

$$\sum_i P_i dQ_i - \sum_i p_i dq_i + F dt = -dW \tag{10.86}$$

From Eq. (10.84), W depends on q_3 only through $\int a_3 dq_3 = a_3 q_3$. The only term, dq_3, in dW is $a_3 dq_3$. The terms in dq_3 from Eq. (10.86) are $-p_3 = -a_3$, or

$$a_3 = q_1^2 \cos^2 q_2 q_3' = r^2 \cos^2 \varphi \frac{d\theta}{dt} \tag{10.87}$$

from Eqs. (10.71) and (10.73). This is the z component of the angular momentum. The magnitude of the angular momentum is

$$C = na^2\sqrt{1-e^2} = \sqrt{\mu a(1-e^2)} \tag{10.88}$$

Its z component is $\sqrt{\mu a(1-e^2)}\cos i$ and so

$$a_3 = Q_3 = \sqrt{\mu a(1-e^2)}\cos i \tag{10.89}$$

We can identify $-p_2$ with the coefficient of dq_2 in the total differential of dW. Only q_2 and dq_2 appear in dW_2; consequently,

$$-p_2 = -\sqrt{a_2^2 - \frac{a_3^2}{\cos^2 q_2}} = -\sqrt{a_2^2 - \frac{a_3^2}{\cos^2 \varphi}} \tag{10.90}$$

Replacing a_3 by $r^2 \cos^2 \varphi (d\theta/dt)$, we have

$$p_2 = \sqrt{a_2^2 - r^4 \left(\frac{d\theta}{dt}\right)^2 \cos^2 \varphi} \tag{10.91}$$

Then, from Eqs. (10.71) and (10.73)

$$p_2 = q_1^2 q_2' = r^2 \frac{d\varphi}{dt} \tag{10.92}$$

Combining these two equations,

$$a_2^2 = r^4 \left[\left(\frac{d\varphi}{dt}\right)^2 + \left(\frac{d\theta}{dt}\right)^2 \cos^2 \varphi \right] \tag{10.93}$$

This is the square of the angular momentum magnitude. So

$$Q_2 = a_2 = \sqrt{\mu a(1 - e^2)} \tag{10.94}$$

(Kovalevsky 1963, pp. 33–34).

10.9.2 Variables Conjugate to Q_i

When W is given by Eq. (10.84), the variables P_j take the constant values given by $b_j = -\partial W/\partial a_j$, according to Jacobi's theorem. This W is defined up to an arbitrary constant and the sign of the integrals. We take a different system of variables P_j,

$$W = \int_{q_1(t_0)}^{q_1(t)} \varepsilon_1 \left(2a_1 + \frac{2\mu}{q_1} - \frac{a_2^2}{q_1^2}\right)^{1/2} dq_1 + \int_0^{\varphi} \varepsilon_2 \left(a_2^2 - \frac{a_3^2}{\cos^2 q_2}\right)^{1/2} dq_2 + \int_0^{\theta} a_3 dq_3 \tag{10.95}$$

where t_0 is the instant of periapsis passage; $\varepsilon_1 = +1$; if $q_1 = r$ is increasing, $\varepsilon_1 = -1$. The derivative of the function of the first integral is continuous and is zero at passages through apoapsis and periapsis. The quantity in the second integral has a continuous derivative. ε_2 is $+1$, when at the discontinuity $q_2 = \varphi = i$ and φ is increasing, or when the argument of latitude, $u = \omega + f$, is between $-\pi/2$ and $+\pi/2$, so that $\cos u > 0$. $\varepsilon_2 = -1$, if $\cos u < 0$ (recall Figs. 5.1, 5.7, 5.8).

The variable conjugate to Q_1 is

$$P_1 = -\frac{\partial W}{\partial a_1} = \int_{q_1(t_0)}^{q_1(t)} \varepsilon_1 \left(2a_1 + \frac{2\mu}{q_1} - \frac{a_2^2}{q_1^2}\right)^{-1/2} dq_1 \tag{10.96}$$

q_1 is r; $a_1 = h = -\mu/2a$, and $a_2^2 = \mu a(1-e^2)$. Multiplying above and below by $r(>0)$, we have

$$P_1 = -\varepsilon_1 \int_{r(t_0)}^{r(t)} \frac{rdr}{\sqrt{-(\mu/a)r^2 + 2\mu r - \mu a(1-e^2)}} \tag{10.97}$$

For the integration, these quantities are given in terms of the eccentric anomaly, E, for which $r = a(1 - e\cos E)$, $dr = ae\sin EdE$. Let E be the eccentric anomaly at t. It is zero at t_0. Let $J = 1 - e\cos E$, then

$$P_1 = -\int_0^E \frac{\varepsilon_1 a(1-e\cos E)ae\sin E\, dE}{\sqrt{\mu a}[-J^2 + 2J - 1 + e^2]^{1/2}}$$

$$= -\varepsilon_1 \int_0^E \frac{a^2(1-e\cos E)e\sin E\, dE}{\sqrt{\mu a}\, e|\sin E|} \tag{10.98a}$$

From the definition of ε_1, we have that

$$\varepsilon_1 \frac{\sin E}{|\sin E|} = +1 \tag{10.99}$$

From Eq. (10.96)

$$\left(2a_1 + \frac{2\mu}{q_1} - \frac{2a_2^2}{q_1^2}\right)^{1/2} = \frac{|\sin E|}{r} \tag{10.100}$$

is zero on passage through perifocus. From Eqs. (10.98) that

$$P_1 = -\int_0^E \frac{a\sqrt{a}}{\sqrt{\mu}}(1-e\cos E)dE$$

$$= -\frac{1}{n}(E - e\sin E) = -(t - t_0) \tag{10.101}$$

from Kepler's third law (see Sect. 4.6). The final Hamiltonian is $h = Q_1$, as per Jacobi's theorem. The equation giving P_1 is

$$\frac{dP_1}{dt} = -\frac{\partial F}{\partial Q_1} = -1 \tag{10.102}$$

which integrates to give $P_1 = -t + b_1$.

The constant of integration b_1 is t_0, the instant of passage through the perihelion.

Similar calculations give $P_3 = -\Omega$ and $P_2 = -\omega$, where we must take precautions in defining the signs and the end points of integration; Ω and ω are, respectively, the argument of the ascending node and the argument of periapsis (see Sect. 5.2) (Kovalevsky 1963, pp. 34–36).

10.9.3 Application to the General Problem

We have established a new system of conjugate variables,

$$Q_1 = -\frac{\mu}{2a}, \; Q_2 = \sqrt{\mu a(1-e^2)}, \; Q_3 = \sqrt{\mu a(1-e^2)}\cos i$$
$$P_1 = -t + t_0, \; P_2 = -\omega, \; P_3 = -\Omega \tag{10.103}$$

whose characteristic function for the two-body problem reduces to $Q_1 = -\mu/2a$. We could make this Hamiltonian zero by a suitable canonical transformation, but such a transformation is not desirable for the rest of the calculation (because we want the Hamiltonian to be able to provide the perturbations for the computations).

Consider the three-body problem and the equations of one of the bodies, where the equations can be extended to the other bodies. The system of equations was

$$\frac{dx_j}{dt} = \frac{\partial F}{\partial y_j}, \; \frac{dy_j}{dt} = -\frac{\partial F}{\partial x_j}, \; j = 1, 2, 3 \tag{10.104}$$

F is in the form of $F = T - V$, and, as we have seen, V contains the term μ/r. Put $V = \mu/r + R$, where R is the disturbing function. In the Hamiltonian $F = (T - \mu/r) - R$, R accounts for perturbations in two-body problem motion. $F^* = T - \mu/r$ is the Hamiltonian of the two-body problem, just discussed. Equations (10.104) become

$$\frac{dx_j}{dt} = \frac{\partial(F^* - R)}{\partial y_j}, \; \frac{dy_j}{dt} = -\frac{\partial(F^* - R)}{\partial x_j}, \; j = 1, 2, 3 \tag{10.105}$$

Let us now transform the variables of this system to the new variables, $P_1, P_2, P_3, Q_1, Q_2, Q_3$, as defined in Eq. (10.103). System (10.104) is not the two-body problem, and its solution's new variables will no longer be constants. We called these variables osculating elements in Sect. 5.2.1. The osculating elements are defined in the set of axes x_1, x_2, x_3 relative to the body in question. The change in variables is such that the new Hamiltonian F_1 is

$$F_1 = F - \left(F^* + \frac{\mu}{2a}\right) = -R - \frac{\mu}{2a} \tag{10.106}$$

Consequently, the system (10.104) is equivalent to the system

$$\frac{dQ_j}{dt} = \frac{\partial(-R - (\mu/2a))}{\partial P_j} \tag{10.107}$$

$$\frac{dP_j}{dt} = -\frac{\partial(-R - (\mu/2a))}{\partial Q_j}, \; j = 1, 2, 3 \tag{10.108}$$

We can improve the appearance of this system by changing all the signs of the Hamiltonian, and putting $P'_j = -P_j$

$$P'_1 = t - t_0, \; P'_2 = \omega, \; P'_3 = \Omega \tag{10.109}$$

We have the following system:

$$\frac{dQ'_j}{dt} = \frac{\partial(R + (\mu/2a))}{\partial P'_j} \tag{10.110}$$

$$\frac{dP'_j}{dt} = -\frac{\partial(R + (\mu/2a))}{\partial Q'_j}, \; j = 1, 2, 3 \tag{10.111}$$

(Kovalevsky 1963, pp. 36–37).

10.10 The Delaunay Variables

The variables P'_2 and P'_3 now represent two classical orbital elements. We can attempt to transform P'_1 such that it represents the mean anomaly. We denote by L, G, H, l, g, h, six new canonical variables, which would be obtained after this transformation. We seek a transformation such that the characteristic function remains unchanged, as well as P'_2 and P'_3, which must be equal to g and h, respectively. The condition for this transformation to be canonical and for the Hamiltonian to remain unchanged is

$$l\,dL + gdG + hdH - P'_1dQ_1 - P'_2dQ_2 - P'_3dQ'_3 = dW \tag{10.112}$$

We want

$$P'_2 = g, \; P'_3 = h, \; l = n(t - t_0) = nP'_1 = \sqrt{\mu}a^{-3/2}P'_1 \tag{10.113}$$

These conditions are fulfilled if

$$Q_2 = G, \; Q_3 = H, \; ldL - P'_1dQ_1 = dW \tag{10.114}$$

and

$$P_1' \left(\sqrt{\mu}\, a^{-3/2} dL - \frac{\mu da}{2a^2} \right) = dW \tag{10.115}$$

A possible solution is $dW = 0$, then

$$\frac{\sqrt{\mu} da}{2\sqrt{a}} = dL \tag{10.116}$$

so that $L = \sqrt{\mu a}$. If we put again

$$\phi = \frac{\mu}{2a} + R = \frac{\mu^2}{2L^2} + R \tag{10.117}$$

the system of equations given by Eq. (10.104) is equivalent to the system

$$\frac{dL}{dt} = \frac{\partial \phi}{\partial l}, \quad \frac{dG}{dt} = \frac{\partial \phi}{\partial g}, \quad \frac{dh}{dt} = \frac{\partial \phi}{\partial h} \tag{10.118a}$$

$$\frac{dl}{dt} = -\frac{\partial \phi}{\partial L}, \quad \frac{dg}{dt} = -\frac{\partial \phi}{\partial G}, \quad \frac{dh}{dt} = -\frac{\partial \phi}{\partial H} \tag{10.118b}$$

ϕ is expressed as a function of the variables L, G, H, l, g, h, whose relation to the classical elements is

$$L = \sqrt{\mu a}, \; G = \sqrt{\mu a(1 - e^2)}, \; H = \sqrt{\mu a(1 - e^2)} \cos i \tag{10.119a}$$

$$l = M = n(t - t_0), \; g = \omega, \; h = \Omega \tag{10.119b}$$

These canonical variables are known as the *Delaunay variables*. They were used by Delaunay for his theory of the Moon, and continue in use for perturbation problems (Kovalevsky 1963, pp. 38–39).

The same reasoning of the osculating classical elements mentioned in Sect. 5.2.1 can be applied to Delaunay's variables. When the perturbations disappear at an instant t, ϕ becomes $\mu/2a$ $(R = 0)$ and the solution of the equations are L, G, H, g, h (constants) and $l = n(t - t_0)$ a variable. Thus, we see that in the general case Delaunay's variables are also osculating values in the sense mentioned in Sect. 5.2.1. They are connected with the classical osculating elements by Eqs. (10.119).

10.11 The Lagrange Equations

The osculating elements are important as variables both in celestial mechanics and in astrodynamics, so we shall establish the differential equations equivalent to the systems already given, but where the variables are classical osculating elements. Starting with the Delaunay equations, Eqs. (10.118), with the six variables L, G, H, l, g, h, we effect a change of variables defined by the relations (10.119), written in the differential form

$$dL = \frac{\sqrt{\mu}}{2\sqrt{a}}da \tag{10.120a}$$

$$dG = \frac{\sqrt{\mu}\sqrt{1-e^2}}{2\sqrt{a}}da - \frac{\sqrt{\mu a}e}{\sqrt{1-e^2}}de \tag{10.120b}$$

$$dH = \frac{\sqrt{\mu}\sqrt{1-e^2}\cos i}{2\sqrt{a}}da - \frac{\sqrt{\mu a}e\cos i}{\sqrt{1-e^2}}de - \sqrt{\mu a(1-e^2)}\sin i di \tag{10.120c}$$

$$dl = dM \tag{10.120d}$$

$$dg = d\omega \tag{10.120e}$$

$$dh = d\Omega \tag{10.120f}$$

From Eqs. (10.120), noting that $\phi = \mu/2a + R$, we obtain

$$\frac{da}{dt} = \frac{2\sqrt{a}}{\sqrt{\mu}}\frac{dL}{dt} = \frac{2\sqrt{a}}{na^{3/2}}\frac{\partial\phi}{\partial l} = \frac{2}{na}\frac{\partial R}{\partial M} \tag{10.121a}$$

$$\frac{de}{dt} = -\frac{\sqrt{1-e^2}}{\sqrt{\mu a}e}\frac{dG}{dt} = \frac{\sqrt{\mu}(1-e^2)}{2\sqrt{a}\sqrt{\mu a}e}\frac{da}{dt} = \frac{1\sqrt{1-e^2}}{na^2e}\frac{\partial R}{\partial\omega} + \frac{(1-e^2)}{na^2e}\frac{\partial R}{\partial M} \tag{10.121b}$$

$$\frac{di}{dt} = \frac{-1}{\sqrt{\mu a}\sqrt{1-e^2}\sin i}\frac{dH}{dt} + \frac{\sqrt{\mu}\sqrt{1-e^2}\cos i}{2\sqrt{a}\sqrt{\mu a}\sqrt{1-e^2}\sin i}\frac{da}{dt} - \frac{\sqrt{\mu a}e\cos i}{\sqrt{1-e^2}\sqrt{\mu a}\sqrt{1-e^2}\sin i}\frac{de}{dt} = \frac{-1}{na^2\sqrt{1-e^2}\sin i}\frac{\partial R}{\partial\Omega} + \frac{\cos i}{na^2\sqrt{1-e^2}\sin i}\frac{\partial R}{\partial\omega} \tag{10.121c}$$

Rearrangement of Eqs. (10.119) gives

$$a = \frac{L^2}{\mu}, \ \sqrt{1-e^2} = \frac{G}{L} \tag{10.122}$$

where $e = \sqrt{1-(G^2/L^2)}$ and $\cos i = H/G$. The last three differential equations of (10.120) give

$$\frac{dh}{dt} = \frac{d\Omega}{dt} = -\frac{\partial R}{\partial H} = -\frac{\partial R}{\partial i}\frac{\partial i}{\partial H} = \frac{1}{na^2\sqrt{1-e^2}\sin i}\frac{\partial R}{\partial i} \tag{10.123a}$$

$$\begin{aligned}\frac{dg}{dt} = \frac{d\omega}{dt} &= -\frac{\partial R}{\partial G} = -\frac{\partial R}{\partial e}\frac{\partial e}{\partial G} - \frac{\partial R}{\partial i}\frac{\partial i}{\partial G} \\ &= \frac{\partial R}{\partial e}\left(\frac{-G}{L^2}\right)\frac{1}{\sqrt{1-(G^2/L^2)}} - \frac{\partial R}{\partial i}\left(\frac{-1}{\sin i}\right)\left(\frac{-H}{G^2}\right) \\ &= \frac{\sqrt{1-e^2}}{na^2 e}\frac{\partial R}{\partial e} - \frac{\cos i}{na^2\sqrt{1-e^2}\sin i}\frac{\partial R}{\partial i}\end{aligned} \tag{10.123b}$$

$$\begin{aligned}\frac{dl}{dt} = \frac{dM}{dt} &= -\frac{\partial}{\partial L}\left(\frac{\mu}{2a}\right) - \frac{\partial R}{\partial L} \\ &= -\frac{\partial}{\partial L}\left(\frac{\mu^2}{2L^2}\right) - \frac{\partial R}{\partial a}\frac{\partial a}{\partial L} - \frac{\partial R}{\partial e}\frac{\partial e}{\partial L} \\ &= \frac{\mu^2}{L^3} - \frac{\partial R}{\partial a}\left(\frac{2L}{\mu}\right) - \frac{\partial R}{\partial e}\left(\frac{G^2}{L^3}\right)\frac{1}{\sqrt{1-(G^2/L^2)}} \\ &= n - \frac{2}{na}\frac{\partial R}{\partial a} - \frac{1-e^2}{na^2 e}\frac{\partial R}{\partial e}\end{aligned} \tag{10.123c}$$

This system of equations is equivalent to the Delaunay system, and constitutes the *Lagrange equations*, also commonly referred to as the *Lagrange planetary equations* (LPE):

$$\frac{da}{dt} = \frac{2}{na}\frac{\partial R}{\partial M} \tag{10.124a}$$

$$\frac{de}{dt} = \frac{-\sqrt{1-e^2}}{na^2 e}\frac{\partial R}{\partial \omega} + \frac{1-e^2}{na^2 e}\frac{\partial R}{\partial M} \tag{10.124b}$$

$$\frac{di}{dt} = \frac{-1}{na^2\sqrt{1-e^2}\sin i}\frac{\partial R}{\partial \Omega} + \frac{\cos i}{na^2\sqrt{1-e^2}\sin i}\frac{\partial R}{\partial \omega} \tag{10.124c}$$

$$\frac{d\Omega}{dt} = \frac{1}{na^2\sqrt{1-e^2}\sin i}\frac{\partial R}{\partial i} \tag{10.124d}$$

$$\frac{d\omega}{dt} = \frac{\sqrt{1-e^2}}{na^2 e}\frac{\partial R}{\partial e} - \frac{\cos i}{na^2\sqrt{1-e^2}\sin i}\frac{\partial R}{\partial i} \tag{10.124e}$$

$$\frac{dM}{dt} = n - \frac{2}{na}\frac{\partial R}{\partial a} - \frac{1-e^2}{na^2 e}\frac{\partial R}{\partial e} \tag{10.124f}$$

Note that in these equations, n represents $\sqrt{\mu}/a^{3/2}$. It is no longer a constant, since a is no longer a constant. In Eq. (10.124f), the n term will be obtained, after integration of the first equation, with the same approximation to the small quantities in R as the other terms. A double integration of the semimajor axis equation is necessary to obtain the mean anomaly. A double integration is always required in celestial mechanics and astrodynamics to solve the problem of a perturbed trajectory. This is an important consequence, when long-period terms and numerical integrations are involved (Kovalevsky 1963, pp. 40–42).

10.12 Small Eccentricity and Small Inclination

Since e and i appear in denominators of some equations, when they are zero the Lagrange formulae are not valid. These singularities are related to the definition of the classical elements discussed in Sect. 5.2. Similarly, when using Delaunay canonical equations, small e and i lead to problems. This is due to the choice of variables.

10.12.1 Small Eccentricity

If there is an elliptical orbit with a small eccentricity, which is subject to perturbations, these perturbations can shorten the semimajor axis, a, and increase the semiminor axis, hence decreasing the eccentricity. Then, the periapsis (and location of the semimajor axis) will become uncertain. The eccentricity can decrease to zero, and then increase again. The periapsis will change, and the mean anomaly will change by the same quantity in the opposite direction. The solutions of the osculating elements will be discontinuous, but $\omega + M$ will remain continuous. Other variables can be selected that are continuous when e passes through zero. For example

$$\eta_1 = e\sin\omega, \ \zeta_1 = e\cos\omega, \ \bar{u} = \omega + M \tag{10.125}$$

where $\bar{u}$ is the *mean argument of latitude*. This change of variables can be effected in Eqs. (10.124).

10.12.2 Small Inclination

Similarly, perturbations cause the orbital plane to change such that the inclination can go through zero, the two nodes are reversed, and the longitude of the ascending node changes by 180°. So we need to change the variables. We can introduce

$$p_1 = \tan\frac{i}{2}\sin\Omega,\ q_1 = \tan\frac{i}{2}\cos\Omega,\ \varpi = \Omega + \omega \tag{10.126}$$

where ϖ is the *longitude of the periapsis* (Kovalevsky 1963, pp. 42–43).

10.12.3 Universal Variables

The cases of zero eccentricity and zero inclination have led to *universal variables*. These are nonsingular variables that replace the classical elements, as briefly mentioned in Sect. 5.2. The universal variables do not have discontinuities for all practical elliptic orbits. One choice of such variables, which is always defined for $i \neq \pi$, $e < 1$, is the *equinoctial orbital elements*, proposed by Broucke and Cefola (1972):

$$\begin{aligned} &a,\ \hbar = e\sin(\Omega + \omega),\ k = e\cos(\Omega + \omega) \\ &p_2 = \tan\frac{i}{2}\sin\Omega,\ q_2 = \tan\frac{i}{2}\cos\Omega,\ \lambda_0 = M_0 + \omega + \Omega \end{aligned} \tag{10.127}$$

where λ_0 is the *mean longitude at epoch.* A different variation of the equinoctial elements was used by Giacaglia (1977) and Nacozy and Dallas (1977):

$$\begin{aligned} &a,\ \zeta = e\cos(\Omega + \omega),\ \eta = e\sin(\Omega + \omega) \\ &p = \sin\frac{i}{2}\cos\Omega,\ q = \sin\frac{i}{2}\sin\Omega,\ \lambda = M + \omega + \Omega \end{aligned} \tag{10.128}$$

where λ is the *mean longitude*. Alternatively, some use the *true longitude*, defined as $\ell = f + \omega + \Omega$, in Eqs. (10.127) and (10.128).

Lagrange's planetary equations for the set of elements (10.128) can be written as (Giacaglia 1977; Nacozy and Dallas 1977)

$$\dot{a} = \frac{2}{na}\frac{\partial R}{\partial \lambda} \tag{10.129a}$$

$$\dot{\lambda} = n - \frac{2}{na}\frac{\partial R}{\partial a} + \frac{\gamma}{2na^2}\left(\zeta\frac{\partial R}{\partial \zeta} + \eta\frac{\partial R}{\partial \eta}\right)$$

$$+ \frac{1}{2na^2\gamma}\left(p\frac{\partial R}{\partial p} + q\frac{\partial R}{\partial q}\right) \tag{10.129b}$$

$$\dot{\zeta} = -\frac{\zeta\gamma}{na^2(1+\gamma)}\frac{\partial R}{\partial \lambda} - \frac{\gamma}{na^2}\frac{\partial R}{\partial \eta} - \frac{\eta}{2na^2\gamma}\left(p\frac{\partial R}{\partial p} + q\frac{\partial R}{\partial q}\right) \tag{10.129c}$$

$$\dot{\eta} = -\frac{\eta\gamma}{na^2(1+\gamma)}\frac{\partial R}{\partial \lambda} + \frac{\gamma}{na^2}\frac{\partial R}{\partial \zeta} + \frac{\zeta}{2na^2\gamma}\left(p\frac{\partial R}{\partial p} + q\frac{\partial R}{\partial q}\right) \tag{10.129d}$$

$$\dot{p} = -\frac{p}{2na^2\gamma}\frac{\partial R}{\partial \lambda} - \frac{1}{4na^2\gamma}\frac{\partial R}{\partial q} + \frac{p}{2na^2\gamma}\left(\eta\frac{\partial R}{\partial \zeta} - \zeta\frac{\partial R}{\partial \eta}\right) \tag{10.129e}$$

$$\dot{q} = -\frac{q}{2na^2\gamma}\frac{\partial R}{\partial \lambda} + \frac{1}{4na^2\gamma}\frac{\partial R}{\partial p} + \frac{q}{2na^2\gamma}\left(\eta\frac{\partial R}{\partial \zeta} - \zeta\frac{\partial R}{\partial \eta}\right) \tag{10.129f}$$

where $\gamma \triangleq \sqrt{1-e^2} = \sqrt{1-\zeta^2-\eta^2}$.

References

Broucke, R.A., Cefola, P.J.: On the equinoctial orbital elements. Celest. Mech. **5**, 303–310 (1972)

Giacaglia, G.: The equations of motion of an artificial satellite in nonsingular variables. Celest. Mech. **15**, 191–215 (1977)

Kovalevsky, J.: Introduction to Celestial Mechanics. Springer, New York, D. Reidel Publishing Company, Dordrecht-Holland (1963)

Nacozy, P.E., Dallas, S.S.: The geopotential in nonsingular orbital elements. Celest. Mech. **15**, 453–466 (1977)

Chapter 11
General Perturbations Theory

11.1 Introduction

We have seen the complexity of the problem when more than two gravitating masses are involved. We have seen two methods of determining the orbits, Cowell's and Encke's methods. Now, let us look at the basic mathematical description of the perturbation problem.

Consider the planetary system. The dominant mass is the Sun, and each planet moves in nearly Keplerian two-body motion around the Sun, which is a good first approximation. The other planets do disturb this motion and slowly change the elements in the Keplerian orbit; a, e, i, Ω, ω and T are slowly changing functions of time. If the changes of these elements are known, the planets' orbital characteristics can be predicted. The distances between bodies are so large, compared with their sizes, that the planets can be treated as point masses.

The perturbations in the motion of an artificial or a natural satellite close to a planet are another type of problem. The oblateness of the planet causes large perturbations from the simple two-body orbit. The motion of artificial Earth satellites is characterized by the rotation of the lines of nodes and the perigee due to the non-central force field, which is caused by the oblateness of the Earth. Other perturbations arise due to the drag of the atmosphere acting on an artificial satellite close to the Earth's surface, the tidal effects between massive natural satellites and their primary, and solar radiation pressure. The Earth oblateness is by the far the most dominant perturbation acting on artificial satellites. We will discuss it in Chap. 12.

When perturbations act upon an orbiting mass, the resulting motion is referred to as *non-Keplerian motion*.

P. Gurfil, P.K. Seidelmann, *Celestial Mechanics and Astrodynamics: Theory and Practice*, Astrophysics and Space Science Library 436,
DOI 10.1007/978-3-662-50370-6_11

11.2 Variation of Parameters

To solve for orbits resulting from non-Keplerian motion, Euler and Lagrange developed the variation of parameters (VOP) method, mentioned in Sect. 9.11, which is a method for solving non-homogenous, nonlinear, differential equations. In essence, the VOP method suggests to turn the constants of integration related to the unperturbed motion, resulting from the homogenous solution of a given differential equation, into functions of time.

Let us introduce the concept of the VOP method. The equations of motion of a particle may be put in the form

$$\ddot{x} + \frac{\mu x}{r^3} = X, \; \ddot{y} + \frac{\mu y}{r^3} = Y, \; \ddot{z} + \frac{\mu z}{r^3} = Z \tag{11.1}$$

where the second terms on the left-hand side are the components of the relative acceleration produced by the central mass, which is placed at the origin of the coordinate system. The right-hand side are the perturbing accelerations produced by all the other forces that affect the motion. If the right-hand side terms are equated to zero, that is we have central-force motion, then

$$x = x(c_1, \ldots, c_6, t), \; \dot{x} = \dot{x}(c_1, \ldots, c_6, t) \tag{11.2a}$$

$$y = y(c_1, \ldots, c_6, t), \; \dot{y} = \dot{y}(c_1, \ldots, c_6, t) \tag{11.2b}$$

$$z = z(c_1, \ldots, c_6, t), \; \dot{z} = \dot{z}(c_1, \ldots, c_6, t) \tag{11.2c}$$

These are expressions of the rectangular coordinates of Keplerian motion in terms of time and six constants of integration. For Keplerian motion, the elements are constants. In the method of VOP, the problem is to satisfy Eqs. (11.1) by the formulas of Eqs. (11.2) that apply to Keplerian motion. Obviously, $c_1, c_2, \ldots, c_6$ can no longer be constants; instead, they become functions of time. The c_k are identical in the Keplerian case to a, e, i, ω, Ω and T, or some combination of them. So the problem is to determine the time rates of change of the c_k.

Consider the elements c_k at any fixed instant. They define the osculating orbit at that time. The planet, or the artificial satellite, at that instant has the same coordinates and velocity components in the unperturbed and the perturbed orbit. That is to say, the planet has the position and is moving instantaneously with a purely two-body motion, as given by the elements c_k. A set of elements defining an osculating orbit could be computed at any point in the actual orbit.

We have previously derived the equations of motion of a mass m around a central mass m_1 under the perturbing influence of a mass m' as

$$\ddot{x} = -k^2 \frac{Mx}{r^3} + \frac{\partial R}{\partial x} \tag{11.3}$$

$$\ddot{y} = -k^2 \frac{My}{r^3} + \frac{\partial R}{\partial y} \tag{11.4}$$

$$\ddot{z} = -k^2 \frac{Mz}{r^3} + \frac{\partial R}{\partial z} \tag{11.5}$$

where

$$R = k^2 m' \left[\frac{1}{\rho} - \frac{xx' + yy' + zz'}{r'^3} \right] \tag{11.6}$$

In vector form those are

$$\ddot{\mathbf{r}} + \frac{\mu \mathbf{r}}{r^3} = \nabla R \tag{11.7}$$

where $\mu = k^2(m_1 + m)$ and

$$\nabla R = \frac{\partial R}{\partial x}\hat{\mathbf{i}} + \frac{\partial R}{\partial y}\hat{\mathbf{j}} + \frac{\partial R}{\partial z}\hat{\mathbf{k}} \tag{11.8}$$

Let us consider what is happening for clarity. Suppose m and m' are moving in circular orbits of radii a and $a' > a$ around the central mass, m_1, as shown in Fig. 11.1. We will consider only the x component of motion.

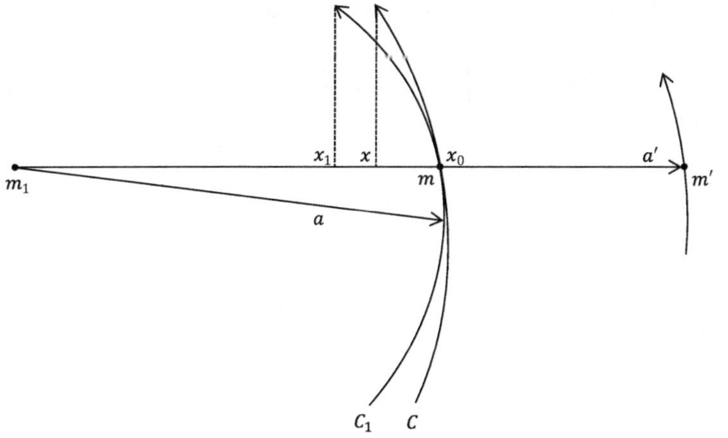

Fig. 11.1 A mass m moving in orbit around a central mass m_1, and perturbed by a mass m'

Assume m' was not present, then the mass m, in an interval of time, τ, would move along the osculating orbit C_1, so that x_0 changes to x_1. Actually, due to the presence of m', the mass m moves along C, the true orbit, so that x_0 changes to x in the same time, τ. For simplicity, we assume that the time interval is small enough so x and x_1 can be written with sufficient accuracy as

$$x = x_0 + \left(\frac{dx}{dt}\right)_0 \tau + \left(\frac{d^2x}{dt^2}\right)_0 \frac{\tau^2}{2} + \ldots \tag{11.9}$$

$$x_1 = x_0 + \left(\frac{\partial x}{\partial t}\right)_0 \tau + \left(\frac{\partial^2 x}{\partial t^2}\right)_0 \frac{\tau^2}{2} + \ldots \tag{11.10}$$

where the subscripts indicate evaluation of the derivatives at the osculating point, x_0. The perturbation due to m' is

$$\begin{aligned} x - x_1 &= \delta x \\ &= \left[\left(\frac{dx}{dt}\right)_0 - \left(\frac{\partial x}{\partial t}\right)_0\right]\tau + \left[\left(\frac{d^2x}{dt^2}\right)_0 - \left(\frac{\partial^2 x}{\partial t^2}\right)_0\right]\frac{\tau^2}{2} \\ &+ \ldots \end{aligned} \tag{11.11}$$

The total derivatives are distinct from the partial derivatives. The total derivative refers to the true orbit and includes the changes in the orbital elements, that is

$$\frac{dx}{dt} = \frac{\partial x}{\partial t} + \sum_{k=1}^{6} \frac{\partial x}{\partial c_k}\dot{c}_k \tag{11.12}$$

The partial derivatives refer to the osculating orbit where the c_k are constants. If the velocity on the osculating orbit at x_0 and that in the true orbit are the same (we will return to this assumption later on), then

$$\left(\frac{dx}{dt}\right)_0 - \left(\frac{\partial x}{\partial t}\right)_0 = 0 \tag{11.13}$$

So, by Eq. (11.7) we have

$$\left(\frac{d^2x}{dt^2}\right)_0 + \frac{\mu x_0}{r_0^3} = \left(\frac{\partial R}{\partial x}\right)_0 \tag{11.14}$$

and, for the osculating orbit

$$\left(\frac{\partial^2 x}{\partial t^2}\right)_0 + \frac{\mu x_0}{r_0^3} = 0 \tag{11.15}$$

Therefore,

$$\left(\frac{d^2x}{dt^2}\right)_0 - \left(\frac{\partial^2 x}{\partial t^2}\right)_0 = \left(\frac{\partial R}{\partial x}\right)_0 \tag{11.16}$$

The perturbation, Eq. (11.11), is

$$\delta x = \frac{1}{2}\tau^2 \left(\frac{\partial R}{\partial x}\right)_0 = \frac{1}{2}\tau^2 k^2 m' \left[\frac{x' - x}{\rho^3} - \frac{x'}{r'^3}\right]_0 \tag{11.17}$$

by Eq. (11.8). For an order of magnitude calculation, assume the masses are approximately aligned so $x' - x \approx a' - a$, and ignore perturbations in the y and z directions. For convenience, express the time interval, τ, in units of the period, P, of m around m_1. From Kepler's third law, the period is approximately

$$P^2 = \frac{4\pi^2 a^3}{k^2 m_1} \quad \text{if} \quad m_1 \gg m \tag{11.18}$$

Introducing this into Eq. (11.17)

$$\delta x = 2\pi^2 a^3 \left(\frac{m'}{m_1}\right) \left(\frac{\tau}{P}\right)^2 \left[\frac{1}{(a' - a)^2} - \frac{1}{a'^2}\right] \tag{11.19}$$

For an order of magnitude calculation of a perturbation, take the effect of Jupiter on an asteroid. Take an asteroid with a period half of Jupiter's period. From Kepler's third law, the asteroid would have mean distance $a = 3.3$ AU. In Eq. (11.19), use $a' = 5.2$ and $m'/m = 0.001$ to find $\delta x \approx 0.16(\tau/P)^2$ AU. Assume the effective contact for the perturbation lasts 0.1 period of the asteroid; then the deviation becomes 0.0016 AU=240,000 km. This is a sizable displacement, reflecting the order of magnitude of perturbations due to Jupiter.

Perturbations of this kind may be involved in the Kirkwood gaps in the minor planets and the Cassini divisions in the rings of Saturn (see also Sect. 1.5.3), which are due in part to the satellite Mimas. These are cases of *secular resonances* and/or chaos.

Now, let us return to the general discussion of the perturbation equation (11.7). From the solution equation in vector form

$$\mathbf{r} = \mathbf{r}(c_1, c_2, c_3, c_4, c_5, c_6, t) \tag{11.20}$$

with $\mathbf{r} = x\hat{\mathbf{i}} + y\hat{\mathbf{j}} + z\hat{\mathbf{k}}$ and the c_k functions of time,

$$\dot{\mathbf{r}} = \frac{\partial \mathbf{r}}{\partial t} + \sum_{k=1}^{6} \frac{\partial \mathbf{r}}{\partial c_k}\dot{c}_k \tag{11.21}$$

The $\dot{c}_k$ are defined such that

$$\sum_{k=1}^{6} \frac{\partial \mathbf{r}}{\partial c_k} \dot{c}_k = \mathbf{0} \tag{11.22}$$

As explained in the simplified example, the velocity in the osculating orbit

$$\frac{\partial \mathbf{r}}{\partial t} = \frac{\partial x}{\partial t}\hat{\mathbf{i}} + \frac{\partial y}{\partial t}\hat{\mathbf{j}} + \frac{\partial z}{\partial t}\hat{\mathbf{k}} \tag{11.23}$$

is the same as in the true orbit, i.e. $\dot{\mathbf{r}} = \dot{x}\hat{\mathbf{i}} + \dot{y}\hat{\mathbf{j}} + \dot{z}\hat{\mathbf{k}}$.

Note, however, that the constraint imposed by Eq. (11.22) is one particular choice, and is also known as the *Lagrange constraint* or *osculation constraint*. Mathematically, this restriction confines the dynamics of the orbital state space to a 9-dimensional submanifold of the 12-dimensional manifold of the orbital elements and their time derivatives. More importantly, this reflects an internal freedom in the mapping $(\mathbf{r}, \dot{\mathbf{r}}) \mapsto (c_k, \dot{c}_k, k = 1, \ldots, 6)$.

Physically, the Lagrange constraint postulates that the trajectory in the inertial space is always tangential to an "instantaneous" ellipse (or hyperbola) defined by the "instantaneous" values of the time-varying orbital elements $c_k(t)$, meaning that the perturbed physical trajectory would coincide with the Keplerian orbit that the body would follow if the perturbing force was to cease instantaneously. As we mentioned previously, this instantaneous orbit is the osculating orbit. Accordingly, the orbital elements which satisfy the Lagrange constraint are osculating orbital elements. The Lagrange constraint , however, is not unique. The generalized form of the Lagrange constraint can be written as

$$\sum_{k=1}^{6} \frac{\partial \mathbf{r}}{\partial c_k} \dot{c}_k = \mathbf{\Phi}(c_k, \dot{c}_k, t) \tag{11.24}$$

where the velocity $\mathbf{\Phi}$ is an arbitrary, user-defined function of the orbital elements, their time derivatives and time. Equation (11.21) then becomes

$$\dot{\mathbf{r}} = \frac{\partial \mathbf{r}}{\partial t} + \mathbf{\Phi}(c_k, \dot{c}_k, t) \tag{11.25}$$

The observation regarding the internal freedom embedded in transforming from the inertial coordinates and velocities to the orbital elements rates of change has been made by Brouwer and Clemence (1961), King-Hele (1958), and Cook (1963). Efroimsky et al. have published key works on planetary equations with a generalized Lagrange constraint (Efroimsky 2002; Newman and Efroimsky 2003; Efroimsky and Goldreich 2003, 2004). They termed the constraint function $\mathbf{\Phi}$ *gauge function* or *gauge velocity*, which are terms taken from the field of electrodynamics. The zero gauge $\mathbf{\Phi} \equiv \mathbf{0}$ was termed the *Lagrange gauge*.

The use of a generalized Lagrange constraint gives rise to *non-osculating orbital elements*. Thus, although the description of the physical orbit in the inertial Cartesian configuration space remains invariant to a particular selection of a gauge velocity, its description in the orbital elements space depends on whether osculating or non-osculating orbital elements are used. However, to avoid complications, we will adhere to the Lagrange gauge, and assume from now on that the orbital elements are osculating, i. e., $\mathbf{\Phi} \equiv 0$.

Proceeding with the constraint (11.22) and differentiating Eq. (11.21) with respect to t, we obtain

$$\ddot{\mathbf{r}} = \frac{\partial^2 \mathbf{r}}{\partial t^2} + \sum_{k=1}^{6} \frac{\partial^2 \mathbf{r}}{\partial t \partial c_k} \dot{c}_k \tag{11.26}$$

Substitution of Eq. (11.26) into Eq. (11.7) yields

$$\frac{\partial^2 \mathbf{r}}{\partial t^2} + \frac{\mu \mathbf{r}}{r^3} + \sum_{k=1}^{6} \frac{\partial^2 \mathbf{r}}{\partial t \partial c_k} \dot{c}_k = \nabla R \tag{11.27}$$

For the osculating orbit $R = 0$ and the c_k are constants, so

$$\frac{\partial^2 \mathbf{r}}{\partial t^2} + \frac{\mu \mathbf{r}}{r^3} = 0 \tag{11.28}$$

and Eq. (11.27) reduces to

$$\sum_{k=1}^{6} \frac{\partial^2 \mathbf{r}}{\partial t \partial c_k} \dot{c}_k = \nabla R \tag{11.29}$$

The time derivatives of the elements can be determined by simultaneously solving Eqs. (11.22) and (11.29) for the $\dot{c}_k$. Before proceeding, we can simplify Eq. (11.29) by noting that

$$\frac{\partial^2 \mathbf{r}}{\partial t \partial c_k} = \frac{\partial}{\partial c_k} \left(\frac{\partial \mathbf{r}}{\partial t} \right) = \frac{\partial \dot{\mathbf{r}}}{\partial c_k} \tag{11.30}$$

This is from the definition, $\dot{\mathbf{r}} = \partial \mathbf{r} / \partial t$, at the osculating orbit. Equation (11.29) can be written

$$\sum_{k=1}^{6} \frac{\partial \dot{\mathbf{r}}}{\partial c_k} \dot{c}_k = \nabla R \tag{11.31}$$

The solution for the $\dot{c}_k$ from Eqs. (11.22) and (11.31) can be performed by a rearrangement that introduces new functions of the parameters c_k, called *Lagrange brackets*. Take the scalar product of Eq. (11.31) by $\partial\mathbf{r}/\partial c_j$, and the scalar product of Eq. (11.22) by $\partial\dot{\mathbf{r}}/\partial c_j$, and subtract to obtain

$$\sum_{k=1}^{6}\left[\frac{\partial\mathbf{r}}{\partial c_j}\cdot\frac{\partial\dot{\mathbf{r}}}{\partial c_k}-\frac{\partial\mathbf{r}}{\partial c_k}\cdot\frac{\partial\dot{\mathbf{r}}}{\partial c_j}\right]\dot{c_k}=\nabla R\cdot\frac{\partial\mathbf{r}}{\partial c_j},\ j=1,\ldots,6 \tag{11.32}$$

The quantity in brackets in Eq. (11.32) is Lagrange's bracket and will be denoted by $[c_j, c_k]$. In terms of its Cartesian components,

$$[c_j,c_k]=\frac{\partial(x,\dot{x})}{\partial(c_j,c_k)}+\frac{\partial(y,\dot{y})}{\partial(c_j,c_k)}+\frac{\partial(z,\dot{z})}{\partial(c_j,c_k)} \tag{11.33}$$

where

$$\frac{\partial(\mathbf{x},\dot{\mathbf{x}})}{\partial(\mathbf{c}_j,\mathbf{c}_k)}\triangleq\begin{vmatrix}\dfrac{\partial x}{\partial c_j} & \dfrac{\partial x}{\partial c_k}\\ \dfrac{\partial\dot{x}}{\partial c_j} & \dfrac{\partial\dot{x}}{\partial c_j}\end{vmatrix} \tag{11.34}$$

with similar expressions for y and z. The determinant in Eq. (11.34) is the Jacobian of x and $\dot{x}$ with respect to c_j and c_k. The right-hand side of Eq. (11.32) is the partial derivative of R with respect to c_j. So, Eq. (11.32) can be written as

$$\sum_{k=1}^{6}[c_j,c_k]\dot{c_k}=\frac{\partial R}{\partial c_j},\ j=1,\ldots,6 \tag{11.35}$$

These six equations are to be solved for $\dot{c_k}$.

To illustrate this method, consider the one-dimensional linear oscillator, whose differential equation of motion is

$$\ddot{x}+x=R(t) \tag{11.36}$$

The solution of this equation when $R=0$ is $x=c_1\sin t+c_2\cos t$. Allow c_1 and c_2 to be functions of time and write

$$\dot{x}=(c_1\cos t-c_2\sin t)+(\dot{c_1}\sin t+\dot{c_2}\cos t) \tag{11.37}$$

The first parenthesis is $\partial x/\partial t$ and the second is the analogue of the summation on the right-hand side of Eq. (11.21), i.e. $\sum_{k=1}^{6}\frac{\partial\mathbf{r}}{\partial c_k}\dot{c_k}$. So similarly to Eq. (11.22) (the Lagrange constraint) we set

$$\dot{c_1}\sin t+\dot{c_2}\cos t=0 \tag{11.38}$$

Differentiating once more, we find

$$\ddot{x} = (-c_1 \sin t - c_2 \cos t) + (\dot{c_1} \cos t - \dot{c_2} \sin t) \tag{11.39}$$

The first parenthesis in Eq. (11.39) is $\frac{\partial^2 x}{\partial t^2}$ and the second is $\dot{c_1}\frac{\partial^2 x}{\partial t \partial c_1} + \dot{c_2}\frac{\partial^2 x}{\partial t \partial c_2}$, which is the analogue of the summation on the right-hand side of Eq. (11.26). Substitution of Eq. (11.39) into Eq. (11.36) yields

$$\dot{c_1} \cos t - \dot{c_2} \sin t = R(t) \tag{11.40}$$

So, we have to solve simultaneously

$$\dot{c_1} \sin t + \dot{c_2} \cos t = 0, \ \dot{c_1} \cos t - \dot{c_2} \sin t = R(t) \tag{11.41}$$

for $\dot{c_1}$ and $\dot{c_2}$. Observe that

$$\frac{\partial}{\partial c_1}\left(\frac{\partial x}{\partial t}\right) = \frac{\partial \dot{x}}{\partial c_1} = \cos t \tag{11.42}$$

$$\frac{\partial}{\partial c_2}\left(\frac{\partial x}{\partial t}\right) = \frac{\partial \dot{x}}{\partial c_2} = -\sin t \tag{11.43}$$

which are analogous to Eq. (11.30). Also,

$$\frac{\partial x}{\partial c_1} = \sin t, \ \frac{\partial x}{\partial c_2} = \cos t \tag{11.44}$$

Thus, Eqs. (11.41) can be written as

$$\dot{c_1}\frac{\partial \dot{x}}{\partial c_1} + \dot{c_2}\frac{\partial \dot{x}}{\partial c_2} = R(t) \tag{11.45}$$

$$\dot{c_1}\frac{\partial x}{\partial c_1} + \dot{c_2}\frac{\partial x}{\partial c_2} = 0 \tag{11.46}$$

Multiplying the first by $\partial x/\partial c_j$ and the second by $\partial \dot{x}/\partial c_j$, $j = 1, 2$ and subtracting, we have

$$\dot{c_1}\left[\frac{\partial x}{\partial c_j}\frac{\partial \dot{x}}{\partial c_1} - \frac{\partial x}{\partial c_1}\frac{\partial \dot{x}}{\partial c_j}\right] + \dot{c_2}\left[\frac{\partial x}{\partial c_j}\frac{\partial \dot{x}}{\partial c_2} - \frac{\partial x}{\partial c_2}\frac{\partial \dot{x}}{\partial c_j}\right]$$

$$= R\frac{\partial x}{\partial c_j}, \ j = 1, 2 \tag{11.47}$$

and introducing Lagrange's brackets,

$$[c_j, c_1]\dot{c_1} + [c_j, c_2]\dot{c_2} = R\frac{\partial x}{\partial c_j},\ j = 1, 2 \tag{11.48}$$

This is analogous to Eq. (11.35). It may be shown then from the values of partials given that

$$[c_1, c_1] = [c_2, c_2] = 0,\ [c_1, c_2] = -1,\ [c_2, c_1] = 1 \tag{11.49}$$

so Eq. (11.48) for $j = 1, 2$ reduces to

$$\dot{c_2} = -R\sin t,\ \dot{c_1} = R\cos t \tag{11.50}$$

$c_1(t)$ and $c_2(t)$ follow by integration, when $R(t)$ is known (McCuskey 1963, pp. 128–134).

11.3 Properties of the Lagrange Brackets

By the definition of $[c_j, c_k]$ given in Eq. (11.33), or, in vector notation,

$$[c_j, c_k] = \left[\frac{\partial \mathbf{r}}{\partial c_j}\cdot\frac{\partial \dot{\mathbf{r}}}{\partial c_k} - \frac{\partial \mathbf{r}}{\partial c_k}\cdot\frac{\partial \dot{\mathbf{r}}}{\partial c_j}\right] \tag{11.51}$$

then we see

$$[c_j, c_j] = 0,\ [c_k, c_j] = -[c_j, c_k] \tag{11.52}$$

Another important property is

$$\frac{\partial}{\partial t}\left[c_j, c_k\right] = 0 \tag{11.53}$$

which implies that the $[c_j, c_k]$ are explicitly independent of the time. This is convenient, because the brackets can be computed for any epoch, such as the perihelion, and they are invariant with respect to t thereafter.

The proof of the property stated in Eq. (11.53) can be given as follows. Denote any one bracket by $[p, q]$ and one of the Jacobians by $J[(x, \dot{x})/(p, q)]$ as defined in

Eq. (11.34). Then

$$\frac{\partial J}{\partial t} = \frac{\partial}{\partial t}\left[\frac{\partial x}{\partial p}\frac{\partial \dot{x}}{\partial q} - \frac{\partial x}{\partial q}\frac{\partial \dot{x}}{\partial p}\right] = \frac{\partial^2 x}{\partial t \partial p}\frac{\partial \dot{x}}{\partial q} - \frac{\partial^2 x}{\partial t \partial q}\frac{\partial \dot{x}}{\partial p}$$
$$+ \frac{\partial^2 \dot{x}}{\partial t \partial q}\frac{\partial x}{\partial p} - \frac{\partial^2 \dot{x}}{\partial t \partial p}\frac{\partial x}{\partial q} \tag{11.54}$$

But $\partial x/\partial t = \dot{x}$ by the definition of the osculating orbit; so, the first two terms become

$$\frac{\partial \dot{x}}{\partial p}\frac{\partial \dot{x}}{\partial q} - \frac{\partial \dot{x}}{\partial q}\frac{\partial \dot{x}}{\partial p} = 0 \tag{11.55}$$

The last two terms of Eq. (11.54) can be written

$$\frac{\partial}{\partial q}\left(\frac{\partial \dot{x}}{\partial t}\right)\frac{\partial x}{\partial p} - \frac{\partial}{\partial p}\left(\frac{\partial \dot{x}}{\partial t}\right)\frac{\partial x}{\partial q} \tag{11.56}$$

By the equations of motion for the osculating orbit

$$\frac{\partial \dot{x}}{\partial t} \equiv \frac{\partial^2 x}{\partial t^2} = -\frac{\mu x}{r^3} = \frac{\partial V}{\partial x} \tag{11.57}$$

where $V = \mu/r$. Equation (11.54) can be in the form

$$\frac{\partial J}{\partial t} = \frac{\partial}{\partial q}\left(\frac{\partial V}{\partial x}\right)\frac{\partial x}{\partial p} - \frac{\partial}{\partial p}\left(\frac{\partial V}{\partial x}\right)\frac{\partial x}{\partial q} \tag{11.58}$$

There are similar expressions with y and z substituted for x. $V(x, y, z)$ is continuous and has continuous derivatives at all points except the origin. So, we can change the order of differentiation and add the terms in y and z,

$$\frac{\partial J}{\partial t} = \frac{\partial [p, q]}{\partial t} = \left[\frac{\partial V_q}{\partial x}\frac{\partial x}{\partial p} + \frac{\partial V_q}{\partial y}\frac{\partial y}{\partial p} + \frac{\partial V_q}{\partial z}\frac{\partial z}{\partial p}\right] \tag{11.59}$$
$$-\left[\frac{\partial V_p}{\partial x}\frac{\partial x}{\partial q} + \frac{\partial V_P}{\partial y}\frac{\partial y}{\partial q} + \frac{\partial V_p}{\partial z}\frac{\partial z}{\partial q}\right] \tag{11.60}$$

where $V_p \equiv \partial V/\partial p$ and $V_q \equiv \partial V/\partial q$. The first group of terms is $\partial V_q/\partial p$ and the second is $\partial V_p/\partial q$. Also, these second partial derivatives are equal. So

$$\frac{\partial [p, q]}{\partial t} = \frac{\partial V_q}{\partial p} - \frac{\partial V_p}{\partial q} = 0 \tag{11.61}$$

Thus, the Lagrange bracket is explicitly independent of the time (McCuskey 1963, pp. 135–136).

11.4 Evaluation of the Lagrange Brackets

The Lagrange brackets $[c_j, c_k]$, defined previously, must be evaluated in terms of the orbital elements, so that Eq. (11.35) may be solved for the $\dot{c}_k$, $k = 1, \ldots, 6$. The procedure will be illustrated and the complete set of values developed. For further details, see Smart (1960).

To begin, recall the definition of the perifocal coordinate system defined in Sect. 5.2, that is, $\hat{\mathbf{P}}$ is a unit vector in the direction of the periapsis, $\hat{\mathbf{Q}}$ is a unit vector in the direction $f = 90°$, and $\hat{\mathbf{R}}$ is a unit vector perpendicular to the orbital plane. The latter is directed in the sense $\hat{\mathbf{R}} = \hat{\mathbf{P}} \times \hat{\mathbf{Q}}$. Then

$$\begin{aligned}\hat{\mathbf{P}} &= (\cos\omega\cos\Omega - \sin\omega\sin\Omega\cos i)\hat{\mathbf{i}} \\ &\quad + (\cos\omega\sin\Omega + \sin\omega\cos\Omega\cos i)\hat{\mathbf{j}} + \sin\omega\sin i\,\hat{\mathbf{k}}\end{aligned} \tag{11.62a}$$

$$\begin{aligned}\hat{\mathbf{Q}} &= (-\sin\omega\cos\Omega - \cos\omega\sin\Omega\cos i)\hat{\mathbf{i}} \\ &\quad + (-\sin\omega\sin\Omega + \cos\omega\cos\Omega\cos i)\hat{\mathbf{j}} + \cos\omega\sin i\,\hat{\mathbf{k}}\end{aligned} \tag{11.62b}$$

$$\hat{\mathbf{R}} = \sin\Omega\sin i\,\hat{\mathbf{i}} - \cos\Omega\sin i\,\hat{\mathbf{j}} + \cos i\,\hat{\mathbf{k}} \tag{11.62c}$$

where $\hat{\mathbf{i}}, \hat{\mathbf{j}}, \hat{\mathbf{k}}$ are unit vectors along x, y, z, respectively (recall Figs. 5.1, 5.7).

The unit vectors $\hat{\mathbf{P}}, \hat{\mathbf{Q}}, \hat{\mathbf{R}}$ depend only on the orbital elements Ω, ω, i, which orient the orbit in space, and not upon a, e, T, which define the shape of the orbit and position of the body on the orbit. The elements c_k fall into two groups of three each. Let us denote the group a, e, T by $\alpha_1, \alpha_2, \alpha_3$, and the group Ω, ω, i by $\beta_1, \beta_2, \beta_3$. The Lagrange brackets are in three categories: $[\alpha_r, \alpha_s]$, $[\alpha_r, \beta_s]$, $[\beta_r, \beta_s]$, where $r \neq s$; $r, s = 1, 2, 3$. We evaluate the first of these to illustrate the procedure. We can simplify Eqs. (11.62) to

$$\hat{\mathbf{P}} = P_1\hat{\mathbf{i}} + P_2\hat{\mathbf{j}} + P_3\hat{\mathbf{k}} \tag{11.63}$$

$$\hat{\mathbf{Q}} = Q_1\hat{\mathbf{i}} + Q_2\hat{\mathbf{j}} + Q_3\hat{\mathbf{k}} \tag{11.64}$$

$$\hat{\mathbf{R}} = R_1\hat{\mathbf{i}} + R_2\hat{\mathbf{j}} + R_3\hat{\mathbf{k}} \tag{11.65}$$

where the components of $\hat{\mathbf{P}}, \hat{\mathbf{Q}}, \hat{\mathbf{R}}$ are the direction cosines of the respective vectors relative to the x, y, z axes. At any instant, the radius vector can be expressed by

$$\mathbf{r} = x\hat{\mathbf{i}} + y\hat{\mathbf{j}} + z\hat{\mathbf{k}} = \xi\hat{\mathbf{P}} + \eta\hat{\mathbf{Q}} \tag{11.66}$$

where ξ, η are Cartesian coordinates in the orbital plane. From Eq. (11.66), the velocity can be written

$$\dot{\mathbf{r}} = \dot{x}\hat{\mathbf{i}} + \dot{y}\hat{\mathbf{j}} + \dot{z}\hat{\mathbf{k}} = \dot{\xi}\hat{\mathbf{P}} + \dot{\eta}\hat{\mathbf{Q}} + \xi\dot{\hat{\mathbf{P}}} + \eta\dot{\hat{\mathbf{Q}}} \tag{11.67}$$

By the definition of the osculating orbit, $\dot{\mathbf{r}} = \partial\mathbf{r}/\partial t$, and $\xi\,\dot{\hat{\mathbf{P}}} + \eta\,\dot{\hat{\mathbf{Q}}} = 0$, so

$$\dot{\mathbf{r}} = \dot{\xi}\,\hat{\mathbf{P}} + \dot{\eta}\,\hat{\mathbf{Q}} \tag{11.68}$$

From Eqs. (11.66)–(11.68), $x, \dot{x}, y, \dot{y}, z, \dot{z}$ can be expressed in terms of the new coordinates and velocity components ξ, η, $\dot{\xi}$, $\dot{\eta}$, whenever desired. From the definition the Lagrangian bracket, $[\alpha_r, \alpha_s]$ is

$$\frac{\partial(x,\dot{x})}{\partial(\alpha_r,\alpha_s)} + \frac{\partial(y,\dot{y})}{\partial(\alpha_r,\alpha_s)} + \frac{\partial(z,\dot{z})}{\partial(\alpha_r,\alpha_s)} =$$

$$\begin{vmatrix} \dfrac{\partial x}{\partial\alpha_r} & \dfrac{\partial x}{\partial\alpha_s} \\ \dfrac{\partial \dot{x}}{\partial\alpha_r} & \dfrac{\partial \dot{x}}{\partial\alpha_s} \end{vmatrix} \begin{vmatrix} \dfrac{\partial y}{\partial\alpha_r} & \dfrac{\partial y}{\partial\alpha_s} \\ \dfrac{\partial \dot{y}}{\partial\alpha_r} & \dfrac{\partial \dot{y}}{\partial\alpha_s} \end{vmatrix} \begin{vmatrix} \dfrac{\partial z}{\partial\alpha_r} & \dfrac{\partial z}{\partial\alpha_s} \\ \dfrac{\partial \dot{z}}{\partial\alpha_r} & \dfrac{\partial \dot{z}}{\partial\alpha_s} \end{vmatrix} \tag{11.69}$$

In place of $\partial x/\partial\alpha_r$, $\partial\dot{x}/\partial\alpha_r$ write by Eqs. (11.66) and (11.68)

$$\frac{\partial x}{\partial\alpha_r} = \frac{\partial x}{\partial\xi}\frac{\partial\xi}{\partial\alpha_r} + \frac{\partial x}{\partial\eta}\frac{\partial\eta}{\partial\alpha_r} \tag{11.70}$$

$$\frac{\partial \dot{x}}{\partial\alpha_r} = \frac{\partial \dot{x}}{\partial\dot{\xi}}\frac{\partial\dot{\xi}}{\partial\alpha_r} + \frac{\partial \dot{x}}{\partial\dot{\eta}}\frac{\partial\dot{\eta}}{\partial\alpha_r} \tag{11.71}$$

ålong with corresponding terms in y and z. Furthermore,

$$\frac{\partial x}{\partial\xi} = P_1, \quad \frac{\partial\dot{x}}{\partial\dot{\xi}} = P_1 \tag{11.72}$$

$$\frac{\partial x}{\partial\eta} = Q_1, \quad \frac{\partial\dot{x}}{\partial\dot{\eta}} = Q_1 \tag{11.73}$$

with similar terms for y and z. Substituting into Eq. (11.69) we have

$$\begin{vmatrix} P_1\dfrac{\partial\xi}{\partial\alpha_r} + Q_1\dfrac{\partial\eta}{\partial\alpha_r} & P_1\dfrac{\partial\xi}{\partial\alpha_s} + Q_1\dfrac{\partial\eta}{\partial\alpha_s} \\ P_1\dfrac{\partial\dot{\xi}}{\partial\alpha_r} + Q_1\dfrac{\partial\dot{\eta}}{\partial\alpha_r} & P_1\dfrac{\partial\dot{\xi}}{\partial\alpha_s} + Q_1\dfrac{\partial\dot{\eta}}{\partial\alpha_s} \end{vmatrix} \tag{11.74}$$

for the first determinant and similar expressions for the second and third determinants.

The direction cosines have the properties

$$P_1^2 + P_2^2 + P_3^2 = 1 \tag{11.75}$$

$$Q_1^2 + Q_2^2 + Q_3^2 = 1 \tag{11.76}$$

$$P_1Q_1 + P_2Q_2 + P_3Q_3 = 0 \tag{11.77}$$

When the determinants are evaluated and the terms collected, we have

$$\begin{aligned}[\alpha_r, \alpha_s] &= \begin{vmatrix} \dfrac{\partial \xi}{\partial \alpha_r} & \dfrac{\partial \xi}{\partial \alpha_s} \\ \dfrac{\partial \dot{\xi}}{\partial \alpha_r} & \dfrac{\partial \dot{\xi}}{\partial \alpha_s} \end{vmatrix} + \begin{vmatrix} \dfrac{\partial \eta}{\partial \alpha_r} & \dfrac{\partial \eta}{\partial \alpha_s} \\ \dfrac{\partial \dot{\eta}}{\partial \alpha_r} & \dfrac{\partial \dot{\eta}}{\partial \alpha_s} \end{vmatrix} \\ &= \frac{\partial(\xi, \dot{\xi})}{\partial(\alpha_r, \alpha_s)} + \frac{\partial(\eta, \dot{\eta})}{\partial(\alpha_r, \alpha_s)} \end{aligned} \tag{11.78}$$

To evaluate these Jacobians, we use the time-invariance property of the bracket, proved previously, and do the evaluation at the periapsis. We know the bracket is constant for all t. We assume a value of t such that the mean anomaly, $M = n(t-T)$, is small, and the eccentric anomaly, $E \approx 0$. Then, we may write to terms of order 3,

$$\sin E = E - \frac{E^3}{3!} + \ldots \tag{11.79}$$

and by Kepler's equation

$$E - e\sin E = n(t - T) \tag{11.80}$$

$$E - eE + e\frac{eE^3}{3!} = n(t - T) \tag{11.81}$$

Neglecting the term eE^3 in comparison with eE, we have

$$E = \frac{n(t - T)}{1 - e} \tag{11.82}$$

and

$$\sin E \approx \frac{n(t - t)}{1 - e}, \quad \cos E \approx 1 - \frac{n^2(t - T)^2}{2(1 - e)^2} \tag{11.83}$$

By definitions of ξ, η and the elliptic equations

$$\xi = r\cos f = acosE - ae, \quad \eta = r\sin f = a\sin E\sqrt{1 - e^2} \tag{11.84}$$

and from Eqs. (11.83)

$$\xi = a\left[1 - \frac{n^2(t-T)^2}{2(1-e)^2}\right] - ae \tag{11.85a}$$

$$\eta = \frac{na\sqrt{1-e^2}(t-T)}{1-e} = na\sqrt{\frac{1+e}{1-e}}(t-T) \tag{11.85b}$$

Differentiating Eqs. (11.85), we obtain

$$\dot{\xi} = -\frac{n^2 a}{(1-e)^2}(t-T), \ \dot{\eta} = na\sqrt{\frac{1+e}{1-e}} \tag{11.86}$$

Before taking partials of $\xi, \eta, \dot{\xi}, \dot{\eta}$ with respect to a, we must express n in terms of a by Kepler's third law, $n^2 a^3 = \mu$. Then we can evaluate the Jacobians in Eq. (11.78). For example, let $[\alpha_r, \alpha_s] = [a, e]$. Then Eqs. (11.85) yield

$$\frac{\partial \xi}{\partial a} = \left[1 - \frac{n^2(t-T)^2}{2(1-e)^2}\right] + a\left[-\frac{n(\partial n/\partial a)(t-T)^2}{(1-e)^2}\right] - e \tag{11.87}$$

and at $t = T$, this becomes

$$\left(\frac{\partial \xi}{\partial a}\right)_T = 1 - e \tag{11.88}$$

In a similar way

$$\left(\frac{\partial \dot{\xi}}{\partial a}\right)_T = \frac{\partial}{\partial a}\left[-\frac{\mu\,(t-T)}{a^2\,(1-e)^2}\right]_T = \left[\frac{2\mu\,(t-T)}{a^3\,(1-e)^2}\right]_T = 0 \tag{11.89}$$

$$\left(\frac{\partial \xi}{\partial e}\right)_T = \left[-\frac{n^2 a\,(t-T)^2}{(1-e)^3} - a\right]_T = -a \tag{11.90}$$

$$\left(\frac{\partial \dot{\xi}}{\partial e}\right)_T = \left[-\frac{2n^2 a\,(t-T)}{(1-e)^3}\right]_T = 0 \tag{11.91}$$

Then the Jacobian

$$\frac{\partial(\xi, \dot{\xi})}{\partial(a, e)} = 0 \tag{11.92}$$

It can be shown that

$$\left(\frac{\partial \eta}{\partial a}\right)_T = 0, \quad \left(\frac{\partial \dot{\eta}}{\partial a}\right)_T = -\frac{n}{2}\sqrt{\frac{1+e}{1-e}} \tag{11.93a}$$

$$\left(\frac{\partial \eta}{\partial e}\right)_T = 0, \quad \left(\frac{\partial \dot{\eta}}{\partial e}\right)_T = \frac{na}{(1-e)\sqrt{1-e^2}} \tag{11.93b}$$

Thus

$$\frac{\partial(\eta, \dot{\eta})}{\partial(a, e)} = 0 \tag{11.94}$$

The Jacobians, Eqs. (11.93) and (11.94), used in Eq. (11.78), yield $[a, e] = 0$. Similarly, all the Lagrangian brackets needed to solve the perturbation equations (11.35) can be evaluated. The non-vanishing brackets are

$$[\Omega, a] = \frac{na \cos i \sqrt{1-e^2}}{2} \tag{11.95}$$

$$[\omega, a] = \frac{na\sqrt{1-e^2}}{2} \tag{11.96}$$

$$[e, \Omega] = \frac{na^2 e \cos i}{\sqrt{1-e^2}} \tag{11.97}$$

$$[e, \omega] = \frac{na^2 e}{\sqrt{1-e^2}} \tag{11.98}$$

$$[i, \Omega] = -na^2 \sin i \sqrt{1-e^2} \tag{11.99}$$

$$[a, T] = \frac{n^2 a}{2} \tag{11.100}$$

and all the others are zero. An element $\sigma = -nT$ can be used in place of T, so

$$[\sigma, a] = \frac{na}{2} \tag{11.101}$$

The brackets in Eqs. (11.95)–(11.98) are of the type $[\alpha_r, \beta_s]$, Eq. (11.99) is $[\beta_r, \beta_s]$, and Eq. (11.100) is $[\alpha_r, \alpha_s]$ (McCuskey 1963, pp. 137–142).

11.5 Solution of the Perturbation Equations

Having evaluated the Lagrangian brackets, we can finish the solution of Eqs. (11.35) for the time derivatives of the orbital elements. Substituting from Eqs. (11.95)–(11.101) into Eq. (11.35) and using the property $[c_j, c_k] = -[c_k, c_j]$, we have

$$\frac{1}{2}na\dot{a} = \frac{\partial R}{\partial \sigma} \tag{11.102}$$

$$\frac{1}{2}na\sqrt{1-e^2}\,\dot{a} - \frac{na^2 e}{\sqrt{1-e^2}}\dot{e} = \frac{\partial R}{\partial \omega} \tag{11.103}$$

$$\frac{1}{2}na\cos i\sqrt{1-e^2}\,\dot{a} - \frac{na^2 e\cos i}{\sqrt{1-e^2}}\dot{e} - na^2\sin i\sqrt{1-e^2}\,\frac{di}{dt} = \frac{\partial R}{\partial \Omega} \tag{11.104}$$

$$na^2\sin i\sqrt{1-e^2}\,\dot{\Omega} = \frac{\partial R}{\partial i} \tag{11.105}$$

$$-\frac{1}{2}na\cos i\sqrt{1-e^2}\,\dot{\Omega} - \frac{1}{2}na\sqrt{1-e^2}\,\dot{\omega} - \frac{1}{2}na\dot{\sigma} = \frac{\partial R}{\partial a} \tag{11.106}$$

$$\frac{na^2 e\cos i}{\sqrt{1-e^2}}\dot{\Omega} + \frac{na^2 e}{\sqrt{1-e^2}}\dot{\omega} = \frac{\partial R}{\partial e} \tag{11.107}$$

The simultaneous solution of these for the time derivatives gives

$$\dot{a} = \frac{2}{na}\frac{\partial R}{\partial \sigma} \tag{11.108a}$$

$$\dot{e} = \frac{1-e^2}{na^2 e}\frac{\partial R}{\partial \sigma} - \frac{\sqrt{1-e^2}}{na^2 e}\frac{\partial R}{\partial \omega} \tag{11.108b}$$

$$\dot{\sigma} = -\frac{1-e^2}{na^2 e}\frac{\partial R}{\partial e} - \frac{2}{na}\frac{\partial R}{\partial a} \tag{11.108c}$$

$$\dot{\Omega} = \frac{1}{na^2\sin i\sqrt{1-e^2}}\frac{\partial R}{\partial i} \tag{11.108d}$$

$$\dot{\omega} = \frac{-\cos i}{na^2\sin i\sqrt{1-e^2}}\frac{\partial R}{\partial i} + \frac{\sqrt{1-e^2}}{na^2 e}\frac{\partial R}{\partial e} \tag{11.108e}$$

$$\frac{di}{dt} = \frac{\cos i}{na^2\sin i\sqrt{1-e^2}}\frac{\partial R}{\partial \omega} - \frac{1}{na^2\sin i\sqrt{1-e^2}}\frac{\partial R}{\partial \Omega} \tag{11.108f}$$

These equations are similar to the Lagrange planetary equations (10.124).

In using the equation for $\dot{\sigma}$, it can be useful to introduce the mean anomaly, $M = n(t-T) = n\tau + \sigma$, in place of σ. The perturbation function can be written in the form $R(a, e, i, \omega, \Omega, M)$, because the Cartesian coordinates , in which R was defined, are functions of time and T, which can be expressed in terms of M. To make this conversion, we note that

$$\dot{M} = n + \dot{n}t + \dot{\sigma} = n + \frac{dn}{da}\dot{a}t + \dot{\sigma} \tag{11.109}$$

and $\dot{\sigma}$ is given in Eq. (11.108c). In using this equation we need $\partial R/\partial a$. However, R depends upon a both explicitly and also through M, because $n = \mu^{1/2}a^{-3/2}$ and, hence, $M = \mu^{1/2}a^{-3/2}t + \sigma$. Therefore,

$$\frac{\partial R}{\partial a} = \left(\frac{\partial R}{\partial a}\right)_M + \frac{\partial R}{\partial M}\frac{\partial M}{\partial a} = \left(\frac{\partial R}{\partial a}\right)_M + \left(\frac{-3n}{2a}t\right)\frac{\partial R}{\partial M} \tag{11.110}$$

where $(\partial R/\partial a)_M$ denotes the derivative taken explicitly with respect to a. $\partial R/\partial e$ is not affected this way. With Eq. (11.110) and $dn/da = -3n/2a$ substituted into Eq. (11.109), we have

$$\dot{M} = n - \frac{3n\dot{a}t}{2a} - \left(\frac{1-e^2}{na^2e}\right)\frac{\partial R}{\partial e} - \frac{2}{na}\left(\frac{\partial R}{\partial a}\right)_M + \frac{3t}{a^2}\frac{\partial R}{\partial M} \tag{11.111}$$

From Eq. (11.108a),

$$\dot{a} = \frac{2}{na}\frac{\partial R}{\partial \sigma} = \frac{2}{na}\frac{\partial R}{\partial M} \tag{11.112}$$

Substituting this into the second term of Eq. (11.111), this cancels the last term, so

$$\dot{M} = n - \left(\frac{1-e^2}{na^2e}\right)\frac{\partial R}{\partial e} - \frac{2}{na}\left(\frac{\partial R}{\partial a}\right)_M \tag{11.113}$$

and

$$\dot{a} = \frac{2}{na}\frac{\partial R}{\partial M} \tag{11.114}$$

These two relations replace the equations for $\dot{a}$ and $\dot{\sigma}$, respectively, when the mean anomaly is used in place of σ as the sixth parameter of the orbit.

In order to apply Eqs. (11.96)–(11.114), it is clear that $R(x, y, z, x', y', z')$ must be expressed in terms of the elements desired. However, in many applications of practical interest, the perturbation forces are expressed in terms of acceleration components along satellite-fixed axes rather than in terms of the perturbation potential; this is the case, for example, for perturbations such as drag, or control forces acting on an artificial satellite. Thus, the perturbing forces are directed along the radius vector, transverse to it, and perpendicular to the plane of the orbit. There are applications where the perturbing force components are tangent to the orbit, normal to the orbit, and perpendicular to the orbital plane. An artificial satellite, moving close to the Earth, experiences atmospheric drag, a perturbing force tangent to the orbit in the direction opposite to the velocity vector.

Let us consider the equations appropriate for the two force systems; radial, transverse, and orthogonal components, and tangential, normal, and orthogonal components. The resulting equations are collectively referred to as the *Gauss variational equations* (GVE). In the derivations, we shall mean force per unit mass, whenever force is used (McCuskey 1963, pp. 142–144).

11.6 Case I: Radial, Transverse, and Orthogonal Components

Let $\hat{\mathbf{u}}_r, \hat{\mathbf{u}}_\theta, \hat{\mathbf{u}}_W$ be a right hand triad of unit vectors with $\hat{\mathbf{u}}_r$ along r, $\hat{\mathbf{u}}_\theta$ perpendicular to r in the orbital plane and with an angle less than 90° from the velocity $\mathbf{v}$, and $\hat{\mathbf{u}}_W$ perpendicular to the orbital plane such that $\hat{\mathbf{u}}_W = \hat{\mathbf{u}}_r \times \hat{\mathbf{u}}_\theta$. This triad forms a coordinate system which is sometimes referred to as an *RSW frame*.

$u = \omega + f$ is the angle from the line of nodes to the radius vector. Then the force under consideration is

$$\mathbf{F} = R'\hat{\mathbf{u}}_r + S'\hat{\mathbf{u}}_\theta + W'\hat{\mathbf{u}}_W \tag{11.115}$$

where the unit vectors in terms of the Cartesian coordinate system are

$$\begin{aligned}
\hat{\mathbf{u}}_r &= (\cos\Omega\cos u - \sin\Omega\sin u\cos i)\hat{\mathbf{i}} \\
&\quad + (\sin\Omega\cos u + \cos\Omega\sin u\cos i)\hat{\mathbf{j}} \\
&\quad + \sin u\sin i\,\hat{\mathbf{k}} &(11.116)\\
\hat{\mathbf{u}}_\theta &= (-\cos\Omega\sin u - \sin\Omega\cos u\cos i)\hat{\mathbf{i}} \\
&\quad + (-\sin\Omega\sin u + \cos\Omega\cos u\cos i)\hat{\mathbf{j}} \\
&\quad + \cos u\sin i\,\hat{\mathbf{k}} \\
\hat{\mathbf{u}}_W &= \sin\Omega\sin i\,\hat{\mathbf{i}} - \cos\Omega\sin i\,\hat{\mathbf{j}} + \cos i\,\hat{\mathbf{k}} &(11.117)
\end{aligned}$$

The partial derivatives $\partial R/\partial x, \partial R/\partial y, \partial R/\partial z$ are the components of the acceleration, when the latter is due to the perturbation function, R. This can be seen from Eq. (11.3). In vector form, this acceleration is

$$\nabla R = \frac{\partial R}{\partial x}\hat{\mathbf{i}} + \frac{\partial R}{\partial y}\hat{\mathbf{j}} + \frac{\partial R}{\partial z}\hat{\mathbf{k}} \tag{11.118}$$

The force components of R enter the perturbation equations only through the partial derivatives, $(\partial R/\partial a), (\partial R/\partial e)$, and alike. The transformation to the new force, $\mathbf{F}$, can be made as follows. If c_j represents any one of the elements, then

$$\frac{\partial R}{\partial c_j} = \nabla R \cdot \frac{\partial \mathbf{r}}{\partial c_j},\ j = 1, \ldots, 6 \tag{11.119}$$

where $\mathbf{r} = x\hat{\mathbf{i}} + y\hat{\mathbf{j}} + z\hat{\mathbf{k}}$. So we need only evaluate

$$\frac{\partial R}{\partial c_j} = \mathbf{F} \cdot \frac{\partial \mathbf{r}}{\partial c_j} \tag{11.120}$$

in terms of the force components R', S', W', and use these partial derivatives in Eqs. (11.108)–(11.114). As an example, we consider $\partial R/\partial a$. Using the notation where $\hat{\mathbf{P}}$ and $\hat{\mathbf{Q}}$ are unit vectors of a perifocal frame (see Sect. 5.2),

$$\mathbf{r} = x\hat{\mathbf{i}} + y\hat{\mathbf{j}} + z\hat{\mathbf{k}} = \xi\hat{\mathbf{P}} + \eta\hat{\mathbf{Q}} \tag{11.121}$$

Thus,

$$\frac{\partial \mathbf{r}}{\partial a} = \hat{\mathbf{P}}\frac{\partial \xi}{\partial a} + \hat{\mathbf{Q}}\frac{\partial \eta}{\partial a} \tag{11.122}$$

Equation (11.120) yields

$$\frac{\partial R}{\partial a} = \mathbf{F} \cdot \hat{\mathbf{P}}\frac{\partial \xi}{\partial a} + \mathbf{F} \cdot \hat{\mathbf{Q}}\frac{\partial \eta}{\partial a} \tag{11.123a}$$

where by Eq. (11.84)

$$\frac{\partial \xi}{\partial a} = \cos E - e = \frac{\xi}{a} \tag{11.124a}$$

$$\frac{\partial \eta}{\partial a} = \sin E\sqrt{1 - e^2} = \frac{\eta}{a} \tag{11.124b}$$

In Eq. (11.115) we defined

$$\mathbf{F} = R'\hat{\mathbf{u}}_r + S'\hat{\mathbf{u}}_\theta + W'\hat{\mathbf{u}}_W \tag{11.125}$$

The unit vectors $\hat{\mathbf{u}}_r$, $\hat{\mathbf{u}}_\theta$ and $\hat{\mathbf{P}}$, $\hat{\mathbf{Q}}$ are related as shown in Fig. 11.2. So

$$\mathbf{F} \cdot \hat{\mathbf{P}} = R' \cos f - S' \sin f \tag{11.126}$$

$$\mathbf{F} \cdot \hat{\mathbf{Q}} = R' \sin f + S' \cos f \tag{11.127}$$

Fig. 11.2 Relation of perifocal and polar unit vectors

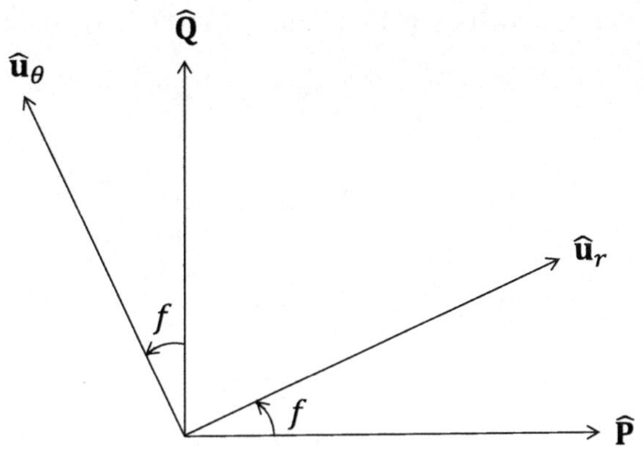

Also, $\xi = r\cos f, \eta = r\sin f$, so Eq. (11.126) becomes

$$\mathbf{F}\cdot\hat{\mathbf{P}} = \frac{1}{r}[R'\xi - S'\eta] \tag{11.128a}$$

$$\mathbf{F}\cdot\hat{\mathbf{Q}} = \frac{1}{r}[R'\eta + S'\xi] \tag{11.128b}$$

Substituting Eqs. (11.128) and (11.128) into (11.123) we have

$$\frac{\partial R}{\partial a} = \frac{1}{ra}R'(\xi^2 + \eta^2) = \frac{r}{a}R' \tag{11.129}$$

This is the partial derivative for the force function postulated. In a similar manner, the other derivatives can be found. These are

$$\frac{\partial R}{\partial a} = R'\frac{r}{a} \tag{11.130a}$$

$$\frac{\partial R}{\partial e} = -R'a\cos f + S'a\sin f\left[1 + \frac{r}{a(1-e^2)}\right] \tag{11.130b}$$

$$\frac{\partial R}{\partial \sigma} = \frac{R'ea\sin f}{\sqrt{1-e^2}} + \frac{S'a^2\sqrt{1-e^2}}{r} \tag{11.130c}$$

$$\frac{\partial R}{\partial \Omega} = S'r\cos i - W'r\sin i\cos u \tag{11.130d}$$

$$\frac{\partial R}{\partial \omega} = S'r \tag{11.130e}$$

$$\frac{\partial R}{\partial i} = W'r\sin u \tag{11.130f}$$

Substituting Eqs. (11.130) into Eqs. (11.108), we have the element rates

$$\dot{a} = \frac{2e\sin f}{n\eta}R' + \frac{2a\eta}{rn}S' \tag{11.131a}$$

$$\dot{e} = \frac{\eta\sin f}{na}R' + \frac{\eta}{a^2ne}\left[\frac{a^2\eta^2 - r^2}{r}\right]S' \tag{11.131b}$$

$$\dot{\sigma} = \left[\frac{\eta^2\cos f}{ane} - \frac{2r}{na^2}\right]R' - \frac{\eta^2\sin f}{a^2ne}\left[1 + \frac{r}{a\eta^2}\right]S' \tag{11.131c}$$

$$\dot{\Omega} = \frac{r\sin u}{a^2n\eta\sin i}W' \tag{11.131d}$$

$$\dot{\omega} = \frac{-\eta\cos f}{ane}R' + \frac{\eta\sin f}{ane}\left[1 + \frac{r}{a\eta^2}\right]S' \tag{11.131e}$$

$$- \frac{r\sin u\cot i}{a^2n\eta}W' \tag{11.131f}$$

$$\frac{di}{dt} = \frac{r\cos u}{a^2n\eta}W' \tag{11.131g}$$

where $\eta \triangleq \sqrt{1-e^2}$. Equations (11.131) are the Gauss variational equations. In a similar manner the equations for $\dot{M}$ and $\dot{a}$, independent of σ, can be transformed. We can see clearly from Eqs. (11.131d) and (11.131g) that the spatial orientation of the orbit will change only when there is a component of specific force W' (McCuskey 1963, pp. 144–147).

11.7 Case II: Tangential, Normal, and Orthogonal Components

We wish to express the radial and transverse components of Case I in terms of the tangential and normal components. Let

$$\mathbf{G} = T'\hat{\mathbf{u}}_T + N'\hat{\mathbf{u}}_N + W'\hat{\mathbf{u}}_W \tag{11.132}$$

define the force. $\hat{\mathbf{u}}_T, \hat{\mathbf{u}}_N, \hat{\mathbf{u}}_W$ are unit vectors along the tangent in the direction of motion, along the normal to the orbit directed toward its concave side (i.e., towards the attraction center), and perpendicular to the orbit in the sense $\hat{\mathbf{u}}_W = \hat{\mathbf{u}}_T \times \hat{\mathbf{u}}_N$, respectively. So W' is the same as W' of Case I. This triad forms a coordinate system which is sometimes referred to as an *NTW frame*.

We are concerned with transforming from R', S' to T', N'. The radial direction component of $\mathbf{G}$ is

$$R' \equiv \mathbf{G} \cdot \hat{\mathbf{u}}_r = T'(\hat{\mathbf{u}}_T \cdot \hat{\mathbf{u}}_r) + N'(\hat{\mathbf{u}}_N \cdot \hat{\mathbf{u}}_r) \tag{11.133}$$

and in the transverse direction it is

$$S' \equiv \mathbf{G} \cdot \hat{\mathbf{u}}_\theta = T'(\hat{\mathbf{u}}_T \cdot \hat{\mathbf{u}}_\theta) + N'(\hat{\mathbf{u}}_N \cdot \hat{\mathbf{u}}_\theta) \tag{11.134}$$

The scalar products are evaluated knowing that $\hat{\mathbf{u}}_T$ and $\hat{\mathbf{u}}_N$ are along and perpendicular, respectively, to the velocity vector. So we may write

$$\mathbf{v} = \dot{r}\hat{\mathbf{u}}_r + r\dot{f}\hat{\mathbf{u}}_\theta = v\hat{\mathbf{u}}_r \tag{11.135}$$

Then the scalar products

$$\frac{\dot{r}}{v} = \hat{\mathbf{u}}_r \cdot \hat{\mathbf{u}}_T = \hat{\mathbf{u}}_\theta \cdot \hat{\mathbf{u}}_N \tag{11.136a}$$

$$\frac{r\dot{f}}{v} = \hat{\mathbf{u}}_\theta \cdot \hat{\mathbf{u}}_r = -\hat{\mathbf{u}}_r \cdot \hat{\mathbf{u}}_N \tag{11.136b}$$

The first parts of Eq. (11.136) can be seen from dot products of Eq. (11.135) by $\hat{\mathbf{u}}_r$ and $\hat{\mathbf{u}}_\theta$, respectively. The second parts are from the unit vectors geometry. From the elliptic motion we have

$$r = \frac{a(1-e^2)}{1+e\cos f} \tag{11.137a}$$

$$r^2\dot{f} = h = \sqrt{k^2Ma(1-e^2)} \tag{11.137b}$$

Evaluation of $\dot{r}$ from the first and elimination of $\dot{f}$ by means of the second yields

$$\dot{r} = \frac{k\sqrt{M}e\sin f}{\sqrt{a(1-e^2)}} \tag{11.138}$$

$$r\dot{f} = \frac{k\sqrt{M}(1+e\cos f)}{\sqrt{a(1-e^2}} \tag{11.139}$$

$$v = \frac{k\sqrt{M}(1+e^2+2e\cos f)^{1/2}}{\sqrt{a(1-e^2)}} \tag{11.140}$$

Then Eqs. (11.136) become

$$\hat{\mathbf{u}}_r \cdot \hat{\mathbf{u}}_T = \hat{\mathbf{u}}_\theta \cdot \hat{\mathbf{u}}_N = \frac{\dot{r}}{v} = \frac{e \sin f}{(1 + e^2 + 2e \cos f)^{1/2}} \tag{11.141a}$$

$$\hat{\mathbf{u}}_\theta \cdot \hat{\mathbf{u}}_T = -\hat{\mathbf{u}}_r \cdot \hat{\mathbf{u}}_N = \frac{1 + e \cos f}{(1 + e^2 + 2e \cos f)^{1/2}} \tag{11.141b}$$

Substituting Eqs. (11.141) into Eqs. (11.133) and (11.134), we have

$$R' = \frac{e \sin f}{\sqrt{1 + e^2 + 2e \cos f}} T' - \frac{1 + e \cos f}{\sqrt{1 + e^2 + 2e \cos f}} N' \tag{11.142}$$

$$S' = \frac{1 + e \cos f}{\sqrt{1 + e^2 + 2e \cos f}} T' + \frac{e \sin f}{\sqrt{1 + e^2 + 2e \cos f}} N' \tag{11.143}$$

With these equations and T', N' given, one can substitute for R' and S' in Eqs. (11.131) and obtain the perturbations in the elements due to a force that has been resolved into tangential, normal, and orthogonal components (McCuskey 1963, pp. 147–149).

We must emphasize that some authors use a different convention for the NTW frame (Vallado 2001, p. 163). In this convention, $\hat{\mathbf{u}}_T$ is parallel to the direction of the velocity vector, the normal axis $\hat{\mathbf{u}}_N$ points away from the attraction center, and $\hat{\mathbf{u}}_W = \hat{\mathbf{u}}_N \times \hat{\mathbf{u}}_T$. In this case, the transformation in Eq. (11.142) is written as (note the sign differences in the N' terms)

$$R' = \frac{e \sin f}{\sqrt{1 + 2e \cos f + e^2}} T' + \frac{1 + e \cos f}{\sqrt{1 + 2e \cos f + e^2}} N' \tag{11.144a}$$

$$S' = \frac{1 + e \cos f}{\sqrt{1 + 2e \cos f + e^2}} T' - \frac{e \sin f}{\sqrt{1 + 2e \cos f + e^2}} N' \tag{11.144b}$$

11.8 Expansion of the Third-Body Potential

To obtain the time rate of change of the orbital elements, the disturbing function R must be expanded as an infinite series, where the orbital elements appear in the coefficients or arguments of trigonometric series. It is also possible to expand the disturbing function as a power series, for artificial satellites, or as a Chebysehev series, as per Carpenter (1966).

Let m and m' be the perturbed and perturbing masses, let r and r' be their distances from the central mass m_1, and let $r' > r$ at all times (recall Fig. 5.7). The analysis is slightly different for $r' < r$, and fails for the Neptune-Pluto case where the values change with respect to each other. The perturbation function can be written as

$$
\begin{aligned}
R &= k^2 m' \left[\frac{1}{\rho} - \frac{xx' + yy' + zz'}{r'^3} \right] \\
&= k^2 m' \left[\left(r^2 + r'^2 - 2rr' \cos\phi \right)^{-1/2} - \frac{rr' \cos\phi}{r'^3} \right] \\
&= k^2 m' \left[\frac{1}{r'} \left[1 + \left(\frac{r}{r'}\right)^2 - 2\left(\frac{r}{r'}\right) \cos\phi \right]^{-1/2} - \frac{rr' \cos\phi}{r'^3} \right] \\
&= \frac{k^2 m'}{r'} \left[\left[1 + \left(\frac{r}{r'}\right)^2 - 2\left(\frac{r}{r'}\right) \cos\phi \right]^{-1/2} - \left(\frac{r}{r'}\right) \cos\phi \right]
\end{aligned}
\tag{11.145}
$$

where ϕ is the angle between the radius vectors. Expanding the first term in the bracket by the binomial theorem, we have

$$
\begin{aligned}
&\left[1 + \left(\frac{r}{r'}\right)^2 - 2\left(\frac{r}{r'}\right) \cos\phi \right]^{-1/2} \\
&= 1 + \left(\frac{r}{r'}\right) \cos\phi + \left(\frac{r}{r'}\right)^2 \left(-\frac{1}{2} + \frac{3}{2} \cos^2\phi \right) \\
&+ \left(\frac{r}{r'}\right)^3 \left(-\frac{3}{2} \cos\phi + \frac{5}{2} \cos^3\phi \right) + \left(\frac{r}{r'}\right)^4 \left(\frac{3}{8} - \frac{15}{4} \cos^2\phi + \frac{35}{8} \cos^4\phi \right) + \ldots
\end{aligned}
\tag{11.146}
$$

The groups of trigonometric functions in Eq. (11.146) that appear due to expanding part of the disturbing function are *Legendre polynomials* of the first kind, $P_n(\cos\phi)$. The Legendre polynomials, generally denoted for some argument x by $P_n(x)$, are useful when expanding functions such as

$$
\frac{1}{\|\mathbf{r} - \mathbf{r}'\|} = \frac{1}{\sqrt{r^2 + r'^2 - 2rr' \cos\phi}} = \sum_{n=0}^{\infty} \frac{(r')^n}{r^{n+1}} P_n(\cos\phi) \tag{11.147}
$$

Each Legendre polynomial, $P_n(x)$, is an nth degree polynomial. It may be expressed using the *Rodrigues formula*:

$$
P_n(x) = \frac{1}{2^n n!} \frac{d^n}{dx^n} \left[(x^2 - 1)^n \right] \tag{11.148}
$$

For example, $P_0(x) = 1, P_1(x) = x$, $P_2(x) = 0.5(3x^2-1)$ and $P_3(x) = 0.5(5x^3-3x)$. In our case, the polynomials are then

$$P_0(\cos\phi) = 1 \tag{11.149}$$

$$P_1(\cos\phi) = \cos\phi \tag{11.150}$$

$$P_2(\cos\phi) = \frac{1}{2}[3\cos^2\phi - 1] = \frac{1}{4}[3\cos 2\phi + 1] \tag{11.151}$$

$$P_3(\cos\phi) = \frac{1}{2}[5\cos^3\phi - 3\cos\phi] = \frac{1}{8}[5\cos 3\phi + 3\cos\phi] \tag{11.152}$$

$$\begin{aligned} P_4(\cos\phi) &= \frac{1}{8}[35\cos^4\phi - 30\cos^2\phi + 3] \\ &= \frac{1}{64}[35\cos 4\phi + 20\cos 2\phi + 9] \end{aligned} \tag{11.153}$$

$$P_5(\cos\phi) = \frac{1}{8}[63\cos^5\phi - 70\cos^3\phi + 15\cos\phi] \tag{11.154}$$

The Legendre polynomials are uniformly bounded, that is

$$|P_n(\cos\phi)| \leq 1,\ n = 0, 1, 2, 3, \ldots \tag{11.155}$$

so the series

$$\sum_{n=0}^{\infty} \left(\frac{r}{r'}\right)^n P_n(\cos\phi) \tag{11.156}$$

is convergent, since $r/r' < 1$ by our choice. Combining Eqs. (11.145) and (11.146), where the $r\cos\phi/r'$ terms cancel, we have

$$R = \frac{k^2 m'}{r'}\left[1 + \left(\frac{r}{r'}\right)^2 P_2 + \left(\frac{r}{r'}\right)^3 P_3 + \left(\frac{r}{r'}\right)^4 P_4 + \ldots\right] \tag{11.157}$$

We must now look at r/r' and the polynomials $P_n(\cos\phi)$ to determine R in terms of the orbital elements. Let us consider the factors in $\left(\frac{r}{r'}\right)^2 P_2(\cos\phi)$ separately.

11.8.1 The Factor $(r/r')^2$

In Sect. 5.9, we saw Kepler's equation and how we can expand E in trigonometric functions of the mean anomaly M. Likewise, we can expand to the e^2 term

$$\frac{r}{a} = 1 + \frac{1}{2}e^2 - e\cos M - \frac{1}{2}e^2\cos 2M + \ldots \tag{11.158}$$

$$\frac{a}{r} = 1 + e\cos M + e^2\cos 2M + \ldots \tag{11.159}$$

$$\cos E = -\frac{1}{2}e + \left(1 - \frac{3}{8}e^2\right)\cos M + \frac{1}{2}e\cos 2M + \frac{3}{8}e^2\cos 3M + \ldots \tag{11.160}$$

$$\sin E = \left(1 - \frac{1}{8}e^2\right)\sin M + \frac{1}{2}e\sin 2M + \frac{3}{8}e^2\sin 3M + \ldots \tag{11.161}$$

From r/a and a/r,

$$\begin{aligned}\frac{r}{r'} &= \left(\frac{r}{a}\right)\left(\frac{a'}{r'}\right)\left(\frac{a}{a'}\right)\\ &= \frac{a}{a'}\left[1 + \frac{1}{2}e^2 - e\cos M - \frac{1}{2}e^2\cos 2M\right]\\ &\times \left[1 + e'\cos M' + e'^2\cos 2M'\right]\end{aligned} \tag{11.162}$$

Squaring Eq. (11.162) yields in terms to order e^2

$$\begin{aligned}\left(\frac{r}{r'}\right)^2 = \left(\frac{a}{a'}\right)^2 &\left[1 + e^2 + e^2\cos^2 M - 2e\cos M - e^2\cos 2M\right.\\ &+ e'\cos^2 M' + 2e'\cos M' + 4ee'\cos M\cos M'\\ &\left.+ 2e'^2\cos 2M' + \ldots\right]\end{aligned} \tag{11.163}$$

Terms such as $\cos^2 M$ and $\cos M\cos M'$ can be transformed into functions of multiple angles or sums and differences of angles,

$$\cos^2 M = \frac{1}{2}(1 + \cos 2M) \tag{11.164}$$

$$\cos M\cos M' = \frac{1}{2}[\cos(M + M') + \cos(M - M')] \tag{11.165}$$

So Eq. (11.162) becomes

$$\begin{aligned}\left(\frac{r}{r'}\right)^2 = \left(\frac{a}{a'}\right)^2 &\left[1 + \frac{3}{2}e^2 + \frac{1}{2}e'^2 - 2e\cos M + 2e'\cos M'\right.\\ &- \frac{1}{2}e^2\cos 2M + \frac{5}{2}e'^2\cos 2M' - 2ee'\cos(M + M')\\ &\left.- 2ee'\cos(M - M')\right]\end{aligned} \tag{11.166}$$

Thus, $(r/r')^2$ is the sum of terms of the form $A_{pq}\cos(pM + qM')$, where p and q are integers, either positive, negative, or zero, and A_{pq} are functions of a, a', e, e'.

11.8.2 The Factor $P_2(\cos\phi)$

We have the polynomial

$$P_2(\cos\phi) = -\frac{1}{2} + \frac{3}{2}\cos^2\phi \tag{11.167}$$

First we want the form of $\cos\phi$ in terms of the orbital elements. From Eq. (11.66),

$$\mathbf{r} = x\hat{\mathbf{i}} + y\hat{\mathbf{j}} + z\hat{\mathbf{k}} = \xi\hat{\mathbf{P}} + \eta\hat{\mathbf{Q}} \tag{11.168}$$

and

$$\xi = r\cos f = a\cos E - ae \tag{11.169}$$

$$\eta = r\sin f = a\sin E\sqrt{1-e^2} \tag{11.170}$$

we have

$$\mathbf{r} = \xi\hat{\mathbf{P}} + \eta\hat{\mathbf{Q}} = a[(\cos E - e)\hat{\mathbf{P}} + (\sqrt{1-e^2}\sin E)\hat{\mathbf{Q}}] \tag{11.171}$$

where $\hat{\mathbf{P}}$ and $\hat{\mathbf{Q}}$ are unit vectors, functions of Ω, ω, i, as given in Eqs. (11.62). An equation similar to Eq. (11.171) can be written expressing $\mathbf{r}'$ to appropriate unit vectors $\hat{\mathbf{P}}'$ and $\hat{\mathbf{Q}}'$, where the prime denotes the disturbing planet. So we can write

$$\begin{aligned}\cos\phi &= \frac{\mathbf{r}\cdot\hat{\mathbf{r}}'}{rr'} = \frac{\left(\xi\hat{\mathbf{P}} + \eta\hat{\mathbf{Q}}\right)\cdot\left(\xi'\hat{\mathbf{P}}' + \eta'\hat{\mathbf{Q}}'\right)}{rr'}\\ &= \frac{\xi\xi'\hat{\mathbf{P}}\cdot\hat{\mathbf{P}}' + \eta\xi'\hat{\mathbf{Q}}\cdot\hat{\mathbf{P}}' + \xi\eta'\hat{\mathbf{P}}\cdot\hat{\mathbf{Q}}' + \eta\eta'\hat{\mathbf{Q}}\cdot\hat{\mathbf{Q}}'}{rr'}\end{aligned} \tag{11.172}$$

Examining a term such as $\xi\xi'\hat{\mathbf{P}}\cdot\hat{\mathbf{P}}'/rr'$, since $\xi = a(\cos E - e)$ and $\xi' = a'(\cos E' - e')$, using the expansion for $\cos E$ in Eq. (11.158) we have

$$\begin{aligned}\frac{\xi\xi'}{rr'} &= \left(\frac{a}{r}\right)\left(\frac{a'}{r'}\right)\left[-\frac{3}{2}e + (1 - \frac{3}{8}e^2)\cos M + \frac{1}{2}e\cos 2M\right.\\ &\left.+\ \frac{3}{8}e^2\cos 3M\right] \times \left[-\frac{3}{2}e' + (1 - \frac{3}{8}e'^2)\cos M' + \frac{1}{2}e'\cos 2M'\right.\\ &\left.+\ \frac{3}{8}e'^2\cos 3M'\right]\end{aligned} \tag{11.173}$$

A typical product from the two brackets is of the form $\cos pM\cos qM'$, which can be transformed into the form $[\cos(pM + qM') + \cos(pM - qM')]/2$.

Products of these terms by series expressions for (a/r) and (a'/r) can be reduced in a similar manner. So the product $\xi\xi'/rr'$ takes the form $B_{pq}\cos(pM + qM')$, where p and q are integers, and B_{pq} are functions of a, a', e, e'. The product $\hat{\mathbf{P}} \cdot \hat{\mathbf{P}}' = P_1P_1' + P_2P_2' + P_3P_3'$, where the components are

$$P_1 = \cos\Omega\cos\omega - \sin\Omega\sin\omega\cos i \tag{11.174}$$

$$P_2 = \sin\Omega\cos\omega + \cos\Omega\sin\omega\cos i \tag{11.175}$$

$$P_3 = \sin\omega\sin i \tag{11.176}$$

and similarly for P_1', P_2', P_3'. We can write $\cos i = 1 - 2\sin^2(i/2)$ and, since i is a small quantity, $\sin(i/2) \approx i/2 = \gamma$. So $\cos i = 1 - 2\gamma^2$. Then $P_1 = \cos(\Omega + \omega) + 2\gamma^2 \sin\omega\sin\Omega$, and, using $2\sin\omega\sin\Omega = [\cos(\Omega + \omega) - \cos(\Omega - \omega)]$,

$$P_1 = (1 - \gamma^2)\cos(\Omega + \omega) + \gamma^2\cos(\Omega - \omega) \tag{11.177a}$$

$$P_1' = (1 - \gamma'^2)\cos(\Omega' + \omega') + \gamma'^2\cos(\Omega' - \omega') \tag{11.177b}$$

Other components of $\hat{\mathbf{P}}$ and $\hat{\mathbf{Q}}$ can be expressed similarly.

We are interested in the form of $\hat{\mathbf{P}} \cdot \hat{\mathbf{P}}'$. From Eqs. (11.177), the product of P_1P_1' will consist of terms of the form $\cos(\Omega+\omega)\cos(\Omega'+\omega')$, which reduce to sums such as $[\cos(\Omega+\omega+\Omega'+\omega')+\cos(\Omega+\omega-\Omega'-\omega')]/2$. The products from $\hat{\mathbf{P}}\cdot\hat{\mathbf{P}}'$ and other scalar products in Eq. (11.172) are of the form $C_j\cos(j_1\Omega + j_2\Omega' + j_3\omega + j_4\omega')$, where $j_i,\ i = 1, 2, 3, 4$ are integers and C_j are functions of γ and γ'. From this analysis of the form of $\xi\xi'P_1P_1'/rr'$, we can see that the form of the perturbation function will be

$$R = k^2m' \sum_p C_p(a, a', e, e', \gamma, \gamma')$$
$$\times \cos\left(p_1M + p - 2M' + p_3\Omega + p_4\Omega' + p_5\omega + p_6\omega'\right) \tag{11.178}$$

where $p_i,\ i = 1, \ldots, 6$ are integers. The expression for R can be used in Eq. (11.108) to obtain the perturbations in the orbital elements. Let $M = nt + \sigma,\ M' = n't + \sigma'$, so

$$p_1M + p_2M' = (p_1n + p_2n')t + p_1\sigma + p_2\sigma' \tag{11.179}$$

Denote the angular argument in R by

$$\theta = (p_1n + p_2n')t + p_1\sigma + p_2\sigma' + p_3\Omega + p_4\Omega' + p_5\omega + p_6\omega' \tag{11.180}$$

If we consider the orbital elements of the perturbing body m' as constants, Eq. (11.180) becomes

$$\theta = (p_1n + p_2n')t + p_1\sigma + p_3\Omega + p_5\omega + \theta_0 \tag{11.181}$$

where θ_0 contains the contributions from $p_2, p_4, p_6, \sigma', \Omega', \omega'$. Then

$$R = k^2 m' \sum_p C_p \cos[(p_1 n + p_2 n')t + p_1\sigma + p_3\Omega + p_5\omega + \theta_0] \tag{11.182}$$

where the summation refers to all $p_i,\ i = 1, \ldots, 6$. Then

$$\frac{\partial R}{\partial \sigma} = \begin{cases} -k^2 m' \sum_p C_p p_1 \sin\theta, & p_1 \neq 0 \\ 0, & p_1 = 0 \end{cases} \tag{11.183a}$$

$$\frac{\partial R}{\partial \Omega} = -k^2 m' \sum_p C_p p_3 \sin\theta \tag{11.183b}$$

$$\frac{\partial R}{\partial \omega} = -k^2 m' \sum_p C_p p_5 \sin\theta \tag{11.183c}$$

$$\frac{\partial R}{\partial e} = k^2 m' \sum_p \frac{\partial C_p}{\partial e} \cos\theta \tag{11.183d}$$

$$\frac{\partial R}{\partial i} = \frac{1}{2} k^2 m' \cos\left(\frac{i}{2}\right) \sum_p \frac{\partial C_p}{\partial \gamma} \cos\theta \tag{11.183e}$$

$$\frac{\partial R}{\partial a} = k^2 m' \sum_p \frac{\partial C_p}{\partial a} \cos\theta - k^2 m' \sum_p C_p \left(p_1 t \frac{\partial n}{\partial a}\right) \sin\theta \tag{11.183f}$$

where $\partial n/\partial a = -3n/2a$. Substituting Eq. (11.183e) into Eq. (11.108d) we have

$$\dot{\Omega} = \frac{1}{na^2 \sin i\sqrt{1-e^2}} \frac{\partial R}{\partial i} = \frac{k^2 m' \cos(i/2)}{2na^2 \sin i\sqrt{1-e^2}} \sum_p \frac{\partial C_p}{\partial \gamma} \cos\theta \tag{11.184}$$

Usually, m' is small compared with the central mass; hence, as a first approximation, the elements on the right of Eq. (11.184) are assumed constant. The derivative of $\dot{\Omega}$ becomes a periodic function of t alone, if p_1 and p_2 are not both zero. If p_1 and p_2 are both zero, $\dot{\Omega}$ is a constant, say A. We can consider $\dot{\Omega}$ separated, so

$$\dot{\Omega} = A + \sum_p B_p \cos[(p_1 n + p_2 n')t + \Omega_1] \tag{11.185}$$

where $\Omega_1 = p_1\sigma_0 + p_3\Omega_0 + p_5\omega_0 + \Omega_0$, and the zero subscript denotes fixed elements. Integrating Eq. (11.185) we have

$$\Omega = \Omega_0 + At + \sum_p \frac{B_p}{(p_1 n + p_2 n')} \sin[(p_1 n + p_2 n')t + \Omega_1] \tag{11.186}$$

where p_1 and p_2 are not both zero. The linear part of the change in Ω, At, is known as a *secular perturbation* term. From Eq. (11.184), we see that, if $p_1 n + p_2 n' = 0$, a secular term will arise. This requires *commensurability* in the periods of the perturbed and perturbing planets. If P and P' are these periods, then

$$\frac{P'}{P} = \frac{n}{n'} = -\frac{p_2}{p_1} \tag{11.187}$$

and p_1 and p_2 are integers. Jupiter and Saturn approach the ratio 5:2. The nature of the periodic perturbations in an element such as Ω depends on the magnitude of B_p, and on $p_1 n + p_2 n'$. B_p is not large in the solar system. If $p_1 n + p_2 n'$ is large for a given pair of values (p_1, p_2), then for these values of p_1, p_2 Ω will have periodic terms of small amplitude and short period. These are *short-period inequalities*. If, for a pair of values p_1, p_2, the quantity $p_1 n + p_2 n'$ is small, then Ω will have a perturbation with a large amplitude and a long period. These are *long-period inequalities*. From the equations for all the elements, it can be seen that all the elements, except a, will exhibit secular and periodic changes, from analysis made to the first order in m', such as that for Ω. For the semimajor axis, however,

$$\dot{a} = \frac{-2k^2 m'}{na} \sum_p C_p p_1 \sin\theta, \qquad p_1 \neq 0 \tag{11.188}$$

$$\dot{a} \equiv 0, \qquad p_1 = 0 \tag{11.189}$$

The semimajor axis oscillates about the mean value a_0, where $a = a_0 + \delta(a)$ with a period

$$P = \frac{2\pi}{p_1 n + p_2 n'}$$

where

$$\delta(a) = \frac{2k^2 m'}{na_0} \sum C_p \left(\frac{p_1}{p_1 n + p_2 n'} \right) \cos\theta \tag{11.190}$$

It has been shown that to the third power in m', there is no secular change in a. This, of course, becomes important when considering the stability of the planetary orbit. The analysis discussed can, of course, be applied to the other elements. The expressions derived are only to first order, so they only contain the factor m'.

We started with the assumption that the elements were constants. If we now were to re-substitute with the secular and periodic terms for the elements, we should obtain the second approximation with terms in t^2, which would all contain the factor m'^2; and thus higher approximations are derived.

The semimajor axis, as mentioned, and the eccentricity, are critical for the stability of the system. If the eccentricity changed sufficiently, we might have a close approach between planets and a radical change in orbits. In 1776, Lagrange showed that the semimajor axis was stable to the first order. In 1809, Poisson showed the same result applied when m'^2 terms are included. The higher order condition is less certain. Hagihara (1957, 1972) has discussed the stability question. Laskar (2008) and Laskar et al. (2011) presented a number of important results on the subject.

There are many more possible approaches to perturbation theory using various coordinates such as rectangular coordinates, Hansen variables, and so on. Likewise, other series such as Chebyshev series can be used in place of trigonometric series (McCuskey 1963, pp. 153–158).

11.9 The Earth-Moon System

The motion of the Moon around the Earth influenced by the Sun is a complex problem. This is due to the large mass ratio of the Moon with respect to the Earth and the proximity of the Moon to the Earth. The primary mass, m_1, is the Earth, m, the perturbed mass, is the Moon, and m' is the Sun. The Moon's orbit is inclined about 5° to the ecliptic; for this discussion we will neglect this inclination. Also we will neglect the eccentricity of the Earth's orbit around the Sun. The Earth's eccentricity $e = 0.016$ has only a second order effect on the analysis. The problem then is reduced to a two-dimensional one.

Let ψ be the celestial longitude of the Moon and ψ' the longitude of the Sun, as shown in Fig. 11.3.

From Eq. (11.157), with $\phi = \psi - \psi'$ the perturbation function becomes

$$R = \frac{k^2 m'}{a'}\left[1 + \left(\frac{r}{a'}\right)^2 P_2 \cos\left(\psi - \psi'\right) + \ldots\right] \tag{11.191}$$

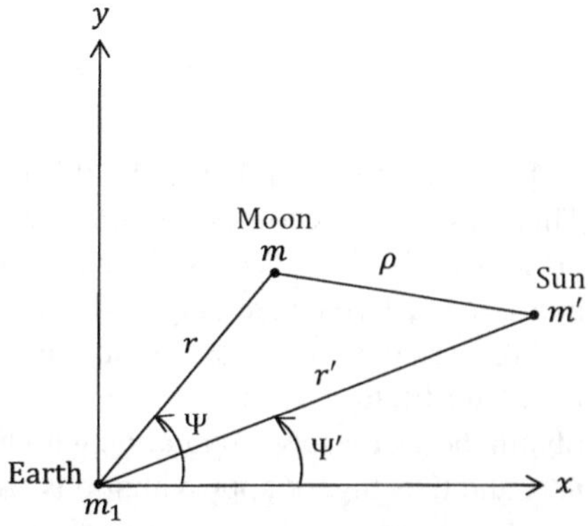

Fig. 11.3 Sun-Earth-Moon geometry

where a' is the assumed circular radius of the Earth's orbit. R will be truncated at second order terms in (r/a'). From the Legendre coefficients we have

$$P_2[\cos(\psi - \psi')] = \frac{1 + 3\cos 2(\psi - \psi')}{4}$$

and since $(r/a')^2 = (a/a')^2(r/a)^2$, R may be written as

$$R = \frac{k^2 m'}{a'}\left[1 + \frac{1}{4}\left(\frac{a}{a'}\right)^2\left(\frac{r}{a}\right)^2 = \frac{3}{4}\left(\frac{a}{a'}\right)^2\left(\frac{r}{a}\right)^2\cos 2(\psi - \psi')\right] \tag{11.192}$$

R must be expressed in terms of the elements of the Moon's orbit. Neglecting the terms in Eq. (11.166) containing e', which has been assumed to be zero, we have

$$\left(\frac{r}{a'}\right)^2 = \left(\frac{a}{a'}\right)^2\left[1 + \frac{3e^2}{2} - 2e\cos M - \frac{1}{2}e^2\cos 2M\right] \tag{11.193}$$

in terms of second order in e. Since the Moon, Sun, and Earth have been assumed to be in the same plane, we take $\psi = \Omega + \omega + f$, where f is the true anomaly of the Moon. Then

$$\left(\frac{r}{a}\right)^2\cos 2\left(\psi - \psi'\right) = \left(\frac{r}{a}\right)^2\cos\left[2f + 2\left(\Omega + \omega - \psi'\right)\right] \tag{11.194}$$

which is expanded to

$$\begin{aligned}&2\left(\frac{r}{a}\right)^2\cos^2 f\cos 2\left(\Omega + \omega - \psi'\right) - \left(\frac{r}{a}\right)^2\cos 2\left(\Omega + \omega - \psi'\right)\\ &- 2\left[\left(\frac{r}{a}\right)\sin f\right]\left[\left(\frac{r}{a}\right)\cos f\right]\sin 2\left(\Omega + \omega - \psi'\right)\end{aligned} \tag{11.195}$$

From $r\cos f = a\cos E - ae$ and $r\sin f = a\sin E\sqrt{1 - e^2}$, and the expansions of $\cos E$ and $\sin E$ in terms of M, we have

$$\begin{aligned}\frac{r}{a}\cos f = &\left(1 - \frac{3}{8}e^2\right)\cos M + \frac{1}{2}e\cos 2M + \frac{3}{8}e^2\cos 3M\\ &- \frac{3}{2}e\end{aligned} \tag{11.196}$$

$$\frac{r}{a}\sin f = \left(1 - \frac{5}{8}e^2\right)\sin M + \frac{1}{2}e\sin 2M + \frac{3}{8}e^2\sin 3M \tag{11.197}$$

Substituting these into Eq. (11.195) and using Eqs. (11.195) and (11.192) in Eq. (11.191), we have for the disturbing function

$$
\begin{aligned}
R = \frac{k^2 m' a^2}{a'^3} \Bigg\{ & \frac{1}{4} + \frac{3}{8}e^2 - \frac{1}{2}e\cos M - \frac{1}{8}e^2\cos 2M + \underbrace{\frac{15}{8}e^2\cos[2(\Omega+\omega-\psi')]}_{(i)} \\
& - \frac{9}{4}e\cos[2(\Omega+\omega-\psi')+M] + \underbrace{\frac{3}{4}\cos[2(\Omega+\omega-\psi)+2M]}_{(ii)} \\
& - \frac{15}{8}e^2\cos[2(\Omega+\omega-\psi')+2M] \\
& + \frac{3}{4}e\cos[2(\Omega+\omega-\psi')+3M] + \frac{3}{4}e^2\cos[2(\Omega+\omega-\psi')+4M] \Bigg\}
\end{aligned}
\tag{11.198}
$$

Terms of order higher than two in e have been ignored. The Moon's mass has also been ignored. A more complete expression of R can be found in Brouwer and Clemence (1961).

While Eq. (11.198) is a simplified version of R, it indicates the complexity of the Moon's motion. Terms in R such as $3k^2m'a^2e^2/(9a'^3)$ cause secular variations in the orbital elements. Terms involving only $\cos M$ and $\cos 2M$ are elliptical terms, similar to terms in a series representation of Keplerian motion. The remaining terms in R depend on the relative positions of the Moon and Sun, and are perturbative terms. Some of these are called (i) *evection* and (ii) *variation* in Eq. (11.198).

Consider the effect of a typical term in R on the orbital elements and the Moon's longitude. The method of analysis for evection will be discussed. Denoting the evection-related term by

$$A = \frac{15}{8}n'^2a^2e^2\cos 2(\Omega+\omega-\psi') \tag{11.199}$$

where $n'^2 = k^2m'/a'^3$, the differential equations for the orbital elements are

$$\dot{a} = \frac{2}{na}\frac{\partial A}{\partial M} = 0 \tag{11.200a}$$

$$\dot{e} = \frac{15}{4}\frac{n'^2 e}{n}\sin 2(\Omega + \omega'_\psi) \tag{11.200b}$$

$$\dot{M} = n - \frac{15}{4}\frac{n'^2}{n}(1+e^2)\cos 2(\Omega+\omega-\psi') \tag{11.200c}$$

$$\dot{\omega} = \frac{d(\Omega+\omega)}{dt} = \frac{15}{4}(1-\frac{1}{2}e^2)\frac{n'^2}{n}\cos 2(\Omega+\omega-\psi') \tag{11.200d}$$

$$\dot{\Omega} = 0 \tag{11.200e}$$

$$\frac{di}{dt} = 0 \tag{11.200f}$$

The last two follow from the assumption that $i = 0$. There is no perturbation in the semimajor axis due to the evection term. Integrating the equations for $\dot{e}, \dot{M}, \dot{\omega}$ yields

$$e = e_0 + \frac{15}{8}\left(\frac{n'}{n}\right) e_0 \cos 2(\Omega_0+\omega_0-\psi') \tag{11.201}$$

$$M = M_0 + \frac{15}{8}\left(\frac{n'}{n}\right)(1+e_0^2)\sin 2(\Omega_0+\omega_0-\psi') \tag{11.202}$$

$$\Omega+\omega = (\Omega+\omega)_0 + \frac{15}{8}(1-\frac{1}{2}e_0^2)\left(\frac{n'}{n}\right)\sin 2(\Omega_0+\omega_0-\psi) \tag{11.203}$$

where the subscripts refer to the osculating orbit of reference.

In the integrations we recall $\psi' = \Omega' + \omega' + n'(t-t_0)$, where n', Ω', ω' refer to the Sun's apparent motion around the Earth. Thus, the perturbations are

$$\delta(e) = \frac{25}{8}\left(\frac{n'}{n}\right) e_0 \cos 2(\Omega_0+\omega_0-\psi') \tag{11.204a}$$

$$\delta(M) = \frac{15}{8}\left(\frac{n'}{n}\right)(1+e_0^2)\sin 2(\Omega_0+\omega_0-\psi') \tag{11.204b}$$

$$\delta(\Omega+\omega) = \frac{15}{8}\left(\frac{n'}{n}\right)(1-\frac{1}{2}e_0^2)\sin 2(\Omega_0+\omega_0-\psi') \tag{11.204c}$$

The first-order perturbations in longitude, $\delta(\psi)$, are determined from $\psi = \Omega+\omega+f$ and $f = M + 2e\sin M$. Then

$$\delta(\psi) = \delta(\Omega+\omega) + (1+2e\cos M)\delta(M) + 2\sin M\delta(e) \tag{11.205}$$

and using Eqs. (11.204)

$$\delta(\psi) = \frac{15}{4}\left(\frac{n'}{n}\right) e_0 \sin[2(\Omega_0+\omega_0-\psi')+M] \tag{11.206}$$

The period of this longitude perturbation caused by the evection term is about one month, with an amplitude of about one degree. Other perturbations would appear if the effects of the Earth's orbit eccentricity and Moon's orbit inclination were included in the analysis. These are perturbations in Ω, and, through Eq. (11.204c), in ω. The appearance of i in the disturbing function and, thus, in $\partial R/\partial i$ leads to

$$\dot{\Omega} = -\frac{3}{4}\frac{n'^2}{n}\frac{i}{\sin i} \approx -\frac{3}{4}\frac{n'^2}{n} \tag{11.207}$$

since i is small. Thus,

$$\delta(\Omega) = -\frac{3}{4}\frac{n'^2}{n}(t - t_0) = -\frac{3}{4}\frac{n'}{n}n'\delta t \tag{11.208}$$

where $n' = 2\pi/P_e$, and P_e is the Earth's orbital period around the Sun. Since for the Moon $n = 2\pi/P_M$, Eq. (11.208) can be written

$$\frac{\delta(\Omega)}{2\pi} = -\frac{3}{4}\frac{P_M}{P_e}\frac{\delta t}{P_e} = -0.056\frac{\delta t}{P_e} \tag{11.209}$$

This indicates that Ω changes by 2π radians, when $\delta t = 6580$ days, approximately 18 years. The minus sign shows the node of the Moon's orbit regresses, i.e. moves west along the ecliptic, with a revolution in about 18 years. This is a secular perturbation, since the change in Ω is a linear function of time.

Hundreds of periodic terms are required to describe the lunar motion, as shown by the exhaustive lunar theory of Brown (1896) and Eckert (Gutzwiller and Schmidt 1986). The principal secular effects are: (i) the line of nodes regresses at an average rate of one revolution in 18.6 years; (ii) the line of apsides (the major axis of the orbit) advances ($\dot{\omega} > 0$) at an average rate of one revolution in 8.9 years (McCuskey 1963, pp. 158–162).

11.10 Expansion of the Gravitational Potential

Under the assumption that the origin of a planet-fixed reference frame coincides with the center of mass of the body, the *gravitational potential* of an arbitrarily-shaped body is given by the *spherical harmonics* expansion (Vallado 2001, pp. 508–517)

$$\begin{aligned} V(r, \phi, \lambda) = \frac{Gm}{r} \Bigg\{ & 1 - \sum_{l=1}^{\infty} J_l \left(\frac{r_{eq}}{r}\right)^l P_l(\cos\phi) \\ & + \sum_{l=1}^{\infty}\sum_{j=1}^{l} J_{lj} \left(\frac{r_{eq}}{r}\right)^n P_{lj}(\cos\phi)\cos\left[j\left(\lambda - \lambda_{lj}\right)\right] \Bigg\} \end{aligned} \tag{11.210}$$

Here λ is the geographic longitude measured eastward from the major axis of the elliptical equatorial cross section that goes through the center of mass of the planet. The angle ϕ is the *colatitude*, i.e. the angle between the body $\hat{\mathbf{z}}$ axis and the radius vector, satisfying

$$\cos\phi = \hat{\mathbf{u}}_r \cdot \hat{\mathbf{u}}_A = \sin i \sin(f + \omega) \tag{11.211}$$

Alternatively, we can use the latitude in the above expression, defined as $\varphi = \pi/2 - \phi$, in which case the terms containing $\cos\phi$ in Eq. (11.210) and in subsequent expressions are replaced by $\sin\varphi$.

We have seen the Legendre polynomials P_l in Sect. 11.8; these polynomials are periodic on the surface of a unit sphere and vanish along l parallels, dividing the surface into $(l + 1)$ zones, thereby earning the name *zonal harmonics* for the coefficients J_l. These harmonics are responsible for the axially-symmetric part of the gravitational potential.

The functions P_{lj} are the *associated Legendre polynomials* of degree j and order l, which are expressed as

$$P_{lj}(x) = \frac{1}{2^l l!}(1 - x^2)^{j/2} \frac{d^{l+j}}{dx^{l+j}}(x^2 - 1)^l \tag{11.212}$$

The terms including the coefficient J_{lj} are collectively referred to as the *tesseral harmonics*. Generally, the tesseral harmonics reflect the triaxiality of the planet. The triaxiality makes itself most evident through the *equatorial ellipticity*.

In Eq. (11.210), the distance from the origin to the orbiting mass is given by the usual expression,

$$r = \frac{a(1 - e^2)}{1 + e\cos f} \tag{11.213}$$

The quantity r_{eq} is the *mean equatorial radius* of the planet, and m is its mass. The sign convention is chosen so that, for a unit mass (cf. Eq. (11.7))

$$\ddot{\mathbf{r}} = \nabla V \tag{11.214}$$

A planet owes its nonsphericity, i.e., its departure from the Newtonian potential, to topographic features, some of which stand testimony to intrinsic geological processes and some are signatures of massive impacts. Oblateness guarantees a noticeable value of the coefficient J_2, an issue which will be discussed in detail in Chap. 12.

In different notations, the potential may be written as

$$V(r,\phi,\lambda) = \frac{Gm}{r}\left[1 - \sum_{l=1}^{\infty} J_l \left(\frac{r_{eq}}{r}\right)^l P_l(\cos\phi)\right.$$
$$\left. + \sum_{l=1}^{\infty}\sum_{j=1}^{l} \left(\frac{r_{eq}}{r}\right)^l P_{lj}(\cos\phi)\left[C_{lj}\cos j\lambda + S_{lj}\sin j\lambda\right]\right] \quad (11.215)$$

or, equivalently, as

$$V(r,\phi,\lambda) = \frac{Gm}{r}\sum_{l=0}^{\infty}\sum_{j=0}^{l}\left(\frac{r_{eq}}{r}\right)^l P_{lj}(\cos\phi)\left[C_{lj}\cos j\lambda + S_{lj}\sin j\lambda\right] \quad (11.216)$$

where the terms including the coefficients C_{lj} are tesseral harmonics, and those including S_{lj} are called *sectorial harmonics*. For any l,

$$P_l(\cos\phi) = P_{l0}(\cos\phi) \quad (11.217)$$

For $l = 2, \ldots, \infty$ and $j = 1, \ldots, l$,

$$C_{lj} \equiv J_{lj}\cos j\lambda_{lj},\; S_{lj} \equiv J_{lj}\sin j\lambda_{lj} \quad (11.218)$$

thus also defining the angle λ_{lj} in Eq. (11.210), while for $l = 2, \ldots, \infty$ and $j = 0$ we have

$$C_{l0} = -J_{l0} \equiv -J_l \quad (11.219)$$

For $l = 0$, we get

$$C_{00} \equiv J_{00} \equiv 1,\; P_0(\cos\phi) = P_{00}(\cos\phi) \equiv 1 \quad (11.220)$$

The $l = 1$ terms need some attention. For a general choice of the origin,

$$C_{11} = \frac{X_{cm}}{r_{eq}},\; S_{11} = \frac{Y_{cm}}{r_{eq}},\; C_{10} \equiv -J_{10} \equiv -J_1 = \frac{Z_{cm}}{r_{eq}} \quad (11.221)$$

X_{cm}, Y_{cm}, Z_{cm} being the Cartesian coordinates of the center of mass (Hobson 1965). The coefficient J_{11} is then related to C_{11} and S_{11} through Eq. (11.218). If, however, we choose to place the origin in the center of mass, we obtain

$$C_{10} \equiv -J_{10} \equiv -J_1 = 0, C_{11} = 0, S_{11} = 0, J_{11} = 0 \quad (11.222)$$

which spares us from all the $l = 1$ terms in Eqs. (11.210), (11.215), and (11.216).

11.11 Atmospheric Drag

Some of the perturbations affecting low Earth orbit (LEO) satellites are atmospheric drag and solar radiation pressure (SRP). Both the translational and rotational satellite dynamics are affected by drag, which depends on the satellite's aerodynamical, thermal, and mechanical properties. The aerodynamical parameters are the density, the drag coefficient, and the lift coefficient, while the mechanical properties are related to the satellite's structure and the cross-sectional area.

Satellite drag modeling has been a vibrant field of study. We shall cover this topic in the current chapter only briefly, with emphasis on modelling drag using analytical, rather than empirical, representations. Additional results, using semianalytical theories, are provided in Sect. 13.3.4.

Satellite aerodynamics have been studied by Fredo and Kaplan (1981), who suggested a numerical procedure for calculating aerodynamical coefficients by basic geometric modeling of the satellite and division into finite elements, and the overall coefficients were determined by integrating the differential force acting on each of the exposed elements. The aerodynamical forces and torques acting on the space shuttle were developed by Stone and Witzgall (2006). Sutton (2009) proposed two analytical methods for calculating the drag and lift coefficients for long satellites, and the statistical results were validated through the CHAMP and GRACE missions. An improved method for aerodynamical model estimation was developed by Fuller and Tolson (2009).

A drag coefficient model for different solar activities was developed by Cook (1965), while the accuracy of the drag model for LEO satellites due to the density uncertainty was examined by Marcos (1990). Calculation of the drag coefficient for different geometrical shapes was performed by Moe et al. (1998), based on orbital measurements. Diverse approaches for the drag effects on LEO satellites were examined, and a high accuracy model for satellite drag was developed by Storz et al. (2005). Most of the available methods for drag coefficient calculation of LEO satellites were reviewed by Prieto et al. (2014).

The specific force due to atmospheric drag can be modeled as

$$\mathbf{F}_{drag} = -\frac{1}{2}\frac{SC_D}{m}\rho\left(\mathbf{v} - \mathbf{v}_{atm}\right)\|\mathbf{v} - \mathbf{v}_{atm}\| \tag{11.223}$$

where m is the satellite mass, S is the cross-sectional reference area, C_D is the drag coefficient defined with respect to the cross-sectional area, $\mathbf{v}$ is the velocity and ρ is the atmospheric density. The vector $\mathbf{v}_{atm}$ is the atmospheric velocity. If the atmosphere is assumed to be spherical and co-rotating with the Earth, then $\mathbf{v}_{atm} = \left[0\ 0\ \omega_e\right]^{\mathrm{T}} \times \mathbf{r}$, where ω_e is the Earth's spin rate and $\mathbf{r}$ is the position vector.

The model used herein approximates the atmospheric density as (Battin 1999, pp. 505–506)

$$\rho = \rho_0 \exp\left(\frac{r_{p0} - r}{H}\right) \tag{11.224}$$

where ρ_0 is the atmospheric density at the initial perigee radius, r_{p_0}, and H is the density scale height of the atmosphere.

The drag specific force vector $\mathbf{F}_{drag}$, written in terms of the osculating elements, can be expressed in the NTW frame (see Sect. 11.7 for a definition and sign conventions) as (Liu and Zhao 1981)

$$\begin{aligned} T' &= -\frac{1}{2}K_1 n^2 a^2 \rho \frac{1+2e\cos f+e^2}{1-e^2} \\ N' &= 0 \\ W' &= -\frac{1}{2}K_2 napr\cos(\omega+f)\sin i\left(\frac{1+2e\cos f+e^2}{1-e^2}\right)^{1/2} \end{aligned} \tag{11.225}$$

where $K_1 = (C_D S_1/m)\,Q$, $K_2 = (C_D S_2/m)\,\omega_e\sqrt{Q}$, and $Q = \left(1 - r_{p_0}\omega_e \cos i/v_{p_0}\right)^2$. The variables S_1 and S_2 are the respective cross-sectional areas perpendicular to the tangential and subnormal directions, and v_{p_0} is the velocity at the initial perigee. The drag components can be transformed from the NTW frame into the RSW frame using the transformation (11.144).

11.12 Regularization of Perturbed Motion

The perturbed two-body problem discussed in this chapter exhibits singularities when the orbiting body approaches the attraction center, i.e. when $r \to 0$. As mentioned in Sect. 1.5.3, this singularity can be eliminated by a proper variable transformation. One possibility is to use the Kustaanheimo-Stiefel (KS) variables (Stiefel and Scheifele 1971). These variables are obtained by transforming the Euclidean space into a *spinor space*, which is a complex vector space, and by using a new independent variable called *universal anomaly* or *fictitious time* (Kustaanheimo and Stiefel 1965). The KS formulation has proven efficient for numerically integrating the equations of motion (Fukushima and Arakida 2000) and for characterizing the long-term dynamics of satellites subject to perturbations (Sharma and Raj 1988, 2009).

Recall that the equations of motion governing the perturbed two-body problem can be written as (cf. Eq. (11.7))

$$\ddot{\mathbf{r}} + \frac{\mu}{r^3}\mathbf{r} = \nabla R + \mathbf{G} \tag{11.226}$$

where $\mathbf{r}$ is the position vector, $r \triangleq \|\mathbf{r}\|$, μ is the gravitational parameter, R is a perturbing potential, and $\mathbf{G}$ denotes any perturbing non-conservative force per unit mass.

The KS transformation maps the three-dimensional space into a four-dimensional space. Given an inertial position vector, more than one spinor can be determined. However, the inverse transformation, from KS variables to the inertial space, is always unique.

The KS variables are denoted by $\mathbf{u} = [u_1, u_2, u_3, u_4]^T$. The injective, nonsingular transformation from the KS variables to inertial coordinates is given by (Stiefel and Scheifele 1971)

$$\tilde{\mathbf{r}} = \begin{bmatrix} x \\ y \\ z \\ 0 \end{bmatrix} = \begin{pmatrix} u_1 & -u_2 & -u_3 & u_4 \\ u_2 & u_1 & -u_4 & -u_3 \\ u_3 & u_4 & u_1 & u_2 \\ u_4 & -u_3 & u_2 & -u_1 \end{pmatrix} \begin{bmatrix} u_1 \\ u_2 \\ u_3 \\ u_4 \end{bmatrix} = \begin{bmatrix} u_1^2 - u_2^2 - u_3^2 + u_4^2 \\ 2(u_1 u_2 - u_3 u_4) \\ 2(u_1 u_3 + u_2 u_4) \\ 0 \end{bmatrix} \tag{11.227}$$

or compactly

$$\tilde{\mathbf{r}} = \mathbf{L}(\mathbf{u})\mathbf{u} \tag{11.228}$$

The symbol $\tilde{(\cdot)}$ denotes the vector $\mathbf{r}$ augmented by a fourth zero component. Equation (11.227) implies that the transformation from inertial coordinates to KS variables is not unique.

Given an initial position vector $\mathbf{r}(0) = [x(0), y(0), z(0)]$, one may transform it into the KS representation as follows (Junkins and Singla 2004). If $x(0) \geq 0$,

$$u_1(0) = \sqrt{\frac{1}{2}[r(0) + x(0)]}, \; u_2(0) = \frac{y(0)}{2u_1(0)} \tag{11.229a}$$

$$u_3(0) = \frac{z(0)}{2u_1(0)}, \; u_4(0) = 0 \tag{11.229b}$$

Otherwise,

$$u_1(0) = \frac{y(0)}{2u_2(0)}, \; u_2(0) = \sqrt{\frac{1}{2}[r(0) - x(0)]} \tag{11.230a}$$

$$u_3(0) = 0, \; u_4(0) = \frac{z(0)}{2u_2(0)} \tag{11.230b}$$

This transformation is used to obtain the initial conditions in KS variables, given an initial position vector in inertial coordinates. It is also seen from Eq. (11.227) that

$$r = u_1^2 + u_2^2 + u_3^2 + u_4^2 = \|\mathbf{u}\|^2 \tag{11.231}$$

As mentioned before, the regularizing transformation derived by Kustaanheimo and Stiefel utilizes an independent variable called universal anomaly or fictitious time. This variable, denoted by s, is defined through the *Sundman transformation*

$$\mathrm{d}t = r\,\mathrm{d}s, \qquad r\frac{\mathrm{d}(\cdot)}{\mathrm{d}t} = \frac{\mathrm{d}(\cdot)}{\mathrm{d}s} \tag{11.232}$$

From Eq. (11.232), it is readily seen that

$$t(s) = \int_{s_0}^{s} r(S)\mathrm{d}S \tag{11.233}$$

Since $0 < r(s) < \infty$, $t(s)$ is monotonically increasing and defined for every s. Hence, one can invert $t(s)$ to determine $s(t)$.

Stiefel and Scheifele (1971) state that the equation governing the dynamics of the perturbed two-body problem, formulated in terms of the fictitious time and the KS variables, is

$$\mathbf{u}'' + \frac{\alpha(\mathbf{u}, \mathbf{u}', s)}{2}\mathbf{u} = \frac{1}{4}\left[\|\mathbf{u}\|^2 \frac{\partial R(\mathbf{u}, s)}{\partial \mathbf{u}} + 2R(\mathbf{u}, s)\mathbf{u}\right] + \frac{\|\mathbf{u}\|^2}{2}\mathbf{L}^T\mathbf{G} \tag{11.234}$$

where $(\cdot)'$ denotes differentiation with respect to the new independent variable s. The perturbing potential R is generally a function of time and the position vector $\mathbf{r}$, or equivalently, a function of $\mathbf{u}$ and s. The parameter α is the total energy, given by

$$\alpha(\mathbf{u}, \mathbf{u}', s) = \frac{\mu - 2\|\mathbf{u}'\|^2}{\|\mathbf{u}\|^2} + R(\mathbf{u}, s) = \frac{\mu}{r} - \frac{\|\dot{\mathbf{r}}\|^2}{2} + R(\mathbf{r}, t) \tag{11.235}$$

In order to integrate (11.234), $\mathbf{u}(0)$ and $\mathbf{u}'(0)$ are needed; $\mathbf{u}(0)$ is computed from Eqs. (11.229) or (11.230), and $\mathbf{u}'(0)$ is obtained as

$$u_1'(0) = \frac{1}{2}\left[u_1(0)\dot{x}_1(0) + u_2(0)\dot{x}_2(0) + u_3(0)\dot{x}_3(0)\right] \tag{11.236a}$$

$$u_2'(0) = \frac{1}{2}\left[-u_2(0)\dot{x}_1(0) + u_1(0)\dot{x}_2(0) + u_4(0)\dot{x}_3(0)\right] \tag{11.236b}$$

$$u_3'(0) = \frac{1}{2}\left[-u_3(0)\dot{x}_1(0) - u_4(0)\dot{x}_2(0) + u_1(0)\dot{x}_3(0)\right] \tag{11.236c}$$

$$u_4'(0) = \frac{1}{2}\left[u_4(0)\dot{x}_1(0) - u_3(0)\dot{x}_2(0) + u_2(0)\dot{x}_3(0)\right] \tag{11.236d}$$

The inertial velocity vector $\dot{\mathbf{r}}(t(s)) = [\dot{x}(t(s))\ \dot{y}(t(s))\ \dot{z}(t(s))]^T$ can be computed from KS variables as (Stiefel and Scheifele 1971):

$$\dot{x}(t(s)) = \frac{2}{r}\left[u_1(s)u_1'(s) - u_2(s)u_2'(s) - u_3(s)u_3'(s) + u_4(s)u_4'(s)\right] \qquad (11.237a)$$

$$\dot{y}(t(s)) = \frac{2}{r}\left[u_2(s)u_1'(s) + u_1(s)u_2'(s) - u_4(s)u_3'(s) - u_3(s)u_4'(s)\right] \qquad (11.237b)$$

$$\dot{z}(t(s)) = \frac{2}{r}\left[u_3(s)u_1'(s) + u_4(s)u_2'(s) + u_1(s)u_3'(s) + u_2(s)u_4'(s)\right] \qquad (11.237c)$$

The solution of Eq. (11.234) represents the actual dynamics of the two-body problem if $\mathbf{u}(0)$ and $\mathbf{u}'(0)$ satisfy the *bilinear relation*, which is stated as

$$u_4(0)u_1'(0) - u_3(0)u_2'(0) + u_2(0)u_3'(0) - u_1(0)u_4'(0) = 0 \qquad (11.238)$$

Once $\mathbf{u}(s)$ is obtained, $\mathbf{r}(s)$ can be determined through Eq. (11.227). To find $\mathbf{r}(t)$, $s(t)$ is substituted into $\mathbf{r}(s)$.

If $\mathbf{G} = \mathbf{0}$ and $\partial R/\partial t = 0$, then $\alpha(\mathbf{u}, \mathbf{u}', s)$ becomes constant, and Eq. (11.234) assumes the form

$$\mathbf{u}'' + \frac{\alpha}{2}\mathbf{u} = \frac{1}{4}\left[\|\mathbf{u}\|^2 \frac{\partial R(\mathbf{u})}{\partial \mathbf{u}} + 2R(\mathbf{u})\mathbf{u}\right] \qquad (11.239)$$

Since α is constant along the trajectory in the phase space, it can be computed at $s = 0$. In such a case, the left-hand side of Eq. (11.239) becomes linear, and, therefore, if the perturbing potential R is identically zero, this representation turns the Keplerian two-body problem into a four-dimensional harmonic oscillator. This observation implies that only perturbations generate nonlinear terms in Eq. (11.239).

References

Battin, R.: An Introduction to the Mathematics and Methods of Astrodynamics. American Institute of Aeronautics and Astronautics, New York (1999)

Brouwer, D., Clemence, G.M.: Methods of Celestial Mechanics. Academic Press, New York (1961)

Brown, E.W.: An Introductory Treatise on the Lunar Theory. The University Press, Cambridge (1896)

Carpenter, L.: Planetary perturbations in Chebyshev series. Tech. Rep. NASA TN D-3168, National Aeronautics and Space Administration, Washington, DC (1966)

Cook, A.H.: The contribution of observations of satellites to the determination of the Earth's gravitational potential. Space Sci. Rev. **2**, 355–437 (1963)

Cook, G.E.: Satellite drag coefficients. Planet. Space Sci. **13**(10), 929–946 (1965)

Efroimsky, M.: Equations for the orbital elements. hidden symmetry. Preprint No 1844 of the Institute of Mathematics and its Applications, University of Minnesota (http://www.ima.umn.edu/preprints/feb02/1844.pdf) (2002)

Efroimsky, M., Goldreich, P.: Gauge symmetry of the n-body problem in the Hamilton-Jacobi approach. J. Math. Phys. **44**, 5958–5977 (2003)
Efroimsky, M., Goldreich, P.: Gauge freedom in the n-body problem of celestial mechanics. Astron. Astrophys. **415**, 1187–1199 (2004)
Fukushima, T., Arakida, H.: Long-term integration error of Kustaanheimo-Stiefel regularized orbital motion. Astron. J. **120**(6), 3333–3339 (2000)
Fuller, J.D., Tolson, R.H.: Improved method for the estimation of spacecraft free-molecular aerodynamic properties. J. Spacecr. Rocket. **46**(5), 938–948 (2009)
Gutzwiller, M.C., Schmidt, D.S.: The motion of the Moon as computed by the method of Hill, Brown, and Eckert. Astron. Papers Part I **23**, 1–272 (1986)
Hagihara, Y.: Stability in celestial mechanics. Scr. Metall. 1 (1957)
Hagihara, Y.: Celestial Mechanics, vol. II. Perturbation Theory. MIT Press, Cambridge (1972)
Hobson, E.W.: The Theory of Spherical and Ellipsoidal Harmonics. Chelsea Publishing, New York (1965)
Junkins, J.L., Singla, P.: How nonlinear is it? a tutorial on nonlinearity of orbit and attitude dynamics. J. Astronaut. Sci. **52**(1 and 2), 7–60 (2004)
King-Hele, D.G.: The effect of the Earth's oblateness on the orbit of a near satellite. Proc. R. Soc. Lond. Sei. A Math. Phys. Sci. **247**(1248), 49–72 (1958)
Kustaanheimo, P., Stiefel, E.: Perturbation theory of Kepler motion based on spinor regularization. J. Reine. Angew. Math. **218**, 204 (1965)
Laskar, J.: Chaotic diffusion in the solar system. Icarus **196**(1), 1–15 (2008)
Laskar, J., Fienga, A., Gastineau, M., Manche, H.: La2010: a new orbital solution for the long-term motion of the Earth. Astron. Astrophys. **532**, A89 (2011)
Liu, L., Zhao, D.: Combined perturbation on near-Earth satellite orbits. Chin. Astron. Astrophys. **5**, 422–433 (1981)
Marcos, F.A.: Accuracy of atmospheric drag models at low satellite altitudes. Adv. Space Res. **10**(3), 417–422 (1990)
McCuskey, S.W.: Introduction to Celestial Mechanics. Addison-Wesley, Reading (1963)
Moe, K., Moe, M.M., Wallace, S.D.: Improved satellite drag coefficient calculations from orbital measurements of energy accommodation. J. Spacecr. Rocket. **35**(3), 266–272 (1998)
Newman, W., Efroimsky, M.: The method of variation of constants and multiple time scales in orbital mechanics. Chaos **13**, 476–485 (2003)
Prieto, D.M., Graziano, B.P., Roberts, P.C.E.: Spacecraft drag modelling. Prog. Aerosp. Sci. **64**, 56–65 (2014)
Sharma, R.K., Raj, M.X.J.: Long-term orbit computations with KS uniformly regular canonical elements with oblateness. Earth Moon Planet. **42**, 163–178 (1988)
Sharma, R.K., Raj, M.X.J.: Contraction of near-Earth satellite orbits using uniformly regular KS canonical elements in an oblate atmosphere with density scale height variation with altitude. Planet. Space Sci. **57**, 34–41 (2009)
Smart, W.M.: Celestial Mechanics. Longmans, London (1960)
Stiefel, E., Scheifele, G.: Linear and Regular Celestial Mechanics. Springer, Berlin, Heidelberg, New York (1971)
Stone, W.C., Witzgall, C.: Evaluation of aerodynamic drag and torque for external tanks in low Earth orbit. J. Res. Natl. Inst. Stand. Technol. **111**(2), 143 (2006)
Storz, M.F., Bowman, B.R., Branson, M.J.I., Casali, S.J., Tobiska, W.K.: High accuracy satellite drag model (HASDM). Adv. Space Res. **36**(12), 2497–2505 (2005)
Sutton, E.K.: Normalized force coefficients for satellites with elongated shapes. J. Spacecr. Rocket. **46**(1), 112–116 (2009)
Fredo, R.M., Kaplan, M.H.: Procedure for obtaining aerodynamic properties of spacecraft. J. Spacecr. Rocket. **18**(4), 367–373 (1981)
Vallado, D.: Fundamentals of Astrodynamics and Applications, 2nd edn. Microcosm Press and Kluwer Academic Publishers, El Segundo (2001)

Chapter 12
Motion Around Oblate Planets

12.1 Introduction

An important perturbation affecting low planetary satellites is the planet's oblateness, approximated mathematically by the second zonal harmonic J_2, discussed in Sect. 11.10. The problem of motion in such a potential field is often referred to as *the main problem* in artificial satellite theory, and has been the subject of many studies over the years. Because the dynamics in this case are generally *nonintegrable dynamics*, and might even exhibit chaos (Celletti and Negrini 1995; Irigoyen and Simo 1993), as discussed in Sects. 1.5 and 1.6, the only possible closed-form solutions may be obtained by approximating and/or averaging the J_2 gravitational potential.

The most important result in the context of the main problem was obtained by Brouwer (1959). By using the von Zeipel method (von Zeipel 1921), integrable approximations of motion under the J_2 perturbing potential were obtained. Brouwer's results were later refined by several other authors, in order to eliminate the singularities occurring at small inclinations and small eccentricities (Lyddane 1963; Izsak 1961). Some solutions were based on the Delaunay variables, discussed in Sect. 10.10.

Another approximation for the gravitational potential of an oblate spheroid, predominant in the Russian school, is that of two fixed attraction centers, which is an integrable problem (Aksenov et al. 1961). Another analytical approach is due to Vinti (1960), who used oblate spheroidal coordinates in order to find an approximating potential for zonal harmonics up to J_4.

A different approach to the main problem belongs to two Spanish scientists, R. Cid Palacios and J. F. Lahulla Fornies. Their main work, published only in Spanish (Cid and Lahulla 1969, 1971), introduced a new way of approximating the potential of an oblate spheroid by using polar-nodal (Whittaker) coordinates. The result reveals a new Hamiltonian, which, unlike the first-order Brouwer model,

P. Gurfil, P.K. Seidelmann, *Celestial Mechanics and Astrodynamics: Theory and Practice*, Astrophysics and Space Science Library 436,
DOI 10.1007/978-3-662-50370-6_12

contains short-periodic terms. Moreover, the Cid-Lahulla model admits a closed-form solution, expressed with the help of *elliptic integrals* (Lara and Gurfil 2012).

In this chapter, we will present analytic solutions to motion about an oblate spheroidal planet. We begin by considering equatorial motion, and then proceed to a more general approach, emanating from the Spanish school.

12.2 Axially-Symmetric Gravitational Field

Let us return to Eq. (11.7), describing the equations of motion of a mass particle under the effect of Newtonian gravity and a perturbing potential R. We rewrite Eq. (11.7) as

$$\ddot{\mathbf{r}} = \nabla W \tag{12.1}$$

where as before $\mathbf{r}$ is the position vector in an inertial reference frame, but here W is the total gravitational potential, or in other words

$$W = \frac{\mu}{r} + R \tag{12.2}$$

If the attracting body has a rotational symmetry, and the origin of the body's reference frame is set at the center of mass, then the potential W can be expanded into a series of the form (see Eq. (11.215))

$$W(r, \phi) = \frac{\mu}{r}\left\{1 - \sum_{n=2}^{\infty} J_n \left(\frac{r_{eq}}{r}\right)^n P_n(\cos\phi)\right\} \tag{12.3}$$

where as seen previously μ is the gravitational parameter, $r = \|\mathbf{r}\|$, r_{eq} is the equatorial radius of the attracting body, J_n, $n \geq 2$ are the zonal harmonics coefficients, P_n, $n \geq 2$ are the Legendre polynomials, which we have already seen in Chap. 11, and ϕ is the colatitude angle, defined as in Eq. (11.211).

The integrability of Eq. (12.1) with the potential (12.3) is a challenging problem. It was proven that when only the J_2 term is retained, the problem is generally non-integrable (Celletti and Negrini 1995; Irigoyen and Simo 1993), except for equatorial motion (for the definition of integrability, see Sect. 1.6.4). However, we see that for $\phi = \pi/2$, i.e in the equatorial plane,

$$W(r) = \frac{\mu}{r}\left[1 + \frac{J_2}{2}\left(\frac{r_{eq}}{r}\right)^2 - \frac{3J_4}{8}\left(\frac{r_{eq}}{r}\right)^4 + \frac{5J_6}{16}\left(\frac{r_{eq}}{r}\right)^6 - \dots\right] \tag{12.4}$$

meaning that the force induced by this potential will be a central force as discussed in Chap. 4. We conclude that equatorial motion will be a central-force motion under any order of even zonal harmonics. This is particularly true when considering only the J_2 zonal harmonic.

12.3 Equatorial Motion

The gravitational potential considered here includes only the first term, J_2, in Eq. (12.4). Considering only J_2 in Eq. (12.4), and adding the specific kinetic energy $v^2/2$, with v being the speed, we obtain the total specific energy

$$\varepsilon = \frac{v^2}{2} - \frac{\mu}{r}\left[1 + \frac{J_2}{2}\left(\frac{r_{eq}}{r}\right)^2\right] \tag{12.5}$$

The total specific energy can be negative, positive, or zero. Similarly to the Keplerian two-body case, discussed in Chap. 5, when the total specific energy is negative, the motion is bounded. In contrast, a positive specific energy gives open orbits, and zero total specific energy represents escape. This notion can be described as follows. Define $U = U(r)$ to be the specific effective potential energy:

$$U(r) \equiv \frac{h^2}{2r^2} - W(r) \tag{12.6}$$

If the effective potential energy U of the radial motion has a local minimum, and the initial conditions are chosen adequately, then the motion is bounded between two concentric circles, of radii $0 < r_{\min} \le r_{\max}$, with $r_{\min}$ and $r_{\max}$ being among the roots of the algebraic equation

$$\varepsilon - U(r) = 0 \tag{12.7}$$

The assumptions made on the effective potential energy U ensure that $r_{\min}$ and $r_{\max}$ are simple roots of Eq. (12.7), and may be equal (circular motion).

If the constant energy line intersects the graph of the potential energy only once, and the initial conditions are adequately chosen, then the motion is unbounded, and Eq. (12.7) has only one simple positive real root, $0 < r_{\min}$. Two subcases of the unbounded motion are possible: (i) the velocity vanishes at infinity, $v_\infty = \lim_{t\to\infty} \dot{r}(t) = 0$, and (ii) the velocity at infinity approaches asymptotically a constant value, $0 < v_\infty < \infty$. Figure 12.1 shows examples associated with the situations described above.

First we examine the case $\varepsilon < 0$. Multiplying Eq. (12.5) by r^3/ε and re-arranging leads to a cubic equation for the variable r,

$$r^3 - \frac{v^2}{2\varepsilon}r^3 + \frac{\mu}{\varepsilon}r^2 + \frac{\mu J_2 {r_{eq}}^2}{2\varepsilon} = 0 \tag{12.8}$$

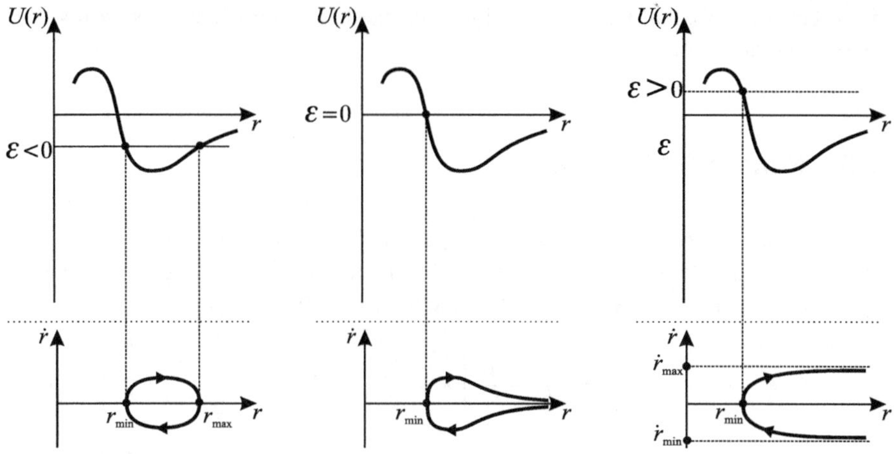

Fig. 12.1 Examples of orbits in the phase plane of the radial motion, associated with constant levels of energy. *Left*: bounded orbit, $\varepsilon < 0$. *Middle*: unbounded orbit, $\varepsilon = 0$. *Right*: unbounded orbit, $\varepsilon > 0$

In the instantaneous periapsis and apoapsis the radial velocity is $\dot{r} = 0$, and the specific angular momentum is $h = rv$. By using this connection, Eq. (12.8) becomes

$$r^3 + \frac{\mu}{\varepsilon}r^2 - \frac{h^2}{2\varepsilon}r + \frac{\mu J_2 {r_{eq}}^2}{2\varepsilon} = 0 \tag{12.9}$$

Equation (12.9) is a cubic equation with 3 real, positive solutions (Martinusi and Gurfil 2011). The solutions of Eq. (12.9) will produce the radii of the circles on which the periapsis and apoapsis lie (for closed orbits), denoted as above by r_{min} and r_{max}, respectively, as well as a third solution r_*, which satisfies $0 < r_* < r_{min} < r_{max}$. Using *Vieta's formulae* for polynomial equations, we can obtain the connections (Martinusi and Gurfil 2011)

$$\frac{r_{min} r_{max}(r_{min} + r_{max})}{J_2 {r_{eq}}^2/2 - r_{min} r_{max}} = \frac{\mu}{\varepsilon} \tag{12.10}$$

$$\left(\frac{J_2 {r_{eq}}^2}{2}\right) \frac{(r_{min} + r_{max})^2}{J_2 {r_{eq}}^2/2 - r_{min} r_{max}} - r_{min} r_{max} = \frac{h^2}{2\varepsilon} \tag{12.11}$$

$$r_* = \left(\frac{J_2 {r_{eq}}^2}{2}\right) \frac{r_{min} + r_{max}}{r_{min} r_{max} - J_2 {r_{eq}}^2/2} \tag{12.12}$$

It is possible to see that the radii of the periapsis and apoapsis depend on the specific angular momentum, specific energy and the values of J_2 and r_{eq}. Due to the laws of conservation these values remain constant, which implies that r_{min}, r_{max} remain

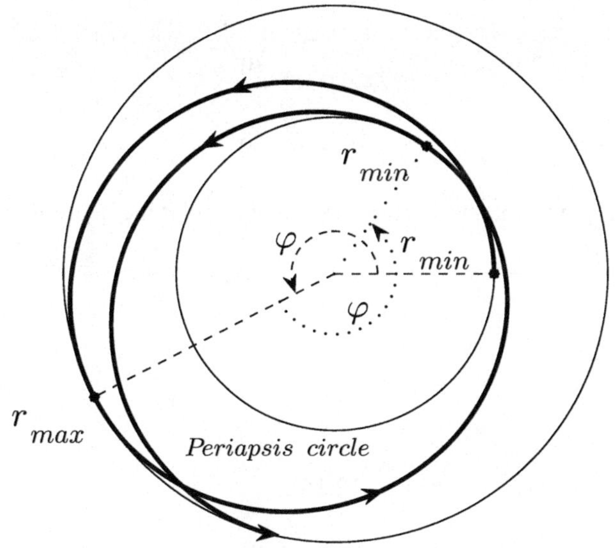

Fig. 12.2 The orbit in the equatorial plane of an oblate planet (bold). The orbital angle φ, and min (periapsis) and max (apoapsis) circles, which define the motion

constant as well. Therefore, the values of r_{min}, r_{max} represent two concentric circles, and the resulting motion can be seen in Fig. 12.2.

12.3.1 The Orbital Angle and Radial Period

Next we calculate the angle between r_{min} and the following r_{max}, which is the orbital angle φ (as shown in Fig. 12.2), and the time between two consecutive r_{min} or two consecutive r_{max}, which is the radial period T.

We begin by writing the position, velocity, and acceleration in polar coordinates, as seen in Chap. 4,

$$\mathbf{r} = r\hat{\mathbf{u}}_r \tag{12.13}$$

$$\dot{\mathbf{r}} = \dot{r}\hat{\mathbf{u}}_r + r\dot{\theta}\hat{\mathbf{u}}_\theta \tag{12.14}$$

$$\ddot{\mathbf{r}} = \left(\ddot{r} - r\dot{\theta}^2\right)\hat{\mathbf{u}}_r + \left(2\dot{r}\dot{\theta} + r\ddot{\theta}\right)\hat{\mathbf{u}}_\theta \tag{12.15}$$

The derivative of the potential in a central force field includes only radial terms, which represent attraction towards the center, and, therefore, the equations of motion are

$$\ddot{r} - r\dot{\theta}^2 = \frac{\partial W}{\partial r} \tag{12.16}$$

$$2\dot{r}\dot{\theta} + r\ddot{\theta} = 0 \tag{12.17}$$

From Eq. (12.17) we obtain the conservation of angular momentum,

$$h \triangleq r^2\dot{\theta} = constant \tag{12.18}$$

Therefore,

$$\dot{\theta} = \frac{h}{r^2} \tag{12.19}$$

By substituting Eq. (12.19) into Eq. (12.16),

$$\ddot{r} = \frac{h^2}{r^3} + \frac{\partial W}{\partial r} \tag{12.20}$$

By defining the effective potential energy as in Eq. (12.6) we can obtain from Eq. (12.20)

$$\ddot{r} = -\frac{\partial U(r)}{\partial r} \tag{12.21}$$

Integrating Eq. (12.21) with respect to time will result in

$$\dot{r} = \sqrt{2\left(\varepsilon - U(r)\right)} \tag{12.22}$$

Next we integrate Eq. (12.22) and obtain the radial period

$$T \triangleq \int_{r_{min}}^{r_{max}} \frac{2}{\sqrt{2\left(\varepsilon - U(s)\right)}} ds \tag{12.23}$$

Using Eqs. (12.19) and (12.22) gives

$$\frac{d\theta}{dr} = \frac{h}{r^2\sqrt{2\left(\varepsilon - U(r)\right)}} \tag{12.24}$$

Thus, the orbital angle is

$$\varphi \triangleq \int_{r_{min}}^{r_{max}} \frac{hs^{-2}}{\sqrt{2\left(\varepsilon - U(s)\right)}} ds \tag{12.25}$$

By using the definition of U while considering only J_2, we obtain the connection

$$\left(\frac{U}{\varepsilon}-1\right)r^3=-1\left(r^3+\frac{\mu}{\varepsilon}r^2-\frac{h^2}{2\varepsilon}r+\frac{\mu J_2 {r_{eq}}^2}{2\varepsilon}\right) \tag{12.26}$$

Since r_{min}, r_{max}, r_* are the solutions of Eq. (12.9), we can write Eqs. (12.23) and (12.25) as

$$T=\frac{2}{\sqrt{-2\varepsilon}}\int_{r_{min}}^{r_{max}}\frac{s^{\frac{3}{2}}}{\sqrt{(s-r_*)(s-r_{min})(r_{max}-s)}}ds \tag{12.27}$$

$$\varphi=\frac{h}{\sqrt{-2\varepsilon}}\int_{r_{min}}^{r_{max}}\frac{s^{-\frac{1}{2}}}{\sqrt{(s-r_*)(s-r_{min})(r_{max}-s)}}ds \tag{12.28}$$

By using elliptic integrals we can write Eqs. (12.27) and (12.28) as

$$T=\frac{2}{\sqrt{-2\varepsilon}}\left[C_1E(w)+C_2F(w)+C_3P\left(\frac{r_{max}-r_{min}}{r_{max}-r_*},\sqrt{w}\right)\right] \tag{12.29}$$

$$\varphi=\frac{2hE(w)}{C_2\sqrt{-2\varepsilon}} \tag{12.30}$$

where

$$w\triangleq\frac{(r_{max}-r_{min})r_*}{(r_{max}-r_*)r_{min}} \tag{12.31}$$

$$C_1\triangleq\frac{{r_*}^2+r_*r_{min}+r_*r_{max}-r_{min}r_{max}}{\sqrt{r_{min}(r_{max}-r_*)}} \tag{12.32}$$

$$C_2\triangleq\sqrt{r_{min}(r_{max}-r_*)} \tag{12.33}$$

$$C_3\triangleq\frac{-{r_*}^2-r_*r_{max}+{r_{min}}^2+r_{min}r_{max}}{\sqrt{r_{min}(r_{max}-r_*)}} \tag{12.34}$$

and E, F, P are *complete elliptic integrals* of the first, second and third kind, respectively, defined as

$$E(m)\triangleq\int_0^1\frac{1}{\sqrt{(1-t^2)(1-mt^2)}}dt \tag{12.35}$$

$$F(m)\triangleq\int_0^1\frac{\sqrt{1-mt^2}}{\sqrt{(1-t^2)}}dt \tag{12.36}$$

$$P(n,k)\triangleq\int_0^1\frac{1}{\sqrt{1-t^2}(1-nt^2)\sqrt{1-k^2t^2}}dt \tag{12.37}$$

12.3.2 New Orbital Elements

Using r_{min}, r_{max} we define new orbital elements for equatorial motion about an oblate planet (Martinusi and Gurfil 2011):

$$a_* \triangleq \frac{r_{max} + r_{min}}{2} \tag{12.38}$$

$$e_* \triangleq \frac{r_{max} - r_{min}}{r_{max} + r_{min}} \tag{12.39}$$

$$p_* \triangleq a_*(1 - {e_*}^2) \tag{12.40}$$

where a_* is equivalent to the classical semimajor axis, e_* is equivalent to the classical eccentricity, and p_* is equivalent to the classical semilatus rectum, but all have a different physical meaning compared to the osculating elements, because the new elements are constants of motion.

By substituting Eqs. (12.38) and (12.39) into Eq. (12.40) we obtain

$$p_* = \frac{2r_{max}r_{min}}{r_{max} + r_{min}} \tag{12.41}$$

We can define the specific energy in terms of the new orbital elements by substituting Eqs. (12.38) and (12.41) into Eq. (12.10)

$$\varepsilon = -\frac{\mu}{2a_*} + \frac{\mu J_2 {r_{eq}}^2}{4p_* {a_*}^2} \tag{12.42}$$

By comparing (12.42) and (12.5) we obtain an expression for the velocity,

$$v = \sqrt{\left(\frac{2\mu}{r} - \frac{\mu}{a_*}\right) + \left(\frac{\mu J_2 {r_{eq}}^2}{r^3} + \frac{\mu J_2 {r_{eq}}^2}{2{a_*}^2 p_*}\right)} \tag{12.43}$$

and by using the definitions from Eqs. (12.38), (12.39) and (12.41) combined with Eq. (12.11), we obtain the expression for the specific angular momentum,

$$h = \sqrt{\mu\left(p_* - \frac{J_2 r_{eq}^2}{2a_*} + \frac{2J_2 r_{eq}^2}{p_*}\right)} \tag{12.44}$$

It is readily seen that the modified values for ε, h and v are larger than the equivalent Keplerian values.

12.3.3 Open Orbits and the Escape Velocity

In previous sections, we examined bounded motion, and assumed that the specific energy is negative (from Eq. (12.8)). Now we obtain expressions for the escape orbit, mentioned in Sect. 4.6, which can be derived by nullifying the specific energy. Thus, from Eq. (12.5)

$$\varepsilon = \frac{v^2}{2} - \frac{\mu}{r}\left[1 + \frac{J_2}{2}\left(\frac{r_{eq}}{r}\right)^2\right] = 0 \tag{12.45}$$

Solving for the velocity,

$$v_{esc} \triangleq \sqrt{\frac{2\mu}{r}\left(1 + \frac{J_2 r_{eq}^2}{2r^2}\right)} \tag{12.46}$$

Comparing the escape velocity (12.46) to the Keplerian model clearly shows that a larger velocity is needed to escape an oblate planet. Therefore, by using the Keplerian value, an escape trajectory cannot be reached. Next, we rewrite Eq. (12.9) as

$$r^3\varepsilon + \mu r^2 - \frac{h^2}{2}r + \frac{\mu J_2 {r_{eq}}^2}{2} = 0 \tag{12.47}$$

Since $\varepsilon = 0$,

$$2\mu r^2 - h^2 r + \mu J_2 {r_{eq}}^2 = 0 \tag{12.48}$$

Solving Eq. (12.48) for r, we obtain two solutions, which represent r_{min} and r_*,

$$r_{min} = \frac{h^2 + \sqrt{h^4 - 8\mu^2 J_2 {r_{eq}}^2}}{4\mu} \tag{12.49}$$

$$r_* = \frac{h^2 - \sqrt{h^4 - 8\mu^2 J_2 {r_{eq}}^2}}{4\mu} \tag{12.50}$$

Finally, we wish to obtain the orbital angle of the escape trajectory. Isolating the specific energy from Eq. (12.10) yields

$$\varepsilon = \frac{\mu\left(J_2 {r_{eq}}^2/2 - r_{min} r_{max}\right)}{r_{min} r_{max}(r_{min} + r_{max})} \tag{12.51}$$

Using the definition for C_2 from Eq. (12.33) and Eq. (12.51) gives

$$C_2\sqrt{-2\varepsilon} = \sqrt{r_{min}(r_{max} - r_*)}\sqrt{\frac{\mu\left(2r_{min}r_{max} - J_2{r_{eq}}^2\right)}{r_{min}r_{max}(r_{min} + r_{max})}} \tag{12.52}$$

Rearranging Eq. (12.52) leads to

$$C_2\sqrt{-2\varepsilon} = \sqrt{\frac{2\mu r_{\min}^2 r_{\max}^2 - \mu J_2 r_{eq}^2 r_{\min} r_{\max} - 2\mu r_{\min}^2 r_{\max} r_* + \mu J_2 r_{eq}^2 r_{\min} r_*}{r_{\min}^2 r_{\max} + r_{\min} r_{\max}^2}} \tag{12.53}$$

For any escape orbit, $r_{max} \to \infty$, and Eq. (12.53) becomes

$$C_2\sqrt{-2\varepsilon} = \sqrt{2\mu r_{min}} \tag{12.54}$$

In addition, for $r_{max} \to \infty$ Eq. (12.31) transforms into

$$w = \frac{r_*}{r_{min}} \tag{12.55}$$

By substituting Eqs. (12.54) and (12.55) into Eq. (12.30) we obtain the orbital angle for escape orbits, which defines the direction at infinity, towards which the orbiting body moves,

$$\varphi_{esc} = \frac{2hE(r_*/r_{min})}{\sqrt{2\mu r_{min}}} \tag{12.56}$$

The terms obtained in this section provide the general description of the escape orbit from the equatorial plane of an oblate planet. A more thorough examination of such orbits was performed by Martinusi and Gurfil (2013).

12.3.4 Circular Orbits

For circular orbits $r_{min} = r_{max} \triangleq r_c$, and, therefore,

$$p_* = a_* = r_c \tag{12.57}$$

By substituting Eq. (12.57) into Eqs. (12.42), (12.43), and (12.44) we obtain

$$v_c = \sqrt{\frac{\mu}{r_c}\left(1 + \frac{3J_2 r_{eq}^2}{2r_c^2}\right)} \tag{12.58}$$

$$\varepsilon_c = \frac{-\mu}{2r_c}\left(1 - \frac{J_2 r_{eq}^2}{2r_c^2}\right) \tag{12.59}$$

$$h_c = \sqrt{\mu r_c\left(1 + \frac{3J_2 r_{eq}^2}{2r_c^2}\right)} \tag{12.60}$$

From Eqs. (12.58), (12.59), and (12.60) it is seen that the velocity, specific energy, and the specific angular momentum of a circular equatorial orbit around an oblate planet are larger than the equivalent values obtained by using Keplerian motion for the same radius r_c.

Another important aspect of the circular equatorial motion is the fact that, since there is no distinction between r_{min} and r_{max}, the orbital angle φ loses its physical meaning, and, therefore, the radial period must be calculated in a different manner. One possible method to calculate the period is by using the basic physical connection

$$h = r^2\dot{\theta} \tag{12.61}$$

which can be re-written as

$$\frac{d\theta}{dt} = \frac{h}{r^2} \tag{12.62}$$

Integrating Eq. (12.62) yields

$$\int_0^T dt = \int_0^\theta \frac{r^2}{h} \tag{12.63}$$

For a circle, $\theta = 2\pi$, so

$$T = 2\pi\frac{r^2}{h} \tag{12.64}$$

Using the expression for the specific angular momentum from Eq. (12.60) will lead to the circular radial period in the presence of J_2,

$$T_c = 2\pi \sqrt{\frac{r_c^3}{\mu}\left(1 - \frac{3J_2 r_{eq}^2}{2r_c^2 + 3J_2 r_{eq}^2}\right)} \tag{12.65}$$

12.4 The Cid-Lahulla Approach

In previous sections we presented solutions for equatorial motion under the effect of the zonal harmonic J_2. We now discuss an alternative approach, the *Cid-Lahulla approach*, which yields approximate analytical solutions for inclined orbits as well, through the concept of *radial intermediaries*.

12.4.1 Polar-Nodal Coordinates

We start by introducing the *polar nodal coordinates* $(r, \theta, \nu, R, \Theta, N)$. We shall use the original notation used by Cid and Lahulla (1969), which is as follows: r is the magnitude of the position vector, θ is the argument of latitude (denoted previously by u in the context of Chap. 5), ν is the right ascension of the ascending node (previously denoted by Ω), R is the radial velocity (not to be confused with the perturbing potential used previously), Θ is the magnitude of the specific angular momentum, and N is the polar component of the specific angular momentum vector, i.e. the projection of the specific angular momentum vector on the inertial z axis.

The transformation between the inertial position and velocity vectors $(\mathbf{r}, \mathbf{v})$ and the polar-nodal coordinates $(r, \theta, \nu, R, \Theta, N)$ is performed as follows. Let the three unit vectors of the inertial planet-centered frame be $\hat{\mathbf{x}}, \hat{\mathbf{y}}, \hat{\mathbf{z}}$. Define the unit vectors $\mathbf{n}_{1,2}$ as

$$\mathbf{n}_1 = \begin{cases} \mathbf{h} \times \hat{\mathbf{z}} / \|\mathbf{h} \times \hat{\mathbf{z}}\| & \mathbf{h} \times \hat{\mathbf{z}} \neq \mathbf{0} \\ \hat{\mathbf{x}} & \mathbf{h} \times \hat{\mathbf{z}} = \mathbf{0} \end{cases} ; \; \mathbf{n}_2 = \hat{\mathbf{h}} \times \mathbf{n}_1 \tag{12.66}$$

Then

$$\begin{gathered} r = \|\mathbf{r}\| ; \quad \begin{cases} \cos\theta = \hat{\mathbf{u}}_r \cdot \mathbf{n}_1 \\ \sin\theta = \hat{\mathbf{u}}_r \cdot \mathbf{n}_2 \end{cases} ; \quad \begin{cases} \cos\nu = \hat{\mathbf{x}} \cdot \mathbf{n}_1 \\ \sin\nu = \hat{\mathbf{y}} \cdot \mathbf{n}_1 \end{cases} ; \\ R = \frac{\mathbf{r} \cdot \mathbf{v}}{r}; \quad \Theta = \|\mathbf{r} \times \mathbf{v}\| ; \qquad N = (\mathbf{r} \times \mathbf{v}) \cdot \hat{\mathbf{z}} \end{gathered} \tag{12.67}$$

The inverse transformation is performed by

$$\mathbf{R}_\nu = \begin{bmatrix} \cos\nu & -\sin\nu & 0 \\ \sin\nu & \cos\nu & 0 \\ 0 & 0 & 1 \end{bmatrix}; \ \mathbf{R}_i = \begin{bmatrix} 1 & 0 & 0 \\ 0 & \cos i & -\sin i \\ 0 & \sin i & \cos i \end{bmatrix}$$

$$\mathbf{R}_\theta = \begin{bmatrix} \cos\theta & -\sin\theta & 0 \\ \sin\theta & \cos\theta & 0 \\ 0 & 0 & 1 \end{bmatrix} \tag{12.68}$$

$$\mathbf{r} = \mathbf{R}_\nu \mathbf{R}_i \mathbf{R}_\theta \left[r \; 0 \; 0 \right]^T, \ \mathbf{v} = \frac{\mathbf{h} \times \mathbf{r}}{r^2} + R\frac{\mathbf{r}}{r} \tag{12.69}$$

where

$$\cos i = \frac{N}{\Theta}; \ \sin i = \sqrt{1 - \left(\frac{N}{\Theta}\right)^2}; \ \mathbf{h} = \mathbf{r} \times \mathbf{v} = \mathbf{R}_\nu \mathbf{R}_i \left[0 \; 0 \; \Theta \right]^T \tag{12.70}$$

Some simplifications may be made to the above formulas. By using some algebraic manipulations, one obtains:

$$\mathbf{Q} \triangleq \mathbf{R}_\nu \mathbf{R}_i \mathbf{R}_\theta \tag{12.71}$$

$$\mathbf{r} = r\mathbf{Q}\hat{\mathbf{x}}; \ \mathbf{v} = \mathbf{Q}\left(R\hat{\mathbf{x}} + \frac{\Theta}{r}\hat{\mathbf{y}} \right), \ \mathbf{h} = \Theta \mathbf{Q}\hat{\mathbf{z}} \tag{12.72}$$

12.4.2 The Cid-Lahulla Radial Intermediary

The Hamiltonian associated with the motion in a Cid-Lahulla potential is obtained by using the von Zeipel method (von Zeipel 1921; Cid and Lahulla 1969). An outline of the method is presented below.

Consider the Hamiltonian associated with the motion in a J_2 potential, written in polar-nodal variables,

$$\mathcal{H} = \frac{1}{2}\left(R^2 + \frac{\Theta^2}{r^2} \right) - \frac{\mu}{r} - \frac{\mu J_2 r_{eq}^2}{r^3} \left(B_{20} + B_{22} \cos 2\theta \right) \tag{12.73}$$

where

$$B_{20} = -\frac{1}{4}\left(1 - 3\frac{N^2}{\Theta^2} \right) \tag{12.74a}$$

$$B_{22} = \frac{3}{4}\left(1 - \frac{N^2}{\Theta^2} \right) \tag{12.74b}$$

Recall the methodology of canonical transformations discussed in Chap. 10 and consider a canonical transformation $(r, \theta, \nu, R, \Theta, N) \rightarrow (r', \theta', \nu', R', \Theta', N')$, given by a generating function $\mathcal{W}$ which satisfies

$$R = \frac{\partial \mathcal{W}}{\partial r} \quad \Theta = \frac{\partial \mathcal{W}}{\partial \theta} \quad N = \frac{\partial \mathcal{W}}{\partial \nu}$$
$$r' = \frac{\partial \mathcal{W}}{\partial R'} \quad \theta' = \frac{\partial \mathcal{W}}{\partial \Theta'} \quad \nu' = \frac{\partial \mathcal{W}}{\partial N'} \tag{12.75}$$

such that the new equations of the phase space are:

$$\frac{dR'}{dt} = -\frac{\partial \mathcal{H}^*}{\partial r'} \quad \frac{d\Theta'}{dt} = -\frac{\partial \mathcal{H}^*}{\partial \theta'} \quad \frac{dN'}{dt} = -\frac{\partial \mathcal{H}^*}{\partial \nu'}$$
$$\frac{dr'}{dt} = \frac{\partial \mathcal{H}^*}{\partial R'} \quad \frac{d\theta'}{dt} = \frac{\partial \mathcal{H}^*}{\partial \Theta'} \quad \frac{d\nu'}{dt} = \frac{\partial \mathcal{H}^*}{\partial N'} \tag{12.76}$$

and the new Hamiltonian $\mathcal{H}^*$ satisfies the equality

$$\mathcal{H}^* \left(r', \theta', \nu', R', \Theta', N'\right) = \mathcal{H}\left(r, \theta, \nu, R, \Theta, N\right) \tag{12.77}$$

In order to determine the generating function $\mathcal{W}$, expand $\mathcal{H}^*$ and $\mathcal{W}$ into a series depending on a small parameter of order J_2 (von Zeipel 1921),

$$\mathcal{W} = \mathcal{W}_0 + \mathcal{W}_1 + \mathcal{W}_2 + \ldots \tag{12.78a}$$

$$\mathcal{H}^* = \mathcal{H}_0^* + \mathcal{H}_1^* + \mathcal{H}_2^* + \ldots \tag{12.78b}$$

and search for $\mathcal{W}_0$ of the form;

$$\mathcal{W}_0 = R'r + \Theta'\theta + N'\nu \tag{12.79}$$

After some manipulations (Cid and Lahulla 1969), the expressions for $\mathcal{H}_0^*$ and $\mathcal{H}_1^*$ are obtained as

$$\mathcal{H}_0^* = \frac{1}{2}\left(R'^2 + \frac{\Theta'^2}{r'^2}\right) - \frac{\mu}{r'} \tag{12.80a}$$

$$\mathcal{H}_1^* = -\frac{\mu J_2 r_{eq}^2}{r'^3} B'_{20} \tag{12.80b}$$

while the partial differential equation for $\mathcal{W}_1$ is written as

$$r^3 R' \frac{\partial \mathcal{W}_1}{\partial r} + \Theta' r \frac{\partial \mathcal{W}_1}{\partial \theta} + \left(\Theta'^2 - \mu r\right) \frac{\partial \mathcal{W}_1}{\partial R'} = \mu J_2 r_{eq}^2 B'_{22} \cos 2\theta \tag{12.81}$$

The associated differential equations are

$$\frac{dr}{R'r^3} = \frac{d\theta}{\Theta' r} = \frac{dR'}{\Theta'^2 - \mu r} = \frac{d\mathcal{W}_1}{\mu J_2 r_{eq}^2 B'_{22} \cos 2\theta} \tag{12.82}$$

and the expression of the generating function $\mathcal{W}_1$ is found to be (Cid and Lahulla 1969)

$$\mathcal{W}_1 = \frac{\mu^2 J_2 r_{eq}^2}{2\Theta'^3} B'_{22} \left[\sin 2\theta + e \sin(2\theta - f) + \frac{e}{3} \sin(2\theta + f)\right] \tag{12.83}$$

where f is the (osculating) true anomaly of the orbit, e its (osculating) eccentricity, both linked to r by

$$r = \frac{\Theta^2}{\mu(1 + e\cos f)} \tag{12.84}$$

By taking into account Eqs. (12.79), (12.83), as well as the equation

$$\begin{aligned} R &= \frac{\partial \mathcal{W}}{\partial r} \quad \Theta = \frac{\partial \mathcal{W}}{\partial \theta} \quad N = \frac{\partial \mathcal{W}}{\partial \nu} \\ r' &= \frac{\partial \mathcal{W}}{\partial R'} \quad \theta' = \frac{\partial \mathcal{W}}{\partial \Theta'} \quad \nu' = \frac{\partial \mathcal{W}}{\partial N'} \end{aligned} \tag{12.85}$$

where $\mathcal{W}$ is approximated as

$$\mathcal{W} \simeq \mathcal{W}_0 + \mathcal{W}_1 \tag{12.86}$$

the contact transformation is explicitly determined by using Eqs. (12.76) as follows (Cid and Lahulla 1969):

$$R = R' - \frac{2\mu J_2 B'_{22} r_{eq}^2}{3r^2\Theta'^2} \cos 2\theta \tag{12.87a}$$

$$\Theta = \Theta' + \frac{\mu^2 J_2 B'_{22} r_{eq}^2}{\Theta'^3} \left[\cos 2\theta + e\cos(2\theta - f) + \frac{e}{3}\cos(2\theta + f)\right] \tag{12.87b}$$

$$N = N' \tag{12.87c}$$

$$r = r' + \frac{\mu J_2 B'_{22} r_{eq}^2}{3\Theta'^2} \cos 2\theta \tag{12.87d}$$

$$\theta = \theta' - \frac{3\mu^2 J_2}{8\Theta'^4}\left[\sin 2\theta + e\sin(2\theta - f) + \frac{e}{3}\sin(2\theta + f)\right]\left(5\cos^2 i' - 3\right) \tag{12.87e}$$

$$+ \frac{\mu J_2 B'_{22} R' r_{eq}^2}{3\Theta'^3}\cos 2\theta - \frac{4\mu J_2 B'_{22} r_{eq}^2}{3\Theta'^2 r}\sin 2\theta$$

$$\nu = \nu' + \frac{3\mu^2 J_2}{4\Theta'^4}\left[\sin 2\theta + e\sin(2\theta - f) + \frac{e}{3}\sin(2\theta + f)\right]\cos i' \tag{12.87f}$$

where for convenience it was denoted

$$\cos i' \triangleq \frac{N'}{\Theta'} \tag{12.88}$$

Equations (12.87) do not yield a closed-form expression for the contact transformation, since Eq. (12.87b) involves solving a 6th order polynomial equation in Θ'. In order to overcome this issue, a first-order approximation of Θ' about Θ is performed. The relationship between Θ and Θ' may be written as

$$\Theta = \Theta' + \frac{\alpha}{\Theta'^3} + \frac{\beta}{\Theta'^5}, \tag{12.89}$$

where

$$\begin{cases} \alpha = \frac{3}{4}\mu^2 J_2 r_{eq}^2\left[\cos 2\theta + e\cos(2\theta - f) + \frac{e}{3}\cos(2\theta + f)\right] \\ \beta = -\frac{3}{4}\mu^2 J_2 N'^2 r_{eq}^2\left[\cos 2\theta + e\cos(2\theta - f) + \frac{e}{3}\cos(2\theta + f)\right] \end{cases} \tag{12.90}$$

Now, consider the function

$$h(\Theta') = \Theta' + \frac{\alpha}{\Theta'^3} + \frac{\beta}{\Theta'^5} \tag{12.91}$$

The approximation is made by considering the Taylor series expansion

$$h(\Theta') = h(\Theta) + (\Theta' - \Theta)\left.\frac{dh}{d\Theta'}\right|_{\Theta} + \frac{(\Theta' - \Theta)^2}{2!}\left.\frac{d^2h}{d\Theta'^2}\right|_{\Theta} + \frac{(\Theta' - \Theta)^3}{3!}\left.\frac{d^3h}{d\Theta'^3}\right|_{\Theta} + \ldots \tag{12.92}$$

which yields

$$0 = \frac{\alpha}{\Theta^3} + \frac{\beta}{\Theta^5} + (\Theta' - \Theta)\left(1 - \frac{3\alpha}{\Theta^4} - \frac{5\beta}{\Theta^6}\right) \tag{12.93}$$
$$+\frac{(\Theta' - \Theta)^2}{2!}\left(\frac{12\alpha}{\Theta^5} + \frac{30\beta}{\Theta^7}\right) - \frac{(\Theta' - \Theta)^3}{3!}\left(\frac{60\alpha}{\Theta^6} + \frac{210\beta}{\Theta^8}\right) + \ldots.$$

The linearized expression of Θ' is

$$\Theta' \simeq \Theta \frac{\Theta^6 - 4\alpha\Theta^2 - 6\beta}{\Theta^6 - 3\alpha\Theta^2 - 5\beta} = \Theta\left(1 - \frac{\alpha\Theta^2 + \beta}{\Theta^6 - 3\alpha\Theta^2 - 5\beta}\right) \tag{12.94}$$

The rest of the equations needed to perform the transformation are

$$N' = N \tag{12.95a}$$

$$R' = R + \frac{2\mu J_2 B'_{22} r^2_{eq}}{3r^2\Theta'^2}\cos 2\theta \tag{12.95b}$$

$$r' = r - \frac{\mu J_2 B'_{22} r^2_{eq}}{3\Theta'^2}\cos 2\theta \tag{12.95c}$$

$$\theta' = \theta + \frac{3\mu^2 J_2}{8\Theta'^4}\left[\sin 2\theta + e\sin(2\theta - f) + \frac{e}{3}\sin(2\theta + f)\right]\left(5\cos^2 i' - 3\right) \tag{12.95d}$$

$$- \frac{\mu J_2 B'_{22} R' r^2_{eq}}{3\Theta'^3}\cos 2\theta + \frac{4\mu J_2 B'_{22} r^2_{eq}}{3\Theta'^2 r}\sin 2\theta$$

$$\nu' = \nu - \frac{3\mu^2 J_2}{4\Theta'^4}\left[\sin 2\theta + e\sin(2\theta - f) + \frac{e}{3}\sin(2\theta + f)\right]\cos i' \tag{12.95e}$$

It is necessary to derive also the inverse transformation, namely from $(r', \theta', \nu', R', \Theta', N')$ to the $(r, \theta, \nu, R, \Theta, N)$ variables. Given the complexity of the system of equations (12.87), a direct approach to solve them for $(r, \theta, \nu, R, \Theta, N)$ seems impossible. Therefore, a classical approximation from perturbation theory is used (Brouwer 1959), and is presented below.

Let $\{c_k, c'_k\}$ be the six-dimensional sets of old and new variables, respectively. Assume that the transformation $c_k \to c'_k$ is given by the equation

$$c_k{}' = c_k + \epsilon f(c_j) \tag{12.96}$$

where ϵ denotes the small perturbing parameter (in the present case, $\epsilon = J_2$). Then the inverse transformation $c'_k \rightarrow c_k$ may be approximated by

$$c_k \approx c'_k - \epsilon f\left(c'_j\right) \tag{12.97}$$

Consequently, the old-to-new contact transformation may be deduced by applying exactly the same procedure as in the old-to-new transformation, by considering Eqs. (12.87) where $J_2 \rightarrow -J_2$ (Brouwer 1959).

12.4.3 Comparison with Brouwer's Approximation

Like the Brouwer (1959) model, the Cid-Lahulla potential was derived based on the classical method of von Zeipel (1921), but starting with a different set of canonical coordinates. While the *Brouwer first-order approximation* is expressed with respect to three constant Delaunay elements, namely

$$\mathcal{H}_1^{*Brouwer}\left(-,-,-,L',G',H'\right) = -\frac{\mu^2}{2L'^2} - \frac{\mu^4 J_2 r_{eq}^2}{L'^3 G'^3}\left(-\frac{1}{2} + \frac{3}{2}\frac{H'^2}{G'^2}\right) \tag{12.98}$$

it contains no short-periodic terms. The Cid-Lahulla potential, to a first-order approximation, is expressed with respect to four polar-nodal variables, of which only two are constant, while the other two, r' and R', contain short-periodic terms,

$$\mathcal{H}_1^{*Cid}\left(r',-,-,R',N',\Theta'\right) = \frac{1}{2}\left(R'^2 + \frac{\Theta'^2}{r'^2}\right) - \frac{\mu}{r'} + \frac{\mu J_2 r_{eq}^2}{4r'^3}\left(1 - 3\frac{N'^2}{\Theta'^2}\right) \tag{12.99}$$

12.5 Solution for Motion in a Cid-Lahulla Potential

Consider the Hamiltonian in Eq. (12.99). As done previously, denote by ε the constant specific energy of the motion, which is equal to the Hamiltonian expressed in Eq. (12.99). For simplicity, the prime symbol will be omitted while carrying on the computations associated with the new Hamiltonian. Hamilton's equations yield

$$\frac{dr}{dt} = R \tag{12.100a}$$

$$\frac{d\theta}{dt} = \frac{\Theta}{r^2} + \frac{3}{2}\frac{\mu J_2 r_{eq}^2}{r^3}\frac{N^2}{\Theta^3} \tag{12.100b}$$

$$\frac{d\nu}{dt} = -\frac{3}{2}\frac{\mu J_2 r_{eq}^2}{r^3}\frac{N}{\Theta^2} \tag{12.100c}$$

$$\frac{dR}{dt} = -\frac{\mu}{r^2} + \frac{\Theta^2}{r^3} + \frac{3}{4}\frac{\mu J_2 r_{eq}^2}{r^4}\left(1 - 3\frac{N^2}{\Theta^2}\right) \tag{12.100d}$$

$$\frac{d\Theta}{dt} = 0 \tag{12.100e}$$

$$\frac{dN}{dt} = 0 \tag{12.100f}$$

The equations are separable, based on the fact that Θ and N are constants. The radial motion $r = r(t)$ satisfies (based on Eq. (12.99), as well as on the fact that $R = \dot{r}$)

$$\dot{r}^2 = 2\varepsilon + \frac{2\mu}{r} - \frac{\Theta^2}{r^2} - \frac{1}{2}\frac{\mu J_2 r_{eq}^2}{r^3}\left(1 - 3\frac{N^2}{\Theta^2}\right) \tag{12.101}$$

The only case of interest is when the total energy ε is strictly negative, which yields bounded orbits. Then Eq. (12.101) may be rewritten as

$$\dot{r}^2 = \frac{2|\varepsilon|}{r^3}\left[-r^3 + \frac{\mu}{|\varepsilon|}r^2 - \frac{\Theta^2}{2|\varepsilon|}r - \frac{\mu J_2 r_{eq}^2}{4|\varepsilon|}\left(1 - 3\frac{N^2}{\Theta^2}\right)\right] \tag{12.102}$$

The algebraic equation

$$r^3 - \frac{\mu}{|\varepsilon|}r^2 + \frac{\Theta^2}{2|\varepsilon|}r + \frac{\mu J_2 r_{eq}^2}{4|\varepsilon|}\left(1 - 3\frac{N^2}{\Theta^2}\right) = 0 \tag{12.103}$$

has always at least one real solution, since it is a cubic equation. Denote by r_1, r_2, r_3 the roots of Eq. (12.103). From the Vieta formulae, it follows

$$\begin{cases} r_1 + r_2 + r_3 = \dfrac{\mu}{|\varepsilon|} \\ r_1 r_2 + r_2 r_3 + r_3 r_1 = \dfrac{\Theta^2}{2|\varepsilon|} \\ r_1 r_2 r_3 = -\dfrac{\mu J_2 r_{eq}^2}{4|\varepsilon|}\left(1 - 3\dfrac{N^2}{\Theta^2}\right) \end{cases} \tag{12.104}$$

Assume that the conditions for Eq. (12.103) to have three real solutions are fulfilled. The last of the Vieta formulae (12.104) shows that one of the roots might be negative, depending on the orbit inclination. In any case, the equation for the radial motion may be written as

$$\dot{r}^2 = \frac{2|\varepsilon|}{r^3}\left(r - r_{\min}\right)\left(r_{\max} - r\right)\left(r - r_*\right) \tag{12.105}$$

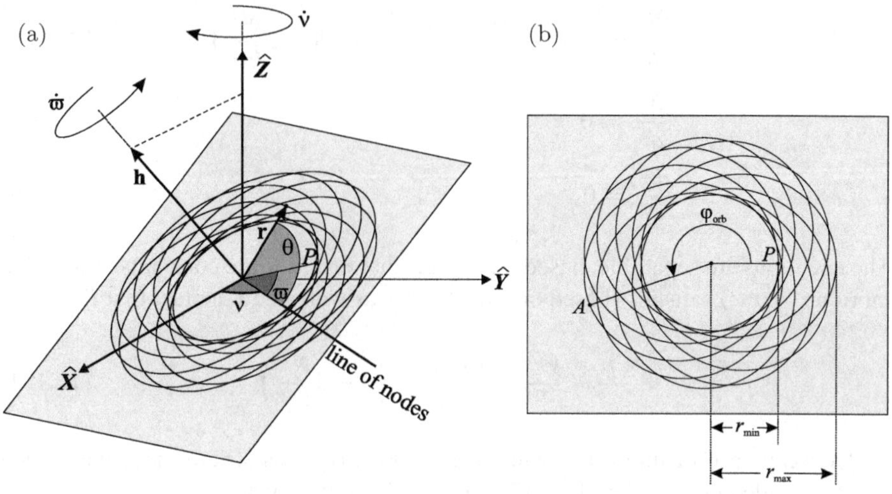

Fig. 12.3 Geometric interpretation of the motion in a Cid-Lahulla potential

where $r_{min,max}$ are the actual bounds of the radial motion (see also Fig. 12.3a), while r_* is the third root of the algebraic equation (12.103), expressed as

$$r_* = \gamma \frac{r_{min} + r_{max}}{r_{min} r_{max} - \gamma}, \; \gamma \triangleq -\frac{J_2 r_{eq}^2}{2}\left(1 - 3\frac{N^2}{\Theta^2}\right) \tag{12.106}$$

The sign of r_* depends on the sign of $1 - 3N^2/\Theta^2 = 1 - 3\cos^2 i$. Three cases are distinguished,

$$\begin{aligned} i &> \cos^{-1}\left(1/\sqrt{3}\right), \; 0 < r_* < r_{min} \\ i &= \cos^{-1}\left(1/\sqrt{3}\right), \; 0 = r_* < r_{min} \\ i &< \cos^{-1}\left(1/\sqrt{3}\right), \; r_* < 0 < r_{min} \end{aligned} \tag{12.107}$$

The equations of motion are then

$$\dot{r}^2 = \frac{2|\varepsilon|}{r^3}(r - r_{min})(r_{max} - r)(r - r_*) \tag{12.108a}$$

$$\dot{\theta} = \frac{\Theta}{r^2} + \frac{3}{2}\frac{\mu J_2 r_{eq}^2}{r^3}\frac{N^2}{\Theta^3} \tag{12.108b}$$

$$\dot{\nu} = -\frac{3}{2}\frac{\mu J_2 r_{eq}^2}{r^3}\frac{N}{\Theta^2} \tag{12.108c}$$

with the initial conditions $r(t_0) = r_0$, $\theta(t_0) = \theta_0$, $\nu(t_0) = \nu_0$, and the values of N and Θ determined based on the initial state $(\mathbf{r}_0, \mathbf{v}_0)$. Solving Eqs. (12.108) leads to the analytical equations of motion.

12.5.1 Main Steps Towards a Solution

By manipulating Eq. (12.108a),

$$dt = \pm \frac{r^{3/2} dr}{\sqrt{2|\varepsilon|(r - r_{\min})(r_{\max} - r)(r - r_*)}} \tag{12.109}$$

If $r_0 \in [r_{\min}, r_{\max}]$, then $r(t) \in [r_{\min}, r_{\max}]$, for all $t \geq t_0$, and, therefore, the motion is bounded and so is the radial velocity $\dot{r}$. The radial motion is periodic (Arnold 1989), and its main period is denoted by T. Its value is determined from Eq. (12.109),

$$T = \frac{2}{\sqrt{2|\varepsilon|}} \int_{r_{\min}}^{r_{\max}} \frac{s^{3/2} ds}{\sqrt{(s - r_*)(s - r_{\min})(r_{\max} - s)}}. \tag{12.110}$$

The closed-form expression for T may take three forms, depending on the orbit inclination, as distinguished in Eqs. (12.107).

Consider the *incomplete elliptic integrals* $F(\cdot,\cdot)$, $E(\cdot,\cdot)$, $\Pi(\cdot,\cdot,\cdot)$ of first, second and third kind, respectively, written by using the Jacobi notation, as follows: let $-1 \leq z \leq 1$, $0 \leq k \leq 1$, then

$$F(z, k) = \int_0^z \frac{du}{\sqrt{1 - u^2}\sqrt{1 - k^2 u^2}} \tag{12.111a}$$

$$E(z, k) = \int_0^z \frac{\sqrt{1 - k^2 u^2}}{\sqrt{1 - u^2}} du \tag{12.111b}$$

$$\Pi(z, \upsilon, k) = \int_0^z \frac{du}{(1 - \upsilon^2 u^2)\sqrt{1 - u^2}\sqrt{1 - k^2 u^2}} \tag{12.111c}$$

and denote by $K(\cdot)$, $E(1, \cdot)$ and $\Pi(1, \cdot, \cdot)$ their complete counterparts.

(i) If $i > \cos^{-1}\left(1/\sqrt{3}\right)$:

$$\begin{aligned} T_{+} &= \frac{\sqrt{2}}{\sqrt{|\varepsilon|}}\left\{\frac{-r_{\min}r_{\max} + r_{*}\left(r_{\max} + r_{\min} + r_{*}\right)}{\sqrt{r_{\min}\left(r_{\max} - r_{*}\right)}}K\left(w_{+}\right)\right. \\ &+ \frac{r_{\min}\left(r_{\max} - r_{*}\right)}{\sqrt{r_{\min}\left(r_{\max} - r_{*}\right)}}E\left(1, w_{+}\right) \\ &+ \left.\frac{\left(r_{\min} - r_{*}\right)\left(r_{*} + r_{\min} + r_{\max}\right)}{\sqrt{r_{\min}\left(r_{\max} - r_{*}\right)}}\Pi\left(1, \frac{r_{\max} - r_{\min}}{r_{\max} - r_{*}}, w_{+}\right)\right\} \end{aligned} \tag{12.112}$$

where

$$w_{+} = \sqrt{\frac{\left(r_{\max} - r_{\min}\right)r_{*}}{\left(r_{\max} - r_{*}\right)r_{\min}}}. \tag{12.113}$$

(ii) If $i = \cos^{-1}\left(1/\sqrt{3}\right)$:

$$T_{0} = \frac{2\pi\mu}{\left(2\left|\varepsilon\right|\right)^{3/2}} \tag{12.114}$$

(iii) If $i < \cos^{-1}\left(1/\sqrt{3}\right)$:

$$\begin{aligned} T_{-} &= \frac{\sqrt{2}}{\sqrt{|\varepsilon|}}\left\{\frac{-r_{\min}r_{\max}}{\sqrt{r_{\max}\left(r_{\min} - r_{*}\right)}}K\left(w_{-}\right)\right. \\ &+ \frac{r_{\max}\left(r_{\min} - r_{*}\right)}{\sqrt{r_{\max}\left(r_{\min} - r_{*}\right)}}E\left(1, w_{-}\right) \\ &+ \left.\frac{r_{\min}\left(r_{*} + r_{\min} + r_{\max}\right)}{\sqrt{r_{\max}\left(r_{\min} - r_{*}\right)}}\Pi\left(1, \frac{r_{\max} - r_{\min}}{r_{\max}}, w_{-}\right)\right\} \end{aligned} \tag{12.115}$$

where

$$w_{-} = \sqrt{\frac{\left(r_{\max} - r_{\min}\right)r_{*}}{r_{\max}\left(r_{\min} - r_{*}\right)}} \tag{12.116}$$

The solution to the system of differential equations (12.108a)–(12.108c) always depends upon three integrals, which are functions of $r \in [r_{\min}, r_{\max}]$,

$$J_r = \int\limits_{r_{\min}}^{r} \frac{s^{3/2} ds}{\sqrt{(s - r_*)(s - r_{\min})(r_{\max} - s)}} \tag{12.117a}$$

$$J_h = \int\limits_{r_{\min}}^{r} \frac{s^{-1/2} ds}{\sqrt{(s - r_*)(s - r_{\min})(r_{\max} - s)}} \tag{12.117b}$$

$$J_\nu = \int\limits_{r_{\min}}^{r} \frac{s^{-3/2} ds}{\sqrt{(s - r_*)(s - r_{\min})(r_{\max} - s)}} \tag{12.117c}$$

Their closed-form expressions may be deduced with the help of the incomplete elliptic integrals, for all three possible cases shown in Eqs. (12.107).

It is useful to point out a few qualitative features of the motion. As assumed previously, the radial motion is periodic, and its main period T is computed as in Eqs. (12.112), (12.114) and (12.115). The motion may be visualized as follows. The satellite moves in a variable plane (see Fig. 12.3a). For an observer for whom this plane is fixed, the motion takes place on a rigid rosette (the trajectory of the motion in a central force field, see Fig. 12.3b), which precesses about the normal to the plane with the rate

$$\dot{\varpi} = \frac{3}{2} \frac{\mu J_2 r_{eq}^2}{r^3} \frac{N^2}{\Theta^3} \tag{12.118}$$

The orbital angle associated with the central motion (from the point of view of another observer, for whom the rosette is fixed in space), denoted by φ_{orb}, (see Fig. 12.3b) is computed from

$$\varphi_{orb} = \int\limits_{r_{\min}}^{r_{\max}} \frac{\Theta}{r^2} \mathrm{d}t = \frac{\Theta}{\sqrt{2|\varepsilon|}} J_h (r_{\max}) \tag{12.119}$$

Its values in the three situations (i)–(iii) are determined as (the superscripts are: "+" for $i > \arccos\left(1/\sqrt{3}\right)$, "(0)" for $i = \arccos\left(1/\sqrt{3}\right)$ and "−" for

$i < \arccos\left(1/\sqrt{3}\right)$):

$$\varphi_{orb}^{+} = \frac{2\Theta}{\sqrt{2|\varepsilon|}} \frac{K(w_+)}{\sqrt{r_{\min}(r_{\max} - r_*)}} \tag{12.120}$$

$$\varphi_{orb}^{(0)} = \pi \tag{12.121}$$

$$\varphi_{orb}^{-} = \frac{2\Theta}{\sqrt{2|\varepsilon|}} \frac{K(w_-)}{\sqrt{r_{\max}(r_{\min} - r_*)}} \tag{12.122}$$

During one radial period, the angle swiped by the position vector for an observer for whom the rosette is fixed, is 2φ. However, for an observer for whom the plane of motion is fixed, the total angle swiped in the interval of time T is different, because of the precession of the rosette about its normal, revealed in Eq. (12.127). The quantity

$$\Delta\omega_{orb} = 2\int_{r_{\min}}^{r_{\max}} \frac{3}{2}\frac{\mu}{r} J_2 \frac{r_{eq}^2}{r^2}\frac{N^2}{\Theta^3} dt = \frac{3\mu J_2 r_{eq}^2}{\sqrt{2|\varepsilon|}}\frac{N^2}{\Theta^3} J_\nu(r_{\max}) \tag{12.123}$$

is the angle of displacement of the rosette about its normal during an entire radial period T, which may also be regarded as the displacement of any pericenter of the rosette in the variable plane of motion. The closed-form expression for $\Delta\omega_{orb}$ is deduced based on the previous considerations, and the three cases are

$$\Delta\omega_{orb}^{(+)} = \frac{6\mu J_2 r_{eq}^2}{\sqrt{2|\varepsilon|}}\frac{N^2}{\Theta^3}\frac{1}{r_*\sqrt{r_{\min}(r_{\max} - r_*)}}\left[K(w_+) - \frac{2(r_{\max} - r_*)}{r_{\max}}E(1, w_+)\right] \tag{12.124a}$$

$$\Delta\omega_{orb}^{(0)} = \frac{\pi\mu^2 J_2 r_{eq}^2}{\Theta^4} \tag{12.124b}$$

$$\Delta\omega_{orb}^{(-)} = \frac{6\mu J_2 r_{eq}^2}{\sqrt{2|\varepsilon|}}\frac{N^2}{\Theta^3}\frac{1}{r_*\sqrt{r_{\max}(r_{\min} - r_*)}}\left[K(w_-) - \frac{2(r_{\min} - r_*)}{r_{\min}}E(1, w_+)\right] \tag{12.124c}$$

The plane of motion precesses about the South-North axis, with the rate described by Eq. (12.108c). The total angular displacement of this plane during a single radial period is computed as

$$\Delta\Omega_{orb} = 2\int_{r_{\min}}^{r_{\max}} \left(-\frac{3}{2}\frac{\mu J_2 r_{eq}^2}{r^3}\frac{N}{\Theta^2}\right) dt = -\frac{3\mu J_2 r_{eq}^2}{\sqrt{2|\varepsilon|}}\frac{N}{\Theta^2} J_\nu(r_{\max}) \tag{12.125}$$

The following additional notations will be useful in deriving the closed-form solution of the motion in a Cid-Lahulla potential,

$$\mathcal{J}_r(r) = \frac{1}{\sqrt{2|\varepsilon|}} J_r(r) \tag{12.126a}$$

$$\mathcal{F}_R = \frac{2}{\sqrt{2|\varepsilon|}} J_r(r_{\max}) = T \tag{12.126b}$$

$$\mathcal{J}_\theta(r) = \frac{\Theta}{\sqrt{2|\varepsilon|}} J_h(r) + \frac{3}{2}\mu J_2 r_{eq}^2 \frac{N^2}{\Theta^3} J_\nu(r) \tag{12.126c}$$

$$\mathcal{F}_\Theta = \frac{2\Theta}{\sqrt{2|\varepsilon|}} J_h(r_{\max}) + 3\mu J_2 r_{eq}^2 \frac{N^2}{\Theta^3} J_\nu(r_{\max}) \tag{12.126d}$$

$$\mathcal{J}_\nu(r) = -\frac{3}{2}\mu J_2 r_{eq}^2 \frac{N}{\Theta^2} J_\nu(r) \tag{12.126e}$$

$$\mathcal{F}_N = -3\mu J_2 r_{eq}^2 \frac{N}{\Theta^2} J_\nu(r_{\max}) \tag{12.126f}$$

12.5.2 New Independent Variable

In Eq. (12.108a), the only case of interest is when

$$r_* < r_{\min} < r < r_{\max}$$

which guarantees that $r - r_* > 0$. The dependence of the radial distance r on time is given by Eq. (12.109), which contains singularities and does not yield a closed-form expression.

A change of variable may be performed as follows. Introduce the independent variable ϵ such that

$$dt = \frac{r^{3/2}}{\sqrt{2|\varepsilon|}(r - r_*)} d\epsilon, \quad \epsilon(t_P) = 0 \tag{12.127}$$

where $t_P \geq 0$ is defined as

$$t_P = \begin{cases} \dfrac{1}{\sqrt{2|\varepsilon|}} \displaystyle\int_{r_{\min}}^{r_0} \frac{s^{3/2} ds}{\sqrt{(s - r_*)(s - r_{\min})(r_{\max} - s)}}, & \mathbf{r}_0 \cdot \mathbf{v}_0 \geq 0 \\ T - \dfrac{1}{\sqrt{2|\varepsilon|}} \displaystyle\int_{r_{\min}}^{r_0} \frac{s^{3/2} ds}{\sqrt{(s - r_*)(s - r_{\min})(r_{\max} - s)}}, & \mathbf{r}_0 \cdot \mathbf{v}_0 < 0 \end{cases} \tag{12.128}$$

Then the initial value problem for the radial motion becomes

$$\left(\frac{d}{d\epsilon}r\right)^2 = (r - r_{\min})(r_{\max} - r), \; r(0) = r_{\min} \tag{12.129}$$

and its closed form solution is expressed as

$$r(\epsilon) = a_*(1 - e_* \cos\epsilon) \tag{12.130}$$

where the following notations were used:

$$a_* \stackrel{\Delta}{=} \frac{r_{\max} + r_{\min}}{2}; \; e_* \stackrel{\Delta}{=} \frac{r_{\max} - r_{\min}}{r_{\max} + r_{\min}} \tag{12.131}$$

The radial motion is expressed exactly as in the Keplerian case, where the function $\epsilon = \epsilon(t)$ is considered as the *pseudo-eccentric anomaly* and it is the solution to the initial value problem (12.127). Unlike the Keplerian case, there does not exist a Kepler-like equation connecting this function and the time variable t.

It is now possible to give the full closed-form solution to the motion in a Cid-Lahulla potential. The solution is explicit with respect to the newly introduced independent variable, the pseudo-eccentric anomaly $\epsilon = \epsilon(t)$, defined with the help of the initial value problem (12.127).

Since the radial distance r is a periodic function of t, with the main period T, it follows from Hamilton's equations (12.108b) and (12.108c) that both functions $\dot{\theta}$ and $\dot{\nu}$ are periodic, with the same main period T. By using Eqs. (12.108), Eqs. (12.126) transform into

$$d\mathcal{J}_r = dt \tag{12.132a}$$

$$\mathcal{F}_R = T \tag{12.132b}$$

$$d\mathcal{J}_\theta = d\theta \tag{12.132c}$$

$$\mathcal{F}_\Theta = \theta(t + T) - \theta(t) \tag{12.132d}$$

$$d\mathcal{J}_\nu = d\nu \tag{12.132e}$$

$$\mathcal{F}_N = \nu(t + T) - \nu(t) \tag{12.132f}$$

Care should be taken when dealing with the differential equalities for θ and ν, since they are valid only for a limited time, given the definitions (12.117) for the integrals J_r, J_h and J_ν. In order to extend these definitions over the entire interval $[t_0, +\infty)$,

$$\dot{\theta} = \begin{cases} \dfrac{d}{dt}\mathcal{J}_\theta(r) & \mathbf{r}\cdot\dot{\mathbf{r}} \geq 0 \\ -\dfrac{d}{dt}\mathcal{J}_\theta(r) & \mathbf{r}\cdot\dot{\mathbf{r}} < 0 \end{cases} \qquad \dot{\nu} = \begin{cases} \dfrac{d}{dt}\mathcal{J}_\nu(r) & \mathbf{r}\cdot\dot{\mathbf{r}} \geq 0 \\ -\dfrac{d}{dt}\mathcal{J}_\nu(r) & \mathbf{r}\cdot\dot{\mathbf{r}} < 0 \end{cases} \tag{12.133}$$

The quantities $\mathcal{F}_R$, $\mathcal{F}_\Theta$, and $\mathcal{F}_N$ are the variations over a period of t, θ, and ν, respectively, while θ and ν are periodic functions of t, with the main period T. Then it can be stated that:

$$\left\lfloor \frac{t-t_0}{\mathcal{F}_R} \right\rfloor = \left\lfloor \frac{\theta(t)-\theta(t_0)}{\mathcal{F}_\Theta} \right\rfloor = \left\lfloor \frac{\nu(t)-\nu(t_0)}{\mathcal{F}_N} \right\rfloor, t \geq t_0, \tag{12.134}$$

where $\lfloor\cdot\rfloor$ denotes the *floor function*, where, for an integer n,

$$\lfloor x \rfloor = n, \; n \leq x < n+1 \tag{12.135}$$

By taking into account Eqs. (12.133), the solution to the Hamilton's equations of motion (12.108a)–(12.108c) is

$$r(t) = a_* \left[1 - e_* \cos\epsilon(t)\right] \tag{12.136}$$

$$\begin{aligned}\theta(r) = {} & \theta_0 + \left\lfloor \frac{\theta-\theta_0}{\mathcal{F}_\Theta} \right\rfloor \mathcal{F}_\Theta \\ & + \left\{\sigma(\dot{r}(t))\, \mathcal{J}_\theta(r(t)) + \left[1-\sigma(\dot{r}(t))\right]\left[\mathcal{F}_\Theta - \mathcal{J}_\theta(r(t))\right]\right\}\Big|_{t_0}^{t}\end{aligned} \tag{12.137}$$

$$\begin{aligned}\nu(r) = {} & \nu_0 + \left\lfloor \frac{\nu-\nu_0}{\mathcal{F}_N} \right\rfloor \mathcal{F}_N \\ & + \left\{\sigma(\dot{r}(t))\, \mathcal{J}_\nu(r(t)) + \left[1-\sigma(\dot{r}(t))\right]\left[\mathcal{F}_N - \mathcal{J}_\nu(r(t))\right]\right\}\Big|_{t_0}^{t}\end{aligned} \tag{12.138}$$

where σ denotes the *Heaviside step function*,

$$\sigma(x) = \begin{cases} 1, & x \geq 0 \\ 0, & x < 0 \end{cases} \tag{12.139}$$

For computational purposes, it is useful to note that

$$\sigma(\dot{r}) = \sigma(\sin\epsilon(t)) \tag{12.140}$$

References

Aksenov, E.P., Grebennikov, E.A., Demin, V.G.: General solution to the problem of motion of a satellite in the normal field of attraction of the Earth. Iskusstvenllye Sputmki Zemli (Artif. Satell. Earth) **8**, 64–72 (1961)

Arnold, V.I.: Mathematical Aspects of Classical Mechanics, 2nd edn. Springer, Berlin (1989)

Brouwer, D.: Solution of the problem of artificial satellite theory without drag. Astron. J. **64**, 378–396 (1959)

Celletti, A., Negrini, P.: Non-integrability of the problem of motion around an oblate planet. Celest. Mech. Dyn. Astron. **61**, 253–260 (1995)

Cid, R., Lahulla, J.: Perturbaciones de corto periodo en el movimiento de un satélite artificial, en función de las variables de Hill. Rev Academia de Ciencias de Zaragoza **XXIV**, 159–165 (1969)

Cid, R., Lahulla, J.: Perturbaciones de segundo orden y corto período, para el movimiento de un satéllite artificial en las variables de Hill. Rev Academia de Ciencias de Zaragoza **XXVI**, 333–343 (1971)

Irigoyen, M., Simo, C.: Nonintegrability of the J_2 problem. Celest. Mech. Dyn. Astron. **55**, 281–287 (1993)

Izsak, I.G.: On satellite orbits with very small eccentricities. Astron. J. **66**, 129–131 (1961)

Lara, M., Gurfil, P.: Integrable approximation of J_2-perturbed relative orbits. Celest. Mech. Dyn. Astron. **114**, 229–254 (2012)

Lyddane, R.H.: Small eccentricities or inclinations in the Brouwer theory of the artificial satellite. Astron. J. **68**, 555–558 (1963)

Martinusi, V., Gurfil, P.: Solutions and periodicity of satellite relative motion under even zonal harmonics perturbations. Celest. Mech. Dyn. Astron. **111**(4), 387–414 (2011)

Martinusi, V., Gurfil, P.: Analytical solutions for J_2-perturbed unbounded equatorial orbits. Celest. Mech. Dyn. Astron. **115**(1), 35–57 (2013)

Vinti, J.P.: Theory of the orbit of an artificial satellite with use of spheroidal coordinates. Astron. J. **65**, 353–354 (1960)

von Zeipel, H.: Recherches sur le mouvement des petites planètes. Almqvist and Wiksells Boktryckeri-A.B. (1921)

Chapter 13
Semianalytical Orbit Theory

13.1 Introduction

Semianalytical theories provide tools for propagating perturbed satellite orbits. The power of these theories lies in the ability to efficiently calculate, and thereby comprehend, the orbital dynamics of satellites. In many cases of practical interest, as we will see in Chap. 14, satellite orbit control laws also utilize mean orbital elements. The *mean elements* are most commonly defined as doubly-averaged elements, obtained from a doubly-averaged Hamiltonian. However, the mean elements in the current chapter will be singly-averaged elements, obtained by removing the short-periodic oscillations only.

There are various alternatives for calculating the mean orbital elements. One possibility is to use an analytical theory, such as the Brouwer (1959) artificial satellite theory, the Kozai (1959a) theory, the Brouwer and Hori (1961) theory, or other theories, including those developed by Lane et al. (1962); Lane (1965); Liu and Alford (1979, 1980); Hoots (1981); Lin and De-zi (1981); Liu (1983); Bezdek and Vokrouhlicki (2004) to name only a few. An alternative is to use batch processing to transform osculating to mean elements using least square-type approaches, such as those that we will discuss in Chap. 16. Brouwer-type theories are usually sensitive to noise and modeling errors and cannot easily accommodate thrust. Also, batch processing is not adequate for real-time on-board implementation as it requires data accumulation for at least a complete orbital period. A different option is to use recursive filtering, to be discussed in Chap. 16, to estimate the mean elements.

The original efforts to write a computationally efficient semianalytical model date back to the early Eighties, with the onset of the *Draper Semianalytical Satellite Theory* (DSST) (Cefola et al. 1980; Zeis and Cefola 1980). DSST contains an extensive treatment of perturbations, expressed in terms of equinoctial elements, and can also be implemented in a recursive form (Taylor and Cefola 1982).

P. Gurfil, P.K. Seidelmann, *Celestial Mechanics and Astrodynamics: Theory and Practice*, Astrophysics and Space Science Library 436,
DOI 10.1007/978-3-662-50370-6_13

In this chapter, we describe an approach for calculating the mean elements from osculating elements. A semianalytical dynamical model that includes zonal/tesseral/sectorial harmonics and drag is formulated to capture the daily, long-periodic, and secular evolution of the mean orbital elements. Because there is a tradeoff between precision and complexity, the semianalytical model is truncated to include the control inputs, the long-periodic and secular terms up to $J_4/C_{33}/S_{33}$, and exponential drag. The mapping from mean to osculating elements is obtained by adding the short-periodic effects of drag and zonal/tesseral/sectorial harmonics to the mean elements.

13.2 Preliminaries

For easier reference, we will repeat some of the coordinate frame definitions made in previous chapters (assuming an Earth satellite), to be used in this chapter. J2000 is an ECI frame. The epoch is January 1, 2000 at noon. The fundamental plane is the mean Earth equator of epoch. The $\hat{\mathbf{x}}_{J2000}$ axis is directed toward the mean vernal equinox at epoch. The $\hat{\mathbf{z}}_{J2000}$ axis lies along Earth's mean rotational axis of epoch, positive northward, and the $\hat{\mathbf{y}}_{J2000}$ axis completes the setup.

The True-of-Date (TOD) frame is an Earth-centered, quasi-inertial coordinate system. The epoch is the time of interest. The fundamental plane is Earth's true equator of epoch. The $\hat{\mathbf{x}}_{TOD}$ axis is directed toward the true vernal equinox of epoch. The $\hat{\mathbf{z}}_{TOD}$ axis lies along Earth's true rotational axis of epoch, positive northward, and the $\hat{\mathbf{y}}_{TOD}$ axis completes the setup.

The distinction between the TOD and J2000 coordinate systems is important for the calculation of mean orbital elements, because the amplitude of the long-periodic variation of the osculating elements depends on Earth's nutation and precession. This effect is particularly significant for the inclination. The long period of the inclination in the TOD frame is around 5 months for low Earth orbits, and the magnitude is slightly varying. However, the long period of the inclination in J2000 is about 3 months, and the magnitude could reach a few tenths of a degree. This phenomenon is caused by precession and nutation. Thus, if measurements are taken with respect to J2000, a transformation to TOD would be required. The transformation of the respective position and velocity vectors can be written as

$$\mathbf{r}_{TOD} = \mathbf{N}^T\mathbf{P}^T\mathbf{r}_{J2000}, \ \mathbf{v}_{TOD} = \mathbf{N}^T\mathbf{P}^T\mathbf{v}_{J2000} \tag{13.1}$$

where $\mathbf{P}$ and $\mathbf{N}$ are Earth's precession and nutation transformation matrices, respectively, as determined by IAU-76/FK5 (Seidelmann 1992).

In an inertial reference frame, the equations of motion for the perturbed Keplerian two-body problem are written as

$$\ddot{\mathbf{r}} + \frac{\mu}{r^3}\mathbf{r} = \mathbf{F} \tag{13.2}$$

where as usual $\mathbf{r}$ is the position vector, $r = \|\mathbf{r}\|$, and $\mathbf{F}$ is the perturbation. As we have seen previously, the position and velocity vectors can be written as functions of time and the classical osculating elements

$$\{a(t),\ e(t),\ i(t),\ \Omega(t)\,, \omega(t)\,, M_0(t)\}$$

with M_0 denoting the mean anomaly at epoch. The variational equations for the classical elements are written either in the Lagrange form (10.124) or the Gauss form (11.131). We will rewrite these equations here for convenience. The LPE for a perturbing potential R are

$$\frac{da}{dt} = \frac{2}{na}\frac{\partial R}{\partial M} \tag{13.3a}$$

$$\frac{de}{dt} = \frac{1-e^2}{na^2 e}\frac{\partial R}{\partial M} - \frac{\sqrt{1-e^2}}{na^2 e}\frac{\partial R}{\partial \omega} \tag{13.3b}$$

$$\frac{di}{dt} = \frac{\cot i}{na^2\sqrt{1-e^2}}\frac{\partial R}{\partial \omega} - \frac{1}{na^2\sqrt{1-e^2}\sin i}\frac{\partial R}{\partial \Omega} \tag{13.3c}$$

$$\frac{d\Omega}{dt} = \frac{1}{na^2\sqrt{1-e^2}\sin i}\frac{\partial R}{\partial i} \tag{13.3d}$$

$$\frac{d\omega}{dt} = \frac{\sqrt{1-e^2}}{na^2 e}\frac{\partial R}{\partial e} - \frac{\cot i}{na^2\sqrt{1-e^2}}\frac{\partial R}{\partial i} \tag{13.3e}$$

$$\frac{dM_0}{dt} = -\frac{1-e^2}{na^2 e}\frac{\partial R}{\partial e} - \frac{2}{na}\frac{\partial R}{\partial a} \tag{13.3f}$$

where $n = \sqrt{\mu/a^3}$, and the Gauss equations, with $\mathbf{F} = R'\hat{\mathbf{u}}_r + S'\hat{\mathbf{u}}_\theta + W'\hat{\mathbf{u}}_W$ as in Eq. (11.115), are written as

$$\frac{da}{dt} = \frac{2}{n\sqrt{1-e^2}}\left(e\sin f\, R' + \frac{p}{r}S'\right) \tag{13.4a}$$

$$\frac{de}{dt} = \frac{\sqrt{1-e^2}}{na}\left[\sin f\, R' + \left(\cos f + \frac{e+\cos f}{1+e\cos f}\right)S'\right] \tag{13.4b}$$

$$\frac{di}{dt} = \frac{r\cos(\omega+f)}{na^2\sqrt{1-e^2}}W' \tag{13.4c}$$

$$\frac{d\Omega}{dt} = \frac{r\sin(\omega + f)}{na^2\sqrt{1-e^2}\sin i} W' \tag{13.4d}$$

$$\frac{d\omega}{dt} = \frac{\sqrt{1-e^2}}{nae}\left[-\cos f\, R' + \sin f\left(1 + \frac{r}{p}\right) S'\right] - \frac{r\cot i \sin(\omega + f)}{\sqrt{\mu p}} W' \tag{13.4e}$$

$$\frac{dM_0}{dt} = \frac{1}{na^2 e}\left[(p\cos f - 2er)\, R' - (p + r)\sin f\, S'\right] \tag{13.4f}$$

where $p = a\left(1 - e^2\right)$ and f is the true anomaly.

A common definition of the mean elements relies on the averaging operator, which, for some given vector-valued periodic function $\mathbf{s}(t)$ with period T is defined by (Battin 1999; pp. 503–504)

$$\bar{\mathbf{s}} = \langle \mathbf{s}(t) \rangle \triangleq \frac{1}{T}\int_0^T \mathbf{s}dt = \frac{1}{2\pi}\int_0^{2\pi} \mathbf{s}\frac{1}{\sqrt{1-e^2}}\left(\frac{r}{a}\right)^2 df = \frac{1}{2\pi}\int_0^{2\pi} \mathbf{s}dM \tag{13.5}$$

where M is the mean anomaly. Thus, any mean element $\bar{c}_i,\ i = 1, 2, \ldots, 6$ is obtained from the osculating element $c_i,\ i = 1, 2, \ldots, 6$ through

$$\bar{c}_i = \langle c_i(t) \rangle \triangleq \frac{1}{T}\int_0^T c_i(t)\, \mathrm{d}t \tag{13.6}$$

where T denotes the orbital period. Thus, mean elements in the current context are obtained by removing the oscillatory motion with a period equal to an orbital period through single averaging. This definition is consistent with other semianalytical theories (Cefola et al. 1980).

An analytical mean-elements calculation scheme was proposed by Kozai (1959a) and Brouwer (1959), who used the averaging technique to isolate the short-periodic terms $c_{i,short}$ in the sense

$$\bar{c}_i(t) = c_i(t) - c_{i,short}(t),\ i = 1, 2, \ldots, 6 \tag{13.7}$$

This implies that

$$\dot{\bar{c}}_i(t) + \dot{c}_{i,short}(t) \equiv g_i = \bar{g}_i + g_{i,short},\ i = 1, 2, \ldots, 6 \tag{13.8}$$

In the Brouwer-Kozai formulation, $\dot{\bar{c}}_i(t)$ consists of secular and long-periodic terms, namely

$$\bar{g}_i = g_{i,sec} + g_{i,long},\ i = 1, 2, \ldots, 6 \tag{13.9}$$

where, written as functions of the orbital elements, for any $i = 1, 2, \ldots, 6$,

$$
\begin{aligned}
g_{sec} &= g_{sec}\left(\bar{a},\, \bar{e},\, \bar{i}\right),\; g_{long} = g_{long}\left(\bar{a},\, \bar{e},\, \bar{i},\, \bar{\Omega},\, \bar{\omega}\right) \\
g_{short} &= g_{short}\left(\bar{a},\, \bar{e},\, \bar{i},\, \bar{\Omega},\, \bar{\omega},\, \bar{M}\right) = g - \bar{g}
\end{aligned}
\tag{13.10}
$$

13.3 Semianalytical Models

In this section, the semianalytical astrodynamical models used for the propagation of the mean elements will be discussed, including zonal, tesseral, and sectorial harmonics, as well as atmospheric drag. We will omit the bar notation of the mean elements for simplicity.

13.3.1 The Zonal Part of the Geopotential

Recalling Eq. (11.210), the perturbing gravitational potential including zonal harmonics only is given by

$$
R_{zonal} = -\frac{\mu}{r}\sum_{n=2}^{\infty} J_n\left(\frac{r_e}{r}\right)^n P_n\left(\sin\varphi\right)
\tag{13.11}
$$

where, as we have seen in Sect. 11.10, J_n, $n = 2, 3, \ldots$ are the zonal gravitational coefficients, φ is the latitude, $\sin\varphi = \sin i \sin u$, $u = \omega + f$ is the argument of latitude, r_e is Earth's mean equatorial radius, and $P_n(x)$ denotes a Legendre polynomial of the first kind of order n, which is expressed as

$$
P_n(x) = \frac{1}{2^n n!}\frac{d^n}{dx^n}\left(x^2 - 1\right)^n
\tag{13.12}
$$

The potential R_{zonal} is averaged prior to substitution into Eqs. (13.3). Applying the averaging operator (13.5) gives

$$
\bar{R}_{zonal} = \frac{1}{2\pi}\int_0^{2\pi} R_{zonal}\frac{1}{\sqrt{1-e^2}}\left(\frac{r}{a}\right)^2 df = R_{sec} + R_{long}
\tag{13.13}
$$

To evaluate the integral in Eq. (13.13), the zonal potential should be written as a function of the classical orbital elements. The resulting variational equations can be written as follows (Danielson et al. 1995), where we use the standard notation for the binomial coefficient,

$$\binom{n}{k} \equiv \frac{n!}{k!(n-k)!}$$

The secular element variations are

$$\dot{a}_{sec} = \dot{e}_{sec} = \mathrm{d}i/\mathrm{d}t_{\,sec} = 0$$

$$\dot{\Omega}_{sec} = n\cos i \sum_{\substack{n\geq 2\\ \text{step } 2}} -J_n \left(\frac{r_e}{p}\right)^n K_{n1}(e) \sum_{q=1}^{n/2} (-1)^{\frac{1}{2}(n+2q)} \frac{2q}{2^{n+2q}} \binom{n}{\frac{n}{2}-q} \binom{n+2q}{n} \binom{2q}{q} (\sin i)^{2q-2}$$

$$\dot{\omega}_{sec} = [\dot{\omega}]_c - \cos i \cdot \dot{\Omega}_{sec}$$

$$[\dot{\omega}]_c = n \sum_{\substack{n\geq 2\\ \text{step } 2}} -J_n \left(\frac{r_e}{p}\right)^n \left[\sum_{q=0}^{n/2} (-1)^{\frac{1}{2}(n+2q)} 2^{-(n+2q)} \binom{n}{\frac{n}{2}-q} \binom{n+2q}{n} \binom{2q}{q} (\sin i)^{2q}\right]$$

$$\cdot \left[(1-e^2)\left(\frac{l}{e^2}K_{n1}(e)\right) + (2n-1)\,K_{n1}(e)\right]$$

$$\dot{M}_{sec} = n\sqrt{1-e^2} \sum_{\substack{n\geq 2\\ \text{step } 2}} -J_n \left(\frac{r_e}{p}\right)^n K_{n1}(e)$$

$$\cdot \sum_{q=0}^{n/2} (-1)^{\frac{1}{2}(n+2q)} \frac{2\,(n+1)}{2^{n+2q}} \binom{n}{\frac{n}{2}-q} \binom{n+2q}{n} \binom{2q}{q} (\sin i)^{2q} - \sqrt{1-e^2}[\dot{\omega}]_c \tag{13.14}$$

where

$$K_{n1}(e) = \sum_{\substack{l=0\\ \text{step } 2}}^{n-2} \binom{n-1}{l} \binom{l}{l/2} \left(\frac{e}{2}\right)^l \tag{13.15}$$

$$\frac{l}{e^2}K_{n1}(e) = \sum_{\substack{l=0\\ \text{step } 2}}^{n-2} \frac{l}{4} \binom{n-1}{l} \binom{l}{l/2} \left(\frac{e}{2}\right)^{l-2} \tag{13.16}$$

The long-periodic terms are

$$\dot{a}_{long} = 0$$

$$\dot{e}_{long} = -\left(\frac{1-e^2}{e}\tan i\right) \mathrm{d}i/\mathrm{d}t_{\,long}$$

$$
\begin{aligned}
\mathrm{d}i/\mathrm{d}t_{\,long} &= n\cos i \sum_{n\geq 3} -J_n \left(\frac{r_e}{p}\right)^n \sum_{p=0}^{(n-2+\delta_1)/2} K_{n1}^p\,(e)\,(n-2p) \\
&\times \left[-\left(1-\delta_1\right)\sin\left(n-2p\right)\omega + \delta_1 \cos\left(n-2p\right)\omega\right] \\
&\times \left[\sum_{q=0}^{p} (-1)^{\frac{1}{2}(n+2q-\delta_1)} 2^{-(2n-2p+2q-1)} \binom{n}{p-q}\binom{2n-2p+2q}{n}\binom{n-2p+2q}{q} (\sin i)^{n-2p+2q-1}\right] \\
\dot{\Omega}_{long} &= n\cos i \sum_{n\geq 3} -J_n \left(\frac{r_e}{p}\right)^n \sum_{p=0}^{(n-2+\delta_1)/2} K_{n1}^p\,(e)\left[\left(1-\delta_1\right)\cos\left(n-2p\right)\omega + \delta_1 \sin\left(n-2p\right)\omega\right] \\
&\times \left[\sum_{q=0}^{p} (-1)^{\frac{1}{2}(n+2q-\delta_1)} \frac{(n-2p+2q)}{2^{(2n-2p+2q-1)}} \binom{n}{p-q}\binom{2n-2p+2q}{n}\binom{n-2p+2q}{q} (\sin i)^{n-2p+2q-2}\right] \\
\dot{\omega}_{long} &= [\dot{\omega}]_l - \cos i \times \dot{\Omega}_{long} \\
[\dot{\omega}]_l &= n \sum_{n\geq 3} -J_n \left(\frac{r_e}{p}\right)^n \\
&\times \sum_{p=0}^{(n-2+\delta_1)/2} \left[\left(1-e^2\right)\left(\frac{l}{e^2} K_{n1}^p\,(e)\right) + (2n-1)\,K_{n1}^p\,(e)\right] \\
&\times \left[\left(1-\delta_1\right)\cos\left(n-2p\right)\omega + \delta_1 \sin\left(n-2p\right)\omega\right] \\
&\times \left[\sum_{q=0}^{p} (-1)^{\frac{1}{2}(n+2q-\delta_1)} 2^{-(2n-2p+2q-1)} \binom{n}{p-q}\binom{2n-2p+2q}{n}\binom{n-2p+2q}{q} (\sin i)^{n-2p+2q}\right] \\
\dot{M}_{long} &= -\sqrt{1-e^2}[\dot{\omega}]_l + n\sqrt{1-e^2} \sum_{n\geq 3} -J_n \left(\frac{r_e}{p}\right)^n \\
&\times \sum_{p=0}^{(n-2+\delta_1)/2} K_{n1}^p\,(e)\left[\left(1-\delta_1\right)\cos\left(n-2p\right)\omega + \delta_1 \sin\left(n-2p\right)\omega\right] \\
&\times \left[\sum_{q=0}^{n} (-1)^{\frac{1}{2}(n+2q-\delta_1)} \frac{(n+1)}{2^{(2n-2p+2q-2)}} \binom{n}{p-q}\binom{2n-2p+2q}{n}\binom{n-2p+2q}{q} (\sin i)^{n-2p+2q}\right]
\end{aligned}
\tag{13.17}
$$

where

$$
\delta_1 = \frac{1}{2}\left[1-(-1)^n\right],\ \delta_2 = \begin{cases} 0 & p = 0 \\ 1 & p \neq 0 \end{cases} \tag{13.18}
$$

$$
K_{n1}^p\,(e) = \delta_2 \sum_{\substack{l=n-2p \\ \text{step } 2}}^{n-2} \binom{n-1}{l}\binom{l}{\frac{1}{2}(l-n+2p)}\left(\frac{e}{2}\right)^l \tag{13.19}
$$

$$\frac{l}{e^2}K_{n1}^{p}(e) = \delta_2 \sum_{\substack{l=n-2p \\ \text{step } 2}}^{n-2} \frac{l}{4}\binom{n-1}{l}\binom{l}{\frac{1}{2}(l-n+2p)}\left(\frac{e}{2}\right)^{l-2} \tag{13.20}$$

13.3.2 Second-Order Effects

As pointed out by Brouwer (1959), due to the nonlinear nature of the variational equations, the short-periodic terms contribute both secular and long-periodic variations of $\mathcal{O}(J_2^2)$.

The resulting short-periodic averaged potential, after transforming into the classical orbital elements, is given by

$$\begin{aligned}\bar{R}_{short} = &\frac{3J_2^2 r_e^4 \mu}{128\eta^7 a^5}\left[5\eta^2 - 4\eta - 5 - \cos^2 i\left(18\eta^2 - 24\eta - 10\right) + \right.\\ &\left.\cos^4 i\left(5\eta^2 + 36\eta + 35\right) + e^2\left(30\cos^4 i - 32\cos^2 i + 2\right)\cos 2\omega\right]\end{aligned} \tag{13.21}$$

where $\eta = \sqrt{1-e^2}$.

Substituting Eq. (13.21) into the LPE (13.3) provides the second-order secular effects, resulting from the short-periodic terms,

$$\begin{aligned}\dot{\Omega}_{sec,2} &= \frac{3J_2^2 r_e^4 n \cos i}{8p^4}\left[-9 - e^2 - 6\sqrt{1-e^2} + \sin^2 i\left(10 - \frac{5e^2}{4} + 9\sqrt{1-e^2}\right)\right]\\ \dot{\omega}_{sec,2} &= \frac{3J_2^2 r_e^4 n}{128p^4}\left[384 + 56e^2 + 192\sqrt{1-e^2} - \sin^2 i\left(824 + 36e^2 + 528\sqrt{1-e^2}\right)\right.\\ &\quad\left. + \sin^4 i\left(430 - 45e^2 + 360\sqrt{1-e^2}\right)\right]\\ \dot{M}_{sec,2} &= \frac{J_2^2 r_e^4 n\sqrt{1-e^2}}{128p^4}\\ &\quad\times\left[192\sqrt{1-e^2} + 120e^2 + 240 - \sin^2 i\left(600 + 120e^2 + 576\sqrt{1-e^2}\right)\right.\\ &\quad\left. + \sin^4 i\left(390 - 75e^2 + 432\sqrt{1-e^2}\right)\right]\end{aligned} \tag{13.22}$$

The expressions for the second-order long-periodic terms are

$$\begin{aligned}
\dot{e}_{long,2} &= -\frac{3J_2^2 r_e^4 ne}{32p^4}(1-e^2)\left(15\cos^2 i - 1\right)\sin^2 i \sin 2\omega \\
\mathrm{d}i/\mathrm{d}t_{\,long,2} &= \frac{3J_2^2 r_e^4 ne^2}{64p^4}\left(15\cos^2 i - 1\right)\sin 2i \sin 2\omega \\
\dot{\Omega}_{long,2} &= -\frac{3J_2^2 r_e^4 ne^2}{16p^4}\left(15\cos^2 i - 8\right)\cos i \cos 2\omega \\
\dot{\omega}_{long,2} &= \frac{3J_2^2 r_e^4 n}{64p^4}\left[30\cos^4 i - 32\cos^2 i + 2 + e^2\left(135\cos^4 i - 112\cos^2 i + 5\right)\right]\cos 2\omega \\
\dot{M}_{long,2} &= -\frac{3J_2^2 r_e^4 n\sqrt{1-e^2}}{64p^4}\left(2-5e^2\right)\left(1-16\cos^2 i + 15\cos^4 i\right)\cos 2\omega
\end{aligned} \tag{13.23}$$

In addition, Kozai (1959a) derived the short-periodic terms of $\mathcal{O}(J_2)$, which are given by

$$\begin{aligned}
a_{short} &= J_2\left(\frac{r_e^2}{a}\right)\left(\frac{a}{r}\right)^3 \\
&\times\left[1-\frac{3}{2}\sin^2 i + \frac{3}{2}\sin^2 i\cos 2(\omega+f) - \left(1-\frac{3}{2}\sin^2 i\right)\left(1-e^2\right)^{-3/2}\right] \\
e_{short} &= \frac{1-e^2}{e}\left(\frac{1}{2a}a_{short} - \tan i\cdot i_{short}\right) \\
i_{short} &= \frac{3J_2\sin 2i}{8}\left(\frac{r_e}{p}\right)^2\left[\cos 2(\omega+f) + e\cos(2\omega+f) + \frac{e}{3}\cos(2\omega+3f)\right] \\
\Omega_{short} &= -\frac{3J_2\cos i}{2}\left(\frac{r_e}{p}\right)^2\left[f - M + e\sin f - \frac{1}{2}\sin 2(\omega+f)\right. \\
&\left. -\frac{e}{2}\sin(2\omega+f) - \frac{e}{6}\sin(2\omega+3f)\right] \\
\omega_{short} &= \frac{3J_2}{4}\left(\frac{r_e}{p}\right)^2\left\{\left(4-5\sin^2 i\right)\left(f-M+e\sin f\right) + \left(2-3\sin^2 i\right)\left[\left(\frac{1}{e}-\frac{e}{4}\right)\sin f\right.\right. \\
&\left. +\frac{1}{2}\sin 2f + \frac{e}{12}\sin 3f\right] - \left[\frac{1}{2e}\sin^2 i + \left(1-\frac{15}{8}\sin^2 i\right)e\right]\sin(f+2\omega) \\
&+\frac{e}{8}\sin^2 i\sin(f-2\omega) - \left(1-\frac{5}{2}\sin^2 i\right)\sin 2(\omega+f) + \frac{3}{4}\sin^2 i\sin(2\omega+4f) \\
&\left. +\left[\frac{7}{6e}\sin^2 i - \left(\frac{1}{3}-\frac{19}{24}\sin^2 i\right)e\right]\sin(2\omega+3f) + \frac{e}{8}\sin^2 i\sin(2\omega+5f)\right\}
\end{aligned}$$

$$
\begin{aligned}
M_{short} = & -\frac{3J_2}{2}\left(\frac{r_e}{p}\right)^2 \frac{\sqrt{1-e^2}}{e} \\
& \times \left\{ \left(1 - \frac{3}{2}\sin^2 i\right)\left(\sin f - \frac{e^2}{4}\sin f + \frac{e}{2}\sin 2f + \frac{e^2}{12}\sin 3f\right) \right. \\
& - \sin^2 i \left[\left(\frac{1}{4} + \frac{5}{16}e^2\right)\sin(2\omega + f) + \frac{e^2}{16}\sin(2\omega - f) \right. \\
& - \left(\frac{7}{12} - \frac{e^2}{48}\right)\sin(2\omega + 3f) \\
& \left.\left. - \frac{3}{8}e\sin(2\omega + 4f) - \frac{e^2}{16}\sin(2\omega + 5f)\right]\right\}
\end{aligned} \tag{13.24}
$$

13.3.3 The Tesseral-Sectorial Part of the Geopotential

The dominant tesseral and sectorial harmonics (degree 2 and order 2) have a period of approximately half a day, which is much longer than the orbital period. Hence, tesseral and sectorial harmonics may have a significant effect on the long-periodic dynamics. The perturbing gravitational potential of the tesseral and sectorial harmonics as given in Sect. 11.10 is

$$
R_{Tesseral} = \frac{\mu}{r}\sum_{n=2}^{\infty}\sum_{m=1}^{n}\left(\frac{r_e}{r}\right)^n P_{nm}(\sin\varphi)\left[C_{nm}\cos m\lambda + S_{nm}\sin m\lambda\right] \tag{13.25}
$$

where λ is the geographic longitude of the satellite measured eastward from the Greenwich meridian, C_{nm}, S_{nm} are harmonic coefficients, and $P_{nm}(x)$ are the associated Legendre polynomials of degree n and order m, which are expressed as

$$
P_{nm}(x) = \frac{1}{2^n n!}\left(1 - x^2\right)^{\frac{m}{2}}\frac{d^{n+m}}{dx^{n+m}}\left(x^2 - 1\right)^n = \left(1 - x^2\right)^{\frac{m}{2}}\frac{d^m}{dx^m}P_n(x) \tag{13.26}
$$

The geopotential up to degree 3 and order 3 can be written based on Eq. (13.25),

$$
\begin{aligned}
R_{22} &= 3\cos^2\varphi\frac{\mu}{r}\left(\frac{r_e}{r}\right)^2\left(C_{22}\cos 2\lambda + S_{22}\sin 2\lambda\right) \\
R_{31} &= \frac{3}{2}\left(5\sin^2\varphi - 1\right)\frac{\mu}{r}\left(\frac{r_e}{r}\right)^3\left(C_{31}\cos\lambda + S_{31}\sin\lambda\right) \\
R_{32} &= 15\sin\varphi\cos^2\varphi\frac{\mu}{r}\left(\frac{r_e}{r}\right)^3\left(C_{32}\cos 2\lambda + S_{32}\sin 2\lambda\right) \\
R_{33} &= 15\cos^3\varphi\frac{\mu}{r}\left(\frac{r_e}{r}\right)^3\left(C_{33}\cos 3\lambda + S_{33}\sin 3\lambda\right)
\end{aligned} \tag{13.27}
$$

Using the averaging method presented in Sect. 13.2, the long-periodic perturbing potentials of the tesseral and sectorial harmonics up to degree 3 and order 3 can be derived,

$$\begin{aligned}
R_{long,22} &= \frac{3}{2}\sin^2 i(1-e^2)^{-\frac{3}{2}}\frac{\mu}{a}\left(\frac{r_e}{a}\right)^2(C_{22}\cos 2\Omega_s + S_{22}\sin 2\Omega_s)\\
R_{long,31} &= \frac{3}{8}e\left(1-e^2\right)^{-\frac{5}{2}}\frac{\mu}{a}\left(\frac{r_e}{a}\right)^3\\
&\quad\times\left\{C_{31}\left[\cos\Omega_s\cos\omega\left(5\sin^2 i-4\right)+\sin\Omega_s\sin\omega\cos i\left(4-15\sin^2 i\right)\right]\right.\\
&\quad\left.+S_{31}\left[\cos\Omega_s\sin\omega\cos i\left(15\sin^2 i-4\right)+\sin\Omega_s\cos\omega\left(5\sin^2 i-4\right)\right]\right\}\\
R_{long,32} &= \frac{15}{4}e\left(1-e^2\right)^{-\frac{5}{2}}\sin^2 i\frac{\mu}{a}\left(\frac{r_e}{a}\right)^3\\
&\quad\times\left\{C_{32}\left[\cos 2\Omega_s\sin\omega\left(3\sin^2 i-2\right)-2\sin 2\Omega_s\cos\omega\cos i\right]\right.\\
&\quad\left.+S_{32}\left[2\cos 2\Omega_s\cos\omega\cos i+\sin\Omega_s\sin\omega\left(3\sin^2 i-2\right)\right]\right\}\\
R_{long,33} &= \frac{45}{4}e\left(1-e^2\right)^{-\frac{5}{2}}\frac{\mu}{a}\left(\frac{r_e}{a}\right)^3\\
&\quad\times\left\{C_{33}\left[2\cos 3\Omega_s\cos\omega\cos i+\sin 3\Omega_s\sin\omega\left(3\sin^2 i-2\right)\right]\right.\\
&\quad\left.+S_{33}\left[2\sin 3\Omega_s\cos\omega\cos i+\cos 3\Omega_s\sin\omega\left(2-3\sin^2 i\right)\right]\right\}
\end{aligned} \tag{13.28}$$

where $\Omega_s = \Omega - \theta$, and θ is the *Greenwich sidereal angle*.

Substituting Eq. (13.28) into Eq. (13.3) yields the rates of change of the long-periodic components contributed by tesseral and sectorial harmonics,

$$\begin{aligned}
\dot{a}_{long,22} &= \dot{e}_{long,22} = \dot{a}_{long,31} = \dot{a}_{long,32} = \dot{a}_{long,33} = 0\\
di/dt_{\ long,22} &= 3\left(\frac{r_e}{p}\right)^2 n\sin i\,(C_{22}\sin 2\Omega_s - S_{22}\cos 2\Omega_s)\\
\dot{\Omega}_{long,22} &= 3\left(\frac{r_e}{p}\right)^2 n\cos i\,(C_{22}\cos 2\Omega_s + S_{22}\sin 2\Omega_s)\\
\dot{\omega}_{long,22} &= \frac{3}{2}\left(\frac{r_e}{p}\right)^2 n\left(5\sin^2 i-2\right)(C_{22}\cos 2\Omega_s + S_{22}\sin 2\Omega_s)\\
\dot{M}_{long,22} &= \frac{9}{2}\left(\frac{r_e}{p}\right)^2 n\sin^2 i(1-e^2)^{\frac{1}{2}}(C_{22}\cos 2\Omega_s + S_{22}\sin 2\Omega_s)
\end{aligned} \tag{13.29}$$

$$
\begin{aligned}
\dot{e}_{long,31} &= \frac{3}{8}\left(\frac{r_e}{p}\right)^3 n\left(1-e^2\right) \\
&\times \left\{C_{31}\left[\sin\omega\cos\Omega_s\left(5\sin^2 i-4\right)+\cos i\cos\omega\sin\Omega_s\left(15\sin^2 i-4\right)\right]\right. \\
&\left. +S_{31}\left[\sin\omega\sin\Omega_s\left(5\sin^2 i-4\right)+\cos i\cos\omega\cos\Omega_s\left(4-15\sin^2 i\right)\right]\right\} \\
di/dt_{\,long,31} &= \frac{3}{8}\left(\frac{r_e}{p}\right)^3 ne\sin i\left\{C_{31}\left[10\cos i\sin\omega\cos\Omega_s+\cos\omega\sin\Omega_s\left(1-15\cos^2 i\right)\right]\right. \\
&\left. +S_{31}\left[10\cos i\sin\omega\sin\Omega_s+\cos\omega\cos\Omega_s\left(15\cos^2 i-1\right)\right]\right\} \\
\dot{\Omega}_{long,31} &= \frac{3}{8}\left(\frac{r_e}{p}\right)^3 ne\left\{C_{31}\left[10\cos i\cos\omega\cos\Omega_s+\sin\omega\sin\Omega_s\left(11-45\cos^2 i\right)\right]\right. \\
&\left. +S_{31}\left[10\cos i\cos\omega\sin\Omega_s+\sin\omega\cos\Omega_s\left(45\cos^2 i-11\right)\right]\right\} \\
\dot{\omega}_{long,31} &= -\frac{3}{8}\left(\frac{r_e}{p}\right)^3 n\left(4e+\frac{1}{e}\right) \\
&\times \left\{C_{31}\left[\cos\omega\cos\Omega_s\left(4-5\sin^2 i\right)+\cos i\sin\omega\sin\Omega_s\left(15\sin^2 i-4\right)\right]\right. \\
&\left. +S_{31}\left[\cos\omega\sin\Omega_s\left(4-5\sin^2 i\right)+\cos i\sin\omega\cos\Omega_s\left(4-15\sin^2 i\right)\right]\right\} \\
&- \dot{\Omega}_{31}\cos i \\
\dot{M}_{long,31} &= \frac{3}{8}\left(\frac{r_e}{p}\right)^3 n(1-e^2)^{\frac{1}{2}} \\
&\times \left\{C_{31}\left[\cos\Omega_s\cos\omega\left(5\sin^2 i-4\right)+\sin\Omega_s\sin\omega\cos i\left(4-15\sin^2 i\right)\right]\right. \\
&\left. +S_{31}\left[\sin\Omega_s\cos\omega\left(5\sin^2 i-1\right)+\cos\Omega_s\sin\omega\cos i\left(15\sin^2 i-4\right)\right]\right\} \\
&\times \left(4e-\frac{1}{e}\right)
\end{aligned}
\tag{13.30}
$$

$$
\begin{aligned}
\dot{e}_{long,32} &= -\frac{15}{4}\left(\frac{r_e}{p}\right)^3 n\left(1-e^2\right)\sin^2 i \\
&\times \left\{C_{32}\left[\cos\omega\cos 2\Omega_s\left(3\sin^2 i-2\right)+2\cos i\sin\omega\sin 2\Omega_s\right]\right. \\
&\left. +S_{32}\left[\cos\omega\sin 2\Omega_s\left(3\sin^2 i-2\right)-2\cos i\sin\omega\cos\Omega_s\right]\right\} \\
di/dt_{\,long,32} &= -\frac{15}{4}\left(\frac{r_e}{p}\right)^3 ne \\
&\times \left\{C_{32}\left[\sin\omega\sin 2\Omega_s\left(2-4\sin^2 i\right)-\cos i\cos\omega\cos 2\Omega_s\left(2+3\sin^2 i\right)\right]\right. \\
&\left. +S_{32}\left[\sin\omega\cos 2\Omega_s\left(4\sin^2 i-2\right)-\cos i\cos\omega\sin 2\Omega_s\left(2+3\sin^2 i\right)\right]\right\}
\end{aligned}
$$

$$
\begin{aligned}
\dot{\Omega}_{long,32} &= \frac{15}{4}\left(\frac{r_e}{p}\right)^3 n \frac{e}{\sin i} \\
&\quad \times \left\{C_{32}\left[\cos 2\Omega_s \sin\omega \cos i\,(7-9\cos^2 i) + \sin 2\Omega_s \cos\omega\,(2-4\cos^2 i)\right]\right. \\
&\quad \left. + S_{32}\left[\sin 2\Omega_s \sin\omega \cos i\,(7-9\cos^2 i) + \cos 2\Omega_s \cos\omega\,(4\cos^2 i - 2)\right]\right\} \\
\dot{\omega}_{long,32} &= \frac{15}{4}\left(\frac{r_e}{p}\right)^3 n \left(4e + \frac{1}{e}\right) \sin i \\
&\quad \times \left\{C_{32}\left[\sin\omega \cos 2\Omega_s\,(3\sin^2 i - 2) - 2\cos i \cos\omega \sin 2\Omega_s\right]\right. \\
&\quad \left. + S_{32}\left[\sin\omega \sin 2\Omega_s\,(3\sin^2 i - 2) + 2\cos i \cos\omega \cos 2\Omega_s\right]\right\} - \dot{\Omega}_{32}\cos i \\
\dot{M}_{long,32} &= -\frac{15}{4}\left(\frac{r_e}{p}\right)^3 n \sin i \left(4e - \frac{1}{e}\right) \\
&\quad \times \left\{C_{32}\left[\cos 2\Omega_s \sin\omega\,(2-3\sin^2 i) + 2\sin 2\Omega_s \cos\omega \cos i\right]\right. \\
&\quad \left. + S_{32}\left[\sin 2\Omega_s \sin\omega\,(2-3\sin^2 i) - 2\cos 2\Omega_s \cos\omega \cos i\right]\right\}\left(1-e^2\right)^{\frac{1}{2}}
\end{aligned}
\tag{13.31}
$$

$$
\begin{aligned}
\dot{e}_{long,33} &= \frac{45}{4}\left(\frac{r_e}{p}\right)^3 n\left(1-e^2\right)\left\{C_{33}\left[2\cos 3\Omega_s \sin\omega \cos i - \sin 3\Omega_s \cos\omega\,(3\sin^2 i - 2)\right]\right. \\
&\quad \left. + S_{33}\left[2\sin 3\Omega_s \sin\omega \cos i - \cos 3\Omega_s \cos\omega\,(2-3\sin^2 i)\right]\right\} \\
di/dt_{\,long,33} &= \frac{45}{4}\left(\frac{r_e}{p}\right)^3 n \frac{e}{\sin i} \\
&\quad \times \left\{C_{33}\left[\cos 3\Omega_s \sin\omega\,(4-7\sin^2 i) + \sin 3\Omega_s \cos\omega \cos i\,(4+3\sin^2 i)\right]\right. \\
&\quad \left. + S_{33}\left[\sin 3\Omega_s \sin\omega\,(4-7\sin^2 i) - \cos 3\Omega_s \cos\omega \cos i\,(4+3\sin^2 i)\right]\right\} \\
\dot{\Omega}_{long,33} &= \frac{45}{2}\left(\frac{r_e}{p}\right)^3 ne\left[C_{33}\left(3\sin 3\Omega_s \sin\omega \cos i - \cos 3\Omega_s \cos\omega\right)\right. \\
&\quad \left. - S_{33}\left(3\cos 3\Omega_s \sin\omega \cos i - \sin 3\Omega_s \cos\omega\right)\right] \\
\dot{\omega}_{long,33} &= \frac{45}{4}\left(\frac{r_e}{p}\right)^3 n\left(4e + \frac{1}{e}\right)\left\{C_{33}\left[2\cos 3\Omega_s \cos\omega \cos i + \sin 3\Omega_s \sin\omega\,(3\sin^2 i - 2)\right]\right. \\
&\quad \left. + S_{33}\left[2\sin 3\Omega_s \cos\omega \cos i + \cos 3\Omega_s \sin\omega\,(2-3\sin^2 i)\right]\right\} - \dot{\Omega}_{33}\cos i \\
\dot{M}_{long,33} &= \frac{45}{4}\left(\frac{r_e}{p}\right)^3 n(1-e^2)^{\frac{1}{2}}\left(4e - \frac{1}{e}\right) \\
&\quad \times \left\{C_{33}\left[2\cos 3\Omega_s \cos\omega \cos i + \sin 3\Omega_s \sin\omega\,(3\sin^2 i - 2)\right]\right. \\
&\quad \left. + S_{33}\left[2\sin 3\Omega_s \cos\omega \cos i + \cos 3\Omega_s \sin\omega\,(2-3\sin^2 i)\right]\right\}
\end{aligned}
\tag{13.32}
$$

By defining

$$\tan m\lambda_{nm} = \frac{S_{nm}}{C_{nm}}$$
$$J_{nm} = -\left(C_{nm}^2 + S_{nm}^2\right)^{\frac{1}{2}} \tag{13.33}$$

The Kozai (1959a) method yields the short-periodic terms due to the tesseral and sectorial harmonics,

$$\begin{aligned} a_{short,22} = & -\frac{3}{2}\frac{J_{22}}{a}r_e^2 \\ & \times \left\{ (1+\cos i)^2 \left[\frac{1}{1-\tau}\cos(2M + 2\omega + 2\Omega_e) - \frac{1}{2}e\left(\frac{1}{1-2\tau}\cos(M + 2\omega + 2\Omega_e) \right.\right.\right. \\ & \left.\left. -\frac{21}{3-2\tau}\cos(3M + 2\omega + 2\Omega_e)\right)\right] + (1-\cos i)^2 \left[\frac{1}{1+\tau}\cos(2m + 2\omega - 2\Omega_e) \right. \\ & \left. -\frac{e}{2}\left(\frac{1}{1+2\tau}\cos(M + 2\omega - 2\Omega_e) - \frac{21}{3-2\tau}\cos(3M + 2\omega - 2\Omega_e)\right)\right] \\ & \left. + 3e\sin^2 i\left[\frac{1}{1-2\tau}\cos(M + 2\Omega_e) + \frac{1}{1+2\tau}\cos(M - 2\Omega_e)\right]\right\} \end{aligned} \tag{13.34}$$

$$\begin{aligned} e_{short,22} = & -\frac{3}{4}J_{22}\left(\frac{r_e}{a}\right)^2 \\ & \times \left\{ (1+\cos i)^2 \left[\frac{1}{2-4\tau}\cos(M + 2\omega + 2\Omega_e) + \frac{7}{6-4\tau}\cos(3M + 2\omega + 2\Omega_e) \right.\right. \\ & \left. +\frac{e}{2}\left(\frac{1}{\tau-1}\cos(2M + 2\omega + 2\Omega_e) + \frac{17}{2-\tau}\cos(4M + 2\omega + 2\Omega_e)\right)\right] \\ & + (1-\cos i)^2 \left[\frac{1}{2+4\tau}(\cos(M + 2\omega - 2\Omega_e)) + \frac{7}{6+4\tau}\cos(3M + 2\omega - 2\Omega_e) \right. \\ & \left. +\frac{e}{2}\left(-\frac{1}{1+\tau}\cos(2M + 2\omega - 2\Omega_e) + \frac{17}{2+\tau}\cos(4M + 2\omega - 2\Omega_e)\right)\right] \\ & + 2\sin^2 i\left[\frac{3}{2-4\tau}\cos(M + 2\Omega_e) + \frac{3}{2+4\tau}\cos(M - 2\Omega_e) \right. \\ & \left.\left. +\frac{9e}{4}\left(\frac{1}{1-\tau}\cos(2M + 2\Omega_e) + \frac{1}{1+\tau}\cos(2M - 2\Omega_e)\right)\right]\right\} \end{aligned} \tag{13.35}$$

$$
\begin{aligned}
i_{short,22} &= -\frac{3}{4}J_{22}\left(\frac{r_e}{a}\right)^2 \sin i \\
&\times \Bigg\{(1+\cos i)\left[\frac{e}{1-2\tau}\cos(M+2\omega+2\Omega_e) - \frac{7e}{3-2\tau}\cos(3M+2\omega-2\Omega_e)\right. \\
&\left. -\frac{1}{1-\tau}\cos(2M+2\omega+2\Omega_e)\right] - (1-\cos i)\left[\frac{e}{1+2\tau}\cos(M+2\omega-2\Omega_e)\right. \\
&\left. -\frac{7e}{3+2\tau}\cos(3M+2\omega-2\Omega_e) - \frac{1}{1+\tau}\cos(2M+2\omega-2\Omega_e)\right] \\
&-6e\left[\frac{1}{1-2\tau}\cos(M+2\Omega_e) - \frac{1}{1+2\tau}\cos(M-2\Omega_e)\right]\Bigg\}
\end{aligned}
\tag{13.36}
$$

$$
\begin{aligned}
\Omega_{short,22} &= -\frac{3}{4}J_{22}\left(\frac{r_e}{a}\right)^2 \\
&\times \Bigg\{(1+\cos i)\left[\frac{e}{1-2\tau}\sin(M+2\omega+2\Omega_e) - \frac{7e}{3-2\tau}\sin(3M+2\omega+2\Omega_e)\right. \\
&\left. -\frac{1}{1-\tau}\sin(2M+2\omega+2\Omega_e)\right] - (1-\cos i)\left[\frac{e}{1+2\tau}\sin(M+2\omega-2\Omega_e)\right. \\
&\left. -\frac{7e}{3+2\tau}\sin(3M+2\omega-2\Omega_e) - \frac{1}{1+\tau}\sin(2M+2\omega-2\Omega_e)\right] \\
&+6e\cos i\left[\frac{1}{1-2\tau}\sin(M+2\Omega_e) + \frac{1}{1+2\tau}\sin(M-2\Omega_e)\right]\Bigg\}
\end{aligned}
\tag{13.37}
$$

$$
\begin{aligned}
\omega_{short,22} &= -\frac{3}{4}\frac{J_{22}}{e}\left(\frac{r_e}{a}\right)^2 \\
&\times \Bigg\{-(1+\cos i)^2\left[\frac{1}{2-4\tau}\sin(M+2\omega+2\Omega_e) - \frac{7}{6-4\tau}\sin(3M+2\omega+2\Omega_e)\right. \\
&\left. +\frac{5e}{2-2\tau}\sin(2M+2\omega+2\Omega_e) - \frac{17e}{4-2\tau}\sin(4M+2\omega+2\Omega_e)\right] \\
&-(1-\cos i)^2\left[\frac{1}{2+4\tau}\sin(M+2\omega-2\Omega_e) - \frac{7}{6+4\tau}\sin(3M+2\omega-2\Omega_e)\right. \\
&\left. +\frac{5e}{2+2\tau}\sin(2M+2\omega-2\Omega_e) - \frac{17e}{4+2\tau}\sin(4M+2\omega-2\Omega_e)\right] \\
&+\frac{9}{2}e\sin^2 i\left[\frac{1}{1-\tau}\sin(2M+2\Omega_e) + \frac{1}{1+\tau}\sin(2M-2\Omega_e)\right] \\
&+2\sin^2 i\left[\frac{3}{2-4\tau}\sin(M+2\Omega_e) + \frac{3}{2+4\tau}\sin(M-2\Omega_e)\right]\Bigg\} \\
&-\Omega_{short,22}\cos i
\end{aligned}
\tag{13.38}
$$

$$
\begin{aligned}
M_{short,22} = & -\frac{9}{4}J_{22}\left(\frac{r_e}{a}\right)^2 \\
& \times \left\{(1+\cos i)^2\left[\frac{1}{1-\tau}\left(1-\frac{1}{2-2\tau}\right)\sin\left(2M+2\omega+2\Omega_e\right)\right.\right. \\
& -\frac{e}{1-2\tau}\left(1-\frac{1}{2-4\tau}\right)\sin\left(M+2\omega+2\Omega_e\right) \\
& \left.-\frac{7e}{3-2\tau}\left(1-\frac{3}{6-4\tau}\right)\sin\left(3M+2\omega+2\Omega_e\right)\right] \\
& -(1-\cos i)^2\left[\frac{e}{1+2\tau}\left(1-\frac{1}{2+4\tau}\right)\sin\left(M+2\omega-2\Omega_e\right)\right. \\
& -\frac{7e}{3+2\tau}\left(1-\frac{3}{6+4\tau}\right)\sin\left(3M+2\omega-2\Omega_e\right) \\
& \left.-\frac{1}{1+\tau}\left(1-\frac{1}{2+2\tau}\right)\sin\left(2M+2\omega-2\Omega_e\right)\right] \\
& +6e\sin^2 i\left[\frac{1}{1-2\tau}\left(1-\frac{1}{2-4\tau}\right)\sin\left(M+2\Omega_e\right)\right. \\
& \left.\left.+\frac{1}{1+2\tau}\left(1-\frac{1}{1+2\tau}\right)\sin\left(M-2\Omega_e\right)\right]\right\} \\
& -\omega_{short,22}-\Omega_{short,22}\cos i
\end{aligned}
\tag{13.39}
$$

where $\Omega_e = \Omega_s - \lambda_{22}$, $\tau = \omega_e/n$, and ω_e is the Earth spin rate.

13.3.4 Atmospheric Drag

Recall the drag modelling done in Sect. 11.11. Substituting Eqs. (11.225) and (11.144) into Eq. (13.4) yields

$$
\begin{aligned}
\dot{a}_{drag} &= -\frac{K_1 n a^2}{\left(1-e^2\right)^{3/2}}\rho\left(1+2e\cos f+e^2\right)^{3/2} \\
\dot{e}_{drag} &= -\frac{K_1 n a}{\left(1-e^2\right)^{1/2}}\rho\left(\cos f+e\right)\left(1+2e\cos f+e^2\right)^{1/2} \\
\mathrm{d}i/\mathrm{d}t_{\,drag} &= -\frac{K_2 a}{4(1+e\cos f)^2}\rho\sin i\left[1+\cos\left(2\omega+2f\right)\right]\left(1-e^2\right)\left(1+2e\cos f+e^2\right)^{1/2} \\
\dot{\Omega}_{drag} &= -\frac{K_2 a}{4(1+e\cos f)^2}\rho\sin\left(2\omega+2f\right)\left(1-e^2\right)\left(1+2e\cos f+e^2\right)^{1/2}
\end{aligned}
$$

$$
\begin{aligned}
\dot{\omega}_{drag} &= -\frac{K_1 na}{e(1-e^2)^{1/2}} \rho \sin f \left(1 + 2e\cos f + e^2\right)^{1/2} - \cos i \cdot \dot{\Omega}_{drag} \\
\dot{M}_{drag} &= -\frac{K_1 na}{e\,(1 + e\cos f)} \rho \sin f \left(1 + e\cos f + e^2\right)\left(1 + 2e\cos f + e^2\right)^{1/2}
\end{aligned}
\tag{13.40}
$$

The atmospheric density can be expanded into a series dependent upon modified Bessel functions (Brouwer and Clemence 1961; Breiter and Metris 1994), a procedure that results in the following variational equations for the secular terms,

$$
\begin{aligned}
\dot{a}_{drag,sec} &= -K_1 \rho_0 n a^2 \left[1 + e^2\left(\frac{3}{4} + \frac{a}{H} + \frac{a^2}{4H^2}\right) + \mathcal{O}\left(e^3\right)\right] \exp\left(\frac{r_{p0} - a}{H}\right) \\
\dot{e}_{drag,sec} &= -K_1 \rho_0 n a \left[\frac{e}{2} + \frac{ae}{2H} + \mathcal{O}\left(e^3\right)\right] \exp\left(\frac{r_{p0} - a}{H}\right) \\
\mathrm{d}i/\mathrm{d}t_{\,drag,sec} &= -\frac{1}{4} K_2 \rho_0 a \sin i \left[1 + e^2\left(\frac{3}{4} - \frac{a}{H} + \frac{a^2}{4H^2}\right) + \mathcal{O}\left(e^3\right)\right] \exp\left(\frac{r_{p0} - a}{H}\right) \\
\dot{\Omega}_{drag,sec} &= 0 \\
\dot{\omega}_{drag,sec} &= 0 \\
\dot{M}_{drag,sec} &= \frac{3}{4} K_1 \rho_0 n^2 a \left[1 + e^2\left(\frac{3}{4} + \frac{a}{H} + \frac{a^2}{4H^2}\right) + \mathcal{O}\left(e^3\right)\right] \exp\left(\frac{r_{p0} - a}{H}\right) \\
&\quad \times (t - t_0)
\end{aligned}
\tag{13.41}
$$

with t_0 denoting the time of epoch, while the long-periodic terms are given by

$$
\begin{aligned}
\dot{a}_{drag,long} &= 0 \\
\dot{e}_{drag,long} &= 0 \\
\mathrm{d}i/\mathrm{d}t_{\,drag,long} &= -\frac{1}{4} K_2 \rho_0 a \sin i \left[\left(\frac{11}{8} - \frac{a}{H} + \frac{a^2}{8H^2}\right) e^2 \cos 2\omega + \mathcal{O}\left(e^3\right)\right] \exp\left(\frac{r_{p0} - a}{H}\right) \\
\dot{\Omega}_{drag,long} &= -\frac{1}{4} K_2 \rho_0 a \left[\left(\frac{11}{8} - \frac{a}{H} + \frac{a^2}{8H}\right) e^2 \sin 2\omega + \mathcal{O}\left(e^3\right)\right] \exp\left(\frac{r_{p0} - a}{H}\right) \\
\dot{\omega}_{drag,long} &= -\cos i \cdot \dot{\Omega}_{drag,long} \\
\dot{M}_{drag,long} &= 0
\end{aligned}
\tag{13.42}
$$

Finally, the effect of the short-periodic terms can be obtained following the method of Kozai (1959a),

$$
\begin{aligned}
a_{drag,short} &= -K_1\rho_0 a^2 \exp\left(\frac{r_{p0}-a}{H}\right)\\
&\quad\times\left[\left(3+\frac{a}{H}\right)e\sin M+\left(\frac{15}{8}+\frac{a}{H}+\frac{a^2}{8H^2}\right)e^2\sin 2M+\mathcal{O}\left(e^3\right)\right]\\
e_{drag,short} &= -K_1\rho_0 a \exp\left(\frac{r_{p0}-a}{H}\right)\\
&\quad\times\left\{\left[1-\left(\frac{1}{2}-\frac{a}{H}-\frac{3a^2}{8H^2}\right)e^2\right]\sin M+\left(\frac{3}{4}+\frac{a}{4H}\right)e\sin 2M\right.\\
&\quad\left.+\left[\frac{2}{3}+\frac{a}{3H}+\frac{a^2}{24H^2}\right]e^2\sin 3M+\mathcal{O}\left(e^3\right)\right\}\\
i_{drag,short} &= -\frac{K_2\rho_0 a\sin i}{4n}\exp\left(\frac{r_{p0}-a}{H}\right)\\
&\quad\times\left\{\left(\frac{a}{H}-1\right)e\sin M+\left(\frac{a}{2H}-\frac{5}{2}\right)e\sin\left(M+2\omega\right)\right.\\
&\quad+\left[\frac{1}{2}+\left(\frac{a^2}{8H^2}-\frac{a}{2H}-\frac{13}{8}\right)e^2\right]\sin\left(2M+2\omega\right)+\left(\frac{a^2}{8H^2}-\frac{3}{8}\right)e^2\sin 2M\\
&\quad+\left(\frac{a}{6H}+\frac{1}{2}\right)e\sin\left(3M+2\omega\right)\\
&\quad\left.+\left(\frac{a^2}{32H^2}+\frac{a}{4H}+\frac{15}{32}\right)e^2\sin\left(4M+2\omega\right)+\mathcal{O}\left(e^3\right)\right\}
\end{aligned}
\tag{13.43}
$$

$$
\begin{aligned}
\Omega_{drag,short} &= \frac{K_2\rho_0 a}{8n}\exp\left(\frac{r_{p0}-a}{H}\right)\\
&\quad\times\left\{\left(\frac{a^2}{16H^2}+\frac{a}{2H}+\frac{15}{16}\right)e^2\cos\left(4M+2\omega\right)+\left(\frac{a}{3H}+1\right)e\cos\left(3M+2\omega\right)\right.\\
&\quad\left.+\left[1+\left(\frac{a^2}{4H^2}-\frac{a}{H}-\frac{13}{4}\right)e^2\right]\cos\left(2M+2\omega\right)+\left(\frac{a}{H}-5\right)e\cos\left(M+2\omega\right)+\mathcal{O}\left(e^3\right)\right\}\\
\omega_{drag,short} &= K_1\rho_0 a\exp\left(\frac{r_{p0}-a}{H}\right)\\
&\quad\times\left\{\left[\frac{1}{e}+\left(\frac{a^2}{8H^2}-1\right)e\right]\cos M+\left(\frac{a^2}{24H^2}+\frac{a}{3H}+\frac{2}{3}\right)e\cos 3M\right.\\
&\quad+\left[\frac{a}{4H}+\frac{3}{4}+\left(\frac{a^3}{48H^3}+\frac{a^2}{16H^2}-\frac{5a}{16H}-\frac{43}{48}\right)e^2\right]\cos 2M\\
&\quad\left.\left(+\frac{a^3}{192H^3}+\frac{5a^2}{64H^2}-\frac{25a}{64H}-\frac{139}{192}\right)e^2\cos 4M+\mathcal{O}\left(e^3\right)\right\}-\cos i\cdot\Omega_{drag,short}
\end{aligned}
$$

$$M_{drag,short} = -K_1 \rho_0 a \left\{ \left[\frac{1}{e} + \left(\frac{a^2}{8H^2} + \frac{3a}{2H} + 4 \right) e \right] \cos M + \left(\frac{a^2}{24H^2} + \frac{a}{3H} + \frac{2}{3} \right) e \cos 3M \right.$$
$$+ \left[\frac{a}{4H} + \frac{3}{4} + \left(\frac{a^3}{48H^3} + \frac{5a^2}{32H^2} + \frac{9a}{16H} + \frac{61}{96} \right) e^2 \right] \cos 2M$$
$$\left. + \left(\frac{a^3}{192H^3} + \frac{5a^2}{64H^2} + \frac{25a}{64H} + \frac{139}{192} \right) e^2 \cos 4M + \mathcal{O}\left(e^3\right) \right\} \exp\left(\frac{r_{p0} - a}{H} \right) \tag{13.44}$$

13.4 Frozen Orbits

An important application of semianalytical orbit theory is the ability to find *frozen orbits*. These orbits have constant mean semimajor axis, eccentricity, inclination, and argument of perigee. For a variety of reasons, among which is precision altitude keeping, these kinds of orbits are advantageous for Earth remote sensing, mapping and observation, and have been used in missions such as RadarSat, Topex, CloudSat, and Aqua (Gurfil and Lara 2013).

The frozen orbit design is commonly done in a J_2-J_3 model by finding the null rate of variation of the eccentricity and argument of periapsis in the Lagrange equations (10.124) for the mean elements (Cutting et al. 1978). However, higher-degree zonals may have a non-negligible influence on the frozen orbit geometry, and, thus, could be needed for a precise orbit design (Rosborough and Ocampo 1992; Shapiro 1995).

The frozen orbit is defined in terms of mean elements, but the computation of the nominal orbit requires the recovery of the short-periodic effects (that are eliminated by averaging). This procedure can be efficiently carried out by means of the perturbation approach relying on Lie transforms (Gurfil and Lara 2013), which shows the important qualitative effects that subsequent zonal harmonics introduce into the frozen orbit problem (Coffey et al. 1994). However, the traditional J_2-J_3 approach may remain valid for low-eccentricity orbits, assuming that they are far enough from the equatorial plane.

Although the frozen orbit definition is tied to an averaging process, these orbits can also be identified as periodic orbits in the meridian plane of the satellite and, therefore, can be computed directly from the non-averaged equations, as shown by Broucke (1994) and Lara et al. (1995). The direct computation of frozen, periodic orbits shows that a subset of frozen orbits are also periodic in the three dimensional space. Indeed, those periodic orbits in the meridian plane of the satellite, whose node rotation rate is commensurate with the orbit mean motion, will close in three-dimensional space. In inertial space, this could happen after very long periods, but in a rotating frame attached to the Earth, these three-dimensional periodic orbits are ideal candidates for *repeat ground-track orbits* (RGT) (Lara 1999), to be discussed in the next section.

In the case of Earth-like bodies, the second-order zonal harmonic dominates over all other harmonics of the expansion of the perturbing gravitational potential. Thus, the main part of these perturbing effects can be studied from the truncation of the perturbing gravitational potential, derived based on Eq. (11.210) up to J_2, which yields

$$R = \frac{\mu}{2r}\left(\frac{r_e}{r}\right)^2 J_2\left(1 - \frac{3}{2}\sin^2 i + \frac{3}{2}\sin^2 i \cos 2u\right) \tag{13.45}$$

where as before μ is the gravitational parameter, r_e is the equatorial radius, and $u = f + \omega$ is the argument of latitude.

As we have seen previously, this simple perturbation model is usually expanded into a trigonometric series with arguments of the type $\beta = jM + k\omega$, with j and k being integers (Kaula 1966), where M is the mean anomaly and ω is the argument of periapsis. From this expansion, one can distinguish among the secular terms $j = k = 0$, the long-periodic terms $j = 0$, $k \neq 0$, and the short-period terms $j \neq 0$. Recall that the disturbing function does not depend on Ω, because of the axial symmetry of the zonal model with respect to the polar axis. The Lagrange equations used in previous sections can be used to show that the arguments of the node and perigee, and also the mean anomaly, undergo secular effects, whereas all the elements exhibit periodic effects.

The transformation from osculating to mean elements can be written to first order in J_2 using a direct application of the averaging operator (13.5). Thus, denoting by primes the mean elements or functions thereof,

$$R = \frac{\mu}{2a'}\left(\frac{r_e}{a'}\right)^2 \frac{J_2}{\eta'^3}\left(1 - \frac{3}{2}\sin^2 i'\right) + \mathcal{O}(J_2^2) \tag{13.46}$$

where $\eta = \sqrt{1 - e^2}$.

This form of the potential reveals a specific inclination where the secular effects on the argument of perigee vanish. From the Lagrange planetary equations, up to $\mathcal{O}(J_2)$,

$$\frac{\mathrm{d}\omega'}{\mathrm{d}t} = n'\left(\frac{r_e}{a'}\right)^2 \frac{3J_2}{4\eta'^4}\left(4 - 5\sin^2 i'\right) \tag{13.47}$$

where a (and, therefore, the mean motion n), e, and i are unaffected by secular variations, and, hence, the corresponding mean values a', n', e', and i' are constant. In consequence, the inclinations satisfying $\sin^2 i' = 4/5$ freeze the mean perigee to a constant value. Each value is called the *critical inclination*,

$$i'_{crit} = 63.435^\circ,\ 116.565^\circ \tag{13.48}$$

where the first value represents a *direct orbit* and the second represents a *retrograde orbit*.

Table 13.1 Inclination polynomials

j	0	1	2
$m_{0,j}$	$\frac{15}{8} - \frac{15}{4}s^2 + \frac{105}{64}s^4$	$\frac{3}{16}(2-3s^2)^2$	$-\frac{3}{8} + \frac{3}{8}s^2 + \frac{15}{64}s^4$
$m_{2,j}$	$-\frac{21}{16}s^2 + \frac{45}{32}s^4$	0	0
$\widetilde{m}_{0,j}$	$-\frac{15}{8} + \frac{75}{8}s^2 - \frac{525}{64}s^4$	0	$\frac{9}{8} - \frac{45}{8}s^2 + \frac{315}{64}s^4$
$\widetilde{m}_{2,j}$	$\frac{45}{16}s^2 - \frac{105}{32}s^4$	0	0
$\gamma_{0,j}$	$\frac{165}{16} - \frac{645}{32}s^2 + \frac{1155}{128}s^4$	$\frac{9}{2} - \frac{99}{8}s^2 + \frac{135}{16}s^4$	$-\frac{21}{16} + \frac{27}{32}s^2 + \frac{135}{128}s^4$
$\gamma_{2,j}$	$\frac{21}{16} - \frac{279}{32}s^2 + \frac{495}{64}s^4$	0	$-\frac{21}{16} + \frac{237}{32}s^2 - \frac{405}{64}s^4$
$\widetilde{\gamma}_{0,j}$	$-\frac{255}{16} + \frac{1875}{32}s^2 - \frac{5775}{128}s^4$	0	$\frac{135}{16} - \frac{945}{32}s^2 + \frac{2835}{128}s^4$
$\widetilde{\gamma}_{2,j}$	$\frac{45}{16} - \frac{615}{32}s^2 + \frac{1155}{64}s^4$	0	$-\frac{45}{16} + \frac{525}{32}s^2 - \frac{945}{64}s^4$

The truncation up to the first-order in J_2 of the transformation from osculating to mean elements is only able to deal with the secular effects. However, since long-periodic effects are related to the motion of the argument of perigee, it seems unrealistic not to deal with them when assessing a frozen perigee condition (Gurfil and Lara 2013). A second-order truncation in the transformation equations from osculating to mean elements enables to cope with long-periodic effects of the disturbing function (Deprit 1981; Coffey et al. 1986),

$$R' = \frac{\mu}{2a'}\left(\frac{r_e}{a'}\right)^2 \frac{J_2}{\eta'^3}\left(1 - \frac{3}{2}s'^2\right) \tag{13.49}$$
$$+\frac{\mu}{2a'}\left(\frac{r_e}{a'}\right)^4 \frac{J_2^2}{\eta'^7}\left(m_{0,0} + m_{0,1}\,\eta' + m_{0,2}\,\eta'^2 + m_{2,0}\,e'^2\cos 2\omega'\right) + \mathcal{O}(J_2^3)$$

where $s \equiv \sin i$, and the inclination polynomials $m_{i,j}$ are given in Table 13.1. Since we only deal here with mean elements, in what follows we drop the prime notation.

Now, the Lagrange planetary equations give

$$\frac{de}{dt} = -n\left(\frac{r_e}{a}\right)^4 \frac{J_2^2}{\eta^6}\left(\frac{21}{16} - \frac{45}{32}s^2\right) s^2\, e \sin 2\omega \tag{13.50}$$

$$\frac{d\omega}{dt} = n\left(\frac{r_e}{a}\right)^2 \frac{J_2}{\eta^4}\frac{3}{4}(4 - 5s^2) \tag{13.51}$$
$$+n\left(\frac{r_e}{a}\right)^4 \frac{J_2^2}{\eta^8}\left[\gamma_{0,0} + \gamma_{0,1}\,\eta + \gamma_{0,2}\,\eta^2 + (\gamma_{2,0} + \gamma_{2,2}\,\eta^2)\cos 2\omega\right]$$

where the inclination polynomials γ are given in Table 13.1.

As opposed to the first-order condition in Eq. (13.47), Eq. (13.50) shows that long-periodic effects in the eccentricity constrain the frozen orbits to exist only when $\omega = k\pi/2$, with k an integer. In addition, the second-order effects in Eq. (13.51) slightly modify the critical inclination value.

The mean elements representation has reduced the phase space of the J_2 problem to a single degree of freedom problem in e and ω, which is made only of closed curves and equilibria. The representation of the reduced phase space can be made without need of integrating the differential equations of the reduced flow, Eqs. (13.50) and (13.51), by means of contour plots of the energy or the disturbing function Eq. (13.49). We see from Eq. (13.50) that circular orbits, up to long-term effects, remain circular. Therefore, although the argument of perigee is undefined for this case, circular orbits are equilibria of the reduced problem and, hence, they are also frozen orbits.

We note that while Eqs. (13.50)–(13.51) fully describe the frozen orbits problem, if second-order effects of J_2 are taken into account, one cannot neglect other effects of the geopotential, such as those of J_3 and J_4. Indeed, J_3 breaks the equatorial symmetry of the problem, and, thus, introduces qualitative changes in the reduced phase space (Coffey et al. 1994). Finally, the inclusion of J_4 in the geopotential makes radical changes in the frozen orbits diagram, stabilizing the behavior of frozen orbits with perigee at $\pm\pi/2$. The perturbing potential is

$$\begin{aligned} R = {} & \frac{\mu}{2a}\left(\frac{r_e}{a}\right)^2 \frac{J_2}{\eta^3}\left(1-\frac{3}{2}s^2\right) + \frac{\mu}{2a}\left(\frac{r_e}{a}\right)^3 \frac{J_3}{\eta^5}\left(3-\frac{15}{4}s^2\right) s\, e \sin\omega \\ & + \frac{\mu}{2a}\left(\frac{r_e}{a}\right)^4 \frac{J_2^2}{\eta^7}\left(m_{0,0} + m_{0,1}\,\eta + m_{0,2}\,\eta^2 + m_{2,0}\, e^2\cos 2\omega\right) \\ & + \frac{\mu}{2a}\left(\frac{r_e}{a}\right)^4 \frac{J_4}{\eta^7}\left(\tilde{m}_{0,0} + \tilde{m}_{0,2}\,\eta^2 + \tilde{m}_{2,0}\, e^2\cos 2\omega\right) + \mathcal{O}(J_2^3) \end{aligned} \tag{13.52}$$

where the inclination polynomials $\tilde{m}_{i,j}$ are given in Table 13.1. From Eq. (13.52) we obtain

$$\begin{aligned} \frac{\mathrm{d}e}{\mathrm{d}t} = {} & -n\left(\frac{r_e}{a}\right)^3 \frac{J_3}{\eta^4}\frac{3}{8}(4-5s^2)\, s\cos\omega \\ & -n\left(\frac{r_e}{a}\right)^4 \frac{e}{\eta^6}\frac{3}{32}\, s^2\left[(14-15s^2)\, J_2^2 + (30-35s^2)\, J_4\right]\sin 2\omega \end{aligned} \tag{13.53}$$

$$\begin{aligned} \frac{\mathrm{d}\omega}{\mathrm{d}t} = {} & n\left(\frac{r_e}{a}\right)^2 \frac{J_2}{\eta^4}\frac{3}{4}(4-5s^2) + n\left(\frac{r_e}{a}\right)^3 \frac{J_3}{\eta^6}\frac{3}{8}\left[\frac{s}{e}(4-5s^2) - \frac{e}{s}(4-35c^2s^2)\right]\sin\omega \\ & + n\left(\frac{r_e}{a}\right)^4 \frac{J_2^2}{\eta^8}\left[\gamma_{0,0} + \gamma_{0,1}\,\eta + \gamma_{0,2}\,\eta^2 + (\gamma_{2,0} + \gamma_{2,2}\,\eta^2)\cos 2\omega\right] \\ & + n\left(\frac{r_e}{a}\right)^4 \frac{J_4}{\eta^8}\left[\tilde{\gamma}_{0,0} + \tilde{\gamma}_{0,2}\,\eta^2 + (\tilde{\gamma}_{2,0} + \tilde{\gamma}_{2,2}\,\eta^2)\cos 2\omega\right] \end{aligned} \tag{13.54}$$

where the inclination polynomials $\tilde{\gamma}_{i,j}$ are given in Table 13.1.

For frozen orbits with low eccentricities, $e = \mathcal{O}(J_2)$, the terms factored by J_2^2 and J_4 are of higher order in Eqs. (13.53) and (13.54) when compared to those factored by J_2 and J_3 and, therefore, can be neglected. Besides, except for the case of orbits that are very close to the equator, one can also neglect the term in Eq. (13.54) that is factored by $J_3\, e/s$, thus leading to the classical equations for determining the (almost circular, non-equatorial) frozen orbit conditions

$$\frac{\mathrm{d}e}{\mathrm{d}t} = -n\left(\frac{r_e}{a}\right)^3 \frac{3J_3}{8\eta^4} \sin i\,(4 - 5\sin^2 i)\cos\omega \tag{13.55}$$

$$\frac{\mathrm{d}\omega}{\mathrm{d}t} = n\,\frac{r_e^2}{a^2}\,\frac{3J_2}{4\eta^4}\,(4 - 5\sin^2 i)\left(1 + \frac{r_e}{a}\,\frac{J_3}{J_2}\,\frac{\sin i}{2e\,\eta^2}\sin\omega\right) \tag{13.56}$$

Equation (13.55) vanishes for $\omega = \pm\pi/2$, which leaves the computation of the frozen orbit's eccentricity limited to the solution of the algebraic equation

$$n\left(\frac{r_e}{a}\right)^2 \frac{3J_2}{4\eta^4}\,(4 - 5\sin^2 i)\left(1 \pm \frac{r_e}{a}\,\frac{J_3}{J_2}\,\frac{\sin i}{2e\,\eta^2}\right) = 0, \tag{13.57}$$

which, for orbits out of the critical inclination, implies that the second parentheses must vanish, and, hence

$$e = -\frac{1}{2}\,\frac{r_e}{a}\,\frac{J_3}{J_2}\sin i + \mathcal{O}(J_2^2) \tag{13.58}$$

13.5 Sun-synchronous and Repeat Ground-track Orbits

In Earth imaging missions, it is often required to have the satellite pass over the same part of the Earth at roughly the same local time each day. The resulting orbit is called a *sun-synchronous orbit* (SSO).

To obtain conditions for sun synchronicity, we use the following notation: Ω_n is the RAAN of the satellite, Ω_s is the right ascension of the Sun, and $\Omega_{ns} = \Omega_n - \Omega_s$ is the *local hour angle*. The *local time of the ascending node* (LTAN) is then

$$\text{LTAN [hr]} = 12 + \frac{\Omega_{ns}\,[\text{deg}]}{15} \tag{13.59}$$

From the Lagrange equations, we know that, up to the order of J_2,

$$\dot{\Omega}_n = -\frac{3}{2}J_2 n\left(\frac{r_e}{p}\right)^2 \cos i \tag{13.60}$$

Sun synchronicity is defined as

$$\dot{\Omega}_n = \dot{\Omega}_s = \frac{2\pi}{1\ [\text{year}]} \tag{13.61}$$

or

$$K_s a^{-7/2} \frac{\cos i}{(1-e^2)^2} = \dot{\Omega}_s, \ K_s = -\frac{3}{2} J_2 \sqrt{\mu} r_e^2 \tag{13.62}$$

Usually, the orbit is near-circular, so the effect of eccentricity is small. Thus, the drift rate can be controlled by modifying a and i, as will be seen in Chap. 14.

However, if the satellite is designed for observing specific regions on Earth, as opposed to global observations, sun synchronicity is not enough. It is then also required that the satellite traces the same track on the ground with a given periodicity pattern. The orbital elements are chosen so that the satellite completes an integer number of revolutions in an integer number of days. This would then constitute an RGT orbit, mentioned in the previous section.

Suppose that the satellite should complete j orbits in k days. The calculation of the required altitude is done iteratively. The first iteration assumes a spherical Earth, and the next iteration uses the obtained result to yield an improved result, taking into account the effect of J_2.

In the first iteration, we calculate the altitude h from the orbital period,

$$T = \frac{kT_{day}}{j} = 2\pi \sqrt{\frac{(r_e+h)^3}{\mu}}, \ T_{day} = 86164 \text{ sec} \tag{13.63}$$

yielding

$$h = \left[\mu \left(\frac{kT_{day}}{2\pi j}\right)^2\right]^{1/3} - r_e = 42164.173 \left(\frac{k}{j}\right)^{2/3} - r_e \text{ [km]} \tag{13.64}$$

Now we can introduce the effect of J_2 by using the semimajor axis $a = r_e + h$ in the Lagrange equations for the mean elements,

$$\dot{\Omega} = K_s a^{-7/2} \left(1-e^2\right)^{-2} \cos i \tag{13.65a}$$

$$\dot{\omega} = -\frac{1}{2} K_s a^{-7/2} \left(1-e^2\right)^{-2} \left(5\cos^2 i - 1\right) \tag{13.65b}$$

$$\dot{M} = -\frac{1}{2} K_s a^{-7/2} \left(1-e^2\right)^{-3/2} \left(3\cos^2 i - 1\right) \tag{13.65c}$$

The amended angular velocity of the orbit is

$$n = \frac{j}{k}\left(\omega_e - \dot{\Omega} - \dot{M} - \dot{\omega}\right) \tag{13.66}$$

where ω_e is the angular velocity of the Earth, which yields the new altitude

$$h = \left(\frac{\mu}{n^2}\right)^{1/3} - r_e \tag{13.67}$$

In addition, we require sun synchronicity according to Eq. (13.62). The semimajor axis has been defined by Eq. (13.67), so Eq. (13.62) provides a single relation for the inclination and eccentricity. Thus, another degree of freedom exists, which may be used to require that the orbit be circular or frozen. Then, another iteration is performed with the set values of e and i to recalculate the semimajor axis.

13.6 Geostationary Orbits

Circular, equatorial, eastward orbits whose periods are equal to the Earth's rotational period, are called *geostationary orbits* (GEO). These orbits are widely used by communication satellites and weather satellites. If the inclination and eccentricity of a GEO are not zero, the orbit becomes a *geosynchronous orbit*. The dynamics in a GEO are affected by perturbations, the most significant of which are the zonal harmonics, the tesseral harmonics, the lunisolar gravitation, and solar radiation pressure. To obtain a semianalytical model of the dynamics in GEO, we will utilize the nonsingular elements defined in Eq. (10.128), and the variational equations (10.129), but with a slight modification: We redefine λ to be the *geographical longitude*, measured along the equator and then along the orbital plane,

$$\lambda = M + \omega + \Omega - \theta \tag{13.68}$$

The variable θ is the Greenwich sidereal angle (recall Eqs. (13.28)), measured from the instantaneous equinox. Hence, Eq. (10.129b) is replaced by

$$\dot{\lambda} = n - \dot{\theta} - \frac{2}{na}\frac{\partial R}{\partial a} + \frac{\gamma}{2na^2}\left(\zeta\frac{\partial R}{\partial \zeta} + \eta\frac{\partial R}{\partial \eta}\right) + \frac{1}{2\,na^2\gamma}\left(p\frac{\partial R}{\partial p} + q\frac{\partial R}{\partial q}\right) \tag{13.69}$$

An expression that relates the motion of the equinox to the variation in $\dot{\theta}$, in units of sidereal days, can be written as (Vallado 2001; p. 184)

$$\dot{\theta} = \omega_e + 5.96006 \times 10^{-11} T_{UT1} - 5.9 \times 10^{-15} T_{UT1}^2$$

where ω_e is the Earth's rotation rate, and T_{UT1} is the number of Julian centuries since the J2000 epoch. The value of ω_e used here is 6.3003880944 rad/day.

For nearly-circular orbits, the values of ζ and η remain small and Eqs. (10.129c)–(10.129d) are decoupled from the rest of the equations. Also, for near-equatorial orbits, the values of p and q are small and the coupling between Eqs. (10.129a)–

(13.69) and Eqs. (10.129e)–(10.129f) can be neglected. Thus, the satellite motion in the orbital plane can be treated separately from the out-of-plane motion.

Typically, the perturbing potential is written as (Belyanin and Gurfil 2009)

$$R = R_e + R_{ls} \tag{13.70}$$

The terms R_e and R_{ls} are generated by Earth's triaxiality and by the lunisolar gravitation, respectively. The expression for Earth's perturbing gravitational potential is taken from Eq. (11.215),

$$\begin{aligned} R_e = \frac{\mu}{r}\Big[& -\sum_{l=2}^{\infty}\left(\frac{r_e}{r}\right)^l J_l P_l(\cos\phi) \\ & + \sum_{l=2}^{\infty}\sum_{m=1}^{l}\left(\frac{r_e}{r}\right)^l P_{lm}(\cos\phi)[C_{lm}\cos m\lambda + S_{lm}\sin m\lambda]\Big] \end{aligned} \tag{13.71}$$

where ϕ is the colatitude angle, satisfying

$$\cos\phi = \sin(\omega + f)\sin i \tag{13.72}$$

and f is the true anomaly; $r_e = 6378.1363$ km is the Earth's equatorial radius, and we recall that P_{lm} is a Legendre polynomial of degree l and order m, J_l are the coefficients of the zonal harmonics, and C_{lm}, S_{lm} are the coefficients of the tesseral harmonics.

For circular orbits, the significant terms of Earth's perturbing potential inducing long-periodic and secular perturbations, written in terms of mean elements as elaborated in Sect. 13.3, are

$$\begin{aligned} R_e &\approx R_{J_2} + R_{J_{22}} \\ &= \frac{1}{4}n^2 J_2 r_e^2(2 - 3\sin^2 i) + 3n^2 r_e^2 J_{22}\cos^2\frac{i}{2}\cos 2(\lambda - \lambda_{22}) \end{aligned} \tag{13.73}$$

where $J_2 = 1082.62668355 \times 10^{-6}$, and

$$\begin{aligned} J_{22} &= \sqrt{C_{22}^2 + S_{22}^2} = 1.81543019 \times 10^{-6} \\ \lambda_{22} &= \frac{1}{2}\tan^{-1}\frac{S_{22}}{C_{22}} = -0.260556 \text{ rad} \end{aligned} \tag{13.74}$$

The perturbing potential due to the Moon and the Sun, approximated up to second order, is given by (Kozai 1959b)

$$\begin{aligned} R_{ls} = {} & \tfrac{1}{4}\mu_m n_m^2 a^2 \big[\big(1-\tfrac{3}{2}\sin^2 i\big)\big(1-\tfrac{3}{2}\sin^2 i_m\big) + \tfrac{3}{4}\sin(2i)\sin(2i_m)\cos(\Omega-\Omega_m) \\ & + \tfrac{3}{4}\sin i^2 \sin^2 i_m \cos[2(\Omega-\Omega_m)]\big] + \tfrac{1}{4}n_s^2 a^2 \big[\big(1-\tfrac{3}{2}\sin^2 i\big)\big(1-\tfrac{3}{2}\sin^2 i_s\big) \\ & + \tfrac{3}{4}\sin(2i)\sin(2i_s)\cos\Omega + \tfrac{3}{4}\sin^2 i \sin^2 i_s \cos(2\Omega)\big] \end{aligned} \tag{13.75}$$

where $\mu_m = 1/82.3$ is the ratio between the mass of the Moon and the sum of the masses of the Earth and the Moon; $n_m = 0.23$ rad/day is the Moon's orbit mean motion; $n_s = 0.017203$ rad/day is the Sun's apparent orbit mean motion; i_m is the Moon's orbit inclination with respect to the equatorial plane; Ω_m is the RAAN of the Moon's orbit with respect to the equatorial plane; and $i_s = \psi = 23.445°$ is the Sun's apparent orbit inclination with respect to the equatorial plane. The above expression is derived based on Kozai (1959b) by setting $e = 0$ and averaging along the period of the third body (Moon and Sun). Thus, the lunar-monthly and solar-yearly oscillations are neglected.

The angles i_m and Ω_m vary with time as a result of lunar regression: $18.3° \leq i_m \leq 28.59°$ and $-13° \leq \Omega_m \leq 13°$. They can be expressed in terms of the Moon's inclination, i_M, and node longitude, Ω_M, referred to the ecliptic plane, which are given by (Kamel and Tibbitts 1973)

$$\begin{aligned} i_M &= 5.145° \\ \Omega_M &= 259.183° - 0.05295°\, t \end{aligned} \tag{13.76}$$

where t is the time in Julian days measured from January 1, 1900, 12:00 hr. Using spherical trigonometry (see Fig. 13.1), i_m and Ω_m are given by

$$\cos i_m = \cos i_M \cos i_s - \sin i_s \sin i_M \cos\Omega_M \tag{13.77}$$

$$\sin\Omega_m = \sin i_M \sin\Omega_M / \sin i_m \tag{13.78}$$

Consequently, the expressions for R_e and R_{ls} in terms of the nonsingular elements (10.128) are

$$\begin{aligned} R_e = {} & \frac{1}{4} n^2 J_2 r_e^2 \left[2 - 12(p^2+q^2)(1-p^2-q^2)\right] \\ & + 3n^2 J_{22} r_e^2 \left(1-p^2-q^2\right)\cos 2(\lambda-\lambda_{22}) \end{aligned} \tag{13.79}$$

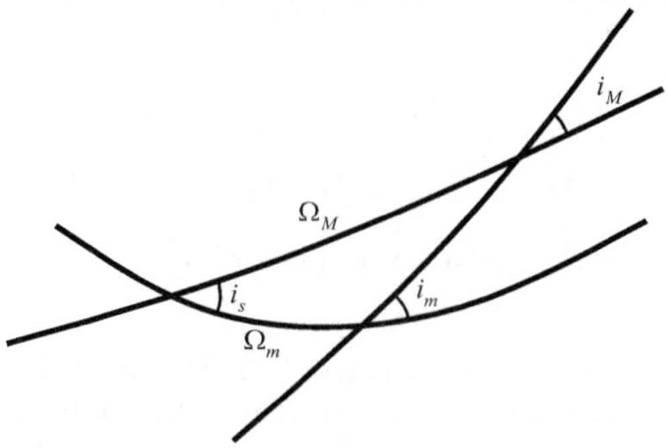

Fig. 13.1 Spherical trigonometry of the lunar orbit

$$\begin{aligned}R_{ls} =& \frac{1}{4}\mu_m n_m^2 a^2\Big[\left(1-\frac{3}{2}\sin^2 i_m\right) - 6\,(p^2+q^2)\cos^2\frac{i}{2}\\&+ 9\,(p^2+q^2)\cos^2\frac{i}{2}\,\sin^2 i_m + 3\,(p^2-q^2)\cos^2\frac{i}{2}\,\cos 2\Omega_m\,\sin^2 i_m\\&+ 6\,p\,q\,\cos^2\frac{i}{2}\,\sin 2\Omega_m\,\sin^2 i_m + 3\,p\,\cos\frac{i}{2}\,\cos i\,\cos\Omega_m\,\sin 2i_m\\&+ 3\,q\,\cos\frac{i}{2}\,\cos i\,\sin\Omega_m\,\sin 2i_m\Big]\\&+ \frac{1}{4}n_s^2 a^2\Big[\left(1-\frac{3}{2}\sin^2 i_s\right) - 6(p^2+q^2)\cos^2\frac{i}{2}\\&+ 6(2p^2+q^2)\cos^2\frac{i}{2}\,\sin^2 i_s + 3p\cos\frac{i}{2}\,\cos i\sin 2i_s\Big]\end{aligned} \tag{13.80}$$

where $\cos\frac{i}{2} = \sqrt{1-p^2-q^2}$ and $\cos i = 1-2p^2-2q^2$.

13.6.1 In-Plane Motion

Assume an ideal geostationary orbit, i.e. $i = e = 0$ and a perturbing potential comprising the J_2 and J_{22} terms only, $R = R_{J_2} + R_{J_{22}}$. The variational equations governing the change in longitude and semimajor axis in this case are

$$\dot{\lambda} = n - \omega_e + 3\frac{n}{a^2}\,r_e^2 J_2 + 18\frac{n}{a^2}\,r_e^2 J_{22}\cos 2(\lambda-\lambda_{22}) \tag{13.81}$$

$$\dot{a} = -12\frac{n}{a}\,r_e^2 J_{22}\sin 2(\lambda-\lambda_{22}) \tag{13.82}$$

By solving $\dot{\lambda} = \dot{a} = 0$ we get 4 equilibrium points, which are called *geostationary points*, 2 of which are stable, and 2 are unstable (Belyanin and Gurfil 2009),

$$\text{stable}: \quad \lambda_s = 75.07°,\ 255.07°; \ a = 42166.237\,\text{km} \tag{13.83}$$

$$\text{unstable}: \quad \lambda_{us} = 165.07°,\ 345.07°; \ a = 42166.279\,\text{km} \tag{13.84}$$

In the nominal gravitation field of the Earth, comprising the J_2 and J_{22} terms, the longitudinal motion of a geosynchronous satellite, initially positioned at a stable geostationary point, is zero. In the presence of perturbations, the equilibria are disturbed and the resulting motion is libration about the stable position with a period of about 2 years (Gedeon 1969; Kamel et al. 1973). The amplitude of librations is dictated by the magnitude of the perturbing potential.

13.6.2 Out-of-Plane Motion

The analysis of *out-of plane motion* is based on a method developed by Kamel and Tibbitts (1973). According to this method, the equations governing the evolution of inclination and ascending node are cast in canonical form, suitable for perturbation analysis. The Hamiltonian of the system then becomes quadratic with coefficients that vary slowly with a period of 18.6 years (same as the lunar nodal regression period). The averaged differential equations are subsequently solved in closed form. The geometry of the averaged solution is used to obtain the initial node location that maximizes the time of staying within a given inclination slot.

We start with the variational equations (11.108) governing the evolution of inclination, i, and ascending node, Ω, written for a near-circular orbit,

$$\frac{di}{dt} = -\frac{1}{na^2 \sin i}\frac{\partial R}{\partial \Omega} \tag{13.85}$$

$$\frac{d\Omega}{dt} = \frac{1}{na^2 \sin i}\frac{\partial R}{\partial i} \tag{13.86}$$

where the perturbing potential satisfies $R = R_{J_2} + R_{ls}$. By introducing new nonsingular variables, representing the Cartesian coordinates of the orbital specific angular momentum unit vector on the equator,

$$h_1 = \sin i \sin \Omega \tag{13.87}$$

$$h_2 = \sin i \cos \Omega \tag{13.88}$$

assuming $\cos i \approx 1$ and differentiating with respect to Ω_M, defined in Eq. (13.76), Eqs. (13.85)–(13.86) can be written in the canonical form

$$\frac{dh_1}{d\Omega_M} = \frac{\partial H}{\partial h_2} \tag{13.89}$$

$$\frac{dh_2}{d\Omega_M} = -\frac{\partial H}{\partial h_1}, \tag{13.90}$$

where H is the Hamiltonian of perturbations. In our case, The Hamiltonian H satisfies

$$H = H_g \tag{13.91}$$

where H_g represents the gravitational perturbations of the Earth, the Moon and the Sun.

By neglecting third-order terms, H_g becomes quadratic in h_1, h_2 with coefficients slightly varying with time (due to the lunar 18.6 years oscillations),

$$H_g = A_{11}h_1^2 + A_{12}h_1h_2 + A_{22}h_2^2 + B_1h_1 + B_2h_2 \tag{13.92}$$

where the coefficients A_{11} to B_2 can be found in Kamel and Tibbitts (1973),

$$A_{11} = \left[1 - \left(1 + \sin^2(\Omega_m)\sin^2(i_m)\right) + \lambda_s \cos^2(i_s)\right]\epsilon + \lambda_{obl} \tag{13.93a}$$

$$A_{12} = -\epsilon \sin^2(i_m)\sin(2\Omega_m) \tag{13.93b}$$

$$A_{22} = \left[1 - \left(1 + \cos^2(\Omega_m)\right)\sin^2(i_m) + \lambda_s \cos(2i_s)\right]\epsilon + \lambda_{obl} \tag{13.93c}$$

$$B_1 = -\epsilon \sin(2i_m)\sin(\Omega_m) \tag{13.93d}$$

$$B_2 = -\epsilon \sin(2i_m)\cos(\Omega_m) - \epsilon\lambda_s \sin(2i_s) \tag{13.93e}$$

$$\lambda_s = \frac{1}{\mu_m}\left(\frac{n_s}{n_m}\right)^2 \tag{13.93f}$$

$$\epsilon = -\frac{3}{8}\,\mu_m\,\dot{\Omega}_M^{-1}\frac{n_m^2}{n} = -\frac{3}{8}\,\mu_m\,\dot{\Omega}_M^{-1}\sqrt{\frac{r_e^3}{\mu}}\left(\frac{a}{r_e}\right)^{3/2} \tag{13.93g}$$

$$\lambda_{obl} = -\frac{3}{4}\,\dot{\Omega}_M^{-1}\sqrt{\frac{\mu}{r_e^3}}\,J_2\left(\frac{r_e}{a}\right)^{7/2} \tag{13.93h}$$

and $\dot{\Omega}_M$ is defined as

$$\dot{\Omega}_M = \frac{d\Omega_M}{dt} = -\frac{0.05295\pi}{180}\ \frac{\text{rad}}{\text{day}} \tag{13.94}$$

The coefficients appearing in Eq. (13.92) are functions of the lunar node and inclination, Ω_m and i_m, which are periodic with a period of 18.6 years. These variables can be expanded into a Fourier series with argument Ω_M (Kamel and Tibbitts 1973),

$$\begin{aligned}\Omega_m &= 0.22578\sin(\Omega_M) - 0.02338\sin(2\Omega_M) + 0.00338\sin(3\Omega_M)\\ &- 0.00055\sin(4\Omega_M) + 0.00010\sin(5\Omega_M) + 0.00010\sin(6\Omega_M)\\ &- 0.00002\sin(7\Omega_M) \end{aligned} \tag{13.95a}$$

$$\begin{aligned} i_m &= 0.41388 + 0.08940\cos(\Omega_M) - 0.00458\cos(2\Omega_M)\\ &+ 0.00052\cos(3\Omega_M) - 0.00006\cos(4\Omega_M) \end{aligned} \tag{13.95b}$$

$$\begin{aligned} A_{11} &= (0.83048 + 0.841637\lambda_s)\,\epsilon + \lambda_{obl} - 0.06507\epsilon\cos(\Omega_M)\\ &+ 0.00466\epsilon\cos(2\Omega_M) \end{aligned} \tag{13.95c}$$

$$A_{12} = -0.07109\epsilon\sin(\Omega_M) - 0.00738\epsilon\sin(2\Omega_M) \tag{13.95d}$$

$$\begin{aligned} A_{22} &= (0.67438 + 0.68327\lambda_s)\,\epsilon + \lambda_{obl} - 0.13032\epsilon\cos(\Omega_M)\\ &- 0.00274\epsilon\cos(2\Omega_M) \end{aligned} \tag{13.95e}$$

$$B_1 = -0.16388\epsilon\sin(\Omega_M) + 0.00319\epsilon\sin(2\Omega_M) \tag{13.95f}$$

$$\begin{aligned} B_2 &= -\,(0.72077 + 0.730162\lambda_s)\,\epsilon - 0.12221\epsilon\cos(\Omega_M)\\ &+ 0.00292\epsilon\cos(2\Omega_M) \end{aligned} \tag{13.95g}$$

Then, the Hamiltonian (13.91) can be rewritten as

$$H = H_0 + \tilde{H} \tag{13.96}$$

where H_0 has constant coefficients, and $\tilde{H}$ depends on Ω_M. In addition, $H_0 \gg \tilde{H}$, which renders Eqs. (13.89)–(13.90) convenient for a perturbation treatment. If $\tilde{H}$ is ignored, then Eqs. (13.89)–(13.90) become linear with constant coefficients. This is equivalent to averaging the equations over 18.6 years.

The averaged solution can be obtained in closed form and will contain two arbitrary constants of integration. Then, the perturbed solution can be found by considering the variations of these "constants" with time in the presence of the perturbation $\tilde{H}$.

13.6.3 Averaged Solution

The averaged equations are given by

$$\frac{dh_1}{d\Omega_M} = 2\bar{A}_{22}h_2 + \bar{B}_2 \tag{13.97}$$

$$\frac{dh_2}{d\Omega_M} = -2\bar{A}_{11}h_1 \tag{13.98}$$

where $\bar{A}_{11}, \bar{A}_{22}, \bar{B}_2$ are the portions of A_{11}, A_{22}, and B_{22} invariant with respect to Ω_M. The equilibrium solution can now be obtained,

$$h_{1e} = 0$$
$$h_{2e} = -\frac{\bar{B}_2}{2\bar{A}_{22}} \tag{13.99}$$

Equations (13.87)–(13.88) imply that the corresponding equilibrium values of the node and inclination are $\Omega_e = 0$, $i_e = \sin^{-1} h_{2e}$. Equations (13.97)–(13.98) can be reduced to the simple harmonic oscillator form,

$$d^2h_1/d\Omega_M^2 + \omega_l^2 h_1 = 0 \tag{13.100}$$

with the normalized period (the normalization factor is $-\dot{\Omega}_M^{-1}$)

$$P = 2\pi/\omega_l \tag{13.101}$$

and

$$\omega_l = 2\sqrt{\bar{A}_{22}\bar{A}_{11}} \tag{13.102}$$

The general solution may be written as

$$h_1 = \sqrt{\delta_1 X}\sin(\omega_l\Omega_M + x) \tag{13.103}$$

$$h_2 = h_{2e} + 2\sqrt{\frac{X}{\delta_1}}\cos(\omega_l\Omega_M + x) \tag{13.104}$$

where x, X are constants of integration, and

$$\delta_1 = 2\sqrt{\frac{\bar{A}_{22}}{\bar{A}_{11}}} \tag{13.105}$$

In the coordinates (h_1, h_2), Eqs. (13.103)–(13.104) describe a counterclockwise ellipse with a center at $(0, h_{2e})$, see Fig. 13.2a. The semimajor axis is $2\sqrt{\frac{X}{\delta_1}}$, lying along the h_2 axis. The semiminor axis is $\sqrt{\delta_1 X}$ and the eccentricity is $e = \frac{1}{2}\sqrt{4-\delta_1^2}$. The ellipse represents the precession cycle of the angular momentum vector about its equilibrium position $(0, h_{2e})$. The period of this precession cycle is given by Eq. (13.101). The numerical values of the constant parameters that

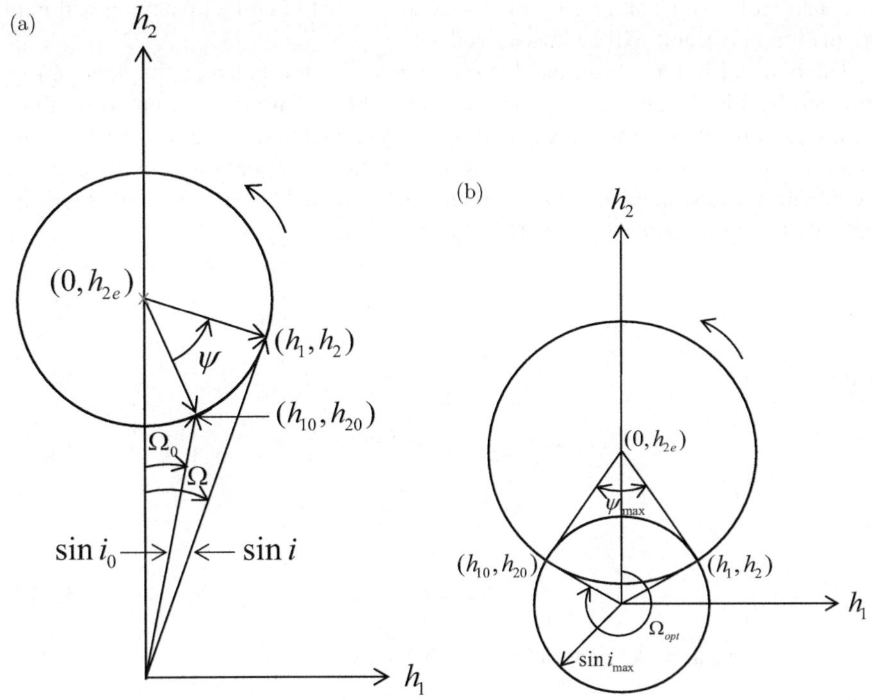

Fig. 13.2 Geometrical interpretation of the averaged solution in $h_1 - h_2$ coordinates. (**a**) Precession cycle of the angular momentum vector. (**b**) Optimal inclination drift cycle

determine the averaged solution are (h_{2e}, δ_1, ω_l are non-dimensional):

h_{2e}	0.13055
ω_l	0.34453
δ_1	1.9457
i_e	7.5001°
P	54.024 yrs

In Fig. 13.2a, the precession time from an initial point (h_{1_0}, h_{2_0}) to a point (h_1, h_2) is proportional to the subtended angle ψ. Due to the fact that $\delta_1 \approx 2$, the precession ellipse can be approximated by a circle. Then, the precession time is given by

$$T = \frac{\psi}{2\pi} P \tag{13.106}$$

In GEO missions, it is generally required to maintain the orbital inclination below a specified maximum value, $i \leq i_{max}$. It is obvious that, if $i_{max} < i_e$, then the inclination constraint cannot be satisfied forever. However, the initial values of the node and inclination can be chosen in such a way that the time of staying within the inclination constraint will be maximized.

The optimal initial conditions for (Ω_0, i_0) can be found using simple geometry, depicted by Fig. 13.2b (see e.g. Kechichian (1997)). The inclination constraint in the coordinates (h_1, h_2) is represented by a circle centered at the origin with radius $\sin i_{max}$. The maximum time of staying within the constraint circle corresponds to the maximum subtended angle ψ_{max}, which is the angle between tangent lines drawn through the equilibrium $(0, h_{2e})$. The maximum time T_{max} is

$$T_{max} = \frac{\psi_{max}}{2\pi} P \tag{13.107}$$

where

$$\psi_{max} = 2 \sin^{-1} \left(\frac{\sin i_0}{h_{2e}} \right) \tag{13.108}$$

and

$$i_0 = i_{max} \tag{13.109}$$

The optimal initial node location is given by

$$\Omega_{opt} = \frac{3\pi}{2} + \frac{\psi_{max}}{2} \tag{13.110}$$

Thus, the optimal inclination drift cycle must be initiated at $i_0 = i_{max}$ and Ω_0 given by Eq. (13.110). At the end of the cycle, the node assumes some final value, Ω_f, and the inclination angle is the same as i_0. If the mission extends beyond T_{max}, then a correction maneuver must be performed. This will be discussed in Sect. 14.8.

13.6.4 The Perturbed Problem

The optimal initial node location, Ω_{opt}, is not constant in time due to the lunar regression. To find how it varies with time, a problem that includes the perturbation $\tilde{H}$ must be solved. Utilizing the variation of parameters technique, the equations of motion for the perturbed problem are found to be

$$\dot{x} = \frac{\partial \tilde{H}}{\partial X} \tag{13.111}$$

$$\dot{X} = -\frac{\partial \tilde{H}}{\partial x}. \tag{13.112}$$

To obtain the Hamiltonian $\tilde{H}$ in terms of x and X, Eqs. (13.103)–(13.104) are substituted into Eq. (13.96). The second harmonics in the Fourier expansions of A_{11}, A_{12}, A_{22}, B_1, B_2 can be ignored, because the coefficients are very small compared to coefficients of the principal harmonics; see Kamel and Tibbitts (1973). Then, the Hamiltonian $\tilde{H}$ is given by

$$\begin{aligned}\tilde{H} = &\sqrt{X}\{\sqrt{\delta_1}\, b_1 \sin\Omega_M \sin(\omega_l\Omega_M + x) + \frac{2}{\sqrt{\delta_1}} b_2 \cos\Omega_M \cos(\omega_l\Omega_M + x)\} \\ &+ \sqrt{X} h_{2e}\{\sqrt{\delta_1}\, a_{12} \sin\Omega_M \sin(\omega_l\Omega_M + x) + \frac{4}{\sqrt{\delta_1}} a_{22} \cos\Omega_M \cos(\omega_l\Omega_M + x)\} \\ &+ X\{a_{11}\,\delta_1 \cos\Omega_M \sin^2(\omega_l\Omega_M + x) + \frac{4}{\sqrt{\delta_1}} a_{22} \cos\Omega_M \cos^2(\omega_l\Omega_M + x) \\ &+ 2\, a_{12} \sin\Omega_M \sin(\omega_l\Omega_M + x)\cos(\omega_l\Omega_M + x)\} \\ &+ a_{22}\cos\Omega_M h_{2e} + b_2 \cos\Omega_M\, h_{2e}\end{aligned} \tag{13.113}$$

where $a_{11}, a_{22}, a_{12}, b_1, b_2$ are the coefficients of the fundamental harmonics in the Fourier expansions of $A_{11}, A_{12}, A_{22}, B_1, B_2$, respectively

$$\begin{aligned}a_{11} &= -0.06507\epsilon,\ a_{12} = -0.07109\epsilon,\ a_{22} = -0.13032\epsilon \\ b_1 &= -0.16388\epsilon,\ b_2 = -0.12221\epsilon\end{aligned} \tag{13.114}$$

The last two terms in Eq. (13.113) can be ignored because they do not depend on x and X and, thus, do not affect the final results. By performing some algebraic and trigonometric manipulations, the expression for $\tilde{H}$ can be written as (Kamel and Tibbitts 1973)

$$\begin{aligned}\tilde{H} =& \sqrt{X}\{(\alpha_1 + \alpha_3)\cos[(\omega_l + 1)\Omega_M + x] + (\alpha_2 + \alpha_4)\cos[(\omega_l - 1)\Omega_M + x]\} \\ &+ X\{\alpha_5 \cos(\Omega_M) + \alpha_6 \cos[(2\omega_l + 1)\Omega_M + 2x] \\ &+ \alpha_7 \cos[(2\omega_l - 1)\Omega_M + 2x]\},\end{aligned} \tag{13.115}$$

where

$$\begin{aligned}\alpha_{1,2} &= \left(b_2 \pm \frac{1}{2} b_1\,\delta_1\right)/\sqrt{\delta_1},\ \alpha_{3,4} = \frac{1}{2} h_{2e}\,(4\,a_{22} \mp a_{12}\,\delta_1)\,/\,\sqrt{\delta_1} \\ \alpha_5 &= \frac{1}{2}\left(a_{11}\,\delta_1^2 + 4\,a_{22}\right)/\delta_1,\ \alpha_{6,7} = \frac{1}{2}\left[\left(2\,a_{22} - \frac{1}{2}\,a_{11}\,\delta_1^2\right)/\delta_1 \mp a_{12}\right]\end{aligned} \tag{13.116}$$

The classical method to solve Eqs. (13.111)–(13.112) is to perform a canonical transformation from the set of variables (x, X) to a new set (y, Y) with corresponding new Hamiltonian K. In our problem, the original Hamiltonian $\tilde{H}$ is time varying. However, it can be seen that $\tilde{H}$ depends on a small parameter: Each of the coefficients α_{1-7} in Eq. (13.116) is a linear combination of the parameters appearing in Eq. (13.114), i.e. has a common factor ϵ.

If we define

$$\bar{\alpha}_i = \alpha_i/\epsilon, \ i = 1, 2 \ldots 7 \tag{13.117}$$

$$\bar{H}_1 = \tilde{H}/\epsilon \tag{13.118}$$

then $\tilde{H}$ may be represented by

$$\tilde{H} = \bar{H}_0 + \epsilon \bar{H}_1, \tag{13.119}$$

where $\bar{H}_0 = 0$.

The canonical transformation for a system described by a Hamiltonian depending on a small parameter can be performed by using Lie transforms. Kamel and Tibbitts (1973) utilized the method developed by Kamel (1970) to obtain a second-order solution for the system defined by Eqs. (13.111)–(13.112), (13.115), (13.116). According to this method, the canonical transformation $(x, X) \rightarrow (y, Y)$ can be constructed recursively to achieve specific requirements in the transformed Hamiltonian K.

To eliminate time-dependence, we seek for a new Hamiltonian

$$K = 0 \tag{13.120}$$

Thus, both Y and y are constants. Following the steps outlined by Kamel (1970), we obtain the first-order expression for the generating function,

$$\begin{aligned} W_1 = & \sqrt{Y}\{\frac{\bar{\alpha}_1 + \bar{\alpha}_3}{(\omega_l + 1)} \sin[(\omega_l + 1)\Omega_M + y] \\ & + \frac{\bar{\alpha}_2 + \bar{\alpha}_4}{(\omega_l - 1)} \sin[(\omega_l - 1)\Omega_M + y]\} \\ & + Y\{\bar{\alpha}_5 \sin \Omega_M + \frac{\bar{\alpha}_6}{(2\omega_l + 1)} \sin[(2\omega_l + 1)\Omega_M + 2y] \\ & + \frac{\bar{\alpha}_7}{(2\omega_l - 1)} \sin[(2\omega_l - 1)\Omega_M + 2y]\} \end{aligned} \tag{13.121}$$

By introducing the new functions

$$\bar{h}_1 = \sqrt{\delta_1 Y} \sin(\omega_l \Omega_M + y) \tag{13.122}$$

$$\bar{h}_2 = h_{2e} + 2\sqrt{\frac{Y}{\delta_1}} \cos(\omega_l \Omega_M + y) \tag{13.123}$$

and assuming W_1 to be quadratic in $\bar{h}_1, \bar{h}_2$, we can write (Kamel and Tibbitts 1973)

$$W_1 = \eta_1 \bar{h}_1^2 + \eta_2 \bar{h}_1 \bar{h}_2 + \eta_3 \bar{h}_2^2 + \eta_4 \bar{h}_1 + \eta_5 \bar{h}_2 \tag{13.124}$$

where

$$\eta_1 = \frac{1}{\delta_1}\left(\bar{\alpha}_5 - \frac{\bar{\alpha}_6}{2\omega_l + 1} + \frac{\bar{\alpha}_7}{2\omega_l - 1}\right) \sin \Omega_M \tag{13.125a}$$

$$\eta_2 = \left(\frac{\bar{\alpha}_6}{2\omega_l + 1} + \frac{\bar{\alpha}_7}{2\omega_l - 1}\right) \cos \Omega_M \tag{13.125b}$$

$$\eta_3 = \frac{\delta_1}{4}\left(\bar{\alpha}_5 + \frac{\bar{\alpha}_6}{2\omega_l + 1} - \frac{\bar{\alpha}_7}{2\omega_l - 1}\right) \sin \Omega_M \tag{13.125c}$$

$$\eta_4 = \left[-h_{2e}\left(\frac{\bar{\alpha}_6}{2\omega_l + 1} + \frac{\bar{\alpha}_7}{2\omega_l - 1}\right)\right. \tag{13.125d}$$

$$\left. + \frac{1}{\sqrt{\delta_1}}\left(\frac{\bar{\alpha}_1 + \bar{\alpha}_3}{\omega_l + 1} + \frac{\bar{\alpha}_2 + \bar{\alpha}_4}{\omega_l - 1}\right)\right] \cos \Omega_M \tag{13.125e}$$

$$\eta_5 = \left[-h_{2e}\frac{\delta_1}{2}\left(\bar{\alpha}_5 + \frac{\bar{\alpha}_6}{2\omega_l + 1} - \frac{\bar{\alpha}_7}{2\omega_l - 1}\right)\right.$$

$$\left. + \frac{\sqrt{\delta_1}}{2}\left(\frac{\bar{\alpha}_1 + \bar{\alpha}_3}{\omega_l + 1} - \frac{\bar{\alpha}_2 + \bar{\alpha}_4}{\omega_l - 1}\right)\right] \sin \Omega_M \tag{13.125f}$$

The coefficients η_{1-5} are found by substituting Eqs. (13.122)–(13.123) into Eq. (13.124) and comparing the coefficients with those in Eq. (13.121). Now, the solutions h_1, h_2 can be found in terms of $\bar{h}_1$, $\bar{h}_2$ (Kamel and Tibbitts 1973; Kamel 1970)

$$h_1 = \bar{h}_1 + \epsilon \frac{\partial W_1}{\partial \bar{h}_2}, \qquad h_2 = \bar{h}_2 - \epsilon \frac{\partial W_1}{\partial \bar{h}_1} \tag{13.126}$$

The parameter ϵ is constant and, thus, can be inserted into the derivatives of W_1,

$$h_1 = \bar{h}_1 + \frac{\partial W}{\partial \bar{h}_2}, \qquad h_2 = \bar{h}_2 - \frac{\partial W}{\partial \bar{h}_1} \tag{13.127}$$

where $W = \epsilon W_1$, i.e. W has the same expression as W_1 with parameters $\bar{\alpha}_{1-7}$ replaced by α_{1-7}.

Finally, after differentiating, we get the approximated solution given by the following transformation:

$$\begin{aligned} h_1 &= \bar{h}_1 + \left(\eta_2 \bar{h}_1 + 2\eta_3 \bar{h}_2 + \eta_5\right) \\ h_2 &= \bar{h}_2 - \left(2\eta_1 \bar{h}_1 + \eta_2 \bar{h}_2 + \eta_4\right) \end{aligned} \tag{13.128}$$

The formulae for the inverse representation are

$$\begin{aligned} \bar{h}_1 &= h_1 - (\eta_2 h_1 + 2\eta_3 h_2 + \eta_5) \\ \bar{h}_2 &= h_2 + (2\eta_1 h_1 + \eta_2 h_2 + \eta_4) \end{aligned} \tag{13.129}$$

The solution in the coordinates (y, Y) can be written in terms of initial conditions. The constants y, Y are evaluated from Eqs. (13.122)–(13.123) in terms of $\bar{h}_{10} = \bar{h}_1(\Omega_{M0})$, $\bar{h}_{20} = \bar{h}_2(\Omega_{M0})$, and are given by

$$\begin{aligned} Y &= \frac{\bar{h}_{10}^2}{\delta_1} + \frac{\delta_1}{4}\left(\bar{h}_{10} - h_{2e}\right)^2 \\ \cos y &= \frac{\bar{h}_{10}}{\sqrt{\delta_1 Y}} \sin(\omega_l \Omega_{M0}) + \frac{\sqrt{\delta_1}(\bar{h}_{20} - h_{2e})}{2\sqrt{Y}} \cos(\omega_l \Omega_{M0}) \\ \sin y &= -\frac{\sqrt{\delta_1}(\bar{h}_{20} - h_{2e})}{2\sqrt{Y}} \sin(\omega_l \Omega_{M0}) + \frac{\bar{h}_{10}}{\sqrt{\delta_1 Y}} \cos(\omega_l \Omega_{M0}) \end{aligned} \tag{13.130}$$

By substituting back into Eqs. (13.122)–(13.123), we obtain

$$\begin{aligned} \bar{h}_1 &= \frac{\delta_1}{2}\left(\bar{h}_{20} - h_{2e}\right) \sin\left[\omega_l\left(\Omega_M - \Omega_{M0}\right)\right] + \bar{h}_{10} \cos\left[\omega_l\left(\Omega_M - \Omega_{M0}\right)\right] \\ \bar{h}_2 &= \left(\bar{h}_{20} - h_{2e}\right) \cos\left[\omega_l\left(\Omega_M - \Omega_{M0}\right)\right] - \frac{2}{\delta_1} \bar{h}_{10} \sin\left[\omega_l\left(\Omega_M - \Omega_{M0}\right)\right] + h_{2e} \end{aligned} \tag{13.131}$$

The resulting solution can now be used to calculate the optimal initial node location, Ω_{opt}, defined in Fig. 13.2b, under the effect of lunar regression.

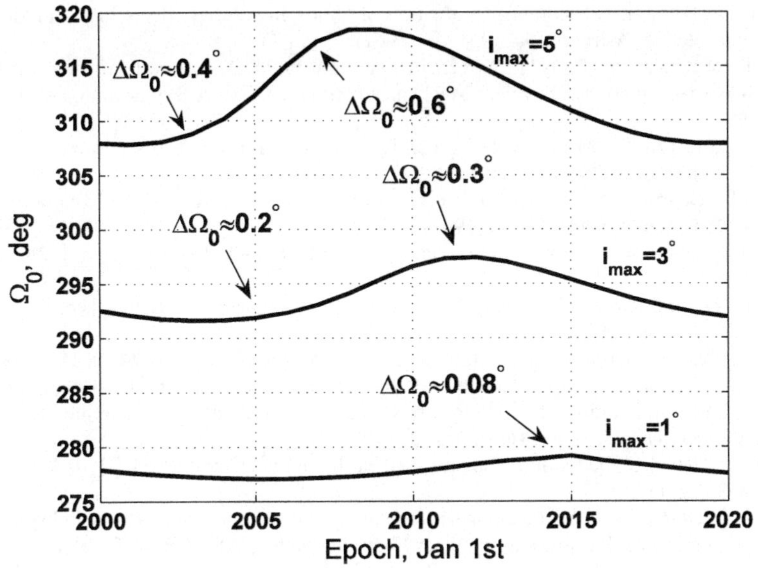

Fig. 13.3 The dependence of Ω_{opt} on time for different inclination constraints

By utilizing an algorithm for calculating Ω_{opt}, detailed in Kamel and Tibbitts (1973), Ω_{opt} and the corresponding time of staying within a given inclination constraint, T_{max}, were calculated over a 20-year period. The dependence of Ω_{opt} on time for different inclination constraints is given in Fig. 13.3. In this figure, the y-axis is the initial node location (defined as the node at epoch, $\Omega_0 = \Omega(t = t_0)$), which is set to be equal to the optimal initial node location, i. e. $\Omega_0 = \Omega_{opt}$. The x-axis is the epoch.

References

Battin, R.: An Introduction to the Mathematics and Methods of Astrodynamics. American Institute of Aeronautics and Astronautics, Reston (1999)

Belyanin, S., Gurfil, P.: Semianalytical study of geosynchronous orbits about a precessing oblate Earth under lunisolar gravitation and tesseral resonance. J. Astronaut. Sci. **57**(3), 517–543 (2009)

Bezdek, A., Vokrouhlickı, D.: Semianalytic theory of motion for close-Earth spherical satellites including drag and gravitational perturbations. Planet. Space Sci. **52**(14), 1233–1249 (2004)

Breiter, S., Metris, G.: Keplerian expansions in terms of Henrard's practical variables. Celest. Mech. Dyn. Astron. **58**(3), 237–244 (1994)

Broucke, R.A.: Numerical integration of periodic orbits in the main problem of artificial satellite theory. Celest. Mech. Dyn. Astron. **58**(2), 99–123 (1994). DOI 10.1007/BF00695787

Brouwer, D.: Solution of the problem of artificial satellite theory without drag. Astron. J. **64**, 378–396 (1959)

Brouwer, D., Clemence, G.M.: Methods of Celestial Mechanics. Academic Press, New York (1961)

Brouwer, D., Hori, G.I.: Theoretical evaluation of atmospheric drag effects in the motion of an artificial satellite. Astron. J. **66**, 193–225 (1961)

Cefola, P., McClain, W., Early, L., Green, A.: A semianalytical satellite theory for weak time-dependent perturbations. In: Flight Mechanics/Estimation Theory Symposium, vol. 1, pp. 67–91 (1980)

Coffey, S.L., Deprit, A., Miller, B.R.: The critical inclination in artificial satellite theory. Celest. Mech. Dyn. Astron. **39**(4), 365–406 (1986)

Coffey, S.L., Deprit, A., Deprit, E.: Frozen orbits for satellites close to an Earth-like planet. Celest. Mech. Dyn. Astron. **59**(1), 37–72 (1994)

Cutting, E., Frautnick, J., Born, G.: Orbit analysis for SEASAT-A. J. Astronaut. Sci. **26**, 315–342 (1978)

Danielson, D.A., Sagovac, C.P., Neta, B., Early, L.W.: Semianalytic satellite theory. Tech. rep., DTIC Document (1995)

Deprit, A.: The elimination of the parallax in satellite theory. Celest. Mech. **24**(2), 111–153 (1981)

Gedeon, G.: Tesseral resonance effects on satellite orbits. Celest. Mech. **1**, 167–189 (1969)

Gurfil, P., Lara, M.: Motion near frozen orbits as a means for mitigating satellite relative drift. Celest. Mech. Dyn. Astron. **116**(3), 213–227 (2013)

Hoots, F.R.: Theory of the motion of an artificial earth satellite. Celest. Mech. Dyn. Astron. **23**(4), 307–363 (1981)

Kamel, A.: Perturbation theory based on Lie transforms and its application to the stability of motion near sun-perturbed Earth-moon triangular libration points. NASA CR-1622 (1970)

Kamel, A., Tibbitts, R.: Some useful results on initial node location for near-equatorial circular satellite orbits. Celest. Mech. **8**, 45–73 (1973)

Kamel, A., Ekman, D., Tibbitts, R.: East-west stationkeeping requirements of nearly synchronous satellites due to Earth's triaxiality and lunisolar effects. Celest. Mech. **8**, 129–148 (1973)

Kaula, W.M.: Theory of Satellite Geodesy. Blaisdell Publishing Company, Waltham, Massachusetts (1966). Reprinted by Dover Publications, Mineola (2000)

Kechichian, J.A.: The analysis of the relative motion in general elliptic orbit with respect to a dragging and precessing coordinate frame. In: Proceedings of the AAS/AIAA Astrodynamics Conference, Sun Valley, ID, pp 2053–2074 (1997)

Kozai, Y.: The motion of a close Earth satellite. Astron. J. **64**, 367–377 (1959a)

Kozai, Y.: On the effects of the Sun and the Moon upon the motion of a close Earth satellite. SAO Spec. Rep. **22**(2), 356–366 (1959b)

Lane, M.H.: The development of an artificial satellite theory using a power-law atmospheric density representation. In: AIAA Second Aerospace Sciences Meeting, New York, NY, USA, AIAA Paper 65-35 (1965)

Lane, M.H., Fitzpatrick, P.M., Murphy, J.J.: On the representation of air density in satellite deceleration equations by power functions with integral exponents. Tech. rep., DTIC, Report No. APGC-TDR-62-15 (1962)

Lara, M.: Searching for repeating ground track orbits: A systematic approach. J. Astronaut. Sci. **47**(3-4), 177–188 (1999)

Lara, M., Deprit, A., Elipe, A.: Numerical continuation of families of frozen orbits in the zonal problem of artificial satellite theory. Celest. Mech. Dyn. Astron. **62**(2), 167–181 (1995)

Lin, L., De-zi, Z.: Combined perturbation on near-Earth satellite orbits. Chin. Astron. Astrophys. **5**(4), 422–433 (1981)

Liu, J.J.F.: Advances in orbit theory for an artificial satellite with drag. J. Astronaut. Sci. **31**(2), 165–188 (1983)

Liu, J.J.F., Alford, R.L.: A semi-analytic theory for the motion of a close-Earth artificial satellite with drag. In: AIAA, Aerospace Sciences Meeting, New Orleans, LA, USA, AIAA Paper 79-0123 (1979)

Liu, J.J.F., Alford, R.L.: Semianalytic theory for a close-Earth artificial satellite. J. Guid. Control Dyn. **3**(4), 304–311 (1980)

Rosborough, G.W., Ocampo, C.: Influence of higher degree zonals on the frozen orbit geometry. In: Kaufman, B., Alfriend, K.T., Roehrich, R.L., Dasenbrock, R.R. (eds.) Astrodynamics 1991,

American Astronautical Society, Univelt, Inc., USA, Advances in the Astronautical Sciences, vol. 76, pp. 1291–1304 (1992)
Seidelmann, P.K.: Explanatory Supplement to the Astronomical Almanac. University Science Books, Washington, DC (1992)
Shapiro, B.E.: Phase plane analysis and observed frozen orbit for the Topex/Poseidon mission. Adv. Astronaut. Sci. **91**, 853–872 (1995)
Taylor, S.P., Cefola, P.J.: Semianalytical satellite theory and sequential estimation, vol. 46, p. 702 (1982)
Vallado, D.: Fundamentals of Astrodynamics and Applications, 2nd edn. Microcosm Press and Kluwer Academic Publishers, London (2001)
Zeis, E., Cefola, P.: Computerized algebraic utilities for the construction of nonsingular satellite theories. J. Guid. Control Dyn. **3**(1), 48–54 (1980)

Chapter 14
Satellite Orbit Control

14.1 Introduction

Satellite orbit control refers to the process of generating thrust for tracking a particular orbit in the presence of orbital perturbations. The particular orbit is mission dependent. It could be a low Earth orbit (LEO), used for Earth imaging, a geostationary orbit (GEO) used for communication and weather monitoring, or an interplanetary orbit.

The control forces can be applied by chemical or electric rocket propulsion systems. Over the past few decades, there has been significant progress in the research of advanced electric propulsion systems for space missions. Propulsion technologies such as Hall thrusters, field emission electric propulsion, and pulsed plasma thrusters have been developed, tested and implemented on board operational spacecraft.

To understand how orbit control systems operate, we start this chapter by introducing some basic concepts in stability and control of dynamical systems. We then distinguish between impulsive and continuous maneuvers, and discuss various problems related to orbit control, including interplanetary travel by means of gravity assists, trajectory optimization, and satellite rendezvous maneuvers.

We devote the last two sections to continuous-thrust orbital maneuvers using orbital elements. Within this context, several methodologies have been proposed over the years, with the use of the variational equations of the orbital elements being the most prevalent approach (Kluever 1998; Gurfil 2007). Other approaches considered a hybrid Cartesian coordinates and orbital element-based control laws (Schaub and Alfriend 2002).

P. Gurfil, P.K. Seidelmann, *Celestial Mechanics and Astrodynamics: Theory and Practice*, Astrophysics and Space Science Library 436,
DOI 10.1007/978-3-662-50370-6_14

14.2 Stability and Control of Dynamical Systems

Let $\mathbf{x}$ be an n-dimensional vector of state variables called a *state vector*. For example, we can include in $\mathbf{x}$ the inertial position and velocity vectors of the satellite, so that $\mathbf{x} = [\mathbf{r}^T \ \mathbf{v}^T]^T$, and $\mathbf{x}$ is a 6-dimensional column vector. We also define an m-dimensional vector of control variables called a *control vector*, denoted by $\mathbf{u}$. In the artificial satellite example, this vector would include the 3 components of the specific thrust vector.

In general, we distinguish between *open-loop control*, where $\mathbf{u}$ is a direct function of time, and *closed-loop control*, or *feedback control*, in which $\mathbf{u} = \mathbf{g}(\mathbf{x})$. For example, the controller $\mathbf{u} = -\mathbf{K}\mathbf{x}$, with $\mathbf{K}$ being an $m \times n$ *gain matrix*, would constitute a *linear feedback controller*.

Using the state and control vectors, we can write the equations of motions as n first-order nonlinear differential equations,

$$\dot{\mathbf{x}} = \mathbf{f}(\mathbf{x}, \mathbf{u}), \ \mathbf{x}(t_0) = \mathbf{x}_0 \tag{14.1}$$

where $\mathbf{f}$ is a vector-valued function, t_0 is the initial time (epoch), and $\mathbf{x}_0$ are the initial conditions.

A fundamental aspect related to the dynamical system (14.1) is stability. We qualitatively discussed this topic in Chap. 1, and will now highlight some qualitative aspects.

First, we consider the uncontrolled version of system (14.1), namely the case wherein $\mathbf{u} \equiv \mathbf{0}$, and define an *equilibrium*, $\mathbf{x}^\star$, for which

$$\mathbf{f}(\mathbf{x}^\star) = \mathbf{0} \tag{14.2}$$

Stability theories analyze the trajectories of system (14.1) with respect to the equilibrium $\mathbf{x}^\star$. We have already seen such an analysis when we introduced the stability of the Lagrangian points in Sect. 8.6. Here, the treatment focuses on the use of the control vector $\mathbf{u}$ for steering the states of the system to the equilibrium $\mathbf{x}^\star$.

A fundamental equilibria stability theory is the *Lyapunov theory*. The equilibrium $\mathbf{x}^\star$ is said to be *stable in the sense of Lyapunov*, if for any $\epsilon > 0$, there exists a $\delta = \delta(\epsilon) > 0$ such that, if $\|\mathbf{x}(t_0) - \mathbf{x}^\star\| \le \delta$, then $\|\mathbf{x}(t) - \mathbf{x}^\star\| \le \epsilon$ for all $t \ge t_0$. The equilibrium is *locally asymptotically stable*, if it is stable in the sense of Lyapunov, and in addition, for $\|\mathbf{x}(t_0) - \mathbf{x}^\star\| \le \delta$,

$$\lim_{t\to\infty} \|\mathbf{x}(t) - \mathbf{x}^\star\| \to 0 \tag{14.3}$$

Finally, the equilibrium is said to be *globally asymptotically stable*, if it is stable in the sense of Lyapunov, and Eq. (14.3) is satisfied for any initial condition $\mathbf{x}(t_0)$.

Lyapunov's theory also provides a method for examining the stability of equilibria. This method, called *Lyapunov's second method*, provides sufficient stability conditions for the equilibrium $\mathbf{x}^\star$ of the system

$$\dot{\mathbf{x}} = \mathbf{f}(\mathbf{x}) \tag{14.4}$$

Suppose there exists a differentiable function $V(\mathbf{x})$, which satisfies $V(\mathbf{x}) > 0$ in some domain D around the equilibrium $\mathbf{x}^\star$, and in addition $V(\mathbf{x}^\star) = 0$. If $\dot{V}(\mathbf{x})$ is continuous both in $\mathbf{x}$ and t, and in addition $\dot{V}(\mathbf{x}) \leq 0$ everywhere except $\mathbf{x} = \mathbf{x}^\star$ in the same domain D, then the equilibrium $\mathbf{x}^\star$ is stable in the sense of Lyapunov. If $\dot{V}(\mathbf{x})$ is not identically 0 for all $\mathbf{x}$ except $\mathbf{x}^\star$, then the equilibrium $\mathbf{x}^\star$ is locally asymptotically stable. Finally, if in addition to all the previous conditions $\lim_{\|\mathbf{x}\| \to \infty} V(\mathbf{x}) \to \infty$, then $\mathbf{x}^\star$ is globally asymptotically stable.

In many practical applications, the nonlinear equations of motion (14.1) can be linearized about a given point $(\mathbf{x}_p, \mathbf{u}_p)$ or about a trajectory, $(\mathbf{x}_t(t), \mathbf{u}_t(t))$. In the former case, we obtain a system of linear time-invariant (LTI) differential equations of the form

$$\delta\dot{\mathbf{x}} = \mathbf{A}\delta\mathbf{x} + \mathbf{B}\delta\mathbf{u}, \ \delta\mathbf{x}(t_0) = \delta\mathbf{x}_0 \tag{14.5}$$

In Eq. (14.5), we have

$$\delta\mathbf{x} = \mathbf{x} - \mathbf{x}_p, \ \delta\mathbf{u} = \mathbf{u} - \mathbf{u}_p, \ \mathbf{A} = \left.\frac{\partial \mathbf{f}}{\partial \mathbf{x}}\right|_{\mathbf{x}_p, \mathbf{u}_p}, \ \mathbf{B} = \left.\frac{\partial \mathbf{f}}{\partial \mathbf{u}}\right|_{\mathbf{x}_p, \mathbf{u}_p} \tag{14.6}$$

The LTI system (14.5) admits a closed-form solution. Since LTI systems are shift-invariant, we can set $t_0 = 0$ without loss of generality, and the solution is given by

$$\delta\mathbf{x}(t) = \mathrm{e}^{\mathbf{A}t}\delta\mathbf{x}(0) + \int_0^t \mathrm{e}^{\mathbf{A}(t-\tau)}\mathbf{B}\delta\mathbf{u}(\tau)d\tau \tag{14.7}$$

where the notation $\mathrm{e}^{\mathbf{Y}}$ denotes the *matrix exponential*, i.e.

$$\mathrm{e}^{\mathbf{Y}} = \mathbf{I} + \sum_{\ell=1}^{\infty} \frac{1}{\ell!}\mathbf{Y}^\ell \tag{14.8}$$

in which $\mathbf{I}$ denotes the identity matrix. The expression $\mathrm{e}^{\mathbf{A}t}$ in Eq. (14.7) is called the *transition matrix*.

The (internal) stability of the linear system (14.5) is determined by the eigenvalues of the matrix $\mathbf{A}$ for $\delta\mathbf{u} \equiv \mathbf{0}$. Thus, we first solve the characteristic equation

$$|\lambda\mathbf{I} - \mathbf{A}| = 0 \tag{14.9}$$

for the n eigenvalues λ. The system would be asymptotically stable, if all the eigenvalues have negative real parts, unstable, if at least one eigenvalue has a positive real part, and stable in the sense of Lyapunov, if all the eigenvalues located on the imaginary axis are distinct. Even if the linear system is unstable, we can still find initial conditions that would yield bounded state trajectories. In this case, we say that this specific selection of initial conditions is included within a *stable subspace*.

14.3 Impulsive and Continuous Maneuvers

Let us assume that the dynamics of a satellite are determined by the two-body problem, and that the only additional acceleration acting on the satellite is due to the specific thrust, $\mathbf{u}$. Thus, for a thrust vector $\mathbf{T}$, we have

$$\mathbf{u} = \frac{\mathbf{T}}{m} \tag{14.10}$$

where m denotes the satellite mass. In this case, we can write the equations of motion as

$$\ddot{\mathbf{r}} = -\frac{\mu \mathbf{r}}{r^3} + \mathbf{u} \tag{14.11}$$

The thrust magnitude is determined by the relation

$$T = \|\mathbf{T}\| = |\dot{m} U_e| \tag{14.12}$$

where $U_e > 0$ is the exhaust velocity. The magnitude of the thrust acceleration is, therefore,

$$u = \|\mathbf{u}\| = \frac{T}{m} = \frac{|\dot{m} U_e|}{m} \tag{14.13}$$

We now take into account the fact that the mass variation $\dot{m}$ is always negative, since the propellent ejected from the satellite always decreases its mass. Hence, we rewrite Eq. (14.13) into

$$u = -\frac{U_e}{m}\frac{dm}{dt} \tag{14.14}$$

The velocity increment obtained at a total burning time of t_b, reducing the satellite's mass from the initial mass m_0 to the final mass m_f, is calculated based on Eq. (14.14) as

$$\Delta v = \int_{t_0}^{t_b} u dt = -\int_{m_0}^{m_f} \frac{U_e}{m} dm \tag{14.15}$$

The exhaust velocity depends on the *specific impulse*, denoted by I_{sp} (a characteristic of the propellant used), and on the standard gravitational acceleration at sea level, g_0,

$$U_e = I_{sp} g_0 \tag{14.16}$$

In general, I_{sp} depends on the fuel chamber's temperature. However, in many common applications, particulary those that involve chemical propellants, it may be assumed that I_{sp}, and, therefore U_e, are approximately constant, thus providing a simple solution to Eq. (14.15),

$$\Delta v = U_e \ln \frac{m_0}{m_f} \tag{14.17}$$

Equation (14.17) leads to the important conclusion that, in chemical propellants, minimization of the fuel mass, which is equal to $m_0 - m_f$, will be obtained by minimization of Δv.

An approximation is often made to the expression for Δv. In this mathematical abstraction, it is assumed that an "infinitely large" force causes an "instantaneous" change in the velocity of the satellite without affecting its position. For some required velocity corrections, this rationale would be valid if $t_b \ll T$, where T is some characteristic time constant, such as the orbital period. This approximation, called an *impulsive maneuver*, is of great value for preliminary mission design. Typically, the impulsive maneuver model would be valid for chemical thrusters.

However, for low-thrust systems, such as *electric propulsion* systems, the actual time history of the maneuver acceleration should be used for calculating Δv. Electric propulsion systems utilize accelerated plasma to generate an exhaust velocity, which could be up to a few orders of magnitude larger than the exhaust velocity in chemical thrusters. This is achieved by generating electromagnetic fields to ionize and accelerate gas, and, hence, requires additional electric power.

For an electric propulsion system, the required propulsion power would be

$$P = \frac{1}{2}\dot{m}U_e^2 = \frac{1}{2}\dot{m}\frac{T^2}{\dot{m}^2} = \frac{m^2u^2}{2\dot{m}} \tag{14.18}$$

The ratio between P and the supplied electric power, P_e, is the efficiency, i.e.

$$\eta = \frac{P}{P_e} \tag{14.19}$$

which leads to

$$\frac{dm}{m^2} = \frac{1}{2\eta P_e} u^2\, dt \tag{14.20}$$

Hence, upon integrating Eq. (14.20), we obtain

$$\frac{1}{m_f} - \frac{1}{m_0} = \frac{1}{2\eta P_e} \int_{t_0}^{t_f} u^2\, dt \tag{14.21}$$

Consequently, an important observation is that minimization of fuel for power-limited electric propulsion systems would be achieved by minimizing the integral of the maneuver acceleration squared, and not the integral of the maneuver acceleration magnitude, as we have seen in the case of constant exhaust velocity chemical thrusters.

14.4 Gravity Assist Maneuvers

An impulsive variation in velocity can be achieved by chemical thrusters, as we have seen in the previous section, but also by utilizing a *flyby*, which is a close approach to planets moving about the Sun. This process is referred to as a *gravity assist maneuver* (GAM). GAMs were explored already in the nineteenth century by Leverrier and Tisserand, who tried to explain the large perturbation of cometary orbits when passing near Jupiter.

Using GAMs as a supplementary propulsive scheme in spaceflight was developed by Lawden in England and simultaneously by Ehricke in the United States; there is still controversy as to who was the true pioneer of this idea (Flandro 2001). The GAM concept has enabled access to distant targets in the solar system and the successful accomplishment of challenging missions with a significant reduction in fuel mass. GAMs were used by Voyager, Galileo, Cassini, Ulysses, and other missions (Meltzer 2007).

Over the years, automated tools were developed for GAM itinerary design. An example for such an automatic tool is the Satellite Tour Design Program. This tool is able to find all GAM trajectories for a given set of launch dates and escape velocities. It has been extensively used for the preliminary investigation of interplanetary trajectories to Mars and the outer planets (Longuski and Williams 1991; Sims et al. 1997; Petropoulos et al. 2000).

14.4.1 Multiple Gravity Assists

We consider orbits with multiple GAMs in a system comprising the Sun (m_1), the flyby planet (m_2), and the spacecraft (m_3), all considered as point masses governed by Newton's gravitational law. In practice, the influence of m_3 on the motion of the other bodies can be neglected, which implies that the planet follows a Keplerian orbit about the Sun. If this orbit is circular, the resulting setup is the circular restricted three-body problem, which we have seen in Chap. 8.

The scaling of time, length, and mass units uses the planet's orbital radius and the inverse of the angular velocity as the length and time units, respectively, and m_1+m_2 as the mass unit. The only remaining parameter is the mass ratio $\mu = m_2/(m_1+m_2)$.

The problem may be further simplified when $\mu \ll 1$. In this case, the heliocentric spacecraft orbit is nearly Keplerian everywhere except a narrow vicinity of the planet, where the spacecraft experiences a short impulse of gravitational attraction. When $\mu \to 0$ simultaneously with the flyby distance, the impulse duration tends to zero, and the resulting trajectory is similar to an ideal elastic collision. This asymptotic case is called, following Poincaré (1899), the *problem of trajectories with consecutive collisions* (an alternative name is *zero-point patched conic model*). It deals with a chain of collision trajectories, i.e., Keplerian arcs having collisions at one or both ends.

The geometry of this problem is depicted in Fig. 14.1. In this figure, $\mathbf{v}_{pl}$ and $\mathbf{v}$ are the heliocentric velocities of the planet and the spacecraft, respectively. The

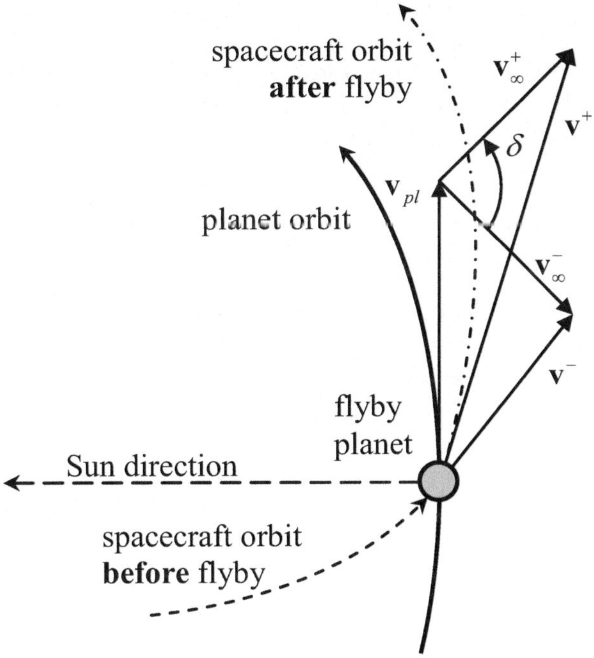

Fig. 14.1 Asymptotic case of flyby: a collision model

hyperbolic excess velocity vector $\mathbf{v}_\infty$ (see Sect. 5.7 for the definition of this velocity) constitutes the velocity of the spacecraft relative to the planet (at infinity). The angle δ is the turning angle of $\mathbf{v}_\infty$. This angle is identical to the angle between the asymptotes of the hyperbolic trajectory, as seen in Fig. 5.6.

Letting the superscripts $(\cdot)^-$ and $(\cdot)^+$ denote the state before and after the flyby, respectively, we can write, based on the conservation of energy along a hyperbolic trajectory,

$$\|\mathbf{v}_\infty^-\| = \|\mathbf{v}_\infty^+\| = v_\infty \tag{14.22}$$

which means that the magnitude of $\mathbf{v}_\infty$ is fixed, although the direction of $\mathbf{v}_\infty$ changes because of the planetary encounter.

14.4.2 Concatenation Rules

In the collision model setup, a multiple gravity assist interplanetary transfer consists of segments, where each segment is patched to the following one by a GAM. In case of an unpowered segment, the transfer is ballistic, and the spacecraft flies on a Keplerian arc; in case of a powered segment, the transfer consists of several Keplerian arcs patched by *deep-space maneuvers*. The arcs before and after each of the flybys must obey the following *concatenation rules*: (i) preservation of relative velocity with respect to the flyby planet; (ii) feasible flyby altitude and (iii) synchronization, meaning that at the end of the arc the spacecraft has to encounter the flyby planet.

The first concatenation rule for a circular planet orbit was described by Tisserand. This rule, called *Tisserand's criterion*, can be obtained from the conservation of Jacobi's constant, see Chap. 8. It reads

$$-\frac{v_\infty^2}{\mu_\odot} + \frac{3}{r_{pl}} = 2\sqrt{\frac{p}{r_{pl}^3}}\cos i + \frac{1}{a} \tag{14.23}$$

where r_{pl} is the orbital radius of the planet, $\mu_\odot$ is the gravitational parameter of the Sun, a is the semimajor axis, and p is the parameter. Based on Eq. (14.22), Eq. (14.23) leads to a connection between the orbital elements of the spacecraft before and after the flyby,

$$2\sqrt{\frac{p^-}{r_{pl}^3}}\cos i^- + \frac{1}{a^-} = 2\sqrt{\frac{p^+}{r_{pl}^3}}\cos i^+ + \frac{1}{a^+} \tag{14.24}$$

If the planet's orbit is elliptic, Tisserand's criterion can be rewritten in terms of the radial and transverse velocities of the spacecraft, denoted by v_r and v_θ, respectively, and the radial and transverse velocities of the planet, $v_{pl,r}$ and $v_{pl,\theta}$,

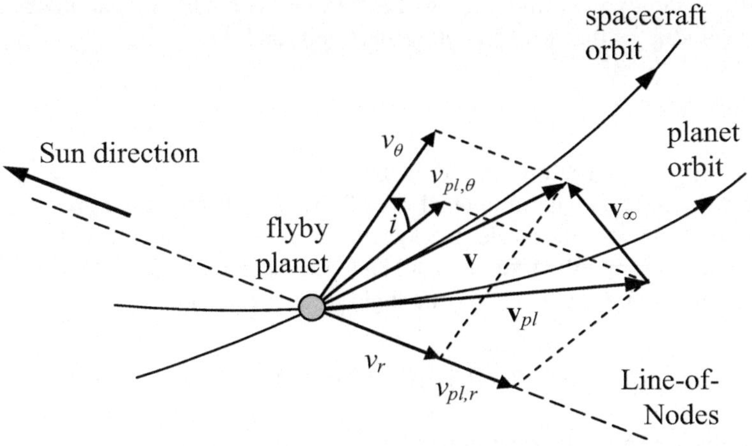

Fig. 14.2 Velocity diagram corresponding to the flyby moment, with the inclination of the spacecraft's orbit, i, measured relative to the planet's orbit

respectively, as seen in Fig. 14.2. The expressions for these velocities are

$$v_r = e\sin f\sqrt{\frac{\mu}{a\left(1-e^2\right)}},\quad v_\theta = \frac{\sqrt{\mu a(1-e^2)}}{r} \tag{14.25}$$

where f is the true anomaly, and r is the orbital radius, given by the usual conic equation

$$r = \frac{a(1-e^2)}{1+e\cos f} \tag{14.26}$$

The modified version of the Tisserand criterion, as obtained by Pisarevsky et al. (2007), is given by

$$v_\infty^2 = \left(v_r - v_{pl,r}\right)^2 + v_\theta^2 + v_{pl,\theta}^2 - 2v_\theta v_{pl,\theta}\cos i \tag{14.27}$$

The second concatenation rule requires that the flyby altitude (the minimal distance to the planet during a hyperbolic flyby) be above some allowed threshold. The flyby altitude, given an elliptic planetary orbit, can be written as (Pisarevsky et al. 2007)

$$h_{fb} = \frac{\mu_{pl}}{v_\infty^2}\left[\frac{2v_\infty}{\sqrt{\Delta v_r^2 + \Delta v_\theta^2 + 4v_\theta^- v_\theta^+ \sin^2(\Delta i/2)}} - 1\right] - R_{pl} \tag{14.28}$$

where h_{fb} is the flyby altitude, μ_{pl} and R_{pl} are the gravitational parameter and the equatorial radius of the flyby planet, respectively, and

$$\Delta v_r \triangleq v_r^+ - v_r^-, \; \Delta v_\theta \triangleq v_\theta^+ - v_\theta^-, \; \Delta i \triangleq i^+ - i^- \tag{14.29}$$

Therefore, two arcs (before and after the flyby, see Fig. 14.1) can be patched if $h_{fb} > h_m$, where h_m is the minimal allowed flyby altitude above the planet.

The third concatenation rule, the synchronization condition, establishes a coupling between the angular size of an arc and the respective flight time. According to Lambert's theorem (see Sect. 6.5), the flight time over an arc is

$$t = \sqrt{a^3/\mu_\odot}\,[(\alpha - \sin\alpha) - (\beta - \sin\beta)] \tag{14.30}$$

The variables α and β are determined by

$$r_1 + r_2 + c = 4a\sin^2(\alpha/2) \tag{14.31}$$

and

$$r_1 + r_2 - c = 4a\sin^2(\beta/2) \tag{14.32}$$

where

$$c \triangleq \|\mathbf{r}_2 - \mathbf{r}_1\| = \sqrt{r_1^2 + r_2^2 - 2r_1r_2\cos\theta} \tag{14.33}$$

The vectors $\mathbf{r}_1$ and $\mathbf{r}_2$ are the initial and final position vectors for the arc, $r_1 = \|\mathbf{r}_1\|$, $r_2 = \|\mathbf{r}_2\|$, and θ is the transfer angle of the arc, as shown in Fig. 14.3.

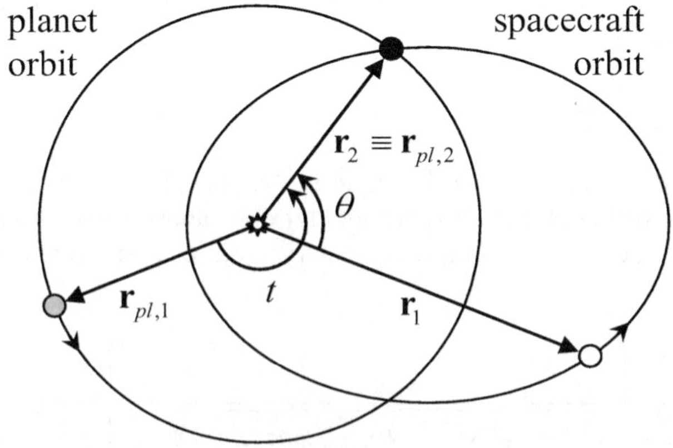

Fig. 14.3 A geometric representation of the synchronization condition

The procedure of obtaining the arc parameters guaranteeing an encounter between the spacecraft and flyby planet is as follows. For a given initial position of the spacecraft, $\mathbf{r}_1$, (see Fig. 14.3), the position of the next flyby planet, $\mathbf{r}_{pl,1}$, and the time until the encounter, t, calculate the final position vector, $\mathbf{r}_2 \equiv \mathbf{r}_{pl,2}$, using the known orbital parameters of the planet. Then, from $\mathbf{r}_1$ and $\mathbf{r}_2$, calculate the transfer angle, θ, using Eq. (14.33). Finally, using Eqs. (14.31) and (14.32), calculate all the remaining transfer parameters. A more detailed analysis of the multiple GAM problem was performed by Pisarevsky and Gurfil (2009).

14.5 Optimization of Orbits

The term *optimization of orbits* refers to designing maneuvers that are efficient in some sense. The most common performance measure in space systems design is the consumed fuel mass. Since launch and operation cost are directly related to the satellite mass, minimization of fuel consumption is vital. Indeed, most of the orbit optimization problems have focused on fuel minimization (Ross 2006; Conway 2010). Other performance metrics could be the transfer time from one orbit to another. In this section, we will outline some basic optimization tools. A more thorough discussion of optimal orbit transfers is provided in Chap. 15.

14.5.1 Static Optimization

If the control of orbits is applied using impulsive maneuvers, the orbit equations can often be described by algebraic relations. In this case, derivation of optimized orbits is based on *static optimization*. This means that differential equations are not used in solving for the optimal maneuvers. The purpose of the following discussion is to introduce a method for solving static optimization problems using the formalism of *Lagrange multipliers*. The method of Lagrange multipliers has been generalized by the *Karush-Kuhn-Tucker conditions* (Haykin 2008), which can also take into account inequality constraints. However, here we will use the classical formalism. To that end, consider the following problem:

$$\begin{aligned}
&\text{minimize} \quad J(\mathbf{x}) \\
&\text{subject to} \\
&g_k(\mathbf{x}) = 0, \; k = 1, \ldots, m
\end{aligned} \tag{14.34}$$

In Eq. (14.34), J is the *cost function*, $\mathbf{x}$ is an n-dimensional vector of optimization variables, $[x_1, x_2, \ldots, x_n]^T$, and g_k are m equality constraints. Now, we define an augmented cost function, J_A, as

$$J_A(\mathbf{x}, \boldsymbol{\lambda}) = J(\mathbf{x}) + \sum_{k=1}^{m} \lambda_k g_k(\mathbf{x}) \tag{14.35}$$

where $\boldsymbol{\lambda} = [\lambda_1, \lambda_2, \ldots, \lambda_m]^T$, $m < n$, is a vector of Lagrange multipliers. If J and g_k have continuous first partial derivatives and the gradients of g_k do not vanish on the domain of J (the domain of J is assumed an open set containing all points satisfying the constraints), then the stationary points of J_A are determined by

$$\nabla J_A = \left[\left(\frac{\partial J_A}{\partial \mathbf{x}}\right)^T, \left(\frac{\partial J_A}{\partial \boldsymbol{\lambda}}\right)^T \right] = \mathbf{0} \tag{14.36}$$

Equation (14.36) gives $m + n$ unique equations for the unknowns $\mathbf{x}^\star$ and $\boldsymbol{\lambda}^\star$, constituting the solutions for the optimization parameters and Lagrange multipliers at the critical point.

To determine whether some stationary point $\{\mathbf{x}^\star, \boldsymbol{\lambda}^\star\}$ is a minimum, one must examine the *Hessian* matrix of the Lagrangian. If there exist vectors $\mathbf{x}^\star$ and $\boldsymbol{\lambda}^\star$ such that

$$\nabla J_A(\mathbf{x}^\star, \boldsymbol{\lambda}^\star) = \mathbf{0} \tag{14.37}$$

and if, for twice continuously-differentiable J and g_k,

$$(-1)^m \det \begin{bmatrix} \frac{\partial^2 J_A(\mathbf{x}^\star, \boldsymbol{\Lambda}^\star)}{\partial x_1 \partial x_1} & \cdots & \frac{\partial^2 J_A(\mathbf{x}^\star, \boldsymbol{\Lambda}^\star)}{\partial x_1 \partial x_p} & \frac{\partial g_1(\mathbf{x}^\star)}{\partial x_1} & \cdots & \frac{\partial g_m(\mathbf{x}^\star)}{\partial x_1} \\ \vdots & \vdots & \vdots & \vdots & \vdots & \vdots \\ \frac{\partial^2 J_A(\mathbf{x}^\star, \boldsymbol{\Lambda}^\star)}{\partial x_p \partial x_1} & \cdots & \frac{\partial^2 J_A(\mathbf{x}^\star, \boldsymbol{\Lambda}^\star)}{\partial x_p \partial x_p} & \frac{\partial g_1(\mathbf{x}^\star)}{\partial x_p} & \cdots & \frac{\partial g_m(\mathbf{x}^\star)}{\partial x_p} \\ \frac{\partial g_1(\mathbf{x}^\star)}{\partial x_1} & \cdots & \frac{\partial g_1(\mathbf{x}^\star)}{\partial x_p} & 0 & \cdots & 0 \\ \vdots & \vdots & \vdots & \vdots & \vdots & \vdots \\ \frac{\partial g_m(\mathbf{x}^\star)}{\partial x_1} & \cdots & \frac{\partial g_m(\mathbf{x}^\star)}{\partial x_p} & 0 & \cdots & 0 \end{bmatrix} > 0 \tag{14.38}$$

for $p = m + 1, \ldots, n$, then J has a local minimum at $\mathbf{x}^\star$ such that (Alfriend et al. 2010; pp. 46–47)

$$g_k(\mathbf{x}^\star) = 0, \quad k = 1, \ldots, m \tag{14.39}$$

An application of static optimization tools for impulsive orbit transfer design is discussed in Chap. 15.

14.5.2 Dynamic Optimization

When orbital maneuvers are performed using *continuous thrust*, as discussed in Sect. 14.3, orbit optimization problems can no longer be treated as static problems. In this case, we must take the differential equations of motion into account, and treat them as constraints, giving rise to *dynamic optimization*.

The underlying dynamical model is that of system (14.1). It is required to find an m-dimensional particular control $\mathbf{u}^\star$, such that the system will satisfy a set of q final conditions at the final time t_f, formulated as

$$\boldsymbol{\psi}\left[\mathbf{x}\left(t_f\right), t_f\right] = \mathbf{0} \tag{14.40}$$

while at the same time minimizing a cost function of the form

$$J = \Phi\left[\mathbf{x}(t_f), t_f\right] + \int_{t_0}^{t_f} L(\mathbf{x}, \mathbf{u}, t)dt \tag{14.41}$$

where L is commonly referred to as the *Lagrangian*. For $\Phi \equiv 0$ and $L \equiv 1$, the cost function J represents a *minimum-time problem*.

System (14.1) is written as the equality constraint

$$\mathbf{f}(\mathbf{x}, \mathbf{u}) - \dot{\mathbf{x}} = \mathbf{0} \tag{14.42}$$

Because this constraint has to be satisfied for the interval $[t_0, t_f]$, it is added to the integrand of J, and the augmented cost becomes

$$J_A = \Phi\left[\mathbf{x}(t_f), t_f\right] + \int_{t_0}^{t_f} \left\{L(\mathbf{x}, \mathbf{u}, t) + \boldsymbol{\lambda}^T \left[\mathbf{f}(\mathbf{x}, \mathbf{u}) - \dot{\mathbf{x}}\right]\right\} dt \tag{14.43}$$

where as previously $\boldsymbol{\lambda}$ are the Lagrange multipliers, also referred to as the *co-states*. We can omit the subscript A, because J and J_A are equal when the dynamic constraints are satisfied. To obtain the necessary conditions for a minimum, we define a Hamiltonian as

$$H(\mathbf{x}, \mathbf{u}, \boldsymbol{\lambda}, t) = L(\mathbf{x}, \mathbf{u}, t) + \boldsymbol{\lambda}^T \mathbf{f}(\mathbf{x}, \mathbf{u}) \tag{14.44}$$

so that

$$J = \Phi\left[\mathbf{x}(t_f), t_f\right] + \int_{t_0}^{t_f} \left[H(\mathbf{x}, \mathbf{u}, \boldsymbol{\lambda}, t) - \boldsymbol{\lambda}^T \dot{\mathbf{x}}\right] dt \tag{14.45}$$

A necessary condition for a minimum is that the first variation of J is nullified, that is

$$\delta J = \frac{\partial J}{\partial \mathbf{u}}\delta\mathbf{u} + \frac{\partial J}{\partial \mathbf{x}}\delta\mathbf{x}(\delta\mathbf{u}) = 0 \tag{14.46}$$

where $\delta\mathbf{x}(\delta\mathbf{u})$ is a *functional.* i.e., the function $\delta\mathbf{x}$ depends on the function $\delta\mathbf{u}$.

Using the technique of integration by parts, Eq. (14.46) leads to conditions for local stationary points known as the *Euler-Lagrange equations* (Stengel 1994; pp. 202–208),

$$\dot{\boldsymbol{\lambda}}^T = -\frac{\partial H}{\partial \mathbf{x}}, \quad \boldsymbol{\lambda}^T(t_f) = \left.\frac{\partial \Phi\left[\mathbf{x}(t), t\right]}{\partial \mathbf{x}}\right|_{t=t_f} \tag{14.47a}$$

$$\frac{\partial H}{\partial \mathbf{u}} = \mathbf{0} \tag{14.47b}$$

which constitute a *two-point boundary value problem* when appended with the initial value problem (14.1).

14.6 Linear Orbit Control

In some cases, orbit control is performed with respect to some given reference. For example, an approach of a transfer vehicle to the ISS requires that the relative motion is controlled to guarantee a safe approach. When the two objects are relatively close, i.e their distance is a few orders of magnitude smaller than the orbital radius, the relative dynamics can be linearized to simplify the maneuver design. Other examples in which proximity plays a major role include *satellite formation flying*, *satellite rendezvous*, and *satellite docking* (Alfriend et al. 2010; pp. 1–11).

Satellite *rendezvous and docking* (RVD) have been playing an important role in space activities since 1966, when Gemini 8 docked on an unmanned Agena Target Vehicle. In the following year, the Soviets carried out the first automated, unmanned RVD mission (Lamkin and Mccandless 1990).

The ability to approach an orbital target has been attracting increasing attention, as the concepts of in-orbit servicing, refueling, repairing, and de-orbiting have emerged. This technology has been demonstrated in several recent missions, including the Engineering Test Satellite VII (Kawano et al. 2001), Orbital Express (Weismuller and Leinz 2006), and Demonstration of Autonomous Rendezvous Technology (Rumford 2003).

Rendezvous is usually divided into two phases: A preparatory phase, often called *closing phase*, and a *final approach phase*, leading to the mating conditions (Fehse 2005). The closing phase reduces the range to the target from several kilometers to a few hundreds of meters using impulsive maneuvers; each trajectory segment

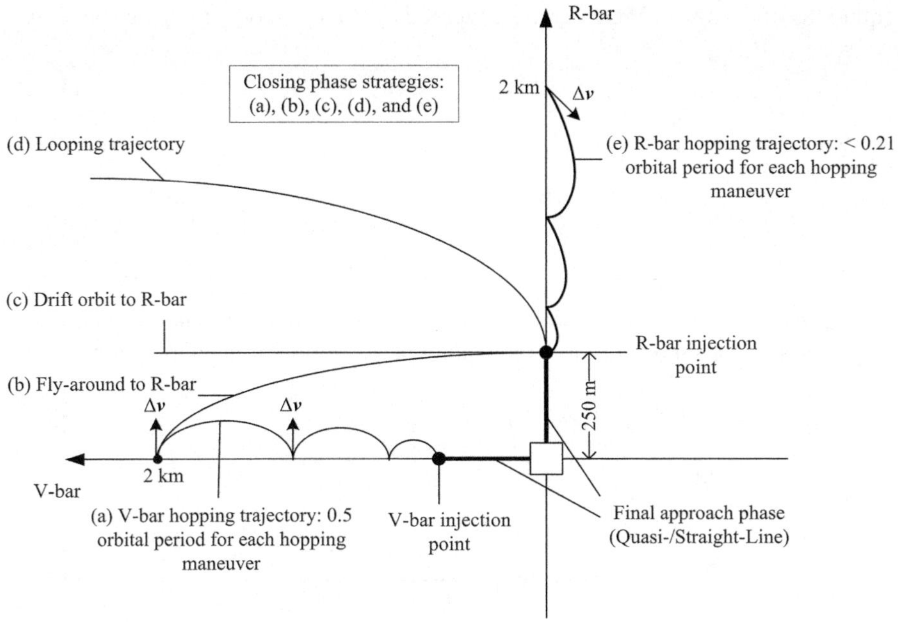

Fig. 14.4 Closing-phase strategies for rendezvous

between two impulsive maneuvers is usually referred to as *hopping*. Various closing phase strategies are depicted in Fig. 14.4.

The closing phase can be executed by approaching the target satellite either from the radial direction, called *R-bar*, or the transverse direction, referred to as *V-bar*, as shown in Fig. 14.4. Hopping along the V-bar to the injection point for the final approach phase can be achieved by applying radial maneuvers, with the duration being half an orbital period (Fehse 2005; Wenfei et al. 2012). This is denoted as Case (a) in Fig. 14.4.

Alternative strategies lead to an injection point for the final approach phase which is located along the R-bar. These strategies include performing an additional fly-around maneuver from V-bar to R-bar (Case (b) in Fig. 14.4); utilizing a natural drift orbit (Case (c) in Fig. 14.4); or using a looping trajectory (Case (d) in Fig. 14.4) (Fehse 2005). A typical example of the fly-around R-bar injection is the Shuttle rendezvous with the ISS (Woffinden and Geller 2007).

In this section, we develop the necessary modelling tools for designing a single-hop rendezvous maneuver. The first step is to develop linearized equations of motion.

To that end, consider two satellites orbiting the same primary. One of the satellites will be termed *chief*; all variables related to the chief will be denoted by $(\cdot)_0$. The other would be the *deputy*; all variables related to the deputy will be denoted by $(\cdot)_1$. The chief serves as the non-maneuvering reference point. The equations of motion of the chief are governed by the two-body dynamics, namely

$$\ddot{\mathbf{r}}_0 = -\frac{\mu}{r_0^3}\mathbf{r}_0 \tag{14.48}$$

where as usual

$$r_0 = \|\mathbf{r}_0\| = \frac{a_0(1-e_0^2)}{(1+e_0\cos f_0)} \tag{14.49}$$

and a_0, e_0, f_0 are the chief's orbit semimajor axis, eccentricity, and true anomaly, respectively. In a similar fashion, the equations of motion of the deputy are

$$\ddot{\mathbf{r}}_1 = -\frac{\mu}{r_1^3}\mathbf{r}_1 \tag{14.50}$$

where

$$r_1 = \|\mathbf{r}_1\| = \frac{a_1(1-e_1^2)}{(1+e_1\cos f_1)} \tag{14.51}$$

and a_1, e_1, f_1 are the deputy's orbit semimajor axis, eccentricity, and true anomaly, respectively. Let

$$\mathbf{r} = \mathbf{r}_1 - \mathbf{r}_0 \tag{14.52}$$

denote the position of the deputy relative to the chief. Subtracting Eq. (14.48) from Eq. (14.50) yields

$$\ddot{\mathbf{r}} = -\frac{\mu(\mathbf{r}_0+\mathbf{r})}{\|\mathbf{r}_0+\mathbf{r}\|^3} + \frac{\mu}{r_0^3}\mathbf{r}_0 \tag{14.53}$$

We now define a rotating frame centered at the chief, so that the x axis points radially outward, i.e. $\hat{\mathbf{i}} = \mathbf{r}_0/\|\mathbf{r}_0\|$, the z axis coincides with the chief's angular momentum vector, that is $\hat{\mathbf{k}} = \mathbf{h}_0/|\mathbf{h}_0\|$, and $\hat{\mathbf{j}} = \hat{\mathbf{k}} \times \hat{\mathbf{i}}$. This rotating frame is called a *local-vertical local-horizontal* (LVLH) frame. In the LVLH frame, the equation of motion of the deputy is (compare to Eq. (8.1))

$$\mathbf{a} + 2\boldsymbol{\omega} \times \mathbf{v} + \boldsymbol{\omega} \times (\boldsymbol{\omega} \times \mathbf{r}) + \dot{\boldsymbol{\omega}} \times \mathbf{r} = -\frac{\mu(\mathbf{r}_0+\mathbf{r})}{\|\mathbf{r}_0+\boldsymbol{\rho}\|^3} + \frac{\mu}{r_0^3}\mathbf{r}_0 \tag{14.54}$$

where $\boldsymbol{\omega} = \dot{f}_0\hat{\mathbf{k}}$ is the angular velocity vector of the LVLH frame with respect to the inertial reference. Expressing the vectors in Eq. (14.54) in terms of their Cartesian components, we have

$$\mathbf{r} = x\hat{\mathbf{i}} + y\hat{\mathbf{j}} + z\hat{\mathbf{k}} \tag{14.55}$$

$$\mathbf{v} = \dot{x}\hat{\mathbf{i}} + \dot{y}\hat{\mathbf{j}} + \dot{z}\hat{\mathbf{k}} \tag{14.56}$$

$$\mathbf{a} = \ddot{x}\hat{\mathbf{i}} + \ddot{y}\hat{\mathbf{j}} + \ddot{z}\hat{\mathbf{k}} \tag{14.57}$$

$$\boldsymbol{\omega} \times \mathbf{v} = -\dot{f}_0\left[\dot{y}\hat{\mathbf{i}} - \dot{x}\hat{\mathbf{j}}\right] \tag{14.58}$$

$$\boldsymbol{\omega} \times (\boldsymbol{\omega} \times \mathbf{r}) = -\dot{f}_0^2\left[x\hat{\mathbf{i}} + y\hat{\mathbf{j}}\right] \tag{14.59}$$

$$\dot{\boldsymbol{\omega}} \times \mathbf{r} = -\ddot{f}_0\left[y\hat{\mathbf{i}} - x\hat{\mathbf{j}}\right] \tag{14.60}$$

which, upon substitution into Eq. (14.54) yields

$$\ddot{x} - 2\dot{f}_0\dot{y} - \ddot{f}_0 y - \dot{f}_0^2 x = -\frac{\mu(r_0 + x)}{[(r_0 + x)^2 + y^2 + z^2]^{\frac{3}{2}}} + \frac{\mu}{r_0^2} \tag{14.61}$$

$$\ddot{y} + 2\dot{f}_0\dot{x} + \ddot{f}_0 x - \dot{f}_0^2 y = -\frac{\mu y}{[(r_0 + x)^2 + y^2 + z^2]^{\frac{3}{2}}} \tag{14.62}$$

$$\ddot{z} = -\frac{\mu z}{[(r_0 + x)^2 + y^2 + z^2]^{\frac{3}{2}}} \tag{14.63}$$

Equations (14.61)–(14.63) together with the two-body relations

$$\ddot{r}_0 = r_0\dot{f}_0^2 - \frac{\mu}{r_0^2}, \; \ddot{f}_0 = -\frac{2\dot{r}_0\dot{f}_0}{r_0} \tag{14.64}$$

constitute a 10-dimensional system of nonlinear differential equations. For $\ddot{f}_0 \neq 0$, these equations admit a single relative equilibrium at $x = y = z = 0$, meaning that the deputy spacecraft will appear stationary in the chief's frame, if and only if their positions coincide on a given elliptic orbit. We will later see that the single relative equilibrium is transformed into infinitely many relative equilibria, if the chief is assumed to follow a circular reference orbit.

If there are external (differential) perturbations, denoted by $\mathbf{d} = [d_x, d_y, d_z]^T$, and (differential) control forces, $\mathbf{u} = [u_x, u_y, u_z]^T$, they are introduced into Eqs. (14.61)–(14.63) in the following manner:

$$\ddot{x} - 2\dot{f}_0\dot{y} - \ddot{f}_0 y - \dot{f}_0^2 x = -\frac{\mu(r_0 + x)}{[(r_0 + x)^2 + y^2 + z^2]^{\frac{3}{2}}} + \frac{\mu}{r_0^2} + d_x + u_x \tag{14.65}$$

$$\ddot{y} + 2\dot{f}_0\dot{x} + \ddot{f}_0 x - \dot{f}_0^2 y = -\frac{\mu y}{[(r_0 + x)^2 + y^2 + z^2]^{\frac{3}{2}}} + d_y + u_y \tag{14.66}$$

$$\ddot{z} = -\frac{\mu z}{[(r_0 + x)^2 + y^2 + z^2]^{\frac{3}{2}}} + d_z + u_z \tag{14.67}$$

A simpler form of the relative motion equations can be derived, if we assume that the chief follows a circular orbit. In many practical cases this is a realistic assumption. In this case, $\dot{f}_0 = n_0$ is constant, $r_0 = a_0$, and, hence, $\ddot{f}_0 = 0$. Substituting into Eqs. (14.61)–(14.63) results in

$$\ddot{x} - 2n_0\dot{y} - n_0^2 x = -\frac{\mu(a_0 + x)}{[(a_0 + x)^2 + y^2 + z^2]^{\frac{3}{2}}} + \frac{\mu}{a_0^2} \tag{14.68}$$

$$\ddot{y} + 2n_0\dot{x} - n_0^2 y = -\frac{\mu y}{[(a_0 + x)^2 + y^2 + z^2]^{\frac{3}{2}}} \tag{14.69}$$

$$\ddot{z} = -\frac{\mu z}{[(a_0 + x)^2 + y^2 + z^2]^{\frac{3}{2}}} \tag{14.70}$$

The equilibria ($\dot{x} = \ddot{x} = \dot{y} = \ddot{y} = \dot{z} = \ddot{z} = 0$) satisfy

$$z = 0,\ (x + a_0)^2 + y^2 = a_0^2 \tag{14.71}$$

Equation (14.71) defines a circle that coincides with the chief's orbit. This result reflects the fact that the deputy will appear stationary in a chief-fixed frame, if the deputy is co-located on the circular orbit of the chief. This type of relative motion is referred to as *co-orbital motion*.

A straightforward approach to obtain the linearized equations is to expand the right-hand side of Eqs. (14.68)–(14.70) into a Taylor series about the origin of the LVLH frame. Taking only the first-order terms in x/a_0, y/a_0, z/a_0, and denoting $n_0 = \sqrt{\mu/a_0^3}$, we get

$$-\frac{\mu(a_0 + x)}{[(a_0 + x)^2 + y^2 + z^2]^{\frac{3}{2}}} \approx n_0^2(2x - a_0) \tag{14.72}$$

$$-\frac{\mu y}{[(a_0 + x)^2 + y^2 + z^2]^{\frac{3}{2}}} \approx -n_0^2 y \tag{14.73}$$

$$-\frac{\mu z}{[(a_0 + x)^2 + y^2 + z^2]^{\frac{3}{2}}} \approx -n_0^2 z \tag{14.74}$$

Rearranging and omitting the subscript 0 (so that $n \equiv n_0$ and $a \equiv a_0$) yields

$$\ddot{x} - 2n\dot{y} - 3n^2 x = 0 \tag{14.75}$$

$$\ddot{y} + 2n\dot{x} = 0 \tag{14.76}$$

$$\ddot{z} + n^2 z = 0 \tag{14.77}$$

Equations (14.75)–(14.77) are called the *Hill-Clohessy-Wiltshire* (HCW) equations. The nonhomogeneous forms of Eqs. (14.75)–(14.77) are

$$\ddot{x} - 2n\dot{y} - 3n^2x = d_x + u_x \tag{14.78}$$

$$\ddot{y} + 2n\dot{x} = d_y + u_y \tag{14.79}$$

$$\ddot{z} + n^2z = d_z + u_z \tag{14.80}$$

where $[d_x, d_y, d_z]^T$ and $[u_x, u_y, u_z]^T$ are, respectively, the vectors of environmental perturbations and control accelerations.

We often normalize the relative coordinates by the radius of the reference orbit, a, and the angular velocities by n, so that in normalized form the unforced HCW equations (14.75)–(14.77) become

$$\bar{x}'' - 2\bar{y}' - 3\bar{x} = 0 \tag{14.81}$$

$$\bar{y}'' + 2\bar{x}' = 0 \tag{14.82}$$

$$\bar{z}'' + \bar{z} = 0 \tag{14.83}$$

where $(\cdot)'$ denotes differentiation with respect to normalized time.

Choosing the state vector $\mathbf{x} = [x, y, z, \dot{x}, \dot{y}, \dot{z}]^T$, Eqs. (14.75)–(14.77) assume the form

$$\dot{\mathbf{x}}(t) = A\mathbf{x}(t) \tag{14.84}$$

where

$$A = \begin{bmatrix} 0 & 0 & 0 & 1 & 0 & 0 \\ 0 & 0 & 0 & 0 & 1 & 0 \\ 0 & 0 & 0 & 0 & 0 & 1 \\ 3n^2 & 0 & 0 & 0 & 2n & 0 \\ 0 & 0 & 0 & -2n & 0 & 0 \\ 0 & 0 & -n^2 & 0 & 0 & 0 \end{bmatrix}, \tag{14.85}$$

and the initial conditions are $\mathbf{x}(0) = [x(0), y(0), z(0), \dot{x}(0), \dot{y}(0), \dot{z}(0)]^T$. The eigenvalues of A are $\{\pm nj, \pm nj, 0, 0\}$, so a secular mode is expected to appear in the solution according to the discussion in Sect. 14.2.

Solving the HCW equations is straightforward. We follow Eq. (14.7) and formulate the solution in terms of the transition matrix,

$$\mathbf{x}(t) = \mathrm{e}^{At}\mathbf{x}(0) \tag{14.86}$$

where

$$e^{At} = \begin{bmatrix} 4-3\,c_{nt} & 0 & 0 & \frac{s_{nt}}{n} & \frac{2}{n}-\frac{2c_{nt}}{n} & 0 \\ -6\,nt+6\,s_{nt} & 1 & 0 & -\frac{2}{n}+\frac{2c_{nt}}{n} & \frac{4s_{nt}}{n}-3\,t & 0 \\ 0 & 0 & c_{nt} & 0 & 0 & \frac{s_{nt}}{n} \\ 3\,ns_{nt} & 0 & 0 & c_{nt} & 2\,s_{nt} & 0 \\ -6\,n+6\,nc_{nt} & 0 & 0 & -2\,s_{nt} & -3+4\,c_{nt} & 0 \\ 0 & 0 & -ns_{nt} & 0 & 0 & c_{nt} \end{bmatrix} \tag{14.87}$$

and $c_{nt} \equiv \cos(nt)$, $s \equiv \sin(nt)$. We can now determine, by substituting Eq. (14.87) into Eq. (14.86), the solutions to the relative position and velocity components,

$$x(t) = \left[4x(0) + \frac{2\dot{y}(0)}{n}\right] + \frac{\dot{x}(0)}{n}\sin(nt) - \left[3x(0) + \frac{2\dot{y}(0)}{n}\right]\cos(nt) \tag{14.88}$$

$$y(t) = -[6nx(0) + 3\dot{y}(0)]t + \left[y(0) - \frac{2\dot{x}(0)}{n}\right] + \left[6x(0) + \frac{4\dot{y}(0)}{n}\right]\sin(nt) + \frac{2\dot{x}(0)}{n}\cos(nt), \tag{14.89}$$

$$z(t) = \frac{\dot{z}(0)}{n}\sin(nt) + z(0)\cos(nt), \tag{14.90}$$

$$\dot{x}(t) = \dot{x}(0)\cos(nt) + [3x(0)n + 2\dot{y}(0)]\sin(nt) \tag{14.91}$$

$$\dot{y}(t) = -[6nx(0) + 3\dot{y}(0)] + [6x(0)n + 4\dot{y}(0)]\cos(nt) - 2\dot{x}(0)\sin(nt) \tag{14.92}$$

$$\dot{z}(t) = \dot{z}(0)\cos(nt) - z(0)n\sin(nt) \tag{14.93}$$

It is now possible to find the impulsive velocity corrections required for rendezvous. This will be performed in two stages: We will initially find an impulsive maneuver that nullifies the final relative distance between the chief and the deputy. This process is termed *targeting* or *guidance*. At the final time, once the relative distance has been nullified, we will determine the additional impulsive maneuver required to cancel the relative velocity and complete the rendezvous.

Designating the moment of the targeting maneuver by $t_0 = 0$, the velocity impulse, $\Delta\mathbf{v}_1$, satisfies

$$\Delta\mathbf{v}_1 = \mathbf{v}^+(0) - \mathbf{v}^-(0) \tag{14.94}$$

where $\mathbf{v}^-(0)$ and $\mathbf{v}^+(0)$ are the relative velocity vectors before and after application of the targeting impulse, respectively. The vector $\mathbf{v}^+(0)$, which is the required initial velocity for targeting, can be found by substituting $x(t_f) = y(t_f) = z(t_f) = 0$ into Eqs. (14.88)–(14.90) and solving for $\dot{x}(0)$, $\dot{y}(0)$, $\dot{z}(0)$, given the initial relative position components $x(0)$, $y(0)$, $z(0)$:

$$\mathbf{v}^+(0) = \begin{bmatrix} \dot{x}(0) \\ \dot{y}(0) \\ \dot{z}(0) \end{bmatrix}^+ = \begin{bmatrix} -\dfrac{n\left[-4x(0)\sin(nt_f) + 3x(0)\,nt_f\cos(nt_f) + 2y(0) - 2y(0)\cos(nt_f)\right]}{-8 + 8\cos(nt_f) + 3nt_f\sin(nt_f)} \\ -\dfrac{n\left[-14x(0) + 14x(0)\cos(nt_f) + 6x(0)\,nt_f\sin(nt_f) - y(0)\sin(nt_f)\right]}{-8 + 8\cos(nt_f) + 3nt_f\sin(nt_f)} \\ -z(0)n\cot(nt_f) \end{bmatrix} \tag{14.95}$$

$\mathbf{v}^+(0)$ does not exist for all flight times, because $\dot{z}^+(0)$ is singular for $nt_f = k\pi$, $k = 0, 1, \ldots$, and the in-plane components $\dot{x}^+(0)$ and $\dot{y}^+(0)$ are singular at $nt_f = 2k\pi$, $k = 0, 1, \ldots$ and at additional points satisfying $8c_{nt_f} + 3nt_f s_{nt_f} = 8$, such as $nt_f = 2.8135\pi$ and $nt_f = 4.8906\pi$.

At t_f, when the deputy satellite reaches the chief satellite, its relative velocity must be nullified to guarantee rendezvous. Therefore, the required velocity impulse at impact must be equal in magnitude and opposite in sign to the final relative velocity,

$$\Delta\mathbf{v}_2 = -\mathbf{v}(t_f) \tag{14.96}$$

where $\mathbf{v}(t_f)$ is the final relative velocity, obtained by substituting $t = t_f$ into Eqs. (14.91)–(14.93),

$$\mathbf{v}(t_f) = \begin{bmatrix} \dot{x}(t_f) \\ \dot{y}(t_f) \\ \dot{z}(t_f) \end{bmatrix} = \begin{bmatrix} \dot{x}^+(0)\cos(nt_f) + [3x(0)n + 2\dot{y}^+(0)]\sin(nt_f) \\ -[6nx(0) + 3\dot{y}^+(0)] + [6x(0)n + 4\dot{y}^+(0)]\cos(nt_f) - 2\dot{x}^+(0)\sin(nt_f) \\ \dot{z}^+(0)\cos(nt_f) - z(0)n\sin(nt_f) \end{bmatrix} \tag{14.97}$$

The total velocity change required for a two-impulse rendezvous, i.e. a single hop, is

$$\Delta v = \|\Delta\mathbf{v}_1\| + \|\Delta\mathbf{v}_2\| \tag{14.98}$$

14.7 Low Earth Orbit Control

Low Earth orbits (LEO) are satellite orbits up to an altitude of about 1000 km. Satellites in LEO perform mainly Earth observation missions. Observation satellites are required to maintain their altitude throughout their lifetime. Other mission parameters of importance are Sun synchronization, required to maintain the same lighting conditions, and the ground trace, required for tracking the same regions on Earth in each pass. The orbital elements of the nominal orbit are chosen to satisfy these constraints. Active control, commonly referred to as *orbit keeping*, is required to keep the satellite on the nominal orbit under the effect of orbital perturbations. We will discuss some common impulsive maneuver methods for orbit keeping.

14.7.1 Altitude Correction

A basic orbital maneuver is keeping a fixed altitude, without controlling the particular location of the satellite on the orbit. The altitude tends to decrease due to the effect of atmospheric drag. When the altitude decreases below some pre-specified threshold, maneuvers are performed to increase the altitude.

Consider a circular orbit of radius r_1. It is required to change the radius to $r_2 > r_1$, as shown in Fig. 14.5. Both orbits are coplanar. Because there is no intersection

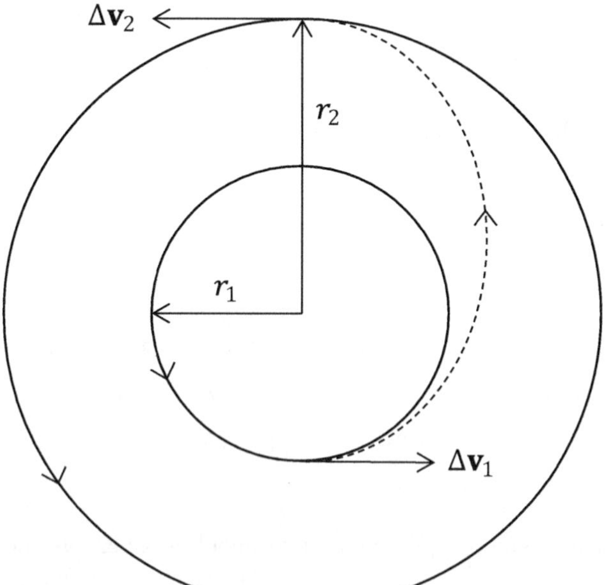

Fig. 14.5 Hohmann transfer for altitude correction

between the initial and final orbits, at least 2 impulsive maneuvers are needed. A maneuver sequence comprised of 2 impulses can be designed as follows. The first impulse, $\Delta\mathbf{v}_1$, is applied in the direction of the velocity of the initial orbit, $\hat{\mathbf{u}}_{\theta_1}$, transferring the satellite into an elliptic orbit whose perigee radius is equal to r_1 and apogee radius is equal to r_2. The second impulse, $\Delta\mathbf{v}_2$, is applied in the direction of the velocity at apogee, $\hat{\mathbf{u}}_{\theta_2}$, transferring the satellite to the higher final orbit. This sequence of maneuvers is called an *Hohmann transfer*. We will prove in Chap. 15 that the Hohmann transfer is in fact the fuel-optimal bi-impulsive transfer between 2 coplanar circular orbits.

The total required velocity increment can be calculated as follows. We first calculate the magnitude of the velocity at the perigee of the transfer ellipse, v_p, which can be obtained from the specific energy equation (4.51),

$$v_p = \sqrt{2\mu\left(\frac{1}{r_1} - \frac{1}{r_1 + r_2}\right)} \tag{14.99}$$

where we substituted $a = (r_1 + r_2)/2$. The magnitude of the first velocity impulse is the difference between v_p and the velocity on the initial circular orbit, and the direction is $\hat{\mathbf{u}}_{\theta_1}$,

$$\Delta\mathbf{v}_1 = \left(v_p - \sqrt{\frac{\mu}{r_1}}\right)\hat{\mathbf{u}}_{\theta_1} \tag{14.100}$$

In an equivalent manner, the magnitude of the velocity at the apogee of the transfer ellipse is

$$v_a = \sqrt{2\mu\left(\frac{1}{r_2} - \frac{1}{r_1 + r_2}\right)} \tag{14.101}$$

and the second velocity change, calculated as the difference between the velocity on the target orbit and the velocity at apogee, becomes

$$\Delta\mathbf{v}_2 = \left(\sqrt{\frac{\mu}{r_2}} - v_a\right)\hat{\mathbf{u}}_{\theta_2} \tag{14.102}$$

In both Eqs. (14.100) and (14.102), the velocity is increased. The direction of both impulses would be in the opposite direction if the altitude is to be reduced from r_2 to r_1. In both cases, however, the overall magnitude of the required maneuver is calculated as

$$\Delta v = \|\Delta\mathbf{v}_1\| + \|\Delta\mathbf{v}_2\| \tag{14.103}$$

Defining $\xi \triangleq r_1/r_2 < 1$ and substituting Eqs. (14.100) and (14.102) into Eq. (14.103) yields

$$\Delta v = \sqrt{\frac{\mu}{r_1}}\left(\sqrt{\frac{2}{\xi+1}}\,(1-\xi) - 1 + \sqrt{\xi}\right) \tag{14.104}$$

For small altitude corrections, namely $\Delta h = r_2 - r_1 \ll r_1$, Eq. (14.104) can be approximated as

$$\Delta v \approx \sqrt{\frac{\mu}{r_1}}\frac{\Delta h}{2r_1} \tag{14.105}$$

14.7.2 Frozen Orbit Control

As mentioned in Sect. 13.4, frozen orbits have constant mean semimajor axis, eccentricity, inclination, and argument of perigee. These orbits, whose mean element evolution is determined by Eqs. (13.55)–(13.56), can be obtained either at the critical inclination values, given by Eq. (13.48), or at $\omega = \pm\pi/2$, together with the eccentricity given by Eq. (13.58).

However, atmospheric drag and solar radiation pressure affect the semimajor axis, eccentricity, and argument of perigee. The semimajor axis can be corrected by using a velocity impulse in the direction of the orbital velocity. At the same time, a radial velocity impulse can correct both the eccentricity and argument of perigee. By using a proper maneuver timing along the orbit, a single velocity impulse can simultaneously correct all three elements.

To obtain the required maneuver, we consider a linearized form of the Gauss equations for near-circular orbits and impulsive maneuvers (Alfriend et al. 2010; p. 256), written in the NTW frame, using the nonsingular elements introduced in Eq. (10.125), namely $\eta_1 = e\sin\omega, \zeta_1 = e\cos\omega, \bar{u} = \omega + M$,

$$\Delta a \approx \frac{2}{n}\Delta v_T \tag{14.106a}$$

$$\Delta \bar{u} \approx -2\gamma\Delta v_N - \gamma\sin u(\cot i)\Delta v_W \tag{14.106b}$$

$$\Delta i \approx \gamma(\cos u)\Delta v_W \tag{14.106c}$$

$$\Delta\zeta_1 \approx \gamma(\sin u)\Delta v_N + 2\gamma(\cos u)\Delta v_T \tag{14.106d}$$

$$\Delta\eta_1 \approx -\gamma(\cos u)\Delta v_N + 2\gamma(\sin u)\Delta v_T \tag{14.106e}$$

$$\Delta\Omega \approx \frac{\gamma\sin u}{\sin i}\Delta v_W \tag{14.106f}$$

where $\Delta v_N, \Delta v_T$, and Δv_W are the magnitudes of the impulse components in the radial, tangential, and out-of-plane directions, respectively, which are small compared to the orbital velocity, $n = \sqrt{\mu/a^3}$, and $\gamma = \sqrt{a/\mu} = 1/v$. Thus, we obtain 2 equations with 2 unknowns: The magnitude of the radial pulse, Δv_N, and its location along the orbit, u,

$$\Delta e \cos u + e\Delta\omega = \frac{\Delta a}{a}, \quad \Delta e \sin u - e\Delta\omega \cos u = \frac{\Delta v_N}{na} \tag{14.107}$$

14.7.3 Sun-synchronous Orbit Control

To maintain the SSO as defined in Sect. 13.5, the nodal drift of the satellite orbit should match the Sun's right ascension rate, and the relative angular position should be equal to a given local hour angle (LHA), i.e.

$$\dot{\Omega}_{ns} = 0 \tag{14.108}$$

and

$$\Omega_{ns} = \text{LHA} \tag{14.109}$$

We first discuss Eq. (14.108). Correcting the drift rate is possible using a combined correction of the semimajor axis and the inclination, using a single impulsive maneuver.

Consider a circular orbit, which is close to sun synchronicity, with the orbital elements a_0 and i_0. The nodal drift of this orbit is denoted by $\dot{\Omega}_0$. A single impulsive maneuver, which includes an along-track component and an out-of-plane component will simultaneously modify the semimajor axis and inclination so that the nodal drift becomes equal to $\dot{\Omega}_s$. The required correction in the nodal drift is

$$\Delta\dot{\Omega} = \dot{\Omega}_0 - \dot{\Omega}_s \tag{14.110}$$

The relation between the required correction and the variations of the semimajor axis and inclination is given by

$$\frac{\Delta\dot{\Omega}}{\dot{\Omega}_s} = -\frac{7}{2}\frac{\Delta a}{a_0} - (\tan i_0)\Delta i \tag{14.111}$$

From the Gauss equations, we realize that

$$\frac{\Delta a}{a_0} = 2\sqrt{\frac{a_0}{\mu}}\,\Delta v_T, \quad \Delta i = \sqrt{\frac{a_0}{\mu}}(\cos u)\,\Delta v_W \tag{14.112}$$

The variation in the inclination depends on u. Hence, the timing in which the impulse is applied along the orbit is important. Combining Eqs. (14.111) and (14.112) we have

$$\frac{\Delta v_T}{\sqrt{\mu/a_0}} = -\frac{7}{49+\tan^2 i_0 \cos^2 u}\left(\frac{\Delta\dot{\Omega}}{\dot{\Omega}_s}\right) \tag{14.113}$$

$$\frac{\Delta v_W}{\sqrt{\mu/a_0}} = -\frac{\tan i_0 \cos u}{49+\tan^2 i_0 \cos^2 u}\left(\frac{\Delta\dot{\Omega}}{\dot{\Omega}_s}\right) \tag{14.114}$$

and the magnitude of the combined impulse, applied as a single impulsive correction, is given by

$$\Delta v = \sqrt{\Delta v_T^2 + \Delta v_W^2} \tag{14.115}$$

The minimum of Δv is obtained when the maneuver is performed at one of the nodes ($u = 0$ or $u = \pi$). If the velocity impulse is applied away from the nodes, the RAAN is affected and, hence, the LHA.

The scheme described so far deals with the adjustment of the nodal drift only. To adjust the LHA as well, i.e. Eq. (14.109), the following procedure is used.

Assume that the LHA should be maintained within the tolerance $\Omega_{ns} \pm \Delta\Omega_{ns}$. The main reason for the violation of the sun synchronicity condition is the decrease in altitude due to the effect of atmospheric drag. Orbit keeping in this case will be performed by applying velocity corrections in the direction of the orbital velocity, so that the satellite remains within an altitude band wherein the LHA tolerance is satisfied.

Assume that the initial altitude is Δa_0 above the nominal altitude, and that the initial node is located at $\Delta\Omega_0$ eastward to the nominal node. According to the drag model (11.223) (neglecting the atmosphere's velocity) and the Gauss equations, the altitude decreases according to

$$\dot{a} = -\rho\frac{SC_D}{m}\sqrt{\mu a} \tag{14.116}$$

In the short term, it may be assumed that the altitude decay rate is constant, so that

$$\Delta a \approx \Delta a_0 + \dot{a}t \tag{14.117}$$

The motion of the node, assuming that the inclination does not change, is given by

$$\Delta\Omega(t) = \Delta\Omega_0 - \frac{7}{2}\frac{\dot{\Omega}_s}{a_0}\left(\Delta a_0 t + \frac{1}{2}\dot{a}t^2\right) \tag{14.118}$$

We will require that when the node reaches the westernmost limit of the tolerance, the altitude reaches its nominal value. The elapsed time is

$$t_1 = -\frac{\Delta a_0}{\dot{a}} \tag{14.119}$$

which, upon substitution into Eq. (14.118) yields the required additional altitude,

$$\Delta a_0 = \sqrt{-\frac{8\dot{a}a_0\Delta\Omega_0}{7\dot{\Omega}_s}} \tag{14.120}$$

The altitude continues decreasing below the nominal altitude, causing an eastward nodal drift. When the satellite reaches the lower altitude limit $-\Delta a_0$, the node assumes its initial value at the easternmost tolerance bound. At this point, a velocity impulse is applied in the direction of the orbital velocity, increasing the semimajor axis to its initial value, i.e. $a_0 + \Delta a_0$. The required impulse is given by

$$\Delta v_T = v\frac{\Delta a_0}{a_0} \tag{14.121}$$

This process entails only tangential velocity impulses, and does not require a specific impulse application timing along the orbit.

14.7.4 Repeat Ground-track Orbit Control

As mentioned in Sect. (13.5), in RGT orbits the satellite tracks the same trace on the ground with a given periodicity pattern. Let ΔG denote the angular deviation of the equatorial ground track with respect to the reference ground track. The rate of the ground track drift depends on altitude. This drift can be determined by calculating the angular separation between 2 subsequent equatorial passes, given by

$$S = -(\omega_e - \dot{\Omega})T_N \tag{14.122}$$

where $\omega_e \approx 7.2921 \times 10^{-5}$ rad/s is Earth's spin rate, and T_N is the nodal period, given approximately by (Vallado 2001; pp. 788–789)

$$T_N = T\left[1 - \frac{3}{2}J_2\left(\frac{r_e}{a}\right)^2(3 - 4\sin^2 i)\right] \tag{14.123}$$

where $T = 2\pi\sqrt{a^3/\mu}$ is the orbital period. Neglecting the Earth oblateness,

$$S = -\omega_e T \tag{14.124}$$

so that, approximately, the ground track depends only upon the semimajor axis a. Variations in a due to orbital decay, given by Eqs. (14.116) and (14.117), cause variations in the ground track pattern according to

$$\dot{G} = -\frac{3}{2}\omega_e \frac{\Delta a}{a_0} \tag{14.125}$$

Thus, the ground track motion is parabolic, similarly to the motion of the LHA. The adjustment of the ground track is, therefore, performed in a similar manner to the method described in the previous subsection, namely, the satellite is initially located at a higher altitude, and the ground track will then pass eastward with respect to the reference value. The decay will shift the track westward. Once the satellite reaches below the nominal altitude, the track will be shifted eastward again and will resume its original value. At this point a tangential velocity impulse is given, which can be determined by

$$\Delta a_0 = \sqrt{-\frac{8a_0\dot{a}\Delta G}{3\omega_e}} \tag{14.126}$$

The remaining expressions for the time between subsequent impulses and the magnitude of the velocity impulse are identical to the expressions obtained in the previous subsection.

14.8 Geostationary Orbit Control

In general, GEO control is divided into inclination corrections, referred to as *north-south stationkeeping*, and longitude corrections, called *east-west stationkeeping*. Other orbital elements, such as the eccentricity, should also be periodically corrected.

14.8.1 North-South Stationkeeping

As mentioned in Sect. 13.6.2, the inclination drift cycle is initiated at $i_0 = i_{max}$, which is the maximum allowed inclination deviation, and at the node location Ω_0 given by Eq. (13.110). At the end of the cycle, the node assumes some final value, Ω_f, and the inclination angle is the same as i_0. Then, an impulsive correction maneuver is performed. The magnitude of the impulsive maneuver is given by

$$\Delta v = 2\sqrt{\frac{\mu}{a}}\sin\frac{i^\star}{2}, \tag{14.127}$$

where $i^\star$ is the wedge angle. If a time-maximizing strategy is used, then $i^\star$ can be computed from (Kechichian 1997)

$$\cos i^\star = \cos(\Omega_{opt} - \Omega_f) \sin^2 i_{max} + \cos^2 i_{max} \tag{14.128}$$

where Ω_{opt} is the initial node location computed for a new inclination drift cycle (Kamel and Tibbitts 1973) (see Sect. 13.6.2).

14.8.2 East-West Stationkeeping

As discussed in Sect. 13.6.1, Earth's triaxiality generates 4 geostationary equilibrium points, 2 of which are stable and 2 are unstable (see Eq. (13.84)). The satellite will tend to approach the nearest stable equilibrium point. Substituting numerical values into Eq. (13.82) will lead to

$$\dot{a} = -0.132 \sin 2(\lambda - \lambda_s) \text{ km/day} \tag{14.129}$$

where λ_s is the nearest stable geostationary point. Differentiating Eq. (13.81) yields

$$\ddot{\lambda} = -\frac{3}{2}\frac{n}{a}\dot{a} \tag{14.130}$$

which, upon substitution of numerical values yields

$$\ddot{\lambda} = -K \sin 2(\lambda - \lambda_s), \; K = 0.00168 \text{ deg/day}^2 \tag{14.131}$$

Assume that the nominal longitude is λ_n. East-west stationkeeping will keep the satellite within a small longitude tolerance, typically fraction of a degree. Hence, we may treat $\ddot{\lambda}$ as constant, and write

$$\ddot{\lambda} = -K \sin 2(\lambda_n - \lambda_s) = \ddot{\lambda}_n \tag{14.132}$$

Integrating Eq. (14.132) twice with respect to time, with the initial conditions $\lambda(0) = \lambda_0$ and $\dot{\lambda}(0) = \dot{\lambda}_0$, provides the relation

$$\lambda = \frac{1}{2}\ddot{\lambda}_n t^2 + \dot{\lambda}_0 t + \lambda_0 \tag{14.133}$$

Thus,

$$\dot{\lambda} = \ddot{\lambda}_n t + \dot{\lambda}_0 \tag{14.134}$$

and the time is expressed as

$$t = \frac{\dot{\lambda} - \dot{\lambda}_0}{\ddot{\lambda}_n} \tag{14.135}$$

By substituting Eq. (14.135) into Eq. (14.133) we have

$$\lambda = \frac{1}{2\ddot{\lambda}_n}(\dot{\lambda} - \dot{\lambda}_0)^2 + \frac{\dot{\lambda}_0}{\ddot{\lambda}_n}(\dot{\lambda} - \dot{\lambda}_0) + \lambda_0 \tag{14.136}$$

Assume that the allowed longitude is $\lambda_n \pm \Delta\lambda$. The satellite is stationed initially at $\lambda_0 = \lambda_n - \Delta\lambda$, so that the perturbation will decrease the initial longitudinal deviation until it reaches zero, and then will grow again. By using an initial altitude offset, $\dot{\lambda}_0$ can be determined so that the angular rate of the longitude deviation will reach zero when the longitude deviation reaches the edge of the allowed limit. In other words, at some time t_f, we require that $\dot{\lambda}(t_f) = 0$, so at that time Eq. (14.136) is evaluated as

$$\lambda_n + \Delta\lambda = -\frac{\dot{\lambda}_0^2}{2\ddot{\lambda}_n} + \lambda_n - \Delta\lambda \tag{14.137}$$

Thus, the initial conditions in which the satellite should be stationed are

$$\lambda_0 = \lambda_n - \Delta\lambda, \; \dot{\lambda}_0 = -2\text{sign}(\ddot{\lambda}_n)\sqrt{-\ddot{\lambda}_n\Delta\lambda} \tag{14.138}$$

The sign of $\Delta\lambda$ is determined according to the direction of the perturbing acceleration, so the argument of the square root in Eq. (14.138) is positive.

The longitude will move from $\lambda_n - \Delta\lambda$ to $\lambda_n + \Delta\lambda$. At this point, the longitudinal drift will change sign, and the longitude will move again towards $\lambda_n - \Delta\lambda$. Once reaching this value, an impulsive maneuver will be used to change the drift rate from $-\dot{\lambda}_0$ to $\dot{\lambda}_0$. The required semimajor axis change is calculated so that the mean motion is changed by

$$\Delta n = 2\dot{\lambda}_0 \tag{14.139}$$

Since $n = \sqrt{\mu/a^3}$, then

$$\Delta a = -\frac{2}{3}\frac{\Delta n}{n}a \tag{14.140}$$

and hence

$$\Delta v = \left|\frac{\mu}{2a^2 v}\Delta a\right| = \frac{a}{3}\left|\Delta n\right| = \frac{2a}{3}\left|\dot{\lambda}_0\right| \tag{14.141}$$

The time between subsequent corrections is determined according to

$$t_m = 2\left|\frac{\dot{\lambda}_0}{\ddot{\lambda}_n}\right| \tag{14.142}$$

which leads to the altitude change given by (assuming t_m is measured in days)

$$\Delta a = -0.132 \sin 2(\lambda - \lambda_s) t_m \text{ km} \tag{14.143}$$

The annual required velocity correction is then

$$\Delta v = 1.75 \sin 2(\lambda - \lambda_s) \text{ m/sec/year} \tag{14.144}$$

14.8.3 Eccentricity Correction

Equations (14.106d)–(14.106e) relate the eccentricity correction to the required velocity impulses. One possibility is to use the east-west stationkeeping tangential impulse and apply it at a specific location along the orbit to correct the eccentricity. For a tangential impulse only, dividing Eq. (14.106e) by Eq. (14.106d) will lead to the required location,

$$\tan u = \frac{\Delta \eta_1}{\Delta \zeta_1} \tag{14.145}$$

It is not guaranteed that the tangential impulse will achieve the required eccentricity correction, and, thus, there are 2 equations with 2 unknowns: u and Δv_N.

14.9 Nonlinear Feedback Control of Orbits

In previous sections, we considered impulsive maneuvers for orbit control. In this section, we will show how continuous thrust can be used to steer the orbital elements towards given desired values based on Lyapunov's second method, developed in Sect. 14.2.

The acceleration caused by thrust is expressed in the RSW frame (see Sect. 11.6) as

$$\mathbf{u} = u_r \hat{\mathbf{u}}_r + u_\theta \hat{\mathbf{u}}_\theta + u_W \hat{\mathbf{u}}_W \tag{14.146}$$

The state vector comprises the classical elements, namely $\mathbf{x} = [a, e, i, \Omega, \omega, M]^T$. With these definitions, we can write the Gauss equations (11.131) using the compact notation

$$\dot{\mathbf{x}} = \mathbf{g}_0(\mathbf{x}) + \mathbf{G}(\mathbf{x})\mathbf{u} \tag{14.147}$$

where

$$\mathbf{g}_0(\mathbf{x}) \triangleq \begin{bmatrix} 0 \\ 0 \\ 0 \\ 0 \\ 0 \\ n \end{bmatrix} \tag{14.148}$$

$$\mathbf{G}(\mathbf{x}) \triangleq \frac{1}{h} \begin{bmatrix} 2a^2 e \sin f & \dfrac{2a^2 p}{r} & 0 \\ p \sin f & (p+r)\cos f + re & 0 \\ 0 & 0 & r\cos(f+\omega) \\ 0 & 0 & \dfrac{r\sin(f+\omega)}{\sin i} \\ \dfrac{-p\cos f}{e} & \dfrac{(p+r)\sin f}{e} & \dfrac{-r\sin(f+\omega)\cos i}{\sin i} \\ \dfrac{b[p\cos f - 2re]}{ae} & \dfrac{-b(p+r)\sin f}{ae} & 0 \end{bmatrix} \tag{14.149}$$

In Eqs. (14.148) and (14.149), we have used the notation

$$n = \sqrt{\mu/a^3},\ p = a(1-e^2),\ h = \sqrt{\mu p},\ b = a\sqrt{1-e^2} \tag{14.150}$$

As previously discussed, the true anomaly f is an implicit function of the mean anomaly M. Based on the solution of Kepler's equation (see Sect. 5.9), f can be expended into the infinite power series (Battin 1999; pp. 210–212)

$$f(M,e) = M + 2\sum_{l=1}^{\infty} \frac{1}{l}\left[\sum_{s=-\infty}^{\infty} J_s(-le)\left(\frac{1-\sqrt{1-e^2}}{e}\right)^{|l+s|}\right]\sin(lM) \tag{14.151}$$

where $J_s(\cdot)$ is a Bessel function of the first kind of order s.

The orbital control problem is to find a continuous feedback control law $\mathbf{u}(\mathbf{x})$ that steers a satellite from an initial elliptic orbit to a desired elliptic orbit. To find such a feedback control law, we will utilize a Lyapunov-based approach.

The first step is to define a subset of orbital elements that we wish to control. These are

$$\mathbf{z} \triangleq [a,\ e,\ i,\ \Omega,\ \omega]^T,\ \mathbf{z}_d \triangleq [a_d,\ e_d,\ i_d,\ \Omega_d,\ \omega_d]^T \tag{14.152}$$

where $\mathbf{z}_d$ denotes the classical orbital elements of some desired target orbit. The mean anomaly of the target orbit is not controlled according to this formulation. Thus, we can write

$$\dot{\mathbf{z}} = \mathbf{Z}(\mathbf{z}, M)\mathbf{u} \tag{14.153}$$

where

$$\mathbf{Z}(\mathbf{z}, M) \triangleq \frac{1}{h}\begin{bmatrix} 2a^2 e \sin[f(M,e)] & \dfrac{2a^2 p}{r} & 0 \\ p\sin[f(M,e)] & (p+r)\cos[f(M,e)] + re & 0 \\ 0 & 0 & r\cos[f(M,e)+\omega] \\ 0 & 0 & \dfrac{r\sin[f(M,e)+\omega]}{\sin i} \\ \dfrac{-p\cos[f(M,e)]}{e} & \dfrac{(p+r)\sin[f(M,e)]}{e} & \dfrac{-r\sin[f(M,e)+\omega]\cos i}{\sin i} \end{bmatrix} \tag{14.154}$$

Now, consider the quadratic form

$$V(\mathbf{z},\ \mathbf{z}_d) = \frac{1}{2}(\mathbf{z} - \mathbf{z}_d)^T \mathbf{P}(\mathbf{z} - \mathbf{z}_d) \tag{14.155}$$

where $\mathbf{P}$ is some positive definite matrix, and a related control law, written as

$$u = -[\nabla V(\mathbf{z},\ \mathbf{z}_d) \cdot \mathbf{Z}(\mathbf{z}, M)]^T = -\left[(\mathbf{z} - \mathbf{z}_d)^T \mathbf{P}\mathbf{Z}(\mathbf{z}, M)\right]^T \tag{14.156}$$

The control law appearing in Eq. (14.155) is known as the *Jurdjevic-Quinn damping feedback* (Jurdjevic and Quinn 1978). To show that the control law (14.156) can indeed render the desired orbital elements set $\mathbf{z}_d$ asymptotically stable in the sense defined in Sect. 14.2, we evaluate the time derivative of V along the trajectories of the system in Eq. (14.153),

$$\dot{V} = \nabla V(\mathbf{z},\ \mathbf{z}_d) \cdot \dot{\mathbf{z}} = \nabla V(\mathbf{z},\ \mathbf{z}_d) \cdot \mathbf{Z}(\mathbf{z}, M)\mathbf{u} \tag{14.157}$$

Utilizing Eq. (14.156) we have

$$\dot{V} = -(\mathbf{z} - \mathbf{z}_d)^T \mathbf{P} \cdot \mathbf{Z}(\mathbf{z}, M)\left[(\mathbf{z} - \mathbf{z}_d)^T \mathbf{P} \cdot \mathbf{Z}(\mathbf{z}, M)\right]^T \leq 0 \tag{14.158}$$

According to Lyapunov's second method, in order to prove asymptotic stability, we need to show that $\dot{V} = 0$, if and only if $\mathbf{z} = \mathbf{z}_d$. To this end, we note that $\dot{V} = 0$ in the following cases:

$$\mathbf{z} = \mathbf{z}_d \tag{14.159}$$

$$a \equiv 0 \tag{14.160}$$

$$a = a_d,\ e = e_d,\ i = i_d,\ \omega = \omega_d, f + \omega_d = k\pi,\ k = 0,\ 1, \ldots \tag{14.161}$$

$$a = a_d,\ e = e_d,\ \Omega = \Omega_d,\ \omega = \omega_d, f + \omega_d = k\pi/2,\ k = 1,\ 3, \ldots \tag{14.162}$$

Equation (14.160) represents the degenerate case of zero angular momentum (rectilinear motion) and is, hence, ruled out. Equations (14.161) and (14.162) are ruled out because they cannot be satisfied for all f. Thus, Eq. (14.159) is the only case in which $\dot{V} = 0$ and the controller (14.156) is a globally asymptotically stabilizing controller, assuring that $\lim_{t\to\infty} \|\mathbf{z}(t) - \mathbf{z}_d(t)\| \to 0$.

By taking a diagonal matrix $\mathbf{P} = \text{diag}[\lambda_a, \lambda_e, \lambda_i, \lambda_\Omega, \lambda_\omega]$, closed-form component-wise expressions for the continuous feedback control law can be obtained by substituting Eq. (14.154) into Eq. (14.156),

$$u_r = -\frac{1}{h}\left[2\Delta a \lambda_a a^2 e \sin f + \Delta e \lambda_e p \sin f - \frac{\Delta\omega\lambda_\omega p \cos f}{e}\right] \tag{14.163a}$$

$$u_\theta = -\frac{1}{h}\left[\frac{2\Delta a \lambda_a a^2 p}{r} + \Delta e \lambda_e[(p + r)\cos f + re] + \frac{\Delta\omega\lambda_\omega (p + r)\sin f}{e}\right] \tag{14.163b}$$

$$u_W = -\frac{1}{h}\left[\Delta i \lambda_i r \cos u + \frac{\Delta\Omega\lambda_\Omega r \sin u}{\sin i} - \frac{\Delta\omega\lambda_\omega r \sin u \cos i}{\sin i}\right] \tag{14.163c}$$

where $u = f + \omega$, $\Delta a = a - a_d$, $\Delta e = e - e_d$, $\Delta\omega = \omega - \omega_d$, $\Delta\Omega = \Omega - \Omega_d$, $\Delta i = i - i_d$.

The continuous thrust acceleration components given in Eqs. (14.163) provide a variable-magnitude thrust, and, hence, can be applied only on throttable propulsion systems. In many practical cases, however, the thrust magnitude is fixed, i.e. unthrottable. This case is discussed in the next section.

14.10 Fixed-Magnitude Continuous-Thrust Orbit Control

Let $\mathbf{x} \triangleq [a, e, i]$ be the state vector, so only the semimajor axis, eccentricity, and inclination are controlled. The control vector is expressed in the NTW frame as per the discussion in Sect. 11.7, so that

$$\mathbf{u} = u_T \hat{\mathbf{u}}_T + u_N \hat{\mathbf{u}}_N + u_W \hat{\mathbf{u}}_W \tag{14.164}$$

In this case, we can write the GVE as $\dot{\mathbf{x}} = \mathbf{G}\mathbf{u}$, where

$$\mathbf{G} = \begin{bmatrix} 0 & \dfrac{2}{n\eta\chi} & 0 \\ -\dfrac{\eta^2\chi}{na}\dfrac{\eta \sin f}{1 + e\cos f} & \dfrac{2\eta\chi}{na}(e + \cos f) & 0 \\ 0 & 0 & \dfrac{r\cos u}{na^2\eta} \end{bmatrix} \tag{14.165}$$

The matrix $\mathbf{G}$ is determined by the GVE as explained in Sect. 11.7. In Eq. (14.165), $\chi = \left(1 + 2ec_f + e^2\right)^{-1/2}$, $\eta = \sqrt{1 - e^2}$, and $n = \sqrt{\mu/a^3}$.

We assume here that the satellite is equipped with a single thruster producing a constant magnitude thrust T_0. Based on Eq. (14.165), the dynamics of the orbital elements are given by

$$\dot{\mathbf{x}} = \frac{T_0}{m}\mathbf{G}\hat{\mathbf{u}} \tag{14.166}$$

where $\hat{\mathbf{u}}$ is a unit vector constituting the control direction, and m is the mass of the satellite. The mass flow rate is given by

$$\dot{m} = -\frac{T_0}{U_e} \tag{14.167}$$

where as in Eq. (14.16), $U_e = I_{sp}g_0$ is the exhaust velocity, I_{sp} is the specific impulse, and g_0 is the gravitational acceleration at sea level.

The control objective is to drive the state $\mathbf{x}$ from $\mathbf{x}_0$ to $\mathbf{x}_d$. Let $\Delta\mathbf{x} = \mathbf{x} - \mathbf{x}_d$. Assuming $\mathbf{G}$ is nonsingular, namely, $f \neq k\pi$, $\theta \neq k\pi/2$, $k = 1, 2, \ldots$, a possible control law to determine the commanded control direction can be written as

$$\hat{\mathbf{u}} = -\frac{\mathbf{G}^{-1}\Delta\mathbf{x}}{\left\|\mathbf{G}^{-1}\Delta\mathbf{x}\right\|} \tag{14.168}$$

The control law in Eq. (14.168) utilizes an inversion of the matrix $\mathbf{G}$ and was, hence, called by Zhang and Gurfil (2014) an *inverse-dynamics controller*. It is a special case of the more general control proposed by Schaub et al. (2000). It can also be interpreted as a Jurdjevic-Quinn controller, discussed in the previous section, with a time-varying gain matrix $\mathbf{P}(t)$, i.e.

$$\hat{\mathbf{u}} = -\frac{\mathbf{G}^T\mathbf{P}(t)\,\Delta\mathbf{x}}{\left\|\mathbf{G}^T\mathbf{P}(t)\,\Delta\mathbf{x}\right\|} \tag{14.169}$$

where

$$\mathbf{P}(t) = \mathbf{G}^{-T}\mathbf{G}^{-1} \tag{14.170}$$

The gain matrix $\mathbf{P}$ is positive definite, but not necessarily diagonal.

To show stability, the following Lyapunov function candidate is considered,

$$V = \frac{1}{2}\Delta\mathbf{x}^T\Delta\mathbf{x} \tag{14.171}$$

The time derivative is

$$\begin{aligned}\dot{V} &= -\frac{T_0}{m}\frac{\Delta\mathbf{x}^T\Delta\mathbf{x}}{\left\|\mathbf{G}^{-1}\Delta\mathbf{x}\right\|}\\ &\leq -\frac{T_0}{m}\frac{\|\Delta\mathbf{x}\|}{\sigma_{\max}\left(\mathbf{G}^{-1}\right)} = -\frac{T_0}{m}\sigma_{\min}(\mathbf{G})\,\|\Delta\mathbf{x}\| \leq 0\end{aligned} \tag{14.172}$$

where $\sigma_{\max}$ and $\sigma_{\min}$ denote the maximum and minimum *singular values* of a matrix.

The inequality in Eq. (14.172) is obtained by using the fact that, for a matrix $\mathbf{M}$ and a vector $\mathbf{a}$, the following relations are satisfied:

$$\sigma_{\min}(\mathbf{M})\|\mathbf{a}\| \leq \|\mathbf{Ma}\| \leq \sigma_{\max}(\mathbf{M})\|\mathbf{a}\|,\ \sigma_{\max}\left(\mathbf{M}^{-1}\right) = \frac{1}{\sigma_{\min}(\mathbf{M})} \tag{14.173}$$

Hence, the control law is asymptotically stable, because $\dot{V} = 0$ only when $\|\Delta\mathbf{x}\| = 0$. Note also that the magnitude of thrust and the mass have no effect on the sign of $\dot{V}$, and, thus, on the stability of the controller.

Since a thruster with a fixed magnitude thrust is assumed, a time-optimal controller, i.e., a controller which achieves the desired state in minimum time, also implies fuel-optimality. To show time optimality we approximate the matrix $\mathbf{G}$ in Eq. (14.166) by substituting the values of some reference state $\mathbf{x}_r$, so that the new matrix is denoted by $\mathbf{G}_r(t) \approx \mathbf{G}(\mathbf{x}_r(t))$. This manipulation removes the state dependence. Thus, the dynamics can be represented as

$$\dot{\mathbf{x}} = \frac{T_0}{m}\mathbf{G}\hat{\mathbf{u}} \approx \frac{T_0}{m}\mathbf{G}_r\hat{\mathbf{u}} \tag{14.174}$$

A constrained time-optimal control problem, seeking to minimize the control operation time t_f, is then formulated as

$$\begin{aligned}&\text{Minimize} && J = t_f,\\ &\text{Subject to} && \begin{cases}\dot{\mathbf{x}} = \dfrac{T_0}{m}\mathbf{G}_r\hat{\mathbf{u}}\\ \mathbf{x}(0) = \mathbf{x}_0, \mathbf{x}\left(t_f\right) = \mathbf{x}_d\\ \|\hat{\mathbf{u}}\| = 1\end{cases}\end{aligned} \tag{14.175}$$

According to the optimal control theory presented in Sect. 14.5, the Hamiltonian is defined as

$$H = 1 + \frac{T_0}{m}\boldsymbol{\lambda}^T\mathbf{G}_r\hat{\mathbf{u}} \tag{14.176}$$

where $\boldsymbol{\lambda}$ is the co-state vector corresponding to $\mathbf{x}$. The necessary condition as given in Eqs. (14.47) becomes

$$\dot{\boldsymbol{\lambda}} = -\left[\frac{\partial H}{\partial \mathbf{x}}\right]^T = 0 \tag{14.177}$$

which gives a constant solution,

$$\boldsymbol{\lambda} = \boldsymbol{\lambda}_c \tag{14.178}$$

After substituting Eq. (14.178) into Eq. (14.176), the optimal control $\hat{\mathbf{u}}^\star$ is found to be the solution to the equation

$$\hat{\mathbf{u}}^\star = \min_{\|\hat{\mathbf{u}}\|=1} H = \min_{\|\hat{\mathbf{u}}\|=1} \boldsymbol{\lambda}_c^T\mathbf{G}_r\hat{\mathbf{u}} \tag{14.179}$$

Equation (14.179) can be solved by using the method of Lagrange multipliers discussed in Sect. 14.5, yielding the optimal control

$$\hat{\mathbf{u}}^\star = -\frac{\mathbf{G}_r^T\boldsymbol{\lambda}_c}{\left\|\mathbf{G}_r^T\boldsymbol{\lambda}_c\right\|} \tag{14.180}$$

The state dynamics are obtained by substituting Eq. (14.180) into Eq. (14.174),

$$\dot{\mathbf{x}}^\star = -\frac{T_0}{m}\frac{\mathbf{G}_r\mathbf{G}_r^T\boldsymbol{\lambda}_c}{\left\|\mathbf{G}_r^T\boldsymbol{\lambda}_c\right\|} \tag{14.181}$$

Considering that the Hamiltonian $H\left(t_f\right) = 0$, another relation can be found,

$$\frac{T_0}{m\left(t_f\right)} \left\| \mathbf{G}_r^T\left(t_f\right) \boldsymbol{\lambda}_c \right\| = 1 \tag{14.182}$$

It follows that the original optimization problem is transformed into a set of nonlinear equations,

$$\begin{cases} \mathbf{x}_d - \mathbf{x}_0 = -\displaystyle\int_0^{t_f} \frac{T_0}{m} \frac{\mathbf{G}_r(t)\, \mathbf{G}_r^T(t)\, \boldsymbol{\lambda}_c}{\left\| \mathbf{G}_r^T(t) -_c \right\|} dt \\ \dfrac{T_0}{m\left(t_f\right)} \left\| \mathbf{G}_r^T\left(t_f\right) \boldsymbol{\lambda}_c \right\| = 1 \end{cases} \tag{14.183}$$

with the unknowns being $\boldsymbol{\lambda}_c$ and t_f.

Equation (14.183) is a two-point boundary value problem and does not permit analytical solutions in general. However, when the control time is short compared to the orbital period, some analytical insight may be gained. During the control, the reference state $\mathbf{x}_r$, and also the matrix $\mathbf{G}_r$, can be treated as constants. By utilizing the relationship

$$m = m_0 - \frac{T_0}{U_e} t \tag{14.184}$$

Equation (14.183) admits the solution

$$\begin{cases} \boldsymbol{\lambda}_c = -\dfrac{m}{T_0} \dfrac{\mathbf{G}_r^{-T} \mathbf{G}_r^{-1} \left(\mathbf{x}_d - \mathbf{x}_0\right)}{\left\| \mathbf{G}^{-1} \left(\mathbf{x}_d - \mathbf{x}_0\right) \right\|} \\ \hat{\mathbf{u}} = \dfrac{\mathbf{G}_r^{-1} \left(\mathbf{x}_d - \mathbf{x}_0\right)}{\left\| \mathbf{G}_r^{-1} \left(\mathbf{x}_d - \mathbf{x}_0\right) \right\|} \\ m_f = m_0 \exp\left(-\dfrac{\left\| \mathbf{G}_r^{-1} \left(\mathbf{x}_d - \mathbf{x}_0\right) \right\|}{U_e} \right) \\ \dot{\mathbf{x}} = \dfrac{T_0}{m} \dfrac{\left(\mathbf{x}_d - \mathbf{x}_0\right)}{\left\| \mathbf{G}_r^{-1} \left(\mathbf{x}_d - \mathbf{x}_0\right) \right\|} \end{cases} \tag{14.185}$$

where the expression for final mass m_f is given instead of t_f. After substituting the control in Eq. (14.185) into Eq. (14.174), the state dynamics can be solved in closed-form,

$$\mathbf{x} = \mathbf{x}_0 + \frac{U_e\left(\mathbf{x}_d - \mathbf{x}_0\right)}{\left\|\mathbf{G}_r^{-1}\left(\mathbf{x}_d - \mathbf{x}_0\right)\right\|} \ln\left(\frac{m_0}{m_0 - \dfrac{T_0}{U_e}t}\right) \tag{14.186}$$

Thus, the controller (14.185) can also be expressed as

$$\hat{\mathbf{u}} = -\frac{\mathbf{G}_r^{-1}\left(\mathbf{x} - \mathbf{x}_d\right)}{\left\|\mathbf{G}_r^{-1}\left(\mathbf{x} - \mathbf{x}_d\right)\right\|} \tag{14.187}$$

which is the same as the controller in Eq. (14.168).

Because the entries of $\mathbf{G}_r$ depend on the fast variables f and u, $\mathbf{G}_r$ is constant for short control times, and the control law (14.168) is in fact nearly time-optimal.

14.11 Comparison of Continuous-Thrust Controllers

This section compares the performance of the inverse-dynamics controller (14.168) with a fixed-magnitude variant of the feedback controller presented in Sect. 14.9, i.e.

$$\mathbf{u} = -\frac{\mathbf{G}^T\mathbf{P}\left(\mathbf{x}_{nc} - \mathbf{x}_d\right)}{\left\|\mathbf{G}^T\mathbf{P}\left(\mathbf{x}_{nc} - \mathbf{x}_d\right)\right\|}, \tag{14.188}$$

where $\mathbf{P}$ is a constant positive definite gain matrix, and the states are $\mathbf{x}_{nc} = [a, i]^T$. A length unit $R_0 = 6378.137\,\text{km}$ is used. After the scaling, the constant gain matrix $\mathbf{P}$ in Eq. (14.188) is taken as

$$P = \begin{bmatrix} 1 & 0 \\ 0 & p_I \end{bmatrix} \tag{14.189}$$

where several values of p_I are chosen between 0.1 and 100. A reference orbit with $a = 7000\,\text{km}$, $e = 0.002$, and $i = 51°$ is chosen for the simulation. The mass of the satellite is 8 kg. The initial errors are set as $\Delta a = 4\,\text{km}$, and $\Delta i = 0.03°$. The nominal values for the single thruster are given as $T_0 = 80\,\text{mN}$ and $I_{sp} = 35\,\text{s}$. The controller began to work near the ascending node, until the mean elements errors were driven below the tolerances.

The results for the controllers in Eqs. (14.168) and (14.188) are depicted in Figs. 14.6 and 14.7. It can be seen from the phase portraits in Fig. 14.6 that only the inverse-dynamics controller (14.168) can produce a straight line, i.e. minimum time convergence. It is also important to note that the controller (14.168) is invariant

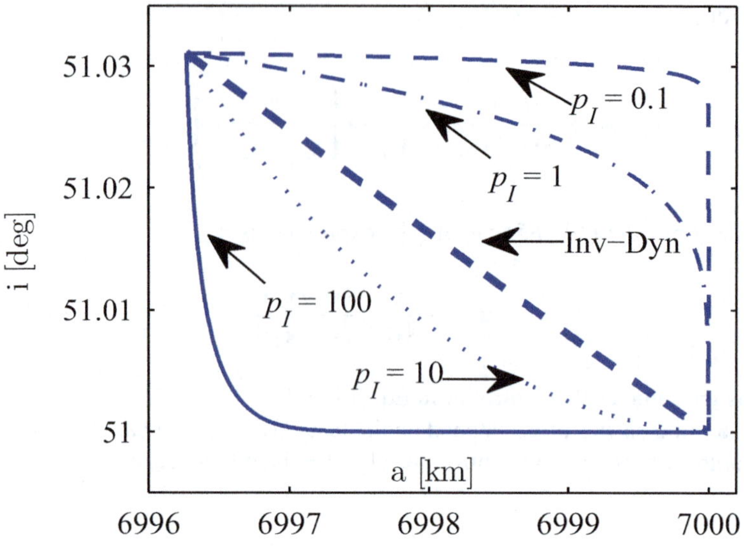

Fig. 14.6 Trajectory evolution of $\{a, i\}$ under different controllers

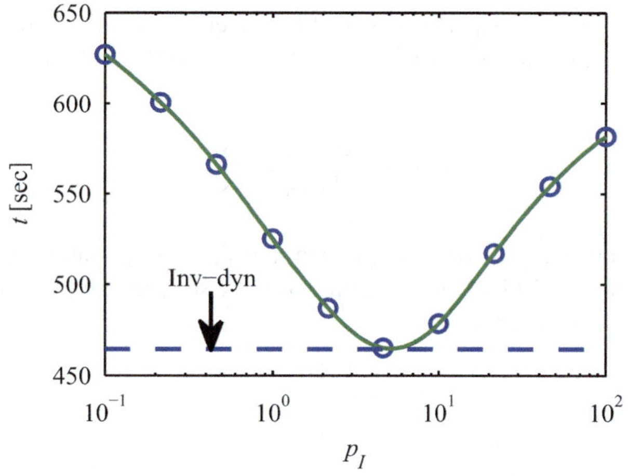

Fig. 14.7 Control time comparison for different controllers

to the scaling. However, for controller (14.188), the scaling plays an important role for the convergence of the algorithm.

The different control times are shown in Fig. 14.7. The control times for controller (14.188) are denoted by circles, while a smooth curve, representing the control time with respect to different gains, is obtained by a cubic spline interpolation. The control time for the constant-gain controller (14.188) varies considerably according to the different gains. Thus, gains should be tuned before the in-flight application. The controller (14.168), however, is relatively straightforward to implement. Furthermore, as can be clearly seen, the control time for controller (14.168) is much shorter than that of controller (14.188), which validates the optimality analysis in the previous section. Specifically, the mass consumption in the current example for the inverse-dynamics controller is 109.3 grams, while the mass consumption of controller (14.188) ranges from 109.4 grams to 144.9 grams. Thus, the fuel consumption can be reduced by up to 33%.

References

Alfriend, K., Vadali, S., Gurfil, P., How, J., Breger, L.: Spacecraft Formation Flying: Dynamics, Control and Navigation. Elseiver, Oxford (2010)

Battin, R.: An Introduction to the Mathematics and Methods of Astrodynamics. American Institute of Aeronautics and Astronautics, Reston (1999)

Conway, B.: Spacecraft Trajectory Optimization. Cambridge University Press, Cambridge (2010)

Fehse, W.: Automated Rendezvous and Docking of Spacecraft. Cambridge University Press, Cambridge (2005)

Flandro, G.A.: From instrumented comets to grand tours: On the history of gravity assist. In: Proceeding of 39th AIAA Aerospace Sciences Meeting and Exhibit, Reno, NV, USA (2001)

Gurfil, P.: Nonlinear feedback control of low-thrust orbital transfer in a central gravitational field. Acta Astronaut. **60**(8), 631–648 (2007)

Haykin, S.: Neural Networks: A Comprehensive Foundation, 3rd edn. Prentice-Hall, Upper Saddle River (2008)

Jurdjevic, V., Quinn, J.P.: Controllability and stability. J. Differ. Equ. **28**(3), 381–389 (1978)

Kamel, A., Tibbitts, R.: Some useful results on initial node location for near-equatorial circular satellite orbits. Celest. Mech. **8**, 45–73 (1973)

Kawano, I., Mokuno, M., Kasai, T., Suzuki, T.: Result of autonomous rendezvous docking experiment of Engineering Test Satellite-VII. J. Spacecr. Rocket **38**(1), 105–111 (2001)

Kechichian, J.A.: The analysis of the relative motion in general elliptic orbit with respect to a dragging and precessing coordinate frame. In: Proceedings of the AAS/AIAA Astrodynamics Conference, Sun Valley, ID, pp. 2053–2074 (1997)

Kluever, C.A.: Simple guidance scheme for low-thrust orbit transfers. J. Guid. Control Dyn. **21**(6), 1015–1017 (1998)

Lamkin, S., Mccandless, W.: Pathfinder autonomous rendezvous and docking project. Tech. Rep. NASA TM-102163, NASA Johnson Space Center (1990)

Longuski, J.M., Williams, S.N.: Automated design of gravity-assist trajectories to Mars and the outer planets. Celest. Mech. Dyn. Astron. **52**(3), 207–220 (1991)

Meltzer, M.: Mission to Jupiter: a history of the Galileo project. Tech. Rep. 7, NASA STI/Recon (2007)

Petropoulos, A.E., Longuski, J.M., Bonfiglio, E.P.: Trajectories to Jupiter via gravity assists from Venus, Earth, and Mars. J. Spacecr. Rocket **37**(6), 776–783 (2000)

Pisarevsky, D.M., Gurfil, P.: Optimizing multiple-flyby orbits for increasing the resolution of space telescopes. J. Spacecr. Rocket **46**(2), 373–380 (2009)

Pisarevsky, D.M., Kogan, A., Guelman, M.: Building interplanetary trajectories with multiple gravity-assisted maneuvers. J. Spacecr. Rocket **44**(4), 985–992 (2007)

Poincaré, H.: Les méthodes nouvelles de la mécanique céleste. Il Nuovo Cimento (1895-1900) **10**(1), 128–130 (1899)

Ross, I.M.: Space trajectory optimization and l^1-optimal control problems. In: Gurfil, P. (Ed.) Modern Astrodynamics, pp. 155–186 (2006)

Rumford, T.E.: Demonstration of Autonomous Rendezvous Technology (DART) project summary. In: Proceedings of SPIE: The International Society for Optical Engineering, International Society for Optics and Photonics, pp. 10–19 (2003)
Schaub, H., Alfriend, K.T.: Hybrid Cartesian and orbit elements feedback law for formation flying spacecraft. J. Guid. Control Dyn. **25**(2), 387–393 (2002)
Schaub, H., Vadali, S.R., Alfriend, K.T.: Spacecraft formation flying control using mean orbital elements. J. Astronaut. Sci. **48**(1), 69–87 (2000)
Sims, J.A., Staugler, A.J., Longuski, J.M.: Trajectory options to Pluto via gravity assists from Venus, Mars, and Jupiter. J. Spacecr. Rocket **34**(3), 347–353 (1997)
Stengel, R.F.: Optimal Control and Estimation. Dover Publications, Mineola (1994)
Vallado, D.: Fundamentals of Astrodynamics and Applications, 2nd edn. Microcosm Press and Kluwer Academic Publishers, London (2001)
Weismuller, T., Leinz, M.: GNC technology demonstrated by the Orbital Express autonomous rendezvous and capture sensor system. In: 29th Annual AAS Guidance and Control Conference, Breckenridge, Colorado, USA, pp. 1–9 (2006)
Wenfei, W., Prathyush, M., Declan, B., Simone, C., Nuno, P., Emanuele, S., Ambroise, B., Alexandre, G., Aymeric, K., Sohrab, S., Samir, B.: GNC technology demonstrated by the Orbital Express autonomous rendezvous and capture sensor system. In: AIAA Guidance, Navigation, and Control Conference, Breckenridge, Colorado, USA, pp. 1–9 (2012)
Woffinden, D.C., Geller, D.K.: Navigating the road to autonomous orbital rendezvous. J. Spacecr. Rocket **44**(4), 898–909 (2007)
Zhang, H., Gurfil, P.: Nanosatellite cluster keeping under thrust uncertainties. J. Guid. Control Dyn. **37**(5), 1406–1414 (2014)

Chapter 15
Optimal Impulsive Orbit Transfers

15.1 Introduction

In Chap. 14, we defined the notion of impulsive maneuvers, used as a common approximation for designing orbital maneuvers. An important orbital maneuver problem is the fuel-optimal impulsive transfer between 2 coplanar circular orbits. Three such transfers are the Hohmann transfer, discussed in Sect. 14.7.1; the *bi-elliptic transfer*; and the *bi-parabolic transfer*. These transfers were originally conceived based on the Keplerian two-body problem, discussed in Chap. 5.

The Hohmann transfer was presented by Walter Hohmann in 1925. The optimality of the Hohmann transfer was proven by Barrar (1962). Other methods for proving the fuel optimality of the Hohmann transfer were developed by Prussing (1992).

The Hohmann transfer is the optimal two-impulse orbit transfer between two circular, coplanar, non-intersecting orbits. The resulting optimal transfer consists of two tangential impulses. The first impulse is given on an arbitrary point of the initial circular orbit, and is collinear with the velocity vector. The transfer orbit is an ellipse, whose perigee is located on the smaller circle. The second impulse, collinear with the velocity vector, is given either at the apogee of the transfer orbit, located on the larger circle, or the perigee, depending on whether the transfer is from the smaller to the larger circle or vice versa.

Hoelker and Silber (1961) suggested that, since for some Hohmann transfers the total impulse becomes greater than the impulse required to escape from the initial orbit, and that the total impulse for Hohmann transfers with some final to initial radii ratio is larger than for Hohmann transfers with an infinite ratio, there exists a more efficient, three-impulse transfer obtained by sending the spacecraft to a transition orbit with a large radius. This maneuver is the bi-elliptic transfer, which consists of two intermediate transfer ellipses. Hoelker and Silber (1961) found the ratio of the final to initial radii above which all bi-elliptic transfers are more efficient than the

P. Gurfil, P.K. Seidelmann, *Celestial Mechanics and Astrodynamics: Theory and Practice*, Astrophysics and Space Science Library 436,
DOI 10.1007/978-3-662-50370-6_15

Hohmann transfer, and showed that the optimal maneuver in terms of fuel cost can be obtained by setting the transition radius to be close to infinity. This maneuver is the bi-parabolic transfer.

In this chapter, we will re-derive the optimal transfer solutions by adopting a unique approach. Instead of assuming that the motion is determined by the two-body problem, we will derive optimal transfer solutions while utilizing the closed-form solutions for J_2-perturbed equatorial motion developed in Chap. 12. We will call these solutions "modified solutions". What we call "classical solutions", can be obtained by simply substituting $J_2 = 0$. Thus, we will obtain more accurate solutions for optimal orbit transfers compared to the classical solutions.

15.2 Modified Hohmann Transfer

As mentioned before, for the two-body problem the optimal two-impulse transfer between two coplanar, non-intersecting circular orbits is the Hohmann transfer (Barrar 1962; Miele and Mathwig 2004; Prussing 1992). Here, we obtain closed-form expressions for the equivalent transfer in the equatorial plane, while taking into account the effect of J_2. This transfer is referred to as the *modified Hohmann transfer*.

Let γ be the angle between the velocity vector and the local horizon (perpendicular to the radius vector), as shown in Fig. 15.1. The notations (1), (2), (m), (A), and (B) denote parameters related to initial orbit, final orbit, transfer orbit, transfer orbit after the first impulse, and transfer orbit before the second impulse, respectively.

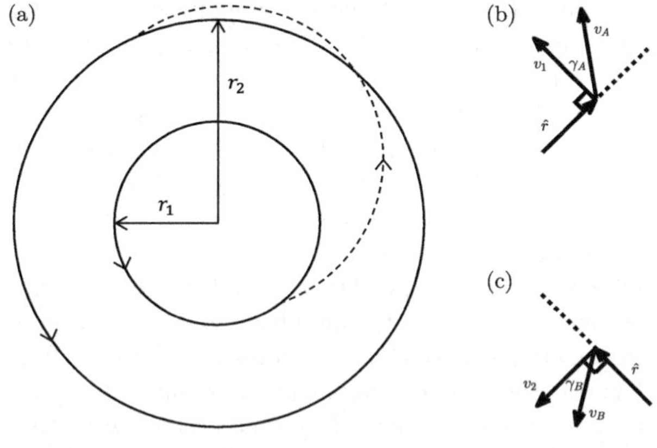

Fig. 15.1 Generic description of a two-impulse transfer. *Left figure*: Initial and final orbits (*solid line*), transfer orbit (*dashed*). *Right figures*: The velocity before and after each impulse. (**a**) The transfer. (**b**) First impulse. (**c**) Second impulse

Figure 15.1 shows a generic two-impulse transfer between two coplanar circular orbits. To optimize the two-impulse maneuver in terms of fuel consumption, the total velocity change is minimized while maintaining the physical constraints of the motion, namely specific energy and angular momentum conservation. Angular momentum conservation means that $h_A = h_B$, where $\mathbf{h} = \mathbf{r} \times \mathbf{v}$. Using the geometry of Fig. 15.1, we obtain the momentum conservation equation

$$r_1 v_A \cos(\gamma_A) = r_2 v_B \cos(\gamma_B) \tag{15.1}$$

The total specific energy of equatorial orbits is given in Eq. (12.5). By using the definition of escape velocity as given in Eq. (12.46), we have

$$\varepsilon = \frac{v^2}{2} - \frac{v_{esc}{}^2(r)}{2} \tag{15.2}$$

Since the impulse is instantaneous, Eq. (15.2) is evaluated for $r = r_A = r_1$ and $r = r_B = r_2$, yielding

$$\varepsilon_A = \frac{v_A{}^2}{2} - \frac{v_{esc}{}^2(r = r_1)}{2} \tag{15.3}$$

$$\varepsilon_B = \frac{v_B{}^2}{2} - \frac{v_{esc}{}^2(r = r_2)}{2} \tag{15.4}$$

Energy conservation suggests that $\varepsilon_A = \varepsilon_B$, and, therefore, we can obtain the energy conservation equation

$$v_A^2 - v_{esc}{}^2(r_1) = v_B^2 - v_{esc}{}^2(r_2) \tag{15.5}$$

By using Eq. (12.58), it is evident that in the equatorial plane the initial and final velocities on the circular orbits are constant, and depend only on the initial and final radii r_1 and r_2,

$$v_1 = \sqrt{\frac{\mu}{r_1}\left(1 + \frac{3J_2 r_{eq}{}^2}{2r_1{}^2}\right)} \tag{15.6}$$

$$v_2 = \sqrt{\frac{\mu}{r_2}\left(1 + \frac{3J_2 r_{eq}{}^2}{2r_2{}^2}\right)} \tag{15.7}$$

Now we split the impulse $\Delta\mathbf{v}$ into radial and transverse components,

$$\Delta\mathbf{v} = \Delta v_r \hat{\mathbf{u}}_r + \Delta v_\theta \hat{\mathbf{u}}_\theta \tag{15.8}$$

In a circular orbit there is no radial velocity, so

$$\Delta v_{\theta 1} = v_A \cos(\gamma_A) - v_1 \tag{15.9}$$

$$\Delta v_{r1} = v_A \sin(\gamma_A) \tag{15.10}$$

$$\Delta v_{\theta 2} = v_2 - v_B \cos(\gamma_B) \tag{15.11}$$

$$\Delta v_{r2} = -v_B \sin(\gamma_B) \tag{15.12}$$

Substituting Eqs. (15.9)–(15.12) into Eq. (15.8) yields

$$\Delta v_1 = \sqrt{v_A^2 + v_1^2 - 2 v_A v_1 \cos(\gamma_A)} \tag{15.13}$$

$$\Delta v_2 = \sqrt{v_B^2 + v_2^2 - 2 v_B v_2 \cos(\gamma_B)} \tag{15.14}$$

To minimize the total velocity impulse while keeping the specific energy and angular momentum constant, we define the augmented cost function (see Sect. 14.5.1)

$$\begin{aligned} L = {} & \Delta v_1 + \Delta v_2 + \lambda_M \left[r_1 v_A \cos(\gamma_A) - r_2 v_B \cos(\gamma_B) \right] \\ & + \lambda_E \left[v_A^2 - v_B^2 - {v_{esc}}^2(r_1) + {v_{esc}}^2(r_2) \right] \end{aligned} \tag{15.15}$$

where λ_M and λ_E are Lagrange multipliers (see Sect. 14.5.1). For finding optimal solutions,

$$\frac{\partial L}{\partial v_A} = \frac{v_A - v_1 \cos(\gamma_A)}{\sqrt{{v_A}^2 + {v_1}^2 - 2\, v_A v_1 \cos(\gamma_A)}} + 2\,\lambda_E v_A + \lambda_M r_1 \cos(\gamma_A) = 0 \tag{15.16}$$

$$\frac{\partial L}{\partial v_B} = \frac{v_B - v_2 \cos(\gamma_B)}{\sqrt{{v_B}^2 + {v_2}^2 - 2\, v_B v_2 \cos(\gamma_B)}} - 2\,\lambda_E v_B - \lambda_M r_2 \cos(\gamma_B) = 0 \tag{15.17}$$

$$\frac{\partial L}{\partial \gamma_A} = \frac{v_A v_1 \sin(\gamma_A)}{\sqrt{{v_A}^2 + {v_1}^2 - 2\, v_A v_1 \cos(\gamma_A)}} - \lambda_M r_1 v_A \sin(\gamma_A) = 0 \tag{15.18}$$

$$\frac{\partial L}{\partial \gamma_B} = \frac{v_B v_2 \sin(\gamma_B)}{\sqrt{{v_B}^2 + {v_2}^2 - 2\, v_B v_2 \cos(\gamma_B)}} + \lambda_M r_2 v_B \sin(\gamma_B) = 0 \tag{15.19}$$

$$\frac{\partial L}{\partial \lambda_E} = v_A^2 - v_B^2 - {v_{esc}}^2(r_1) + {v_{esc}}^2(r_2) = 0 \tag{15.20}$$

$$\frac{\partial L}{\partial \lambda_M} = r_1 v_A \cos(\gamma_A) - r_2 v_B \cos(\gamma_B) = 0 \tag{15.21}$$

Equations (15.16)–(15.21) yield several solutions, shown in Table 15.1.

For a closed orbit, the velocity cannot be zero, and, therefore, Solutions 1–8 are infeasible. A negative velocity magnitude is not physical, so Solutions 9 and 10 are infeasible as well.

Table 15.1 Two-impulse optimal transfer—list of solutions

Solution	v_A	v_B	γ_A [deg]	γ_B [deg]
1	$+\sqrt{v_{esc}{}^2(r_1) - v_{esc}{}^2(r_2)}$	0	+90	± 90
2	$+\sqrt{v_{esc}{}^2(r_1) - v_{esc}{}^2(r_2)}$	0	−90	± 90
3	$-\sqrt{v_{esc}{}^2(r_1) - v_{esc}{}^2(r_2)}$	0	+90	± 90
4	$-\sqrt{v_{esc}{}^2(r_1) - v_{esc}{}^2(r_2)}$	0	−90	± 90
5	0	$+\sqrt{v_{esc}{}^2(r_2) - v_{esc}{}^2(r_1)}$	+90	± 90
6	0	$+\sqrt{v_{esc}{}^2(r_2) - v_{esc}{}^2(r_1)}$	−90	± 90
7	0	$-\sqrt{v_{esc}{}^2(r_2) - v_{esc}{}^2(r_1)}$	+90	± 90
8	0	$-\sqrt{v_{esc}{}^2(r_2) - v_{esc}{}^2(r_1)}$	−90	± 90
9	$\pm\sqrt{\frac{r_2^2(v_{esc}{}^2(r_1)-v_{esc}{}^2(r_2))}{r_2^2-r_1^2}}$	$\mp\sqrt{\frac{r_1^2(v_{esc}{}^2(r_1)-v_{esc}{}^2(r_2))}{r_2^2-r_1^2}}$	180	0
10	$\pm\sqrt{\frac{r_2^2(v_{esc}{}^2(r_1)-v_{esc}{}^2(r_2))}{r_2^2-r_1^2}}$	$\mp\sqrt{\frac{r_1^2(v_{esc}{}^2(r_1)-v_{esc}{}^2(r_2))}{r_2^2-r_1^2}}$	0	180
11	$\pm\sqrt{\frac{r_2^2(v_{esc}{}^2(r_1)-v_{esc}{}^2(r_2))}{r_2^2-r_1^2}}$	$\pm\sqrt{\frac{r_1^2(v_{esc}{}^2(r_1)-v_{esc}{}^2(r_2))}{r_2^2-r_1^2}}$	180	180
12	$\pm\sqrt{\frac{r_2^2(v_{esc}{}^2(r_1)-v_{esc}{}^2(r_2))}{r_2^2-r_1^2}}$	$\pm\sqrt{\frac{r_1^2(v_{esc}{}^2(r_1)-v_{esc}{}^2(r_2))}{r_2^2-r_1^2}}$	0	0

Table 15.2 Two impulse optimal transfer—the feasible solutions

	Δv_1	Δv_2	Result
Solution 11	$\|v_1 + v_A\|$	$\|v_2 + v_B\|$	Maximum
Solution 12	$\|v_1 - v_A\|$	$\|v_2 - v_B\|$	Minimum

To detect which of Solutions 11 and 12 provides a minimum, the total impulse magnitude is calculated, see Table 15.2. It is readily seen that Solution 12 is the optimal solution. From Solution 12 we obtain

$$\Delta v_1 = v_A - v_1 = \sqrt{\frac{r_2^2\left[v_{esc}{}^2\left(r_1\right) - v_{esc}{}^2\left(r_2\right)\right]}{r_2^2 - r_1^2}} - v_1 \tag{15.22}$$

$$\Delta v_2 = v_2 - v_B = v_2 - \sqrt{\frac{r_1^2\left[v_{esc}{}^2\left(r_1\right) - v_{esc}{}^2\left(r_2\right)\right]}{r_2^2 - r_1^2}} \tag{15.23}$$

Figure 15.2 depicts the maneuver resulting from Solution 12. Recall our discussion of the classical Hohmann transfer in the previous chapter. It is evident that, as in the classical Hohmann transfer, the resulting transfer orbit is bounded between the initial and the final circles. But in contrast to the classical transfer, the new transfer orbit is not an ellipse, which is manifested by the fact that the orbital angle φ_m is larger than 180°.

As in the classical Hohmann transfer, the obtained optimal maneuver requires velocity impulses in the direction of the instantaneous velocity, causing the transfer orbit to be between perigee and apogee, and the spacecraft completes half a period

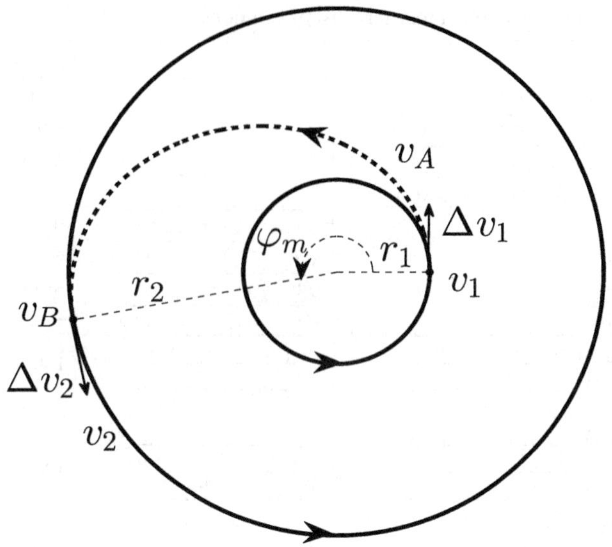

Fig. 15.2 J_2-modified Hohmann transfer: Initial and final orbits (*solid thick line*), transfer orbit (*dashed thick line*)

Table 15.3 Parameter values for J_2-modified and classical Hohmann transfers

Parameter	J_2-modified Hohmann transfer	Classical Hohmann transfer
v_1 [km/s]	7.661688	7.656220
Δv_1 [km/s]	0.303746	0.304338
Δv_2 [km/s]	0.291709	0.292211
v_2 [km/s]	7.062329	7.058686
T_m [s]	3169.51	3167.58
φ_m [deg]	180.2202	180.0000

on the transfer orbit. However, although the maneuvers seem similar, the presence of J_2 affects the optimal maneuver as follows:

1. The impulse size changes due to the different values of $v_{esc}(r_1)$, $v_{esc}(r_2)$, v_1, v_2.
2. The transfer trajectory changes due to the difference in the angle φ_m between the (instantaneous) perigee and apogee, and the consequent change in the orbit shape, as shown in Fig. 15.2.
3. The transfer time is different compared to the classical Hohmann transfer.

As an example, a simulation is performed with the scenario being a transfer between a circular orbit with a radius of $r_1 = 6800$ km and a circular orbit with a radius of $r_2 = 8000$ km. The orbital perturbations include a complete geopotential model, lunisolar perturbations, and drag. The initial impulse is applied at the instantaneous apogee, for the sake of comparison with the classical Hohmann transfer. The parameter values used in the simulation can be seen in Table 15.3.

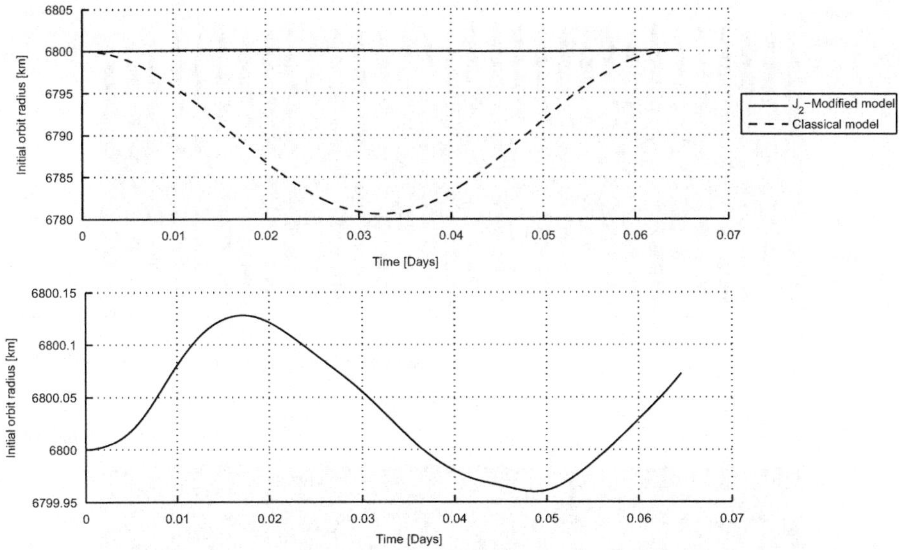

Fig. 15.3 Simulation results of the equatorial Hohmann transfer—showing the initial orbit radius (J_2-modified transfer result magnified at the bottom graph)

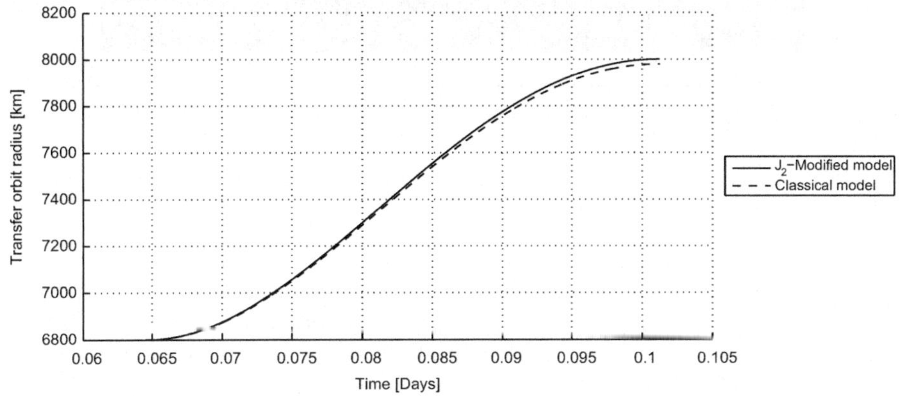

Fig. 15.4 Simulation results of the equatorial Hohmann transfer—showing the transfer orbit radius

The first implementation of the model is an equatorial transfer. The results are shown in Figs. 15.3, 15.4, 15.5, which compare the time histories of the orbital radius for the initial, transfer, and final orbits, respectively, between the classical and modified Hohmann transfers. The circular orbits are generated as explained in Sect. 12.3.4.

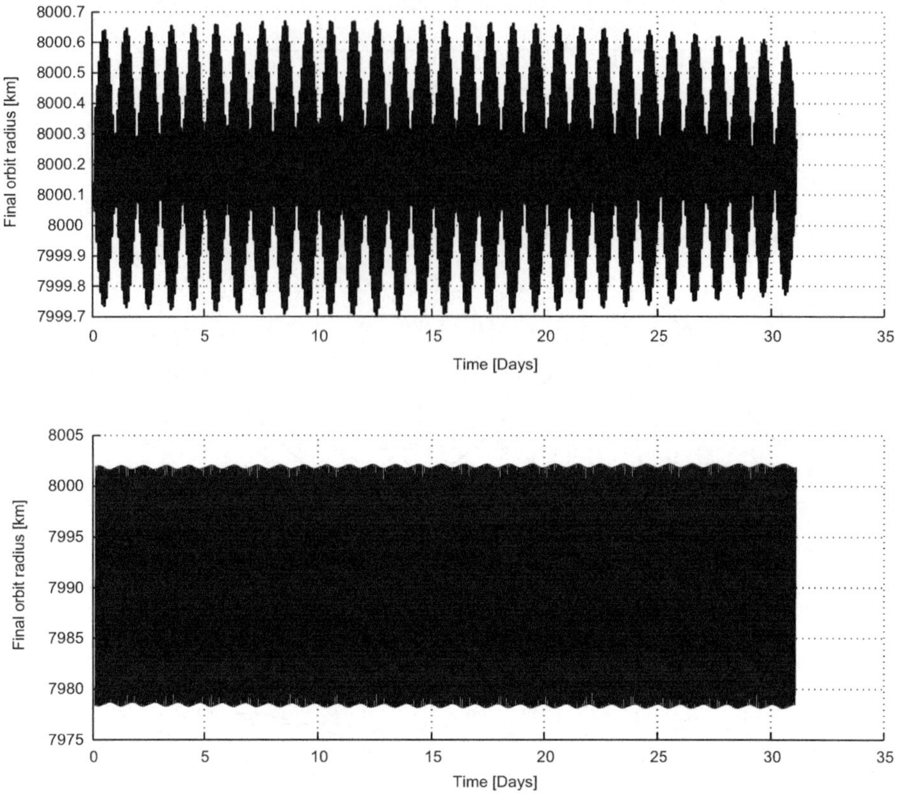

Fig. 15.5 Simulation results of the equatorial Hohmann transfer, performed using the J_2-modified transfer (*top*) and the classical transfer (*bottom*)—showing the final orbit radius

Figure 15.3 shows that the initial orbit obtained using the modified initial conditions, corresponding to the circular equatorial orbit in the presence of J_2, is quite accurate, with the deviation being less than 150 m even in the presence of perturbations other than J_2.

Figure 15.4 shows that the classical Hohmann transfer is unable to converge to the final orbit radius, and the transfer orbit ends 21 km short of r_2.

Figure 15.5 depicts the final orbit radius evolution. Even after a period of one month, the orbital radius remains tightly bounded, with a maximum deviation of less than 1 km, while the classical transfer leads to a deviation of up to 20 km from the target orbit radius.

Consequently, the J_2-modified Hohmann transfer yields improved results compared to the classical transfer even in the presence of other perturbations.

Examining the evolution of the new orbital elements, defined in Eqs. (12.38)–(12.40), provides another insight to the effect of perturbations during the transfer, which may not be visible by examining only the radius of the orbit. Figures 15.6

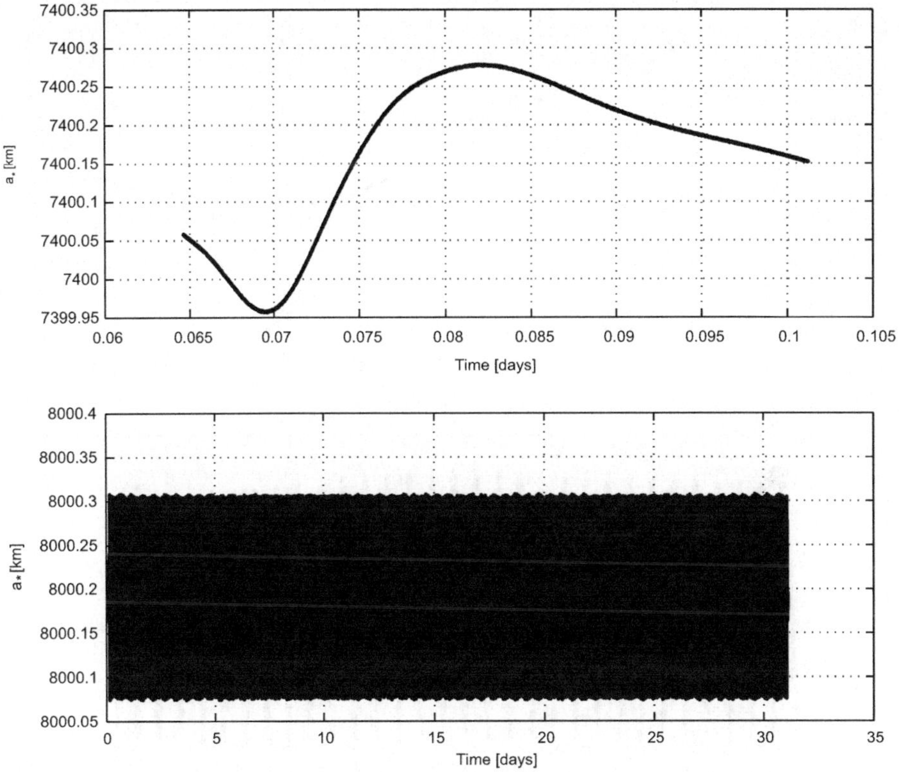

Fig. 15.6 Simulation results of the modified equatorial Hohmann transfer—showing a_* during the transfer (*top*) and final (*bottom*) orbits

and 15.7 show the evolution of the elements a_* and e_* during the transfer and the final orbits of the modified Hohmann maneuver, obtained by using Eqs. (15.22) and (15.23). It is possible to see that the new elements remain almost constant during the motion. Consequently, the cumulative effect of all perturbations except J_2 is small during the transfer.

To extend the possible applications of the modified maneuver, we test the model for a moderately inclined orbit, by performing the maneuver given by Eqs. (15.22)–(15.23) with an initial inclination of 30°.

Figure 15.8 shows the initial orbit radius. The deviation is still greatly reduced—from almost 20 km to less than 5 km—with respect to the Keplerian initialization.

Figure 15.9 shows the transfer orbit radius. As in the equatorial case, the classical Hohmann transfer is unable to achieve the final orbit radius, and the transfer orbit ends 18 km short of r_2. In contrast, the distance from the center of attraction at the end of the modified Hohmann transfer deviates from the nominal value of r_2 by 3 km only.

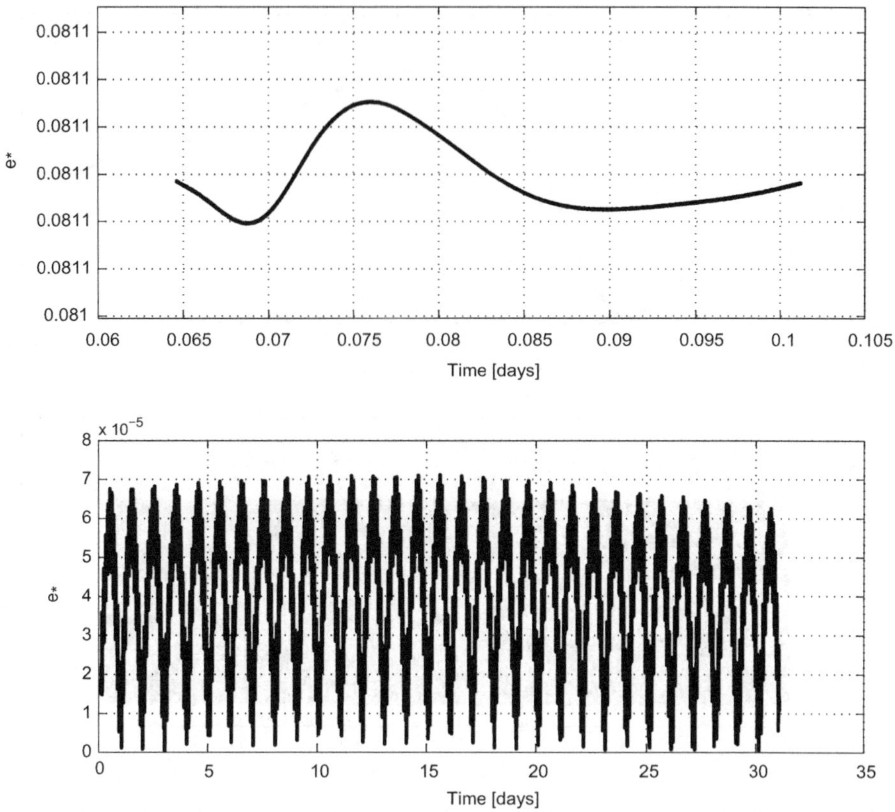

Fig. 15.7 Simulation results of the modified equatorial Hohmann transfer—showing e_* during the transfer (*top*) and final (*bottom*) orbits

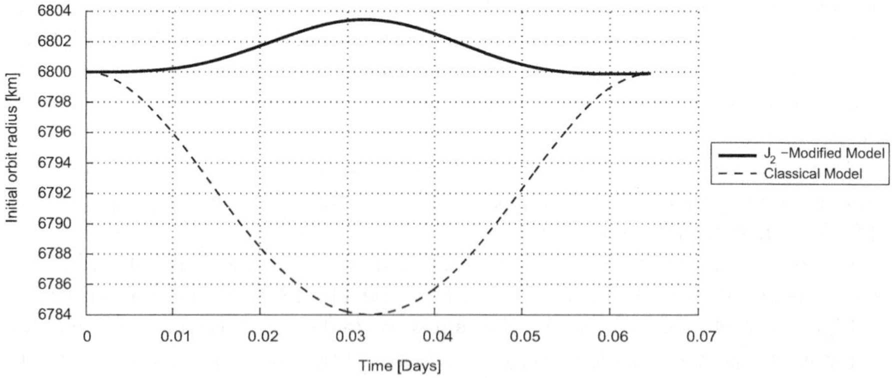

Fig. 15.8 Simulation result of a 30° inclined Hohmann transfer—showing the initial orbit radius

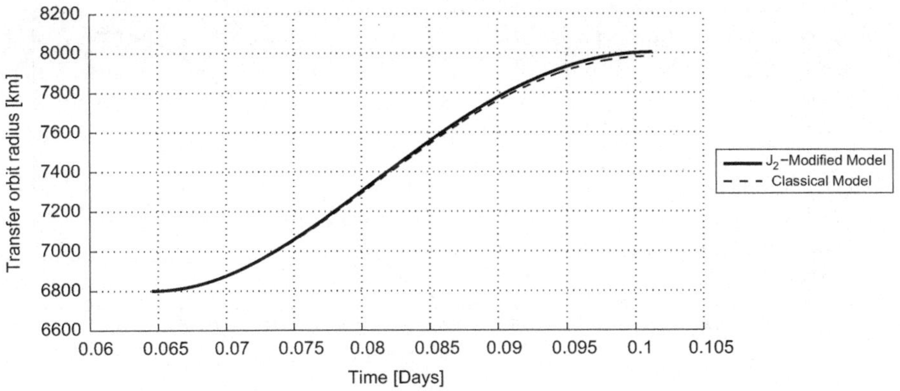

Fig. 15.9 Simulation results of a 30° inclined Hohmann transfer—showing the transfer orbit radius

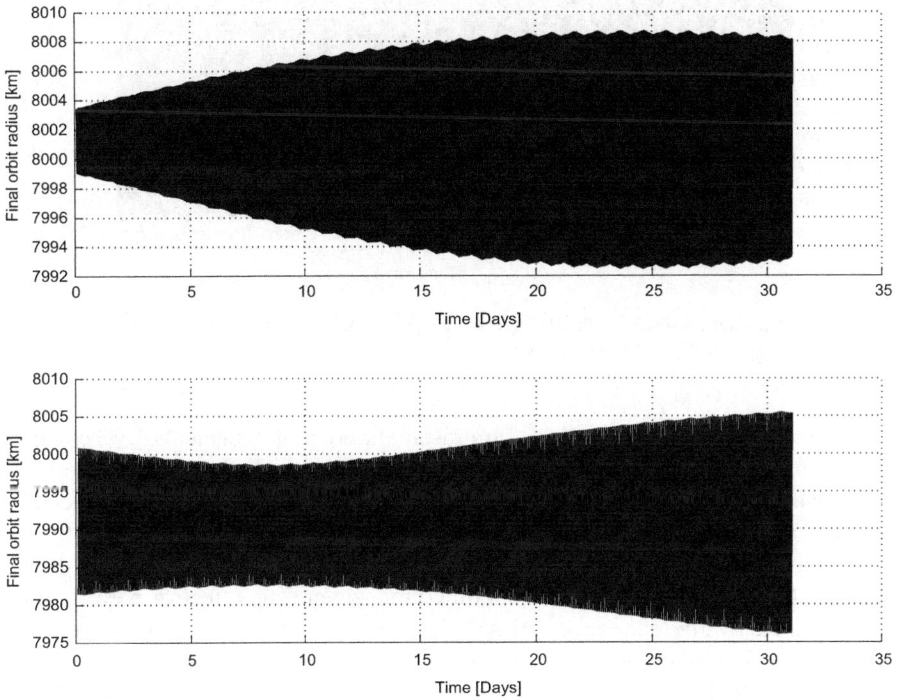

Fig. 15.10 Simulation results of the Hohmann transfer, performed with an initial inclination of 30°, by using a J_2 modified transfer (*top*) and the classical transfer (*bottom*)—showing the final orbit radius

Figure 15.10 shows the final orbit radius. It is possible to see that the error of the modified transfer is significantly smaller than the classical transfer, with an error variation of +8.65 km/−7.48 km compared to +5.45 km/−23.89 km.

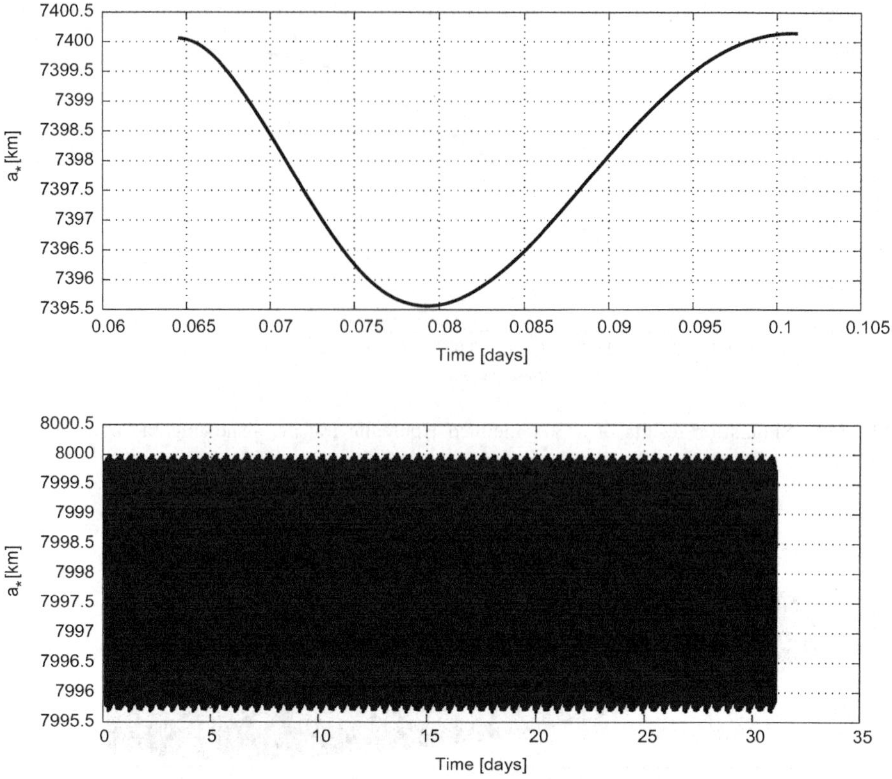

Fig. 15.11 Simulation results of a 30° inclined modified Hohmann transfer—showing a_*, during the transfer (*top*) and final (*bottom*) orbits

Another noticeable fact is that in contrast to the equatorial maneuver, we can now see a significant change in the radial distance amplitude, which is mostly due to J_3. The zonal harmonic J_3 may have a significant effect on the eccentricity, especially in inclined near-circular orbits.

Figures 15.11 and 15.12 show the evolution of a_* and e_*, respectively, during the transfer and final orbits of the modified maneuver. The effect of perturbations is illustrated by the change in orbital elements, and is significantly larger than the minor effect which was seen in the equatorial transfer. This confirms that the equatorial model does not accurately describe the motion outside the equatorial plane, and the change in orbital elements can be seen as an indicator to the ability of the model to accurately describe the motion. However, this does not diminish the improvement in comparison to the classical transfer.

Finally, it is emphasized that for inclinations larger than 60°, the modified solution does not necessarily lead to improved results. In fact, for higher inclinations it can produce larger errors than the classical transfer.

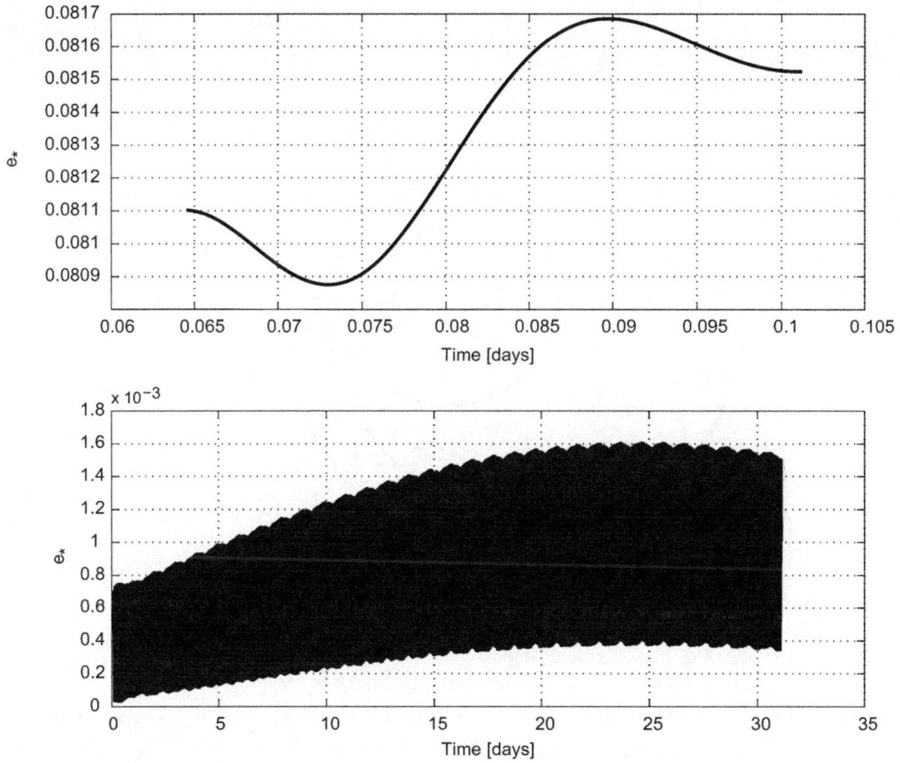

Fig. 15.12 Simulation results of a 30° inclined modified Hohmann transfer—showing e_*, during the transfer (*top*) and final (*bottom*) orbits

15.3 Modified Bi-Elliptic and Bi-Parabolic Transfers

In this section, we will derive a more efficient maneuver than the modified Hohmann transfer, by optimizing the three-impulse transfer between coplanar circular orbits in the equatorial plane, while analytically taking into account the effect of J_2. It is important to note that although the transfers are called "modified bi-elliptic" and "modified bi-parabolic", the orbit shape is neither an ellipse nor a parabola.

15.3.1 Definitions

The classical bi-elliptic transfer consists of two Hohmann transfers: one between the initial orbit with radius r_1 and a user-defined mid-orbit with a radius of r_3, and a second Hohmann transfer between the mid-orbit and the final orbit, with a radius of r_2. Depending on the value of the radius r_3, three distinctive possible transfers can

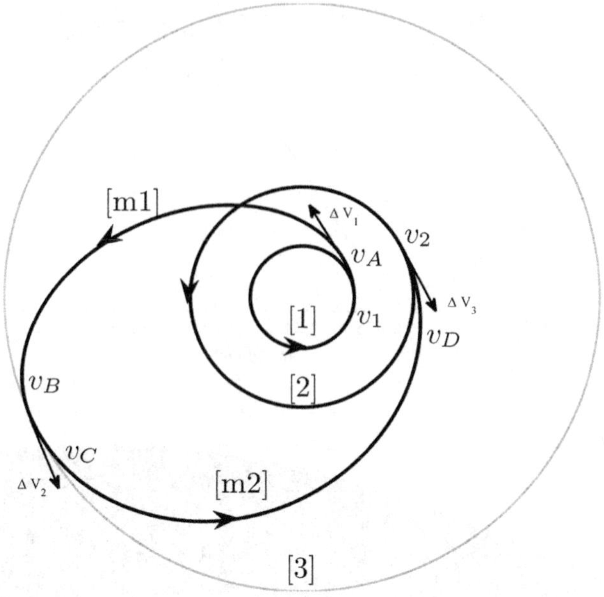

Fig. 15.13 External bi-elliptic transfer

be performed; a transfer where $r_3 < r_1$; a transfer where $r_1 < r_3 < r_2$; and a transfer where $r_2 < r_3$. The first two cases are called "internal transfers", and the third case is referred to as an "external transfer". Hoelker and Silber (1961) showed that for a Keplerian motion, the optimal maneuver is obtained for an external transfer, while internal maneuvers are more fuel-expensive than the Hohmann transfer; therefore, we focus on external maneuvers. An illustration of an external maneuver can be seen in Fig. 15.13.

In our discussion, the following notation will be used: (1)—initial orbit, (2)—final orbit, (3)—mid-orbit, (m1)—first transfer orbit, (m2)—second transfer orbit, (A)—first transfer orbit after the first impulse, (B)—first transfer orbit before the second impulse, (C)—second transfer orbit after the second impulse, (D)—second transfer orbit before the third impulse.

15.3.2 *Modified Bi-Elliptic Transfer*

We can also extend the classical bi-elliptic transfer to the case of motion under J_2. This transfer will be referred to as the *modified bi-elliptic transfer*. The velocity impulses are given by

$$\Delta v_1 = v_A - v_1 \tag{15.24}$$

$$\Delta v_2 = v_C - v_B \tag{15.25}$$

$$\Delta v_3 = v_2 - v_D \tag{15.26}$$

Since the initial, final, and mid-orbits are circular, the velocities are given by

$$v_i = \sqrt{\frac{\mu}{r_i}\left(1 + \frac{3J_2 r_{eq}{}^2}{2r_i{}^2}\right)}, \quad i = 1, 2, 3 \tag{15.27}$$

Next, we use Eqs. (15.22) and (15.23), applied to a modified Hohmann transfer from a circular orbit with $r = r_1$ to a circular orbit with $r = r_3$, and obtain

$$v_A = \sqrt{\frac{r_3^2\left[v_{esc}{}^2(r_1) - v_{esc}{}^2(r_3)\right]}{r_3^2 - r_1^2}} \tag{15.28}$$

$$v_B = \sqrt{\frac{r_1^2\left[v_{esc}{}^2(r_1) - v_{esc}{}^2(r_3)\right]}{r_3^2 - r_1^2}} \tag{15.29}$$

The additional modified Hohmann transfer from a circular orbit with $r = r_3$ to a circular orbit with $r = r_2$ requires that

$$v_C = \sqrt{\frac{r_2^2\left[v_{esc}{}^2(r_3) - v_{esc}{}^2(r_2)\right]}{r_2^2 - r_3^2}} \tag{15.30}$$

$$v_D = \sqrt{\frac{r_3^2\left[v_{esc}{}^2(r_3) - v_{esc}{}^2(r_2)\right]}{r_2^2 - r_3^2}} \tag{15.31}$$

where $v_{esc}(r)$ is given in Eq. (12.46). The total impulse (for each component the positive value for an external transfer is selected) is given by

$$\Delta v_{total} = v_A - v_1 + v_C - v_B - v_2 + v_D \tag{15.32}$$

Next, some algebraic manipulations are performed to simplify Eq. (15.32). First,

$$v_A - v_B = \sqrt{\frac{r_3 - r_1}{r_3 + r_1}\left[v_{esc}{}^2(r_1) - v_{esc}{}^2(r_3)\right]} \tag{15.33}$$

$$v_C + v_D = \sqrt{\frac{r_3 + r_2}{r_3 - r_2}\left[v_{esc}{}^2(r_2) - v_{esc}{}^2(r_3)\right]} \tag{15.34}$$

Now, we define

$$J_e \triangleq \frac{J_2 r_{eq}{}^2}{2r_1{}^2}, \quad X \triangleq \frac{r_3}{r_1}, \quad Y \triangleq \frac{r_2}{r_1}, \quad K \triangleq \sqrt{\frac{2\mu}{r_1}} \tag{15.35}$$

Using this notation, the expressions for the three escape velocities are

$$v_{esc}(r_1) = \sqrt{\frac{2\mu}{r_1}(1+J_e)} \tag{15.36}$$

$$v_{esc}(r_2) = \sqrt{\frac{2\mu}{r_1 Y}\left(1+\frac{J_e}{Y^2}\right)} \tag{15.37}$$

$$v_{esc}(r_3) = \sqrt{\frac{2\mu}{r_1 X}\left(1+\frac{J_e}{X^2}\right)} \tag{15.38}$$

Substituting Eqs. (15.36), (15.37) and (15.38) into Eqs. (15.33) and (15.34) yields

$$v_A - v_B = \sqrt{\left(\frac{2\mu}{r_1}\right)\frac{r_3-r_1}{r_3+r_1}\left[(1+J_e)-\left(\frac{1}{X}+\frac{J_e}{X^3}\right)\right]} \tag{15.39}$$

$$v_C + v_D = \sqrt{\left(\frac{2\mu}{r_1}\right)\frac{r_3+r_2}{r_3-r_2}\left[\left(\frac{1}{Y}+\frac{J_e}{Y^3}\right)-\left(\frac{1}{X}+\frac{J_e}{X^3}\right)\right]} \tag{15.40}$$

Normalizing Eqs. (15.39) and (15.40) by using Eq. (15.35) gives

$$\frac{v_A - v_B}{K} = \sqrt{\frac{X-1}{X+1}\left[(1+J_e)-\left(\frac{1}{X}+\frac{J_e}{X^3}\right)\right]} \tag{15.41}$$

$$\frac{v_C + v_D}{K} = \sqrt{\frac{X+Y}{X-Y}\left[\left(\frac{1}{Y}+\frac{J_e}{Y^3}\right)-\left(\frac{1}{X}+\frac{J_e}{X^3}\right)\right]} \tag{15.42}$$

Differentiating Eqs. (15.41) and (15.42) with respect to X is equivalent to differentiation with respect to r_3 (since r_1 is constant), so

$$\frac{\partial\left(\frac{v_A-v_B}{K}\right)}{\partial X} = \frac{(X-1)(3J_e+5J_eX+2J_eX^2+2J_eX^3+X^2+3X^3)}{2X^4(X+1)^2\sqrt{\frac{(X-1)^2(J_e+J_eX+J_eX^2+X^2)}{X^3(X+1)}}} \tag{15.43}$$

$$\frac{\partial\left(\frac{v_C+v_D}{K}\right)}{\partial X} = \frac{-(X^2Y^2+2J_eX^2+4J_eXY+3J_eY^2)}{2X^4Y^2\sqrt{\frac{(X+Y)(X^2Y^2+J_eX^2+J_eXY+J_eY^2)}{X^3Y^3}}} \tag{15.44}$$

To find the optimal solution, we need to differentiate Eq. (15.32) with respect to r_3. This is equivalent to differentiating Eq. (15.32), normalized by the constant value K, with respect to X. By using Eqs. (15.43) and (15.44), and adopting the notation

$$D_X \triangleq \frac{\partial\left(\frac{\Delta v_{total}}{K}\right)}{\partial X} \tag{15.45}$$

we obtain

$$D_X = \frac{(X-1)(3J_e + 5J_eX + 2J_eX^2 + 2J_eX^3 + X^2 + 3X^3)}{2X^4(X+1)^2\sqrt{\frac{(X-1)^2(J_e+J_eX+J_eX^2+X^2)}{X^3(X+1)}}} - \frac{(X^2Y^2 + 2J_eX^2 + 4J_eXY + 3J_eY^2)}{2X^4Y^2\sqrt{\frac{(X+Y)(X^2Y^2+J_eX^2+J_eXY+J_eY^2)}{X^3Y^3}}} \tag{15.46}$$

Because the value of D_X depends on X, Y and J_e, trying to directly solve Eq. (15.46) will lead to a four dimensional surface, which contains the extremum points. To circumvent the complexity of the direct solution, an alternative approach is used (Hoelker and Silber 1961), containing the following steps:

1. Finding a critical value of Y, denoted by Y_{crit}, for which the modified bi-elliptic transfer is more economic than the modified Hohmann transfer for at least some mid-orbits.
2. Showing that for Y_{crit}, the modified bi-elliptic transfer is more economic than the modified Hohmann transfer for all mid orbits, and that by increasing the value of X, a more economic solution is obtained.
3. Extending Step 2 for all $Y_{crit} < Y$.

15.3.2.1 Calculating Y_{crit}

It is possible to compare the total velocity impulse of the modified bi-elliptic transfer to the total velocity impulse of the modified Hohmann transfer by imposing the mid-orbit radius to be equal to the final orbit radius ($r_3 = r_2$). At the point where $r_3 = r_2$ or $X = Y$, the total impulse of the modified Hohmann transfer equals the total impulse of the modified bi-elliptic transfer. If, for that point, D_X is negative, the total impulse size begins with a negative slope, and will decrease for a positive increment of X. Since $Y < X$, at least some values of X will provide a more economic solution than the transfer with $X = Y$. Thus,

1. For $X = Y$, the total impulse size of the modified bi-elliptic transfer equals the total impulse size of the modified Hohmann transfer.
2. If, for $X = Y$, D_X is negative, then for some values of X the modified bi-elliptic transfer is more economic than the modified Hohmann transfer.

Considering these facts, first we find the values of Y under the condition $Y = X$, for which D_X is zero. This will provide the critical value above which D_X is negative, and the modified bi-elliptic transfer is more economic.

Equating Eq. (15.46) to zero and simplifying, while considering the fact that $1 \leq Y \leq X$, leads to

$$
\begin{aligned}
&(3J_e + 5J_eX + 2J_eX^2 + 2J_eX^3 + X^2 + 3X^3) \\
&\times \sqrt{Y(X+Y)(X^2Y^2 + J_eX^2 + J_eXY + J_eY^2)} \\
&- (X^2Y^2 + 2J_eX^2 + 4J_eXY + 3J_eY^2)\sqrt{(J_e + J_eX + J_eX^2 + X^2)(X+1)^3} \\
&= 0
\end{aligned}
\tag{15.47}
$$

Setting $X = Y$ gives

$$
\begin{aligned}
&\sqrt{2}\sqrt{Y^2 + 3J_e}(3J_e + 5J_eY + 2J_eY^2 + 2J_eY^3 + Y^2 + 3Y^3) \\
&- (Y^2 + 9J_e)\sqrt{(Y+1)^3(J_e + J_eY + J_eY^2 + Y^2)} = 0
\end{aligned}
\tag{15.48}
$$

Equation (15.48) depends on r_1 through J_e, and for $J_e = J_2 = 0$

$$\sqrt{2}(1 + 3Y) - (Y+1)^{\frac{3}{2}} = 0. \tag{15.49}$$

The positive solution of Eq. (15.49) is $Y_{crit} = 15.58171873876$, which corresponds to the classical solution obtained by Hoelker and Silber (1961).

At this point, to solve Eq. (15.48), a numerical procedure will be employed. In order to accommodate possible solutions, we need to approximate the range of the different variables appearing in Eq. (15.48). The sphere of influence radius can be approximated using Eq. (1.19).

For Earth, $r_{SOI} \approx 925,000$ km. Since the initial radius cannot be smaller than r_{eq}, the values of X and Y cannot extend beyond 150 without exceeding the SOI radius. Therefore, after taking some margin of error, the variables can be limited as follows:

$$r_{eq} < r_1 < 1000000 \text{ km} \tag{15.50}$$

$$1 < Y < 156 \tag{15.51}$$

$$Y < X < 156 \tag{15.52}$$

Numerically solving Eq. (15.48) for Y while using the values relevant for Earth, namely

$$J_2 = 0.0010826269, \; r_{eq} = 6378.1363 \text{ km}, \; \mu = 398600.4415 \text{ km}^3/\text{sec}^2 \tag{15.53}$$

and denoting the solution by Y_{crit}, will yield the results shown in Fig. 15.14.

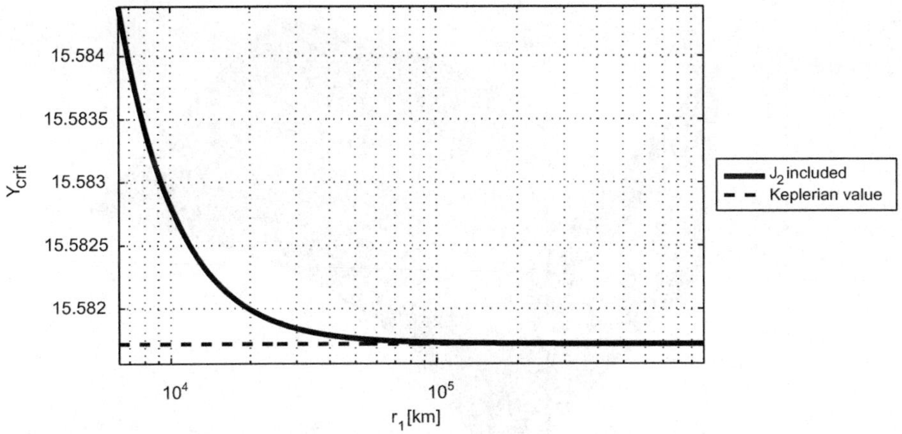

Fig. 15.14 The final-to-initial radii ratio (Y_{crit}) for which the modified bi-elliptic transfer is more economic than the modified Hohmann transfer given an external mid-orbit, as a function of the initial radius

The results shown in Fig. 15.14 are the values of Y for which the normalized total impulse of the modified bi-elliptic transfer is equal to the normalized total impulse of the modified Hohmann transfer, and D_X is zero.

Evaluation of $\partial D_X/\partial X$ shows that this derivative is negative. Therefore, for some values of X that satisfy the condition $Y_{crit} = Y < X$, we obtain $D_X < 0$, and it is possible to conclude that for every value of Y_{crit} there exist at least some values of X providing a modified bi-elliptic transfer, which is more economic than the modified Hohmann transfer.

15.3.2.2 Evaluating the effect of X on maneuvers where $Y = Y_{crit}$

By using the obtained values $Y = Y_{crit}$ and their compatible values of J_e, it is now possible to evaluate the effect of X on the bi-elliptic transfers while considering only two variables (since J_e is directly derived from Y_{crit}). Numerical evaluation of Eq. (15.46), for all values of Y_{crit} (and the corresponding values of J_e) and all values of $Y_{crit} < X$ can be seen in Fig. 15.15.

The derivative is always negative. Therefore, we can conclude that for values of $Y \approx Y_{crit}$, all modified bi-elliptic maneuvers are more economic than the modified Hohmann transfer. Moreover, the maneuver becomes more economic as the value of X becomes larger.

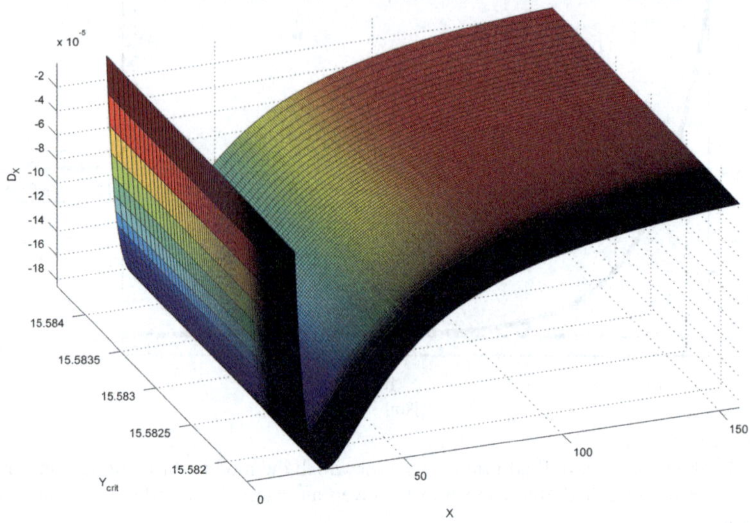

Fig. 15.15 Partial derivative of the normalized total impulse with respect to X for all values of $Y = Y_{crit}$ and $Y_{crit} < X$

15.3.2.3 Extending the evaluation to include all $Y_{crit} < Y$

Finally, we calculate the normalized derivative of the total impulse with respect to X, from Eq. (15.46), for

$$r_{eq} < r_1 < 1000000\,\text{km} \tag{15.54}$$

$$Y_{crit}(r_1) \leq Y < \frac{1000000}{r_1} \tag{15.55}$$

$$Y < X < \frac{1000000}{r_1} \tag{15.56}$$

The result is a 4 dimensional surface. The value of the normalized derivative for all the values listed above is

$$-5.2935\;10^{-4} \leq D_X \leq -1.6044\;10^{-5} \tag{15.57}$$

By evaluating only for values where $Y \geq Y_{crit}$, it is possible to exclude the regions that contain extremum points. By doing so, we obtain a region where the derivative is always negative. Consequently,

1. For $Y \geq Y_{crit}$, all modified bi-elliptic maneuvers are more fuel-efficient than the modified Hohmann transfer.
2. For $Y \geq Y_{crit}$, increasing the value of X will improve the fuel efficiency of the transfer.
3. Under the condition that $Y \geq Y_{crit}$, the optimal modified bi-elliptic transfer is obtained for $X \to \infty$.

15.3.3 Modified Bi-Parabolic Transfer

As shown in the previous section, the optimal modified bi-elliptic transfer is obtained for $X \to \infty$, where the total impulse size is asymptotically converging to a constant value. This value is a particular case of the modified bi-elliptic transfer, called the *modified bi-parabolic transfer*. By setting $r_3 \to \infty$ (which corresponds to $X \to \infty$) in Eqs. (15.27)–(15.31) we obtain

$$v_1 = \sqrt{\frac{\mu}{r_1}\left(1 + \frac{3J_2{R_e}^2}{2{r_1}^2}\right)} \tag{15.58}$$

$$v_2 = \sqrt{\frac{\mu}{r_2}\left(1 + \frac{3J_2{R_e}^2}{2{r_2}^2}\right)} \tag{15.59}$$

$$v_3 = 0 \tag{15.60}$$

$$v_A = v_{esc}(r_1) \tag{15.61}$$

$$v_B = 0 \tag{15.62}$$

$$v_C = 0 \tag{15.63}$$

$$v_D = v_{esc}(r_2) \tag{15.64}$$

The impulses are

$$\Delta v_1 = v_A - v_1 \tag{15.65}$$

$$\Delta v_2 \approx 0 \tag{15.66}$$

$$\Delta v_3 = v_2 - v_D \tag{15.67}$$

This is the optimal maneuver (among all modified bi-elliptic maneuvers) for all values of $Y \geq Y_{crit}$.

15.4 Comparison Between the Modified Bi-Parabolic and the Modified Hohmann Transfers

To compare the modified Hohmann transfer to the modified bi-parabolic transfer, we first calculate the point at which the two transfers yield the same total impulse,

$$(v_{A,Hoh} - v_1) + (v_2 - v_{B,Hoh}) = (v_{A,Bip} - v_1) + (v_{D,Bip} - v_2) \tag{15.68}$$

where the subscripts "Hoh" and "Bip" denote the values for the modified Hohmann and modified bi-parabolic transfers, respectively. Substituting from Eqs. (15.61), (15.64), (15.22) and (15.23) yields

$$\sqrt{\frac{r_2^2\left({v_{esc}}^2(r_1) - {v_{esc}}^2(r_2)\right)}{r_2^2 - r_1^2}} - \sqrt{\frac{r_1^2\left({v_{esc}}^2(r_1) - {v_{esc}}^2(r_2)\right)}{r_2^2 - r_1^2}} + v_2$$
$$= v_{esc}(r_1) + v_{esc}(r_2) - v_2 \tag{15.69}$$

Simplifying Eq. (15.69) gives

$$\sqrt{\frac{r_2 - r_1}{r_2 + r_1}\left({v_{esc}}^2(r_1) - {v_{esc}}^2(r_2)\right)} = v_{esc}(r_1) + v_{esc}(r_2) - 2v_2 \tag{15.70}$$

Normalizing the velocities provides the expressions

$$\frac{v_2}{K} = \sqrt{\frac{1}{2Y}\left(1 + 3\frac{J_e}{Y^2}\right)} \tag{15.71}$$

$$\frac{v_{esc}(r_1)}{K} = \sqrt{1 + J_e} \tag{15.72}$$

$$\frac{v_{esc}(r_2)}{K} = \sqrt{\frac{1}{Y}\left(1 + \frac{J_e}{Y^2}\right)} \tag{15.73}$$

Normalizing Eq. (15.70) by K, and substituting Eqs. (15.71), (15.72) and (15.73), gives

$$\sqrt{\frac{Y-1}{Y+1}\left[1+J_e-\frac{1}{Y}\left(1+\frac{J_e}{Y^2}\right)\right]} = \sqrt{1+J_e} + \sqrt{\frac{1}{Y}\left(1+\frac{J_e}{Y^2}\right)} - \sqrt{\frac{2}{Y}\left(1+3\frac{J_e}{Y^2}\right)} \tag{15.74}$$

Re-arranging Eq. (15.74) leads to

$$\begin{aligned}&\sqrt{(Y-1)\left(Y^3+Y^3J_e-Y^2-J_e\right)}\\&+\sqrt{Y+1}\left(\sqrt{2Y^2+6J_e}-\sqrt{Y^3+Y^3J_e}-\sqrt{Y^2+J_e}\right)=0\end{aligned} \tag{15.75}$$

Solving Eq. (15.75) for Earth with values taken from Eq. (15.53) will yield the value of the final-to-initial radii ratio Y, above which the modified bi-parabolic transfer is more economic than the modified Hohmann transfer. This value of Y is denoted by Y_m.

At this stage, several observations can be made:

1. For values of $Y > 1$ the function in Eq. (15.75) is continuous, and has no singularities.
2. The modified Hohmann transfer equals the modified bi-parabolic transfer only for $Y = Y_m$.
3. It has already been proven that for $Y = Y_{crit} > Y_m$, all modified bi-elliptic transfers (which include the modified bi-parabolic transfer) are more economic than the modified Hohmann transfer.

From these observations, we conclude that since there is at least one value of Y, larger than Y_m, for which the modified bi-parabolic transfer is more economic than the modified Hohmann transfer, Y_m represents the minimal value of Y above which all modified bi-parabolic transfers are more economic than the modified Hohmann transfers. The Keplerian value of $Y_m = 11.9387654726$, obtained by Hoelker and Silber (1961), was re-calculated by setting $J_e = 0$, and degenerating Eq. (15.75) to the equation

$$Y(Y-1)+\sqrt{Y+1}\left(\sqrt{2}Y-\sqrt{Y^3}-Y\right)=0 \tag{15.76}$$

In Fig. 15.16, it is possible to see that for small values of r_1, the difference between the Keplerian case and the J_2-perturbed case is the largest. This corresponds to the fact that the effect of J_2 is more pronounced for low orbits.

To compare the obtained solutions, we test the transfers in simulations. The simulations were performed using a full geopotential model, lunisolar attraction, and drag.

15.4.1 Bi-Elliptic Transfer

The first test case will be an equatorial transfer between a circular orbit with a radius of $r_1 = 6800\,\text{km}$, to a coplanar circular orbit with a radius of $r_2 = 108800\,\text{km}$. The modified bi-elliptic transfer is performed using a mid-orbit with a radius of

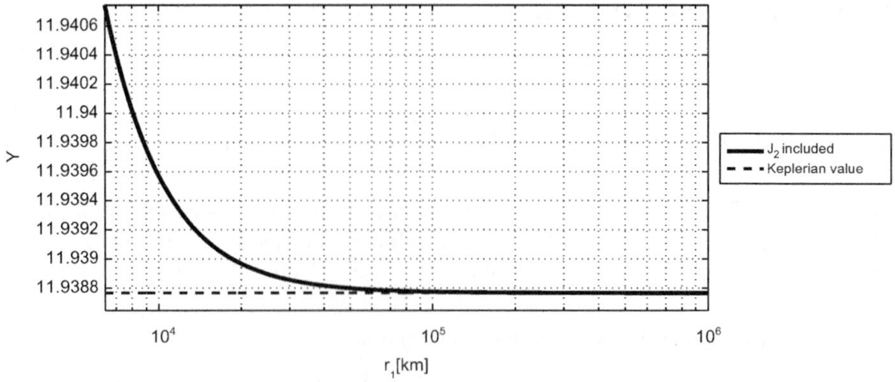

Fig. 15.16 The final-initial radii ratio for which the modified bi-parabolic transfer is more economic than the modified Hohmann transfer

Table 15.4 Parameter values of the J_2 modified bi-elliptic transfer, J_2 modified Hohmann transfer, and the classical bi-elliptic maneuver

Parameter	J_2 modified bi-elliptic transfer	Classical bi-elliptic transfer	J_2 modified Hohmann transfer
v_1 [km/sec]	7.661688	7.656220	7.661688
T_1 [sec]	5576.53	5580.52	5576.53
Δv_1 [km/sec]	2.879713	2.882524	2.845227
T_{m1} [sec]	81704.69	81701.45	69149.99
φ_{m1} [deg]	180.0716	180.0000	180.0726
Δv_2 [km/sec]	1.165079	1.165222	—
T_{m2} [sec]	195577.47	195576.98	—
φ_{m2} [deg]	180.0009	180.0000	—
Δv_3 [km/sec]	0.055491	0.055491	1.257378
v_2 [km/sec]	1.914060	1.914055	1.914060
T_2 [sec]	357152.02	357153.02	357152.02
Δv_{total} [km/sec]	4.100283	4.103238	4.102605
Total transfer time [hr]	77.0228	77.0218	19.2083

$r_3 = 122400$ km. Thus, $Y = 16$ ($Y > Y_{crit}$) and $X = 18$, which suggests that such a transfer is more economic than the modified Hohmann transfer.

We test the improvement obtained by modifying the bi-elliptic transfer by comparing it to the classical bi-elliptic transfer, as well as comparing the modified bi-elliptic transfer to the modified Hohmann transfer. The maneuver parameter values can be seen in Table 15.4.

The modified bi-elliptic transfer is more economic than the modified Hohmann transfer, as expected. However, the small fuel saving comes at the price of a significantly increased maneuver time. It is also interesting to see that the classical

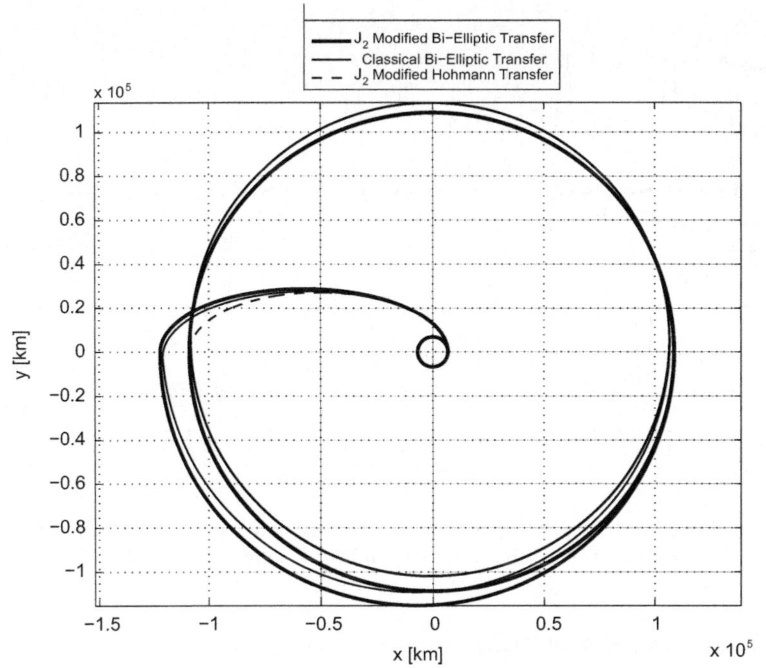

Fig. 15.17 The modified bi-elliptic transfer, the classical bi-elliptic transfer, and the modified Hohmann transfer in the equatorial plane

bi-elliptic transfer is the most expensive maneuver, which leads to the conclusion that by modifying the bi-elliptic transfer we obtain a more accurate maneuver, and at a lower fuel cost. The different transfers can be seen in Fig. 15.17.

The significant deviation of the classical bi-elliptic transfer from the target orbit can be seen in Fig. 15.17. This is a result of an insufficient velocity at the start of the maneuver.

Although it seems as if the orbital angle of the Keplerian transfer is larger than 180°, in reality the insufficient velocity causes the final radius to be smaller than expected, and, therefore, the radial period of the transfer orbit is also smaller than expected. That causes the Keplerian transfer second impulse to be applied after the local apogee, while the radius is decreasing, which leads to additional errors.

The final orbit radius error is depicted in Fig. 15.18. It is possible to see the significant reduction of the final radius error—from more than 6000 km in the classical bi-elliptic maneuver to about 250 km in the modified bi-elliptic transfer. Also, the modified Hohmann transfer produces more accurate results, with a final radius error around 100 km.

In general, for distant orbits, the external perturbations, such as the third-body effect, become significantly larger, and, therefore, the final error remains quite large in all cases.

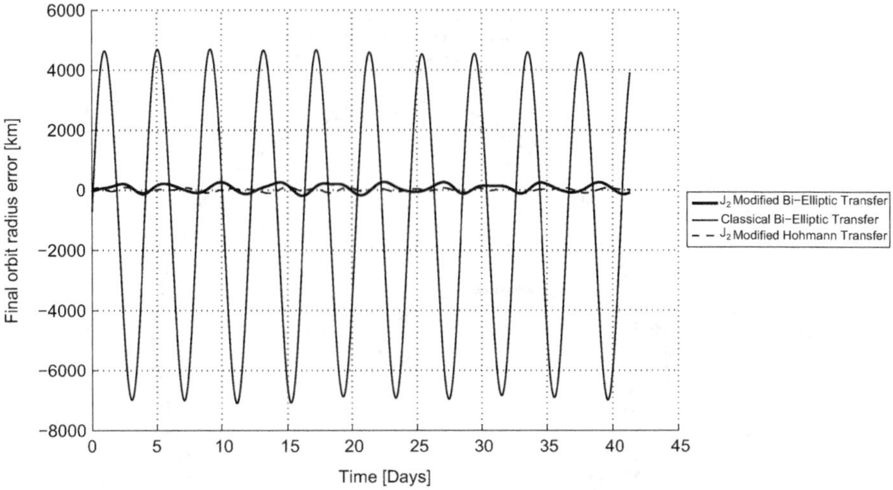

Fig. 15.18 Final orbit radius deviation from the nominal value, obtained using the modified bi-elliptic transfer, the classical bi-elliptic transfer, and the modified Hohmann transfer

15.4.2 Bi-Parabolic Transfer

An orbit with an infinite radius is a theoretical concept, so propagating an ideal bi-parabolic transfer is not possible. Therefore, we should make some approximations. The radial distance, which is considered "infinite", is the SOI border, calculated using Eq. (1.19). The second approximation will be the elimination of the third-body perturbation.

The selected maneuver will be a transfer between an initial circular orbit in the equatorial plane with $r_1 = 6800$ km, to a final circular orbit with $r_2 = 85000$ km, and a mid-orbit with $r_3 = 880000$ km (close to the border of the SOI). In terms of X and Y, the normalized values are $Y = 12.5$ and $X = 129.41$, and since $Y > Y_m$ we can expect the modified bi-parabolic transfer to be more fuel efficient than the modified Hohmann transfer, which will be tested here as well.

The parameter values of the approximated bi-parabolic transfer can be seen in Table 15.5. The resulting transfers can be seen in Fig. 15.19, and the final radius error in Fig. 15.20. Although the modified bi-parabolic transfer is more economic than the modified Hohmann transfer, it considerably extends the transfer time, as expected. In this case, the classical bi-parabolic transfer results in a collision due to the extensive deviation from the target orbit, which is depicted by the straight line trajectory in Fig. 15.19.

The error in the classical bi-parabolic transfer is comprised of three factors: The first is the initialization, which causes the initial velocity to be smaller than required; the second is the transfer orbit, which due to the insufficient initial velocity, reaches

Table 15.5 Parameter values of the approximated J_2 modified bi-parabolic transfer, J_2 modified Hohmann transfer, and the approximated original classical bi-parabolic maneuver

Parameter	J_2 modified bi-parabolic transfer	Classical bi-parabolic transfer	J_2 modified Hohmann transfer
v_1 [km/sec]	7.661687	7.656220	7.661687
T_1 [sec]	5576.53	5580.52	5576.53
Δv_1 [km/sec]	3.126838	3.129718	2.759804
T_{m1} [sec]	1469186.45	1469178.34	48935.59
φ_{m1} [deg]	180.0653	180.0000	180.0750
Δv_2 [km/sec]	0.199115	0.199135	—
T_{m2} [sec]	1667735.42	1667734.69	—
φ_{m2} [deg]	180.0005	180.0000	—
Δv_3 [km/sec]	0.758992	0.758997	1.331797
v_2 [km/sec]	2.165516	2.165506	2.165516
T_2 [sec]	246625.16	246626.29	246625.16
Δv_{total} [km/sec]	4.084945	4.087849	4.091600
Total transfer time [hr]	871.3671	871.3647	13.5932

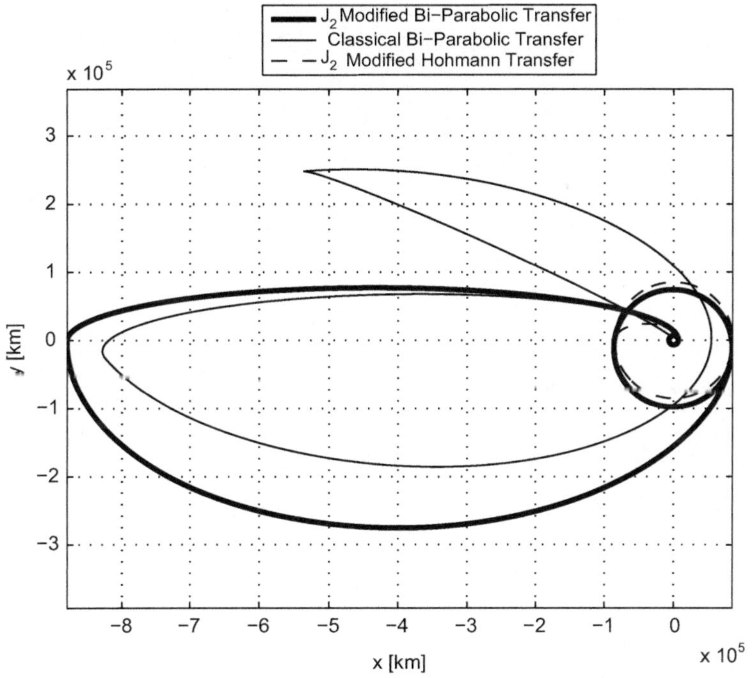

Fig. 15.19 The modified bi-parabolic transfer, the classical bi-parabolic transfer, and the modified Hohmann transfer in the equatorial plane

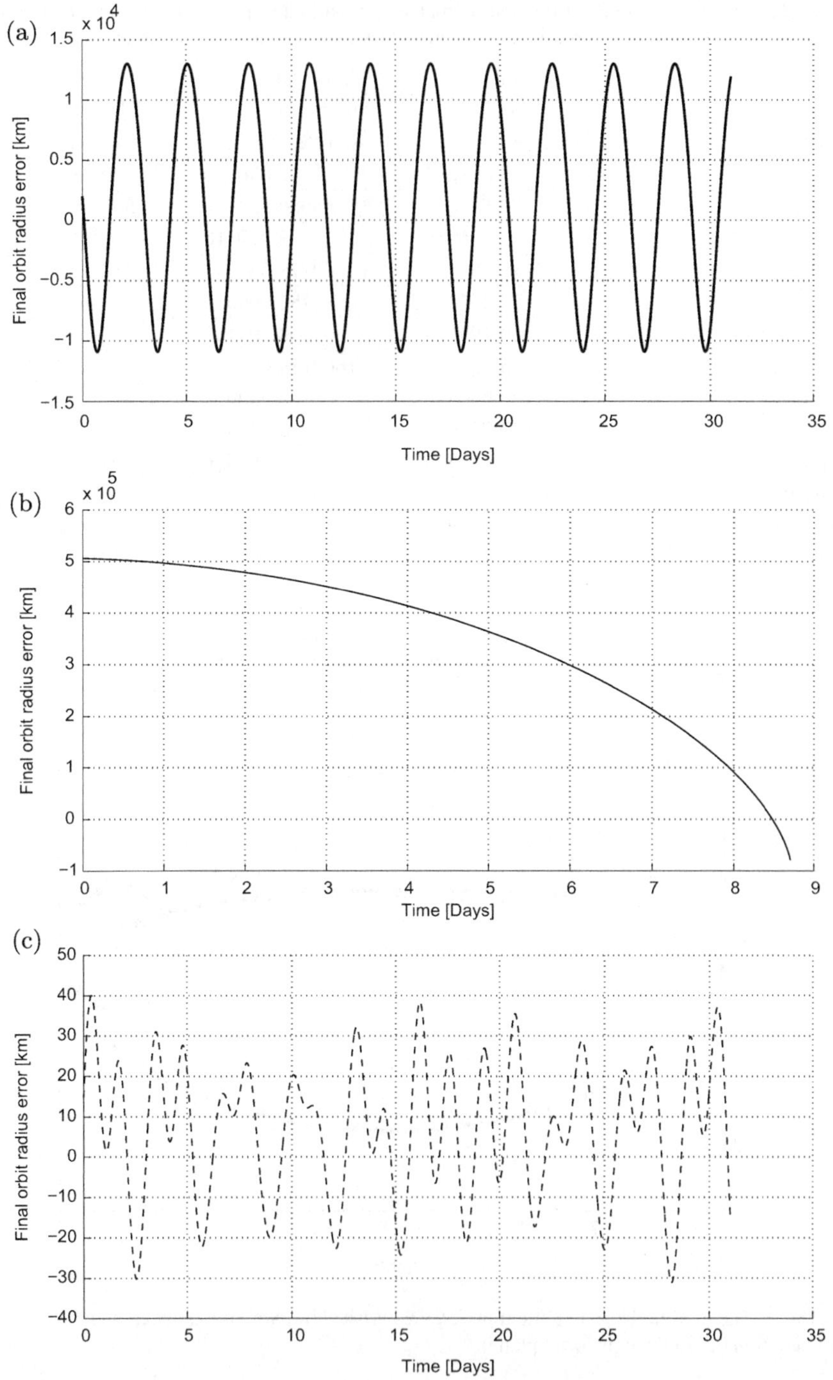

Fig. 15.20 Final orbit radius deviation from the nominal value, obtained using the (**a**) modified bi-parabolic transfer, (**b**) the classical bi-parabolic transfer, and (**c**) the modified Hohmann transfer

a smaller maximal radius; and the third is a smaller radial period, which causes the impulses to miss their target location, thus the transfer fails completely.

It is important to note that the collision of the classical bi-parabolic transfer is a result of the scenario defined here, and for a different scenario design, a different result can be obtained.

References

Barrar, R.B.: An analytic proof that the Hohmann-type transfer is the true minimum two-impulse transfer. Acta Astronaut. **9**(1), 1–11 (1962)

Hoelker, R., Silber, R.: The bi-elliptical transfer between co-planar circular orbits. Planet. Space Sci. **7**, 164–175 (1961)

Miele, C.M.A., Mathwig, J.: Reflections on the Hohmann transfer. J. Optim. Theory Appl. **123**(2), 233–253 (2004)

Prussing, J.: Simple proof of the global optimality of the Hohmann transfer. J. Guid. Control Dyn. **15**(4), 1037–1038 (1992)

Chapter 16
Orbit Data Processing

16.1 Introduction

In Chap. 9 we discussed numerical procedures for computation of orbits. In this chapter we will discuss other numerical procedures as needed for celestial mechanics and astrodynamics. These include the processing and filtering of observations and fitting ephemerides to the observations, determining an approximation for a given set of measurements, the use of polynomials for the computation from numerical integrations of positions of objects for specific times, and the use of different expressions for the series expansions of general perturbations theories. We will also present a *recursive filtering* approach for fitting dynamical models to noisy sets of observations. Whereas a polynomial fit to a given set of observations is usually referred to as *batch processing*, recursive filtering generates an approximation for the state of a system as each measurement is obtained.

There are two cases where determining an approximation by fitting a polynomial to a discrete set of points is inefficient. First, when all values of a function in an interval are specified, and it is desirable to use as many values as possible, rather than the least, to approximate the function. The presence of discontinuities in the function, or its derivative, becomes important. Second, when there are a limited set of approximate values of uncertain accuracies available. It is not desirable to determine a high degree polynomial to fit the data exactly. A high degree polynomial might fit the data exactly, but it would oscillate around the true function. The second case is when there are a large number of observations with observational errors and noise.

The method of *least squares* is designed for both of these cases. In addition to the principle of least squares, we will also introduce some classical sets of *orthogonal polynomials* and some of their properties.

P. Gurfil, P.K. Seidelmann, *Celestial Mechanics and Astrodynamics: Theory and Practice*, Astrophysics and Space Science Library 436,
DOI 10.1007/978-3-662-50370-6_16

16.2 Principle of Least Squares

Instead of seeking a polynomial $y(x)$ of degree n, which agrees exactly with a function $f(x)$ at $n+1$ points, we can seek to have $y(x)$ and $f(x)$ agree in some sense as well as possible over a domain D. The domain can be a continuous interval, or a set of $N+1$ points, $N > n$. A possible measure of agreement would be for the sum of the squared errors to be a minimum. This measure is *Legendre's principle of least squares*. A weighting function, $w(x)$, which is based on the precision of the $f(x)$ value, can be added, so $w(x)$ corresponds to the value of $f(x)$ when $x = x_i$. Thus, the square error at x_i is multiplied by $w(x_i)$ to determine the sum of the squares.

Suppose that values of $f(x)$ are known at points $x_0, x_1, x_2, \ldots, x_n$ in a domain D, or a continuous interval (a, b); then an approximation can be made,

$$f(x) \approx \sum_{k=0}^{n} a_k\phi_k(x) \equiv y(x) \tag{16.1}$$

where $\phi_0(x), \ldots, \phi_n(x)$ are $n+1$ chosen functions. To obtain a polynomial approximation of degree n, we could take $\phi_0 = 1, \phi_1 = x, \phi_2 = x^2, \ldots, \phi_n = x^n$, or a more advantageous function. The nonnegative weighting function is

$$w(x) \geq 0 \tag{16.2}$$

We define the residual $R(x)$ as

$$R(x) = f(x) - \sum_{k=o}^{n} a_k\phi_k(x) \equiv f(x) - y(x) \tag{16.3}$$

Then the best approximation, in a least squares sense, is when the $a's$ are determined so the sum of $w(x)R^2(x)$ over D is a minimum. The sum is $\langle wR^2 \rangle$ and the requirement is

$$\langle wR^2 \rangle \equiv \left\langle w(f - \sum_{k=0}^{n} a_k\phi_k)^2 \right\rangle = \text{minimum} \tag{16.4}$$

which implies the conditions

$$\frac{\partial}{\partial a_r}\left\langle w(f - \sum_{k=0}^{n} a_k\phi_k)^2 \right\rangle = 0, \; r = 0, 1, 2, \ldots, n \tag{16.5}$$

or

$$\left\langle w\phi_r(f - \sum_{k=0}^{n} a_k\phi_k)^2 \right\rangle \equiv \langle w\phi_r(f - y)\rangle = 0 \tag{16.6}$$

or

$$\sum_{k=0}^{n} a_k \langle w\phi_r\phi_k\rangle = \langle w\phi_r f\rangle \,,\; r = 0, 1, 2, \ldots, n \tag{16.7}$$

which leads to $n+1$ simultaneous linear equations in the $n+1$ unknown parameters $a_0, a_1, \ldots, a_n$. These are called the *normal equations*. These can be expressed as

$$\langle w(x)\phi(x)R(x)\rangle = 0,\; r = 0, 1, 2, \ldots, n \tag{16.8}$$

We also have

$$\left\langle wR^2\right\rangle = \langle wR \cdot R\rangle = \left\langle wR(f - \sum_{k=0}^{n} a_k\phi_k)\right\rangle = \langle wRf\rangle - \sum_{k=o}^{n} a_k \langle w\phi_k R\rangle \tag{16.9}$$

When the coefficients $a_0, a_1, \ldots, a_n$ satisfy Eq. (16.7), the squared residual sum reduces to the following, since by Eq. (16.8) $\langle w\phi_r R\rangle = 0$,

$$\left\langle wR^2\right\rangle_{\min} = \langle wRf\rangle \equiv \langle wf(f - y)\rangle \equiv \left\langle wf^2\right\rangle - \sum_{k=0}^{n} a_k \langle w\phi_k f\rangle \tag{16.10}$$

The size of this quantity is a criterion of the approximation over D. If we have $n+1$ points and functions $\phi_0, \phi_1, \ldots, \phi_n$, polynomials of degree less than n, then $R(x)$ can be zero at each point. So the n degree polynomial $y(x)$ agrees with $f(x)$ at $n+1$ points. If we have $N+1$ points, $N > n$, an exact fit is impossible, and the least squares produces a function, which is the best fit under criterion (16.4). From Eq. (16.7), the unknowns in the left hand members of the normal equation are independent of the function $f(x)$ to be approximated. From the coordinate function and weighting function, they can be recalculated. Since $\left\langle w\phi_i\phi_j\right\rangle \equiv \left\langle w\phi_j\phi_i\right\rangle$, the coefficient a_i in the jth equation equals the coefficient a_j in the ith equation, and the coefficients of the $a's$ are symmetrical with respect to the principal diagonal. This reduces the calculations required.

Another simplification is accomplished by choosing coordinate functions such that

$$\left\langle w\phi_i\phi_j\right\rangle = 0,\; i \neq j \tag{16.11}$$

This set of $\phi's$ are an orthogonal set with weighting function $w(x)$ over D. This set of functions, from the normal equations (16.7), are uncoupled, and of the form

$$a_r \left\langle w\phi_r^2 \right\rangle = \left\langle w\phi_r f \right\rangle, \; r = 0, 1, 2, \ldots, n \tag{16.12}$$

$w(x)$ is nonnegative, and it is assumed that none of the functions vanish everywhere over D, so we obtain

$$a_r = \frac{\langle w\phi_r f \rangle}{\langle w\phi_r^2 \rangle}, \; r = 0, 1, 2, \ldots, n \tag{16.13}$$

From Eqs. (16.10) and (16.12), the value of $\left\langle wR^2 \right\rangle$ can be written as

$$\left\langle wR^2 \right\rangle_{\min} = \left\langle wf^2 \right\rangle - \sum_{k=0}^{n} a_k^2 \left\langle w\phi_k^2 \right\rangle \tag{16.14}$$

The root-mean square (RMS) error over D relative to $w(x)$ is defined to be

$$\epsilon_{RMS} \equiv (f - y)_{RMS} = \sqrt{\frac{\langle wR^2 \rangle}{\langle w \rangle}} \tag{16.15}$$

When $w(x) \equiv 1$, the quantity $\langle 1 \rangle$ represents the length of the interval in the continuous case and the number $(N + 1)$ of points in D in the discrete case. In the discrete case, the data are frequently observations, and the function $f(x)$ is not known (Hildebrand 1956; pp. 258–261).

16.3 Least Squares Approximation

Consider applying the least squares method to a discrete set of points in a domain D, or a continuous interval. Take an approximation of the form

$$f(x) \approx \sum_{k=0}^{n} a_k \phi_k(x) \tag{16.16}$$

over $N + 1$ points $x_0, x_1, x_2, \ldots, x_N$, where $N > n$, and the weighted squared error is to be a minimum,

$$\sum_{i=0}^{N} w(x_i) \left[f(x_i) - \sum_{k=0}^{n} a_k \phi_k(x_i) \right]^2 = \text{minimum} \tag{16.17}$$

The $n + 1$ normal equations (16.7) become

$$a_0 \sum_{i=0}^{N} w(x_i)\phi_r(x_i)\phi_0(x_i) + a_1 \sum_{i=0}^{N} w(x_i)\phi_r(x_i)\phi_1(x_i) + \ldots \quad (16.18a)$$

$$a_n \sum_{i=0}^{N} w(x_i)\phi_r(x_i)\phi_n(x_i) = \sum_{i=0}^{N} w(x_i)\phi_r(x_i)f(x_i)$$

$$r = 0, 1, 2, \ldots, n \quad (16.18b)$$

These equations are obtained by writing down the $N + 1$ equations which would require that Eq. (16.16) be an equality at the $N + 1$ points x_i.

$$a_0\phi_0(x_0) + a_1\phi_1(x_0)+, \ldots, +a_n\phi_n(x_0) = f(x_0) \quad (16.19a)$$

$$a_0\phi_0(x_1) + a_1\phi_1(x_1)+, \ldots, +a_n\phi_n(x_1) = f(x_1) \quad (16.19b)$$

$$\vdots \quad (16.19c)$$

$$a_0\phi_0(x_N) + a_1\phi_1(x_N)+, \ldots, +a_n\phi_n(x_N) = f(x_N) \quad (16.19d)$$

The rth normal equation is obtained by multiplying each equation by the coefficient of a_r in that equation, and by the weight associated with that equation, and then summing the results. The weights assigned should be independent of the least squares results, and unity should be used in the absence of a good source. When $N = n$, the problem reduces to Eqs. (16.19) in $n + 1$ unknowns, and the normal equations are the original equations.

As an example, let us fit a line by the least squares method to the data in Table 16.1.

We have

$$a_0 + a_1 x = f(x) \quad (16.20)$$

The coefficients a_0 and a_1 and the data of Table 16.1 give

$$\begin{matrix} 1 & 0 & 1.00 \\ 1 & 1 & 3.85 \\ 1 & 2 & 6.50 \\ 1 & 3 & 9.35 \\ 1 & 4 & 12.05 \end{matrix} \quad (16.21)$$

Table 16.1 Data for fitting a straight line

x	0	1	2	3	4
$f(x)$	1.00	3.85	6.50	9.35	12.05

Table 16.2 Values obtained from a least squares approximation

x	0	1	2	3	4
$f(x)$	1.00	3.85	6.50	9.35	12.05
$y(x)$	1.03	3.79	6.55	9.31	12.07

Taking weights of unity, the first normal equation is formed by adding the columns of (16.21) giving [5, 10, 32.75]. the second normal equation is formed by multiplying the elements of each row by the element in the second column of the row, and adding the results, giving [10, 30, 93.10]. the normal equations are

$$5a_0 + 10a_1 = 32.75 \tag{16.22}$$

$$10a_0 + 30a_1 = 93.10 \tag{16.23}$$

The symmetry is as expected. The solution is $a_0 = 1.03, a_1 = 2.76$, and the approximation is

$$f(x) \approx y(x) = 1.03 + 2.76x \tag{16.24}$$

Table 16.2 contains the given data and the approximation values

The sum of the squared errors is 0.0090 and the RMS of the points is 0.042, which represents the RMS deviation of the approximation function from the true function, if the given values are exact function values. If the given values are observed values, 0.042 is the RMS error of the given values (Hildebrand 1956; pp. 261–263).

16.4 Orthogonal Polynomials

We want a least squares approximation over the interval (a, b) by constructing a set of polynomials $\phi_0(x), \phi_1(x), \ldots, \phi_r(x)$, where each is orthogonal to all others over (a, b), with a nonnegative weighting function $w(x)$. We seek a polynomial $\phi_r(x)$ of degree r, which is orthogonal over (a, b) to all polynomials of degree less than r, such that

$$\int_a^b w(x)\phi_r(x)q_{r-1}(x)dx = 0 \tag{16.25}$$

where w is specified and q_{r-1} is an arbitrary polynomial of degree $r-1$, or less. Put in a more useful form, integrate by parts r times, and use the fact that $q_{r-1}^{(r)} \equiv 0$. Introduce the notation

$$w(x)\phi_r(x) \equiv \frac{d^r U_r(x)}{dx^r} = U_r^{(r)}(x) \tag{16.26}$$

so Eq. (16.25) becomes

$$\int_a^b U_r^{(r)}(x) q_{r-1}(x) dx = 0 \tag{16.27}$$

and after r integrations by parts

$$[U_r^{(r-1)} q_{r-1} - U_r^{(r-2)} q'_{r-1} + U_r^{(r-3)} q_{r-1}'' - \ldots,. + (-1)^{r-1} U_r^{(r-1)} q_{r-1}^{r-1}]_0^b = 0 \tag{16.28}$$

The function $\phi_r(x)$ defined by Eq. (16.26)

$$\phi_r(x) = \frac{1}{w(x)} \frac{d^r U_r(x)}{dx^r} \tag{16.29}$$

is a polynomial of degree r, so $U_r(x)$ satisfies the differential equation

$$\frac{d^{r+1}}{dx^{r+1}} \left[\frac{1}{w(x)} \frac{d^r U_r(x)}{dx^r} \right] = 0 \tag{16.30}$$

in (a, b). For Eq. (16.28) to be satisfied for any values of

$$q_{r-1}(a), q_{r-1}(b), q'_{r-1}(a), q'_{r-1}(b)$$

leads to $2r$ boundary conditions

$$U_r(a) = U'_r(a) = U_r''(a) = \cdots = U_r^{r-1}(a) = 0 \tag{16.31a}$$

$$U_r(b) = U'_r(b) = U_r''(b) = \cdots = U_r^{r-1}(b) = 0 \tag{16.31b}$$

If a solution of Eq. (16.30), which satisfies Eqs. (16.31) for each integer r, can be obtained, the rth member of the set of functions is given by Eq. (16.29). Each solution will contain an arbitrary multiplicative constant. The formulated problem possesses a solution, even when a and/or b is infinite, assuming that $w(x) \geq 0$ in (a, b) and that $\int_a^b x^k w(x) dx$ exists for all nonnegative integral values of k. The coefficients in the expression

$$y(x) = \sum_{r=0}^{n} a_r \phi_r(x) \tag{16.32}$$

are determined from

$$\int_a^b w(x) [f(x) - y(x)]^2 dx = \text{minimum} \tag{16.33}$$

in the form

$$a_r = \frac{\int_a^b wf\phi_r dx}{\int_a^b w\phi_r^2 dx} = \frac{\int_a^b wf\phi_r dx}{\gamma_r} \tag{16.34}$$

where the numerator depends on f, but the denominator γ_r is independent of f, and can be calculated only once. The calculation of γ_r can be simplified. Write

$$\phi_r(x) = A_{r0} + A_{r1}x + \cdots + A_r x^r \tag{16.35}$$

so that A_{rk} is the coefficient of x^k in $\phi_r(x)$, and $A_r \equiv A_{rr}$ is its leading coefficient. It follows

$$\gamma_r = \int_a^b w(x)\phi_r(x)\phi_r(x)dx \equiv \int_a^b w(x)\phi_r(x)[A_{r0} + A_{r1}x + \cdots + A_r x^r]dx \tag{16.36}$$

and from Eq. (16.25)

$$\int_a^b w(x)\phi_r(x)x^i dx = 0, \ i = 0, 1, 2, \ldots, r-1 \tag{16.37}$$

so

$$\gamma_r = A_r \int_a^b x^r w(x)\phi_r(x)dx = A_r \int_a^b x^r U_r^{(r)}(x)dx \tag{16.38}$$

Integrating by parts r times and using Eqs.(16.31), this relation takes the form

$$\gamma_r \equiv \int_a^b w(x)\phi_r^2(x)dx = (-1)^r r! A_r \int_a^b U_r(x)dx \tag{16.39}$$

where A_r is the coefficient of x^r in $\phi_r(x)$ (Hildebrand 1956; pp. 269–271).

16.5 Chebyshev Series

A development in celestial mechanics is the introduction of *Chebyshev series* in the 1960s. Historically, Fourier series have been used for planetary and lunar theories. Attempts have been made to utilize power series, and various polynomial approaches to interpolation have been applied. Carpenter (1966) provided the first application of Chebyshev polynomials for celestial mechanics, for development of planetary theories over a specified time period. Now Chebyshev polynomials, or economization of power series from Chebyshev polynomials, are being used to represent and compress ephemeris data over varying periods of time to various accuracies (Deprit et al. 1979; Lee et al. 2003). They provide compact means of providing the data and an accurate method of interpolation at the same time. Whereas Fourier series give accurate representation at the middle of an interval of time and reduced accuracy with greater distance from the middle, Chebyshev polynomials are weighted for increased accuracy at the ends.

16.5.1 Chebyshev Approximation

Chebyshev polynomials have the advantage of minimizing errors at the ends of the interval, so a weighting function $1/\sqrt{(x-a)(b-x)}$ is useful. A linear change of variables has transformed the interval into $(-1, 1)$, so the weighting function is

$$w(x) = \frac{1}{\sqrt{1-x^2}} \tag{16.40}$$

For an orthogonal polynomial of degree r, $\phi_r(x)$, to be orthogonal to all polynomials of degree inferior to r over the interval (a, b), it is required that

$$\int_a^b w(x)\phi_r(x)q_{r-1}(x)dx = 0 \tag{16.41}$$

where $w(x)$ is a specified weighting function, and q_{r-1} is an arbitrary polynomial of degree $r-1$ or less. For the interval $(-1, 1)$ and this weighting function, Eq. (16.41) becomes

$$\int_{-1}^1 \frac{\phi_r(x)q_{r-1}(x)}{\sqrt{1-x^2}}dx = 0 \tag{16.42}$$

The change of variables $x = \cos\theta$ gives

$$\int_0^\pi \phi_r(\cos\theta)q_{r-1}(\cos\theta)d\theta = 0 \tag{16.43}$$

$\cos k\theta$ can be expressed as a polynomial of degree k in $\cos\theta$, and any polynomial of degree k in $\cos\theta$ can be expressed as a linear combination of $1, \cos\theta, \cos 2\theta, \ldots, \cos k\theta$, so Eq. (16.43) will be satisfied, if and only if,

$$\int_0^\pi \phi_r(\cos\theta)\cos k\theta\, d\theta = 0,\ k = 0, 1, \ldots, r-1 \tag{16.44}$$

Thus, the function

$$\phi_r(\cos\theta) = C_r \cos r\theta \tag{16.45}$$

has this property. Using $x = \cos\theta$, we return to the variable x. We establish that the functions

$$\phi_r(x) = C_r \cos(r \cos^{-1} x) \tag{16.46}$$

are the orthogonal polynomials. With $C_r = 1$, these polynomials are the *Chebyshev polynomials*, $T_r(x)$, so

$$\phi_r(x) = T_r(x) = \cos(r \cos^{-1} x) \tag{16.47}$$

These polynomials possess the orthogonal property

$$\int_{-1}^{1} \frac{T_r(x) T_s(x)}{\sqrt{1-x^2}} dx = 0, \; r \neq s \tag{16.48}$$

where r and s are nonnegative integers. The first six of these polynomials are

$$\begin{aligned} &T_0(x) = 1, \quad T_1(x) = x, \quad T_2(x) = 2x^2 - 1, \quad T_3(x) = 4x^3 - 3x \\ &T_4(x) = 8x^4 - 8x^2 + 1, \quad T_5(x) = 16x^5 - 20x^3 + 5x \end{aligned} \tag{16.49}$$

and additional polynomials may be determined from the recurrence formula

$$T_{r+1}(x) = 2xT_r(x) - T_{r-1}(x) \tag{16.50}$$

The product of two Chebyshev polynomials is

$$2T_r(x)T_q(x) = T_{r+q}(x) + T_{|r-q|}(x) \tag{16.51}$$

The integral is

$$\begin{aligned} &\int T_0(x)dx = T_1(x), \quad \int T_1(x)dx = \frac{1}{4}T_2(x) \\ &2\int T_r(x)dx = \frac{1}{r+1}T_{r+1}(x) - \frac{1}{r-1}T_{r-1}(x, \; r > 1 \end{aligned} \tag{16.52}$$

To evaluate the factor

$$\gamma_r = \int_{-1}^{1} \frac{T_r^2(x)}{\sqrt{1-x^2}} dx \tag{16.53}$$

write $x = \cos\theta$ and $T_r(x) = \cos r\theta$, so

$$\gamma_r = \int_0^{\pi} \cos^2 r\theta \, d\theta = \begin{cases} \pi, & r = 0 \\ \frac{\pi}{2}, & r \neq 0 \end{cases} \tag{16.54}$$

With the weighting factor $w(x) = 1/\sqrt{1-x^2}$, the least squares polynomial of degree n approximating $f(x)$ in $(-1, 1)$ is

$$y(x) = \sum_{r=0}^{n} a_r T_r(x), \ -1 < x < 1 \tag{16.55}$$

where

$$a_0 = \frac{1}{\pi}\int_{-1}^{1} \frac{f(x)}{\sqrt{1-x^2}}dx, \ a_r = \frac{2}{\pi}\int_{-1}^{1} \frac{f(x)T_r(x)}{\sqrt{1-x^2}}dx, \ r \neq 0 \tag{16.56}$$

The integrated weighted square error of all polynomials of degree n or less,

$$\int_{-1}^{1} \frac{1}{\sqrt{1-x^2}}[f(x) - y_n(x)]^2 dx \tag{16.57}$$

is least when $y_n(x)$ equals the right hand member of Eq. (16.55) (Hildebrand 1956; pp. 279–281).

16.5.2 Other Polynomial Approximations

Using other weighting functions, many types of least squares polynomial approximations can be formed. For the weighting function over $(-1, 1)$

$$w(x) = (1-x)^{\alpha}(1+x)^{\beta}, \ \alpha > -1, \beta > -1 \tag{16.58}$$

This reduces to the Legendre case, when $\alpha = \beta = 0$, and to the Chebyshev case, when $\alpha = \beta = -1/2$. The rth orthogonal polynomial is

$$\phi_r(x) = C_r(1-x)^{-\alpha}(1+x)^{-\beta}\frac{d^r}{dx^r}\left[(1-x)^{\alpha+r}(1+x)^{\beta+r}\right] \tag{16.59}$$

This can be identified as the rth Jacobi polynomial, when C_r is properly specified. The factor C_r for $T_r(x)$ is $(-2)^r r!/(2r)!$, so Eq. (16.47) can be written

$$T_r(x) = \frac{(-2)^r r!}{(2r)!}(1-x^2)^{1/2}\frac{d^r}{dx^r}(1-x^2)^{r-1/2} \tag{16.60}$$

Similar polynomials, $S_r(x)$, are from the weighting function $w(x) = (1-x^2)^{1/2}$, and can be expressed as

$$S_r(x) = \frac{\sin[(r+1)\cos^{-1}x]}{\sin(\cos^{-1}x)}$$
$$= \frac{(-2)^r(r+1)!}{(2r+1)!}(1-x^2)^{-1/2}\frac{d^r}{dx^r}(1-x^2)^{r+1/2} \tag{16.61}$$

and

$$S_r(x) = \frac{1}{r+1}T'_{r+1}(x) \tag{16.62}$$

which can be expressed as

$$S_r(x) = \frac{1}{\sqrt{1-x^2}}\sin[(r+1)\cos^{-1}x] = \frac{\sin(r+1)\theta}{\sin\theta} \tag{16.63}$$

where $\theta = \cos^{-1}x$. For the weighting function

$$w(x) = x^{\beta}e^{-\alpha x}, \ \beta > -1, \alpha > 0 \tag{16.64}$$

over $(0, \infty)$ there follows

$$\phi_r(x) = C_r x^{-\beta}e^{\alpha x}\frac{d^r}{dx^r}(x^{\beta+r}e^{-\alpha x}) \tag{16.65}$$

These polynomials are called *Sonine polynomials*, or *generalized Laguerre polynomials* (Hildebrand 1956; pp. 281–282).

16.6 Fourier Approximation: Continuous Range

If the function to be approximated is a periodic function $f(x)$ of known period, and the period is 2π after an adjustment of the units, then

$$f(x+2\pi) = f(x) \tag{16.66}$$

A set of coordinate functions is

$$1, \cos x, \cos 2x, \ldots, \cos rx, \ldots$$
$$\sin x, \sin 2x, \ldots, \sin rx, \ldots \tag{16.67}$$

where each member has a period 2π. The following useful properties are present:

1. A product of any two members can be expressed as a linear combination of two members.
2. The derivative of each member is a member.
3. The integral of each member is a member, except for the constant.
4. The set is orthogonal over any period interval, say the period $(-\pi, \pi)$, so that

$$\int_{-\pi}^{\pi} \sin jx \sin kx\, dx = 0,\ j \neq k \tag{16.68a}$$

$$\int_{-\pi}^{\pi} \cos jx \cos kx\, dx = 0,\ j \neq k \tag{16.68b}$$

$$\int_{-\pi}^{\pi} \sin jx \cos kx\, dx = 0 \tag{16.68c}$$

where j and k are nonnegative integers; negative integers are not considered. We want an approximation of the form

$$f(x) \approx a_0 + \sum_{k=1}^{n} (a_k \cos kx + b_k \sin kx) \tag{16.69}$$

where the coefficients are to be such that the integrated square error is a minimum. From the period of $f(x)$ and the sine and cosine harmonics, one period interval of, for example, $(-\pi, \pi)$, might be considered. The requirement

$$\int_{-\pi}^{\pi} \left[f(x) - a_0 - \sum_{k=1}^{n} (a_k \cos kx + b_k \sin kx) \right]^2 dx = \text{minimum} \tag{16.70}$$

leads to

$$\begin{aligned}
&\int_{-\pi}^{\pi} \left[f(x) - a_0 - \sum_{k=1}^{n} (a_k \cos kx + b_k \sin kx) \right] dx = 0 \\
&\int_{-\pi}^{\pi} \cos rx \left[f(x) - a_0 - \sum_{k=1}^{n} (a_k \cos kx + b_k \sin kx) \right] dx = 0,\ r = 1, 2, \ldots, n \\
&\int_{-\pi}^{\pi} \sin rx \left[f(x) - a_0 - \sum_{k=1}^{n} (a_k \cos kx + b_k \sin kx) \right] dx = 0,\ r = 1, 2, \ldots, n
\end{aligned} \tag{16.71}$$

when the partial derivatives of the left-hand member of Eq. (16.70), with respect to a_0, a_r, and b_r, are equated to zero. Equations (16.68) and

$$\int_{-\pi}^{\pi} dx = 2\pi, \int_{-\pi}^{\pi} \cos^2 kx\, dx = \int_{-\pi}^{\pi} \sin^2 kx\, dx = \pi,\ k \neq 0 \tag{16.72}$$

lead to the expressions

$$\begin{aligned} a_0 &= \frac{1}{2\pi}\int_{-\pi}^{\pi} f(x)dx,\ a_k = \frac{1}{\pi}\int_{-\pi}^{\pi} f(x)\cos kx\, dx,\ k \neq 0 \\ b_k &= \frac{1}{\pi}\int_{-\pi}^{\pi} f(x)\sin kx\, dx \end{aligned} \tag{16.73}$$

If $f(x) = f(-x)$, then $b_k = 0$, and Eq. (16.69) reduces to

$$f(x) \approx a_0 + \sum_{k=1}^{n} a_k \cos kx \tag{16.74}$$

where

$$\begin{aligned} a_0 &= \frac{1}{2\pi}\int_{-\pi}^{\pi} f(x)dx = \frac{1}{\pi}\int_{0}^{\pi} f(x)dx \\ a_k &= \frac{1}{\pi}\int_{-\pi}^{\pi} f(x)\cos kx\, dx = \frac{2}{\pi}\int_{0}^{\pi} f(x)\cos kx\, dx,\ k \neq 0 \end{aligned} \tag{16.75}$$

If $f(x) = -f(-x)$, then $a_o = a_k = 0$, and Eq. (16.69) becomes

$$f(x) \approx \sum_{k=1}^{n} b_k \sin kx \tag{16.76}$$

where

$$b_k = \frac{1}{\pi}\int_{-\pi}^{\pi} f(x)\sin kx\, dx = \frac{2}{\pi}\int_{0}^{\pi} f(x)\sin kx\, dx \tag{16.77}$$

If $f(x)$ is bounded and piecewise differentiable, the approximation comes closer to $f(x)$ as $n \to \infty$ at x, where $f(x)$ is continuous, and at discontinuity points to $\frac{1}{2}[f(x+) + f(x-)]$, which is the mean value of the right- and left-hand limits at the discontinuities. Each coefficient is independently determined, and independent of the number of harmonics in the approximation, which is typical of least squares

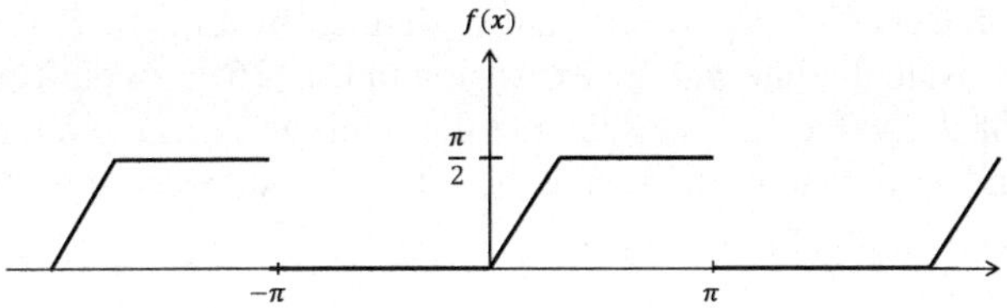

Fig. 16.1 Discontinuous periodic function

approximations by orthogonal functions. Define $f(x)$ over $(-\pi, \pi)$ as

$$\begin{aligned} f(x) &= 0, \quad -\pi < x \leq 0 \\ f(x) &= x, \quad 0 \leq x \leq \pi/2 \\ f(x) &= \pi/2, \quad \frac{\pi}{2} \leq x \leq \pi \end{aligned} \tag{16.78}$$

and it is periodic with period 2π, as shown in Fig. 16.1, elsewhere. Since $f(x)$ is neither even nor odd, the presence of both sine and cosine harmonics may be anticipated. Equation (16.73) gives

$$\begin{aligned} a_0 &= \frac{1}{2\pi}\left[\int_{-\pi}^{0} 0dx + \int_{0}^{\pi/2} xdx + \int_{\pi/2}^{\pi} \frac{\pi}{2}dx\right] = \frac{3\pi}{16} \\ a_k &= \frac{1}{\pi}\left[\int_{-\pi}^{0} 0\cos kx\, dx + \int_{0}^{\pi/2} x\cos kx\, dx + \int_{\pi/2}^{\pi} \frac{\pi}{2}\cos kx\, dx\right] \\ &= \frac{1}{\pi k^2}(1 - \cos\frac{k\pi}{2}), \; k \neq 0 \\ b_k &= \frac{1}{\pi}\left[\int_{-\pi}^{0} 0\sin kx\, dx + \int_{0}^{\pi/2} x\sin kx\, dx + \int_{\pi/2}^{\pi} \frac{\pi}{2}\sin kx\, dx\right] \\ &= \frac{1}{\pi k^2}(\sin\frac{k\pi}{2} - \frac{k\pi}{2}\cos k\pi) \end{aligned} \tag{16.79}$$

Thus,

$$\begin{aligned} f(x) = \frac{3\pi}{16} &- \frac{1}{\pi}\cos x - \frac{1}{2\pi}\cos 2x - \frac{1}{9\pi}\cos 3x - \ldots \\ &+ \frac{2+\pi}{2\pi}\sin x - \frac{1}{4}\sin 2x + \frac{3\pi - 2}{9\pi}\sin 3x - \ldots \end{aligned} \tag{16.80}$$

Since $f(x)$ has discontinuities, a good approximation will require a large number of terms, particularly near discontinuities. In many cases, low harmonic coefficients are adequate. If $f(x)$ is not periodic, but is defined by Eq. (16.78) in the interval $(-\pi, \pi)$, then inside the interval the expansion in Eq. (16.79) would be satisfactory, independent of the values of $f(x)$ outside the interval. A trigonometric expansion will be periodic both inside and outside the interval, independent of the behavior of $f(x)$ (Hildebrand 1956; pp. 369–372).

16.7 Fourier Approximation: Discrete Range

Let $f(x)$ be of period 2π, with values only known at $2N + 1$ equally-spaced discrete points

$$-\pi, -\frac{(N-1)\pi}{N}, \ldots, \frac{-\pi}{N}, 0, \frac{\pi}{N}, \ldots, \frac{(N-1)\pi}{N}, \pi \tag{16.81}$$

of the interval $(-\pi, \pi)$. Periodicity means $f(-\pi) = f(\pi)$, so there are $2N$ independent data, to be used to determine the coefficients of $2N$ terms of an approximation like Eq. (16.69). Denote the rth abscissa as

$$x_r = r\frac{\pi}{N}, \quad r = -N+1, -N+2, -N+3, \ldots, -1, 0, 1, \ldots, N-1, N \tag{16.82}$$

so $2N$ independent values $f_r \equiv f(x_r)$ are determined. Only the $2N$ functions $1, \cos x, \cos 2x, \ldots, \cos Nx, \sin x, \sin 2x, \ldots, \sin(N-1)x$ are independent over the domain comprising this set of abscissas, because $\sin Nx$ vanishes at each of these points, and the functions $\cos(N+1)x, \ldots,$ and $\sin(N+1)x, \ldots,$ take on the same values at the points as does one of the $2N$ functions given above. For example, since $\sin Nx_r = 0$, we have

$$\cos(N+1)x_r = \cos Nx_r \cos x_r = (-1)^r \cos x_r = \cos(N-1)x_r$$

Under summation over the set in Eq. (16.82), this set of functions is orthogonal. With the notation of Eq. (16.82),

$$\sum_{r=-N+1}^{N} \sin jx_r \sin kx_r = 0, \; j \neq k$$

$$\sum_{r=-N+1}^{N} \cos jx_r \cos kx_r = 0, \; j \neq k$$

$$\sum_{r=-N+1}^{N} \sin jx_r \cos kx_r = 0 \tag{16.83}$$

where j and k are integers between 0 and N inclusive, similar to Eq. (16.68). When $j = k$,

$$\sum_{r=-N+1}^{N} \sin^2 kx_r = \sum_{r=-N+1}^{N} \cos^2 kx_r = N, \; k \neq 0, N$$

$$\sum_{r=-N+1}^{N} 1 = 2N, \quad \sum_{r=-N+1}^{N} \cos^2 Nx_r = 2N \tag{16.84}$$

can be established, similar to Eq. (16.72). We assume an approximation

$$f(x) \approx A_0 + \sum_{k=1}^{n} (A_k \cos kx + B_k \sin kx) \tag{16.85}$$

where $n \leq N$, and adopting the least squares criterion

$$\sum_{r=-N+1}^{N} \left[f(x_r) - A_0 - \sum_{k=1}^{n} (A_k \cos kx_r + B_k \sin kx_r) \right]^2 = \text{minimum} \tag{16.86}$$

a derivation similar to Eqs. (16.70)–(16.73), using Eqs. (16.83) and (16.84), yields

$$A_0 = \frac{1}{2N} \sum_{r=-N+1}^{N} f(x_r), \; A_k = \frac{1}{N} \sum_{r=-N+1}^{N} f(x_r) \cos kx_r, \; k \neq 0, N$$

$$A_n = \frac{1}{2N} \sum_{r=-N+1}^{N} f(x_r) \cos Nx_r, \; B_k = \frac{1}{N} \sum_{r=-N+1}^{N} f(x_r) \sin kx_r \tag{16.87}$$

The coefficients in Eq. (16.85) are obtained by summation. The calculation of each coefficient is independent of the others, and independent of n, when $n \leq N$. When $n = N$, the least squares criterion requires that Eq. (16.85) be equal at $2N$ points, specified by Eq. (16.82). Equations (16.87) are in a more symmetrical form

$$A_0 = \frac{1}{2N} \left[\frac{1}{2} f_{-N} + f_{-N+1} + \ldots, + f_{-1} + f_0 + f_1 + \ldots, f_{N-1} + \frac{1}{2} f_N \right]$$

$$A_k = \frac{1}{N} \left[\frac{1}{2} f_{-N} \cos kx_{-N} + f_{-N+1} \cos kx_{-N+1} + \ldots, + f_{-1} \cos kx_{-1} \right.$$

$$\left. + f_0 \cos kx_0 + f_1 \cos kx_1 + \ldots, + f_{N-1} \cos kx_{N-1} + \frac{1}{2} f_N \cos kx_N \right], \; k \neq 0, N$$

$$A_N = \frac{1}{2N}\left[\frac{1}{2}f_{-N}\cos Nx_{-N} + f_{-N+1}\cos Nx_{-N+1} + \ldots, + f_{-1}\cos Nx_{-1}\right.$$
$$\left. + f_0\cos Nx_0 + f_1\cos Nx_1 + \ldots, + f_{N-1}\cos Nx_{N-1} + \frac{1}{2}f_N\cos Nx_N\right]$$
$$B_k = \frac{1}{N}\left[\frac{1}{2}f_{-N}\sin kx_{-N} + f_{-N+1}\sin kx_{-N+1} + \ldots, + f_{-1}\sin kx_{-1}\right.$$
$$\left. + f_0\sin kx_0 + f_1\sin kx_1 + \ldots, + f_{N-1}\sin kx_{N-1} + \frac{1}{2}f_N\sin kx_N\right], \quad k \neq 0, N \tag{16.88}$$

from $f_{-N} = f_N$. It is convenient to separate $f(x)$ into even and odd components, by introducing the auxiliary functions

$$F(x) = \frac{1}{2}[f(x) + f(-x)], \quad G(x) = \frac{1}{2}[f(x) - f(-x)] \tag{16.89}$$

so $f(x) = F(x) + G(x)$. Since $x_{-r} = -x_r$ and $x_0 = 0$, Eqs. (16.87) and (16.88) may be reduced to

$$A_0 = \frac{1}{N}\left[\frac{1}{2}F_0 + F_1 + F_2 + \ldots, + F_{N-1} + \frac{1}{2}F_N\right]$$
$$A_k = \frac{2}{N}\left[\frac{1}{2}F_0 + F_1\cos kx_1 + F_2\cos kx_2 + \ldots, + F_{N-1}\cos kx_{N-1}\right.$$
$$\left. + \frac{1}{2}F_N\cos kx_N\right], \quad k \neq 0, N$$
$$A_N = \frac{1}{N}\left[\frac{1}{2}F_0 - F_1 + F_2 - \ldots, + (-1)^{N-1}F_{N-1} + (-1)^N\frac{1}{2}F_N\right]$$
$$B_k = \frac{2}{N}[G_1\sin kx_1 + G_2\sin kx_2 + \ldots, + G_{N-1}\sin kx_{N-1}] \tag{16.90}$$

To illustrate the use of these formulas, consider the case $N = 6$, corresponding to the use of 12 independent ordinates. The tabular forms are appropriate (though more systematization is possible).

The sum of the products are given at the bottom of each column. For the empirical data shown in Table 16.5 and 16.6, the data columns of Tables 16.3 and 16.4 are

$$\frac{F_0}{2} = 0.605, \; G_1 = 0.075$$
$$F_1 = 1.245, \; G_2 = 0.160$$
$$F_2 = 1.300, \; G_3 = 0.150$$

Table 16.3 Using a discrete Fourier approximation: cosine terms

x	Data	$\cos x$	$\cos 2x$	$\cos 3x$	$\cos 4x$	$\cos 5x$	$\cos 6x$
0	$\frac{1}{2}f_0 = \frac{1}{2}F_0$	1	1	1	1	1	1
$\pi/6$	$\frac{1}{2}(f_1 = f_{-1}) = F_1$	$\frac{1}{2}\sqrt{3}$	$\frac{1}{2}$	0	$-\frac{1}{2}$	$-\frac{1}{2}\sqrt{3}$	-1
$\pi/3$	$\frac{1}{2}(f_2 + f_{-2}) = F_2$	$\frac{1}{2}$	$-\frac{1}{2}$	-1	$-\frac{1}{2}$	$\frac{1}{2}$	1
$\pi/2$	$\frac{1}{2}(f_3 + f_{-3}) = F_3$	0	-1	0	1	0	-1
$2\pi/3$	$\frac{1}{2}(f_4 + f_{-4}) = F_4$	$-\frac{1}{2}$	$-\frac{1}{2}$	1	$-\frac{1}{2}$	$-\frac{1}{2}$	1
$5\pi/6$	$\frac{1}{2}(f_5 + f_{-5}) = F_5$	$-\frac{1}{2}\sqrt{3}$	$\frac{1}{2}$	0	$-\frac{1}{2}$	$\frac{1}{2}\sqrt{3}$	-1
π	$\frac{1}{2}f_6 = \frac{1}{2}F_6$	-1	1	-1	1	-1	1
	$6A_0$	$3A_1$	$3A_2$	$3A_3$	$3A_4$	$3A_5$	$6A_6$

Table 16.4 Using a discrete Fourier approximation: sine terms

x	Data	$\sin x$	$\sin 2x$	$\sin 3x$	$\sin 4x$	$\sin 5x$
$\pi/6$	$\frac{1}{2}(f_1 - f_{-1}) = G_1$	$\frac{1}{2}$	$\frac{1}{2}\sqrt{3}$	1	$\frac{1}{2}\sqrt{3}$	$\frac{1}{2}$
$\pi/3$	$\frac{1}{2}(f_2 - f_{-2}) = G_2$	$\frac{1}{2}\sqrt{3}$	$\frac{1}{2}\sqrt{3}$	0	$-\frac{1}{2}\sqrt{3}$	$-\frac{1}{2}\sqrt{3}$
$\pi/2$	$\frac{1}{2}(f_3 - f_{-3}) = G_3$	1	0	-1	0	1
$2\pi/3$	$\frac{1}{2}(f_4 - f_{-4}) = G_4$	$\frac{1}{2}\sqrt{3}$	$-\frac{1}{2}\sqrt{3}$	0	$\frac{1}{2}\sqrt{3}$	$-\frac{1}{2}\sqrt{3}$
$5\pi/6$	$\frac{1}{2}(f_5 - f_{-5}) = G_5$	$\frac{1}{2}$	$-\frac{1}{2}\sqrt{3}$	1	$-\frac{1}{2}\sqrt{3}$	$\frac{1}{2}$
		$3B_1$	$3B_2$	$3B_3$	$3B_4$	$3B_5$

$$
\begin{aligned}
F_3 &= 1.250, \; G_4 = 0.145 \\
F_4 &= 1.195, \; G_5 = 0.085 \\
F_5 &= 1.095 \\
\frac{F_6}{2} &= 0.535
\end{aligned}
\tag{16.91}
$$

and calculation gives

$$
\begin{aligned}
&A_0 = 1.204, \; A_1 = 0.084, \; A_2 = -0.062, \; A_3 = -0.012, \; A_4 = -0.009 \\
&B_1 = 0.165, \; B_2 = 0.001, \; B_3 = 0.003, \; B_4 = -0.007
\end{aligned}
\tag{16.92}
$$

for the coefficients of harmonics through the fourth. If all the available harmonics are retained, the resultant approximation takes on the prescribed value at each of the points employed in the calculation (Table 16.5 and 16.6). Retention of a smaller number of harmonics leads to the appropriate least-squares approximation relevant to that set of points. Tables similar to those above, but employing larger sets of data and further systematized are in the literature and books of tables (Hildebrand 1956; pp. 373–377).

Table 16.5 Using a discrete Fourier approximation: empirical data

θ	0°	30°	60°	90°	120°	150°	180°	210°	240°
f	1.21	1.32	1.46	1.40	1.34	1.18	1.07	1.01	1.05
x	0	$\pi/6$	$\pi/3$	$\pi/2$	$2\pi/3$	$5\pi/6$	π	$-5\pi/6$	$-2\pi/3$

Table 16.6 Using a discrete Fourier approximation: empirical data (cont.)

θ	270°	300°	330°	360°
f	1.10	1.14	1.17	1.21
x	$-\pi/2$	$-\pi/3$	$-\pi/6$	2π

16.8 Optimum Polynomial Interpolation

Suppose a function $f(x)$ is approximated by the polynomial $y(x)$ of degree n, and agrees with $f(x)$ at $n+1$ points $x_0, x_1, \ldots, , x_n$, then

$$f(x) = y(x) = \pi(x)\frac{f^{n+1}(\xi)}{(n+1)!} \tag{16.93}$$

where $\pi(x) = (x - x_0)(x - x_1), \ldots, (x - x_n)$, and ξ lie in the interval of the largest and smallest values of $x_0, x_1, x_2, \ldots, x_n$. An appropriate change of variables can reduce the interval to $(-1, 1)$. We will consider trigonometric interpolation formulas.

How ξ depends on the $n+1$ values of x depends on the function $f(x)$. If we want the error over $(-1, 1)$ to be as small as possible for the functions having $n+1$ continuous derivatives in $(-1, 1)$, we should make $|\pi(x)|$ as small as possible, knowing that the coefficient of the highest power of x in $\pi(x)$ must be unity from Eq. (16.93). So we require

$$\int_{-1}^{1} w(x)[\pi(x)]^2 dx = \text{minimum} \tag{16.94}$$

where $w(x)$ is the nonnegative weighting function in $(-1, 1)$. $\pi(x)$ can be expressed as

$$\pi(x) = x^{n+1} + C_n x^n + \ldots, + C_2 x^2 + C_1 x + C_0 \tag{16.95}$$

and can be specified by $n+1$ coefficients $C_0, C_1, \ldots, C_n$. Then Eq. (16.94) requires that the partial derivative of the left-hand member with respect to each C_r must be zero. Since

$$\frac{\partial \pi(x)}{\partial C_r} = x^r, \ r = 0, 1, 2, \ldots, n \tag{16.96}$$

the requirement becomes

$$2\int_{-1}^{1} w(x)\frac{\partial \pi(x)}{\partial C_r}\pi(x)dx \equiv 2\int_{-1}^{1} w(x)\pi(x)x^r dx = 0,\ r = 0, 1, 2 \ldots, n \tag{16.97}$$

so $\pi(x)$ is a polynomial of degree $n + 1$, with a unity leading coefficient. It is orthogonal to all polynomials of an inferior degree over $(-1, 1)$ with $w(x)$. The $n + 1$ points where the polynomial $y(x)$ and the function $f(x)$ should agree, are the zeros of the polynomial. The integral approximation

$$\int_{-1}^{1} w(x)f(x)dx \approx \int_{1}^{1} w(x)y(x)dx \tag{16.98}$$

is the corresponding Gaussian quadrature formula. Let $w(x) = 1$ and seek to minimize the integral of the square of the error $E(x)$ over $(-1, 1)$, the $n + 1$ values should be the zeros of the Legendre polynomials $P_{n+1}(x)$ (see Sect. 11.8); i.e.

$$\int_{-1}^{1} P_r(x)P_s(x)dx = 0,\ r \neq s \tag{16.99}$$

where

$$P_0(x) = 1,\ P_1(x) = x,\ P_2(x) = \frac{1}{2}(3x^2 - 1),\ P_3(x) = \frac{1}{2}(5x^3 - 3x)$$
$$P_4(x) = \frac{1}{8}(35x^4 - 30x^2 + 3),\ P_5(x) = \frac{1}{8}(63x^5 - 70x^3 + 15x) \tag{16.100}$$

and

$$P_{r+1}(x) = \frac{2r+1}{r+1}xP_r(x) - \frac{r}{r+1}P_{r-1}(x) \tag{16.101}$$

We take $w(x) = 1/\sqrt{1-x^2}$ and seek to minimize the integral of $[E(x)]^2/\sqrt{1-x^2}$. Then the values are the zeros of the $(n+1)^{th}$ Chebyshev polynomial

$$T_{n+1}(x) = \cos[(n+1)\cos^{-1} x] \tag{16.102}$$

and are given by

$$x_i = \cos\left(\frac{2i+1}{n+1}\frac{\pi}{2}\right),\ i = 0, 1, 2, \ldots, , n \tag{16.103}$$

Since the coefficient of x^r in $T_r(x)$ is 2^{r-1}, then

$$\pi(x) = 2^{-n}T_{n+1}(x) \tag{16.104}$$

The extreme values of $\pi(x)$ in $(-1, 1)$ are then $\pm 2^{-n}$ and have alternating signs at the end points, $x = \pm 1$, and at n interior points, each of which separates a pair of adjacent values. With this choice of values, the coefficient of $f^{(n+1)}(\xi)/(n+1)!$ in the error term of Eq. (16.93) oscillates with a constant amplitude 2^{-n} as x increases from -1 to 1. Since the coefficient of x^r in $P_r(x)$ is $2^{-r}(2r)!/(r!)^2$, the use of the zeros of $P_{n+1}(x)$ as values corresponds to

$$\pi(x) = \frac{2^{n+1}[(n+1)!]^2}{(2n+2)!} P_{n+1}(x) \tag{16.105}$$

The Legendre polynomial has the values +1 at $x = +1$ and $(-1)^{n+1}$ at $x = -1$. $P_{n+1}(x)$ oscillates in $(-1, 1)$, such that the n successive maxima and minima, separating pairs of adjacent zeros, decrease in magnitude toward the center of the interval. The maximum value of $\pi(x)$ in Eq. (16.105) over $(-1, 1)$ is approximated by $2^{-n}\sqrt{\pi n/4}$, when n is large. Where the zeros of $P_{n+1}(x)$ minimize the RMS value of $\pi(x)$ over $(-1, 1)$, the zeros of $T_{n+1}(x)$ lead to a value of $|\pi(x)|_{\max}$, which is smaller by a factor that increases in proportion to $n^{1/2}$. The Chebyshev polynomials have an error oscillating uniformly over the interval, while Legendre polynomial errors oscillate with increasing amplitude towards the interval ends. If the maximum error is to be limited, the Chebyshev polynomial is best (Hildebrand 1956; pp. 386–388).

16.9 Chebyshev Interpolation

A linear combination of Chebyshev polynomials, of degrees zero through n, can approximate any degree n polynomial, $y(x)$, which agrees with $f(x)$, when $x = x_0, x_1, \ldots, x_n$, where x_r is the rth zero of $T_{n+1}(x)$. So

$$f(x) = \sum_{k=0}^{n} C_k T_k(x) + \frac{1}{2^n(n+1)!} T_{n+1}(x) f^{(n+1)}(\xi) \tag{16.106}$$

agreeing with Eq. (16.93), where $|\xi| < 1$ when x is in $(-1, 1)$. The $C's$ are determined such that the error term is suppressed, when $x = x_0, x_1, x_2, \ldots, x_n$, where

$$x_i = \cos\left(\frac{2i+1}{2n+2}\pi\right), \quad i = 0, 1, 2, \ldots, n \tag{16.107}$$

Introduce the change of variables

$$x = \cos\theta, \ 0 \le \theta \le \pi \tag{16.108}$$

then

$$f(x) \approx \sum_{k=0}^{n} C_k T_k(x), \quad -1 < x < 1 \tag{16.109}$$

becomes

$$F(\theta) \approx \sum_{k=0}^{n} C_k \cos k\theta, \quad 0 < \theta < \pi \tag{16.110}$$

with $F(\theta) = f(\cos\theta)$. The $C's$ are determined such that Eq. (16.110) is an equality, when $\theta = \theta_i$ and $F(\theta) = f(\cos\theta)$,

$$\theta_i = \cos^{-1} x_i = \frac{2i+1}{2n+2}\pi, \quad i = 0, 1, \ldots, n \tag{16.111}$$

At the equally spaced points $\pi/(2n+2), 3\pi/(2n+2), \ldots, (2n+1)\pi/(2n+2)$ agreement occurs. $\cos j\theta$ and $\cos k\theta$ are orthogonal under summation over the $(n+1)$ points defined by Eq. (16.111),

$$\sum_{r=0}^{n} \cos j\theta_r \cos k\theta_r = \begin{cases} 0, & j \neq k \\ \frac{n+1}{2}, & j = k \neq 0 \\ n+1, & j = k = 0 \end{cases} \tag{16.112}$$

where j and k are nonnegative integers not exceeding n. The left-hand side of Eq. (16.112) is identical to $\sum_{r=0}^{n} T_j(x_r) T_k(x_r)$; then, since $T_0(x), T_1(x) \ldots$ are orthogonal under integration over $(-1, 1)$ relative to $w(x) = 1/\sqrt{1-x^2}$, the $T_0(x), T_1(x), \ldots, T_n(x)$, with a unit weighting function, are orthogonal under summation over the zeros of $T_{n+1}(x)$. From Eq. (16.112), the $C's$ are

$$C_0 = \frac{1}{n+1} \sum_{r=0}^{n} F(\theta_r)$$

$$C_k = \frac{2}{n+1} \sum_{r=0}^{n} F(\theta_r) \cos k\theta_r \tag{16.113}$$

where θ_r is given by Eq. (16.111), or

$$C_0 = \frac{1}{n+1} \sum_{r=0}^{n} f(x_r)$$

$$C_k = \frac{2}{n+1} \sum_{r=0}^{n} f(x_r) T_k(x_r) \tag{16.114}$$

Table 16.7 Chebyshev interpolation

θ	$f(x)$ $F(\theta)$	$x = T_1(x)$ $\cos\theta$	$T_2(x)$ $\cos 2\theta$	$T_3(x)$ $\cos 3\theta$	$T_4(x)$ $\cos 4\theta$	$T_5(x)$ $\cos 5\theta$
$\pi/12$	$F_1 = f_1$	A	$\frac{1}{2}\sqrt{3}$	$\frac{1}{2}\sqrt{2}$	1/2	B
$\pi/4$	$F_2 = f_2$	$\frac{1}{2}\sqrt{2}$	0	$-\frac{1}{2}\sqrt{2}$	-1	$-\frac{1}{2}\sqrt{2}$
$5\pi/12$	$F_3 = f_3$	B	$-\frac{1}{2}\sqrt{3}$	$-\frac{1}{2}\sqrt{2}$	1/2	A
$7\pi/12$	$F_4 = f_4$	-B	$-\frac{1}{2}\sqrt{3}$	$\frac{1}{2}\sqrt{2}$	1/2	-A
$3\pi/4$	$F_5 = f_5$	$-\frac{1}{2}\sqrt{2}$	0	$\frac{1}{2}\sqrt{2}$	-1	$\frac{1}{2}\sqrt{2}$
$11\pi/12$	$F_6 = f_6$	-A	$\frac{1}{2}\sqrt{3}$	$-\frac{1}{2}\sqrt{2}$	1/2	-B
	$6C_0$	$3C_1$	$3C_2$	$3C_3$	$3C_4$	$4C_5$

where x_i is given by Eq. (16.107). For $n = 5$

$$A = \cos\frac{\pi}{12} = \frac{1}{2}\sqrt{2+\sqrt{3}} \doteq 0.96593$$

$$B = \cos\frac{5\pi}{12} = \frac{1}{2}\sqrt{2-\sqrt{3}} \doteq 0.25882 \tag{16.115}$$

Table 16.7, with dual headings, can be used either with the function $f(x)$, over $-1 \leq x \leq 1$, with the unequally spaced values in the third column, or with the function, $F(\theta)$, over $0 \leq \theta \leq \pi$, with the equally spaced values in the first column. Thus, the coefficient of $\cos 4\theta$ in Eq. (16.110) is from Table 16.7 and Eq. (16.114)

$$C_4 = \frac{1}{3}\left(\frac{1}{2}F_1 - F_2 + \frac{1}{2}F_3 + \frac{1}{2}F_4 - F_5 + \frac{1}{2}F_6\right) \tag{16.116}$$

while the coefficient of $T_4(x)$ in Eq. (16.109) from Table 16.7 and Eq. (16.113) is

$$C_4 = \frac{1}{3}\left(\frac{1}{2}f_1 - f_2 + \frac{1}{2}f_3 + \frac{1}{2}f_4 - f_5 + \frac{1}{2}f_6\right) \tag{16.117}$$

When the $C's$ are determined, the right-hand side of the approximation is evaluated as

$$f(x) \approx \sum_{k=0}^{n} C_k T_k(x) \tag{16.118}$$

at intermediate points using available tables of Chebyshev polynomials (Hildebrand 1956; pp. 389–391).

16.10 Economization of Power Series

When the integral of the product of $1/\sqrt{1-x^2}$ and the square of the error is to be minimized, the nth degree least-squares polynomial approximation to a function $f(x)$ over $(-1, 1)$, is

$$f(x) \approx y(x) = \sum_{k=0}^{n} a_k T_k(x), \ |x| < 1 \tag{16.119}$$

where

$$a_0 = \frac{1}{\pi} \int_{-1}^{1} \frac{f(x)}{\sqrt{1-x^2}} dx, \ a_k = \frac{2}{\pi} \int_{-1}^{1} \frac{f(x) T_k(x)}{\sqrt{1-x^2}} dx, \ k \geq 1 \tag{16.120}$$

Since the coefficients determined by summation over a discrete set, or by integration over an interval, are generally unequal, the approximation will not usually be that of Eq. (16.118). The two approximations may be of similar nature, in that each error will tend to oscillate with uniform amplitude over $(-1, 1)$. The finite Legendre series least squares approximation, with uniform weighting, will tend to oscillate with amplitudes increasing towards the interval ends. So an approximation may be achieved with fewer Chebyshev series than Legendre series, if smallness of the maximum error is the criterion.

The evaluation of the integrals of Eq. (16.120) is not readily effected for many functions. An alternative method is useful, if a function $f(x)$ is a power series for sufficiently small values of $|x|$. If one has the relation

$$f(x) = \sum_{k=0}^{n} A_k x^k + E_n(x) \tag{16.121}$$

where

$$|E_n(x)| < \epsilon_1, \ -1 \leq x \leq 1 \tag{16.122}$$

and ϵ_1 is smaller than the error tolerance ϵ, while $|A_n| + \epsilon_1$ is not smaller, then the last term in the approximation

$$f(x) \approx \sum_{k=0}^{n} A_k x^k \tag{16.123}$$

cannot be neglected. The right-hand side of Eq. (16.123) can be expanded in Chebyshev polynomials. Since it is a degree n polynomial, the series will end with the term involving $T_n(x)$, and be

$$\sum_{k=0}^{n} A_k x^k \equiv \sum_{k=0}^{n} a_k T_k(x) \tag{16.124}$$

The terms of highest degree in $T_r(x)$ are given by

$$T_r(x) = 2^{r-1}\left(x^r - \frac{r}{4}x^{r-2} + \dots\right) \tag{16.125}$$

and expressing the two members of Eq. (16.124) in terms of decreasing powers of x

$$A_n x^n + A_{n-1}x^{n-1} + A_{n-2}x^{n-2} + \cdots = 2^{n-1}a_n\left(x^n - \frac{n}{4}x^{n-2} + \dots\right)$$
$$+2^{n-2}a_{n-1}\left(x^{n-1} - \frac{n-1}{4}x^{n-3} + \dots\right) + 2^{n-3}a_{n-2}\left(x^{n-2} - \dots\right) + \dots \tag{16.126}$$

it follows that

$$a_n = 2^{-(n-1)}A_n,\ a_{n-1} = 2^{-(n-2)}A_{n-1},\ a_{n-2} = 2^{-(n-3)}(A_{n-2} + \frac{n}{4}A_n), \dots \tag{16.127}$$

If n is sufficiently large, the coefficients of

$$T_n(x), T_{n-1}(x), \dots, T_{n-m+1}(x)$$

in Eq. (16.124) will be small compared to the coefficients of

$$x^n, x^{n-1}, \dots, x^{n-m+1}$$

in Eq. (16.121) for some m, and it could be that

$$(|a_{n-m+1}| + |a_{n-m+2}| +, \dots, |a_n|) + \epsilon_1$$

is smaller than ϵ, thus a tolerable error in the approximation to $f(x)$. Since $|T_r(x)| \leq 1$ in $(-1, 1)$, then the last m terms in the right-hand member of Eq. (16.124) are negligible. The approximation in Eq. (16.123) can be replaced by

$$f(x) \approx \sum_{k=0}^{n-m} a_k T_k(x) \tag{16.128}$$

Table 16.8 Process of power series economization

$1 = T_0$	$T_0 = 1$
$x = T_1$	$T_1 = x$
$x^2 = \frac{1}{2}(T_0 + T_2)$	$T_2 = 2x^2 - 1$
$x^3 = \frac{1}{4}(3T_1 + T_3)$	$T_3 = 4x^3 - 3x$
$x^4 = \frac{1}{8}(3T_0 + 4T_2 + T_4)$	$T_4 = 8x^4 - 8x^3 + 1$
$x^5 = \frac{1}{16}(10T_1 + 5T_3 + T_5)$	$T_5 = 16x_5 - 20x^3 + 5x$
$x^6 = \frac{1}{32}(10T_0 + 15T_2 + 6T_4 + T_6)$	$T_6 = 32x^6 - 48x^4 + 18x^2 - 1$
$x^7 = \frac{1}{64}(35T_1 + 21T_3 + 7T_5 + T_7)$	$T_7 = 64x^7 - 112x^5 + 56x^3 - 7x$
$x^8 = \frac{1}{128}(35T_0 + 56T_2 + 28T_4 + 8T_6 + T_8)$	$T_8 = 128x^8 - 256x^6 + 160x^4 - 32x^2 + 1$
$x^9 = \frac{1}{256}(126T_1 + 84T_3 + 36T_5 + 9T_7 + T_9)$	$T_9 = 256x^9 - 576x^7 + 432x^5 - 120x^3 + 9x$

where $m > 0$. This approximation can be transformed to an expression of the form

$$f(x) \approx \sum_{k=0}^{n-m} \bar{A}_k x^k \tag{16.129}$$

if desired. Thus, a polynomial approximation to $f(x)$ over $(-1, 1)$ is obtained with fewer terms than required by a truncated power series, and involving the smallest possible number of polynomial terms, which will supply an accuracy within the prescribed tolerance limits. The error $E_n(x)$ in Eq. (16.121) is accepted as a fixed error, and an efficient approximation to $f(x) - E_n(x)$ is sought. The approximation obtained may not be the best possible one, if $|E_n(x)|$ is small relative to ϵ. The transformations can be facilitated using the following two sets of relations; the second being the Chebyshev polynomials, and the first being obtained by successively inverting the second set members, as shown in Table 16.8.

To illustrate, if a polynomial approximation to e^x is required for the interval $(-1, 1)$ with a tolerance of 0.01, the truncation of a Maclaurin series gives a polynomial approximation of degree five,

$$e^x \approx 1 + x + \frac{1}{2}x^2 + \frac{1}{6}x^3 + \frac{1}{24}x^4 + \frac{1}{120}x^5 \equiv y(x) \tag{16.130}$$

with an error

$$|E(x)| = \left|\frac{e^\xi}{720}x^6\right| < \frac{e}{720} < 0.0038 \tag{16.131}$$

Neglecting the $x^5/120$ term would permit a possible error exceeding the prescribed tolerance. Using the first set of relations in Table 16.8 transforms Eq. (16.130) into

$$y(x) = \frac{81}{64}T_0 + \frac{217}{192}T_1 + \frac{13}{48}T_2 + \frac{17}{384}T_3 + \frac{1}{192}T_4 + \frac{1}{1920}T_5 \tag{16.132}$$

where $T_r = T_r(x)$. If the last two terms are neglected, an additional error not exceeding $11/1920 < 0.0058$ for all x in $(-1, 1)$ would be introduced. Thus, a total error smaller than $|0.0096|$ is achieved with

$$e^x \approx \frac{81}{64}T_0 + \frac{217}{192}T_1 + \frac{13}{48}T_2 + \frac{17}{384}T_3 \tag{16.133}$$

Using the second set of Table 16.8,

$$e^x \approx \frac{1}{384}(382 + 382x + 208x^2 + 68x^3), \ |x| \leq 1 \tag{16.134}$$

For comparison, in Legendre polynomials,

$$y(x) = \frac{47}{40}P_0 + \frac{309}{280}P_1 + \frac{5}{14}P_2 + \frac{19}{270}P_3 + \frac{1}{105}P_4 + \frac{1}{945}P_5 \tag{16.135}$$

Since the last term could be neglected, a fourth degree polynomial would satisfy the approximation. This procedure, *economization of power series*, is useful when minimization of numerical operations is desirable. It can be applied to any polynomial once the interval has been transformed to $(-1, 1)$ (Hildebrand 1956; pp. 391–396).

16.11 Recursive Filtering

Whereas in previous sections we described model fitting procedures based on batch processing of data, in this section we will discuss a method for estimating the state vector of a dynamical system (see Sect. 14.2) based on measurements obtained at each time instant. This process is referred to as *filtering*. The most notable result in this context is the *Kalman filter* (KF) (Kalman 1960; Kalman and Bucy 1961). The KF was further developed by NASA and successfully implemented for the first time in the Apollo mission in the late 1960s. The KF provided a solution to the Apollo spacecraft trajectory estimation and control problem, and was vital for the mission success (Grewal and Andrews 2010). Kalman filters are widely used in orbit determination, guidance, and navigation systems of satellites.

The KF is the optimal filter for a linear system with linear measurements and additive Gaussian noise. However, most physical processes are nonlinear. To deal with nonlinearities, the system can be linearized about the estimated trajectory. The resulting filtering scheme is known as the *extended Kalman filter* (EKF). The EKF has become one of the most widely used nonlinear filters, and was implemented in navigation systems in low Earth orbits (Busse 2003).

There are a number of variants for the EKF formulation. The one that best suits astrodynamical applications uses a continuous-time physical model and a discrete-time (sampled) measurement model. In between measurement samples, the state of the satellite is estimated based on the model only. The physical time is denoted by t and the sampled time instances are denoted by t_k. As in Sect. 14.2, the state vector is denoted by $\mathbf{x}$. The estimated state will be denoted by $\hat{\mathbf{x}}$.

Looking at Eq. (14.1), the continuous dynamics are modelled as

$$\dot{\mathbf{x}}(t) = \mathbf{f}(\mathbf{x}, \mathbf{u}, t) + \mathbf{w}(t) \tag{16.136}$$

Here, we added the term $\mathbf{w}(t)$ to the model appearing in Eq. (14.1). This term, commonly referred to as the *process noise*, is an additive white noise, which is assumed to be zero-mean and Gaussian, and is used for modelling uncertainties. The *power spectral density* of the process noise is

$$E[\mathbf{w}(t)\mathbf{w}^T(\tau)] = \mathbf{Q}(t)\delta(t-\tau) \tag{16.137}$$

where $E(\cdot)$ denotes the *expected value* operator, and $\delta(\cdot)$ is the *Dirac delta function*. Now, we distinguish between two estimates. The first, is an estimate of the state at some time τ before a measurement is obtained. This estimate is denoted by $\hat{\mathbf{x}}(\tau^-)$. The second, is an estimate updated by a measurement, which will be denoted by $\hat{\mathbf{x}}(\tau^+)$. In between measurements, i.e. between the times t_{k-1} and t_k, the estimated state is propagated by using numerical integration, e.g. the Runge-Kutta method (See Sect. 9.9) to obtain (Stengel 1994; p. 388)

$$\hat{\mathbf{x}}(t_k^-) = \hat{\mathbf{x}}(t_k^+) + \int_{t_{k-1}}^{t_k} \mathbf{f}[\hat{\mathbf{x}}(\tau^-), \mathbf{u}(\tau), \tau] d\tau \tag{16.138}$$

where the initialization is written as

$$\hat{\mathbf{x}}_0 = E(\mathbf{x}_0) \tag{16.139}$$

The EKF approximates the nonlinear dynamics by linearizing it about the estimate $\hat{\mathbf{x}}(\tau^-)$. The corresponding Jacobian is given by

$$\mathbf{F} = \left. \frac{\partial \mathbf{f}}{\partial \mathbf{x}} \right|_{\hat{\mathbf{x}}(\tau^-)} \tag{16.140}$$

In addition to the state estimate, the EKF propagates a covariance estimate of the state. Still without considering any measurements, the covariance estimate is propagated as (Stengel 1994; p. 388)

$$\mathbf{P}(t_k^-) = \mathbf{P}(t_{k-1}^+) + \int_{t_{k-1}}^{t_k} \left[\mathbf{F}(\tau)\mathbf{P}(\tau) + \mathbf{P}(\tau)\mathbf{F}^T(\tau) + \mathbf{Q}(\tau)\right] d\tau \tag{16.141}$$

and the initialization is performed according to

$$\mathbf{P}_0 = E\left[(\mathbf{x}_0 - \hat{\mathbf{x}}_0)(\mathbf{x}_0 - \hat{\mathbf{x}}_0)^T\right] \tag{16.142}$$

The vector of sampled measurements, $\mathbf{y}$, is expressed through the nonlinear measurement equation

$$\mathbf{y}(t_k) = \mathbf{h}[\mathbf{x}(t_k), t_k] + \boldsymbol{v}(t_k) \tag{16.143}$$

where $\boldsymbol{v}$ denotes the *measurement noise*, which is assumed to be a white, zero-mean Gaussian random sequence, so that

$$E[\boldsymbol{v}(t_k)] = \mathbf{0}, \ E[\boldsymbol{v}(t_k)\boldsymbol{v}^T(t_k)] = \mathbf{R}(t_k) \tag{16.144}$$

It is also assumed that the measurement noise is uncorrelated with the process noise. Now, the nonlinear measurement equation is also linearized about the current estimate,

$$\mathbf{H}(t_k) = \left.\frac{\partial \mathbf{h}}{\partial \mathbf{x}}\right|_{\hat{\mathbf{x}}(t_k^-)} \tag{16.145}$$

and the *Kalman gain* is calculated as

$$\mathbf{K}(t_k) = \mathbf{P}(t_k^-)\mathbf{H}^T(t_k)\left[\mathbf{H}(t_k)\mathbf{P}(t_k^-)\mathbf{H}^T(t_k) + \mathbf{R}(t_k)\right]^{-1} \tag{16.146}$$

This gain is used for calculating a state estimate update as (Stengel 1994; p. 388)

$$\hat{\mathbf{x}}(t_k^+) = \hat{\mathbf{x}}(t_k^-) + \mathbf{K}(t_k)\left\{\mathbf{y}(t_k) - \mathbf{h}[\hat{\mathbf{x}}(t_k^-), t_k]\right\} \tag{16.147}$$

and the covariance estimate update

$$\mathbf{P}(t_k^+) = [\mathbf{I} - \mathbf{K}(t_k)\mathbf{H}(t_k)]\mathbf{P}(t_k-) \tag{16.148}$$

The process is then repeated for the next index k.

16.12 Mean Elements Estimator

We will now illustrate the application of the EKF to an astrodynamical problem. The goal is to obtain estimated mean orbital elements by utilizing instantaneous measurements of the osculating elements (or functions thereof). To that end, we write the fully-assembled semianalytical model described in Sect. 13.3 as

$$\begin{aligned}\dot{\bar{\mathbf{c}}} &= \dot{\mathbf{c}}_{sec}(\bar{\mathbf{c}}) + \dot{\mathbf{c}}_{long}(\bar{\mathbf{c}}) + \dot{\mathbf{c}}_{sec,2}(\bar{\mathbf{c}}) + \dot{\mathbf{c}}_{long,2}(\bar{\mathbf{c}}) \\ &\quad + \dot{\mathbf{c}}_{long,nm}(\bar{\mathbf{c}}) + \dot{\mathbf{c}}_{drag,sec}(\bar{\mathbf{c}}) + \dot{\mathbf{c}}_{drag,long}(\bar{\mathbf{c}}) + \mathbf{u} + \mathbf{w}\end{aligned} \tag{16.149}$$

where $\mathbf{c} \triangleq [a, e, i, \Omega, \omega, M_0]$; $\bar{\mathbf{c}}$ denotes the mean elements; $\mathbf{u}$ are the control inputs (either impulsive or continuous), whose effect is modeled using the GVE (13.4); and $\mathbf{w}$ is the process noise, satisfying Eq. (16.137). As in Sect. 13.3, the subindices are used for denoting secular terms of the first (*sec*) and second (*sec*, 2) order that are due to the gravitational harmonics, long-periodic terms of the first order (*long*) and second order (*long*, 2) that are due to zonal harmonics and tesseral harmonics (*long*, *nm*), and the secular (*drag*, *sec*) and long-periodic (*drag*, *long*) effects of drag.

The incorporation of the control forces using Eqs. (13.4) involves an inherent approximation, since the GVE are written in osculating, not mean, elements. However, as previous studies have indicated, this approximation is adequate (Schaub and Alfriend 2001).

The satellite on-board sensors—e.g., a GPS receiver—provide measurements of the inertial position and velocity or unfiltered outputs of the osculating orbital elements. Thus, following the discussion in Sect. 13.3, it is possible to write the following relation between the osculating and mean elements,

$$\mathbf{c}_{osc} = \bar{\mathbf{c}} + \mathbf{c}_{short}(\bar{\mathbf{c}}) + \mathbf{c}_{drag,short}(\bar{\mathbf{c}}) + \mathbf{c}_{short,nm}(\bar{\mathbf{c}}) + \boldsymbol{\upsilon} \tag{16.150}$$

where $\boldsymbol{\upsilon}$ is the measurement noise, a white, zero-mean Gaussian random sequence, satisfying Eq. (16.144). It is readily seen that Eqs. (16.149) and (16.150) constitute nonlinear process and measurement equations, respectively, which correspond to Eqs. (16.136) and (16.143), namely

$$\dot{\mathbf{x}}(t) = \mathbf{f}(\mathbf{x}(t), \mathbf{u}, t) + \mathbf{w}(t) \tag{16.151}$$

$$\mathbf{y}(t_k) = \mathbf{h}(\mathbf{x}(t_k)) + \boldsymbol{\upsilon}(t_k) \tag{16.152}$$

where in this case the state vector to be estimated is $\mathbf{x} \equiv \bar{\mathbf{c}}$, and $\mathbf{y} \equiv \mathbf{c}_{osc}$ is the measurement vector. To obtain the estimated mean elements $\hat{\bar{\mathbf{c}}}$ from the osculating elements measurements, a nonlinear estimation algorithm, such as the EKF, can be used (Zhong and Gurfil 2013).

The mean elements estimation algorithm will be illustrated using three test cases: Orbits with no control inputs; impulsive orbital corrections; and continuous-thrust orbital transfer. The "true" orbits are generated based on a high precision orbit propagator, including a 21×21 gravity model, drag according to the ISA-1976 model (Noaa and Usaf 1976) and other perturbations, including solar radiation pressure and lunisolar attraction. For the first example, a simulation time of one year is chosen so that the long-periodic terms, whose period is around five months, can be seen in the orbital dynamics.

16.12.1 Initial Conditions and Parameter Values

The mission epoch is 1 Mar 2012 10:00:00.000 UTC, and it lasts until 28 Feb 2013 10:00:00.000 UTC. Table 16.9 gives the initial conditions in terms of osculating orbital elements.

The simulated observation data are generated by contaminating the simulation-generated position and velocity vectors with measurement errors. The standard deviations of the position and velocity errors, for each axis, are 5 m and 2 cm/s, respectively.

The transformation of the position and velocity measurement noise covariance into osculating orbital elements can be done by means of a *Monte-Carlo simulation*, i.e., repeated random sampling. The results are presented in Table 16.10, where the measurement noise covariance matrix **R** is given by

$$\mathbf{R} = \text{diag}\left[\text{cov}_a, \text{cov}_e, \text{cov}_i, \text{cov}_\Omega, \text{cov}_\omega, \text{cov}_M\right] \tag{16.153}$$

and the notation cov_x denotes the covariance of x. The covariance is calculated as follows: Since the common formulation of recursive filtering assumes additive noise, it is required to calculate the covariance of the orbital elements starting from the covariance of position and velocity. Since the mapping from inertial position and velocity to orbital elements is nonlinear, a Monte-Carlo simulation is used to evaluate the probability density function of the elements and the related noise statistics. The covariance is time dependent, but is averaged to yield constant values. It is approximated as a diagonal matrix, since the cross-correlation terms practically vanish.

Table 16.9 Initial osculating orbital elements values

Parameter	Numerical value
a	7000 km
e	0.01
i	55°
Ω	10°
ω	10°
M	10°

Table 16.10 Measurement noise covariance of the osculating orbital elements

Parameter	Numerical value
cov_a	500.4 m^2
cov_e	8.966×10^{-12}
cov_i	1.886×10^{-12} rad^2
cov_Ω	9.020×10^{-13} rad^2
cov_ω	3.179×10^{-8} rad^2
cov_M	3.086×10^{-8} rad^2

Table 16.11 Geopotential and drag model coefficients

Parameter	Numerical value
J_2	$1082.62668355 \times 10^{-6}$
J_3	$-2.53265648533 \times 10^{-6}$
J_4	$-1.61962159137 \times 10^{-6}$
C_{22}	$1.57446037456 \times 10^{-6}$
S_{22}	$-9.03803806639 \times 10^{-7}$
C_{31}	$2.19263852917 \times 10^{-6}$
S_{31}	$2.68424890297 \times 10^{-7}$
C_{32}	$3.08989206881 \times 10^{-7}$
S_{32}	$-2.11437612437 \times 10^{-7}$
C_{33}	$1.00548778064 \times 10^{-7}$
S_{33}	$1.97222559006 \times 10^{-7}$
ω_e	$7.2921158553 \times 10^{-5}$ rad/s
r_e	6378.137 km
μ	$3.98600436 \times 10^{14}$ m^3/s^2
H	68.7 km
ρ_0	2.34×10^{-13} kg/m^3
C_D	2.2
m	6 kg
S	0.2474 m^2

The estimated state propagation was performed based on the model of Eq. (16.149), which included zonal harmonics up to J_4 and tesseral/sectorial terms up to C_{33}/S_{33}. Table 16.11 lists the numerical values of the geopotential coefficients and other parameters related to the astrodynamical models.

To obtain faster convergence, a judicious initialization of the estimated mean elements is required. To that end, the following initialization is used:

$$\hat{\bar{\mathbf{c}}}(t_0) = \mathbf{c}_{osc}(t_0) - \mathbf{c}_{short}(\mathbf{c}_{osc}(t_0)) - \mathbf{c}_{drag,short}(\mathbf{c}_{osc}(t_0)) \tag{16.154}$$

Equation (16.154) generates an approximation of the initial estimated values by replacing the mean elements with the (measured) osculating elements in the expressions for the short-periodic variations.

16.12.2 Uncontrolled Orbits, Single Run

The simulation study starts by examining the long-periodic and secular mean element evolution for a one-year mission. The results are presented in Figs. 16.2, 16.3, 16.4, which compare the osculating and mean values of the semimajor axis (Fig. 16.2a), eccentricity (Fig. 16.2b), inclination (Fig. 16.3a), RAAN (Fig. 16.3b), and argument of perigee (Fig. 16.4). It is evident that the filter captures the long-

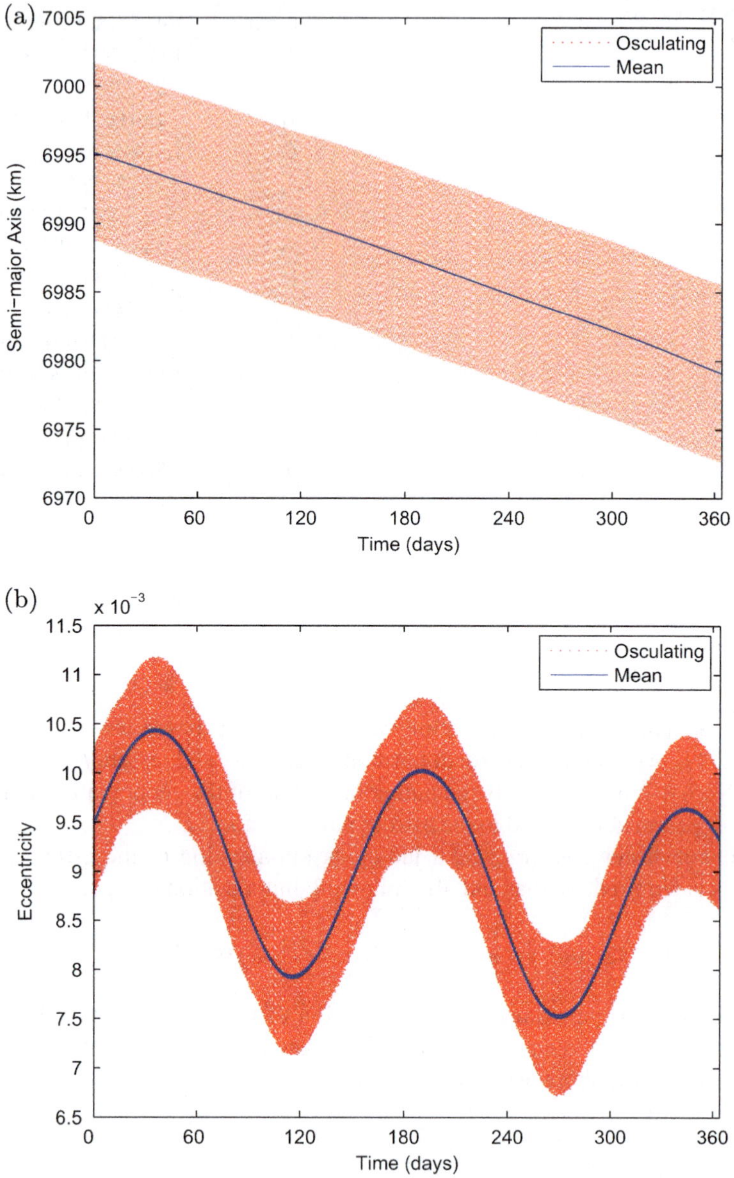

Fig. 16.2 A 1-year simulation run comparing the osculating and estimated mean (**a**) semimajor axis and (**b**) eccentricity

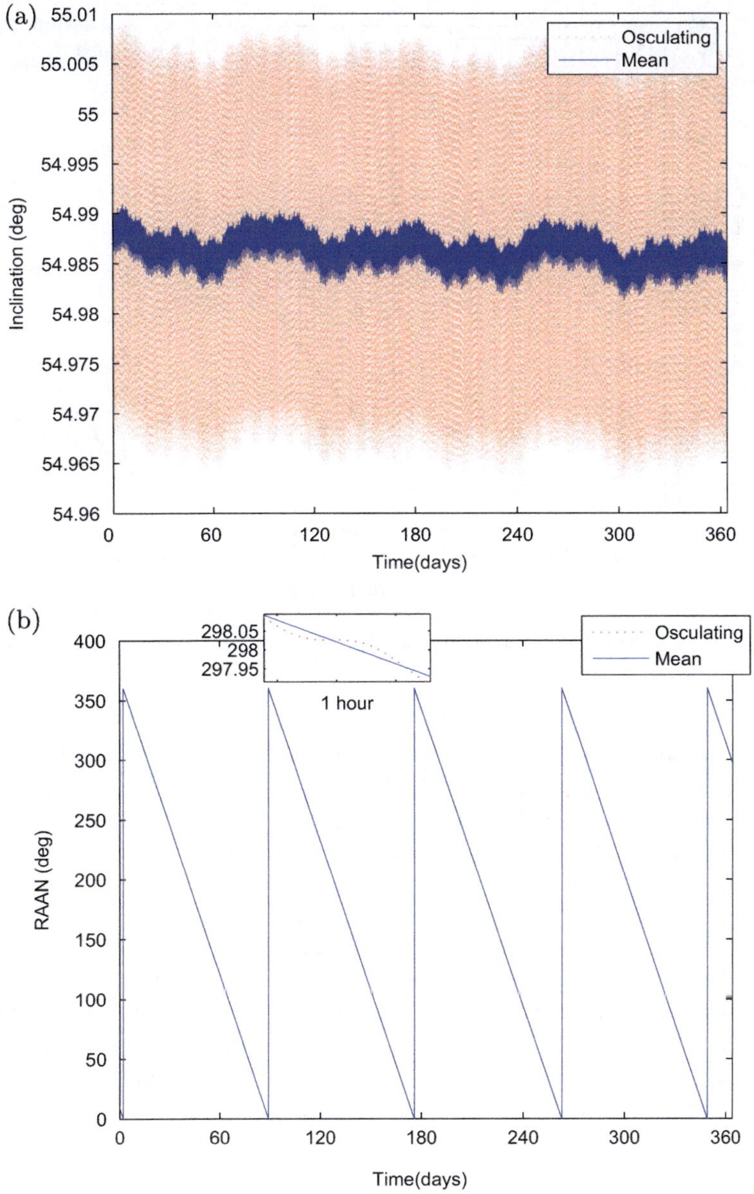

Fig. 16.3 A 1-year simulation run comparing the osculating and estimated mean (**a**) inclination and (**b**) RAAN

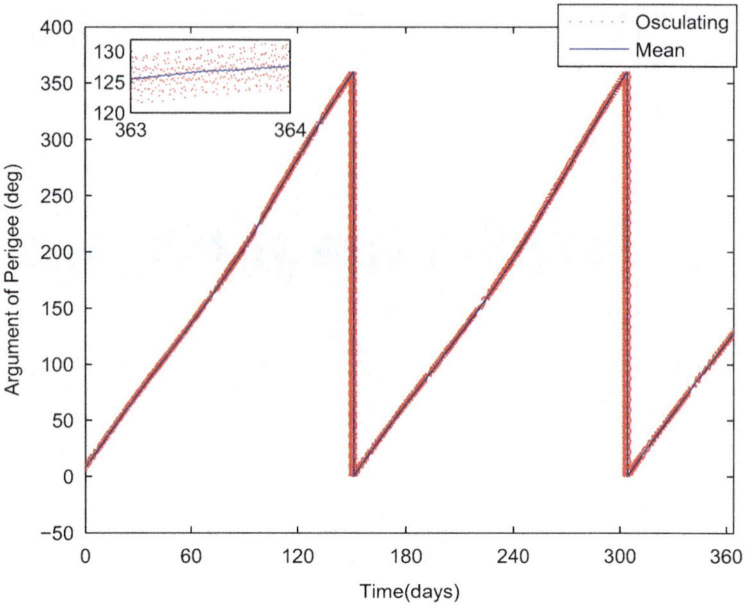

Fig. 16.4 A 1-year simulation run comparing the osculating and estimated mean argument of perigee

periodic evolution of the eccentricity and inclination as well as the secular evolution of the semimajor axis (due to atmospheric drag).

The next step is to evaluate the mean elements estimation errors. However, as opposed to standard filtering problems, in which the states of the process model are used as reference for comparing the estimated states, in the case at hand the semianalytical model of the mean elements, given by Eq. (16.149), is truncated on purpose, and does not include effects such as lunisolar attraction, gravitational perturbations beyond order 4 and solar radiation pressure, for the sake of computational efficiency. Thus, it makes little sense to use it as a reference for evaluating the estimation errors. Instead, a batch numerical averaging procedure is carried out to evaluate the "true" mean elements per the definition in Eq. (13.6). The batch averaging relies on the extended Simpson quadrature rule, given by (see Sect. 9.5) where the alternation of $2/3$ and $4/3$ continues throughout the interior of the parentheses in Eq. (16.156). The vector of estimation errors is then defined as

$$\mathbf{e_c} \triangleq \bar{\mathbf{c}} - \hat{\bar{\mathbf{c}}} \tag{16.155}$$

The estimation errors of Eq. (16.155) are compared to a direct calculation of the mean elements using the Brouwer artificial satellite theory (Schaub and Junkins 2003; pp. 693–696). The purpose of the comparison between the Brouwer and EKF-based estimation of the mean elements is to examine whether a direct application of the Brouwer theory may be used as a substitute for recursive filtering; as will be seen shortly, the answer to this question is negative.

A comparison of estimation errors between the EKF and the Brouwer theory is depicted in Figs. 16.5, 16.6, 16.7. In this figure, the estimation errors of the semimajor axis (Fig. 16.5a), eccentricity (Fig. 16.5b), inclination (Fig. 16.6a), RAAN (Fig. 16.6b), and argument of perigee (Fig. 16.7) are shown for a 24-h period to better illustrate the quantitative aspects of the differences between the Brouwer theory and the recursive filter. It is evident that the Brouwer errors are noisier than the filter-based estimation errors. In addition, the eccentricity estimation through Brouwer's theory is slightly biased, whereas in the filter-based estimation it is unbiased.

16.12.3 Orbits with No Control Inputs, Monte-Carlo Runs

The observations made previously can be substantiated using a long-term Monte-Carlo simulation. Figures 16.8–16.10 depict the results of 100 Monte-Carlo simulation runs, each propagated for 175 days. In each of Figs. 16.8, 16.9, 16.10, the single standard deviation ($\pm\sigma$) curves as obtained by the filter and through Brouwer's theory are shown. The numerical values of the standard deviations are summarized in Table 16.12. It is evident that the filter yields at least an order of magnitude better accuracy than the Brouwer theory. The difference is particularly dramatic in the estimation of the semimajor axis, for which the filter provides a 1-σ value of about 60 cm, compared to a Brouwer 1-σ value of about 22 m.

16.12.4 Impulsive Maneuvers

We can now examine how the filter behaves when the control term in Eq. (16.149) becomes active. The first test is conducted for an impulsive maneuver. The direction of the thrust is along the velocity vector, with a magnitude of 1 N. The estimated mean semimajor axis and eccentricity in the presence of this maneuver are shown in Fig. 16.11a and b, respectively. The filter keeps tracking the mean elements, which indicates that it can be combined with impulsive maneuvers for correcting the mean elements, as explained in Sect. 14.3.

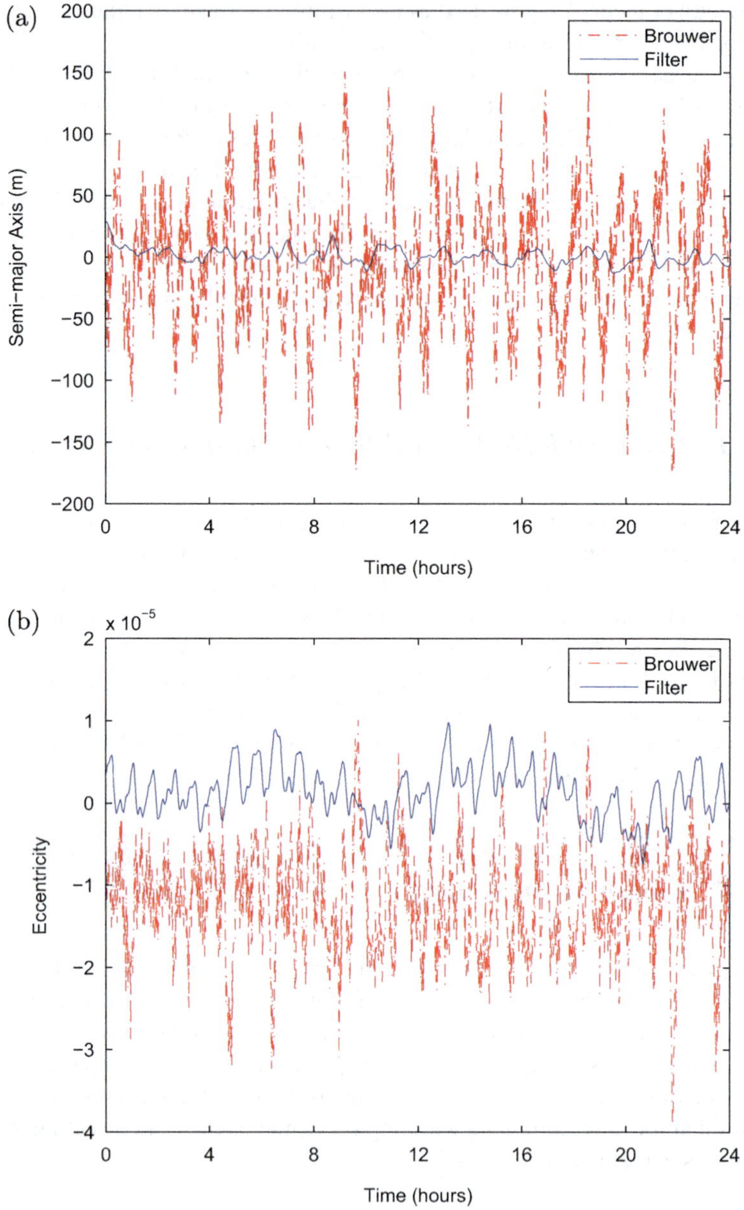

Fig. 16.5 Simulation run comparing the mean (**a**) semimajor axis and (**b**) eccentricity estimation errors as obtained from the Brouwer theory and the EKF

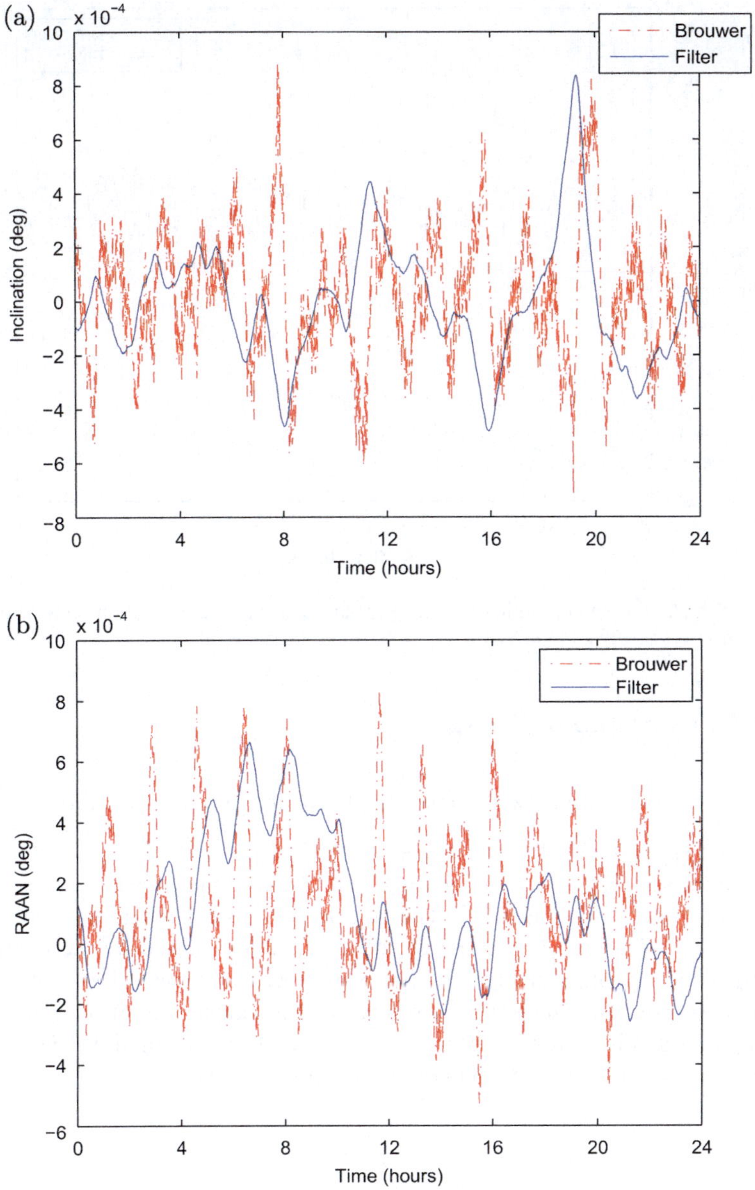

Fig. 16.6 Simulation run comparing the mean (**a**) inclination and (**b**) RAAN estimation errors as obtained from the Brouwer theory and the EKF

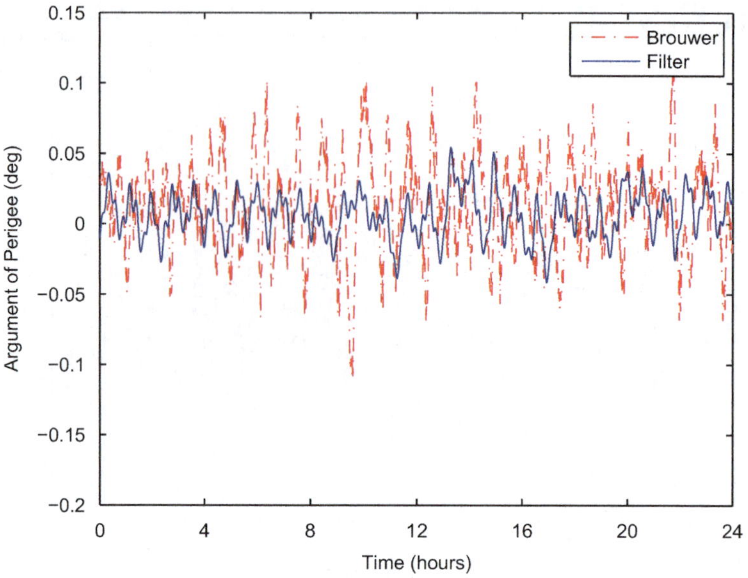

Fig. 16.7 Simulation run comparing the mean argument of perigee estimation error as obtained from the Brouwer theory and the EKF

16.12.5 Continuous Thrust

The final test case involves continuous thrust. The magnitude of the thrust is $0.08\,N$ and the direction is determined by the unit vector $\left[\sqrt{3}/3, \sqrt{3}/3, \sqrt{3}/3\right]^T$ in the NTW frame (see Sect. 11.7,) indicating that all three maneuver axes are active. A thrusting time of 3 h is used. The results for a 12-h integration are shown in Figs. 16.12, 16.13, 16.14. Although the acceleration caused by the continuous thrusting induces both secular and periodic variations in the mean elements, the filter is capable of providing an unbiased estimation of the mean elements, including the semimajor axis (Fig. 16.12a), eccentricity (Fig. 16.12b), inclination (Fig. 16.13a), RAAN (Fig. 16.13b), and argument of perigee (Fig. 16.14).

$$\begin{aligned}\bar{\mathbf{c}} &= \frac{1}{T}\int_0^T \mathbf{c}(t)\,\mathrm{d}t = \frac{1}{N}\int_{t_0}^{t_{N-1}} \mathbf{c}(t)\,\mathrm{d}t \\ &= \frac{1}{N}\left(\frac{1}{3}\mathbf{c}(t_0) + \frac{4}{3}\mathbf{c}(t_1) + \frac{2}{3}\mathbf{c}(t_2) + \frac{4}{3}\mathbf{c}(t_3) + \right. \\ &\quad \left. \cdots + \frac{2}{3}\mathbf{c}(t_{N-3}) + \frac{4}{3}\mathbf{c}(t_{N-2}) + \frac{1}{3}\mathbf{c}(t_{N-1})\right) + \mathcal{O}\left(\frac{1}{N^4}\right)\end{aligned} \tag{16.156}$$

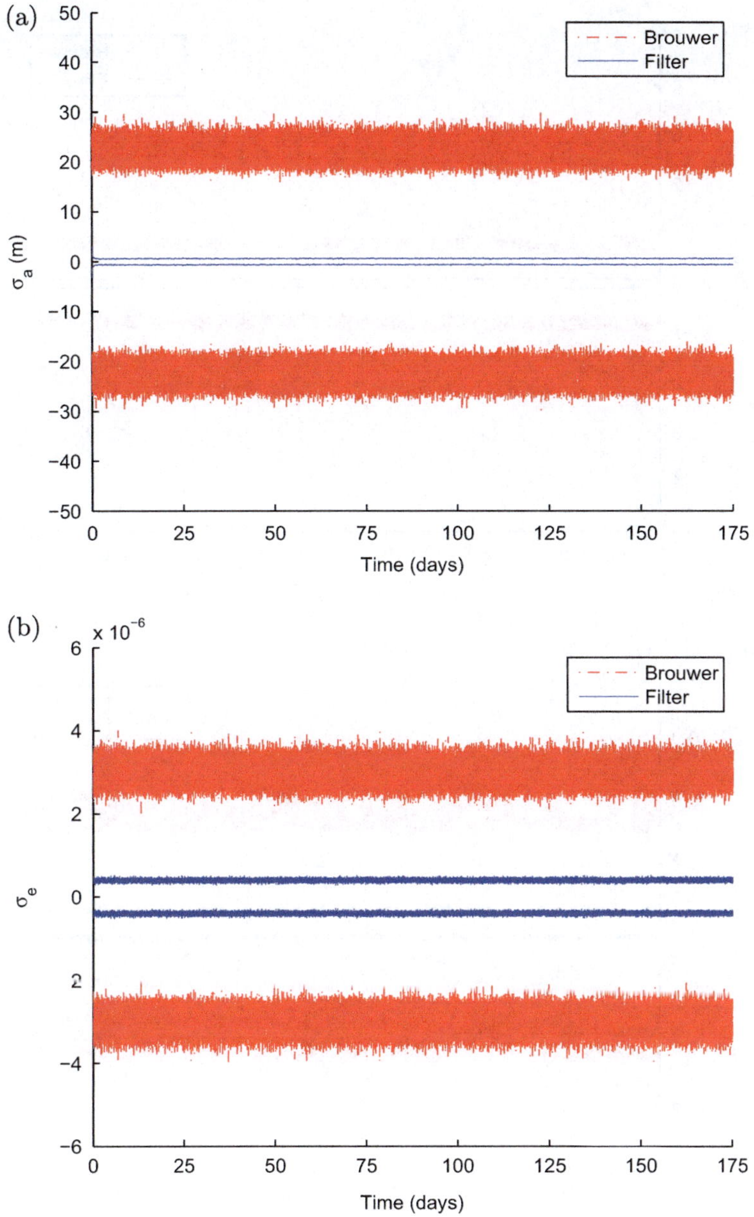

Fig. 16.8 The standard deviation of the mean (**a**) semimajor axis and (**b**) eccentricity estimation errors as obtained from 100 Monte-Carlo simulation runs, compared to the Brouwer theory

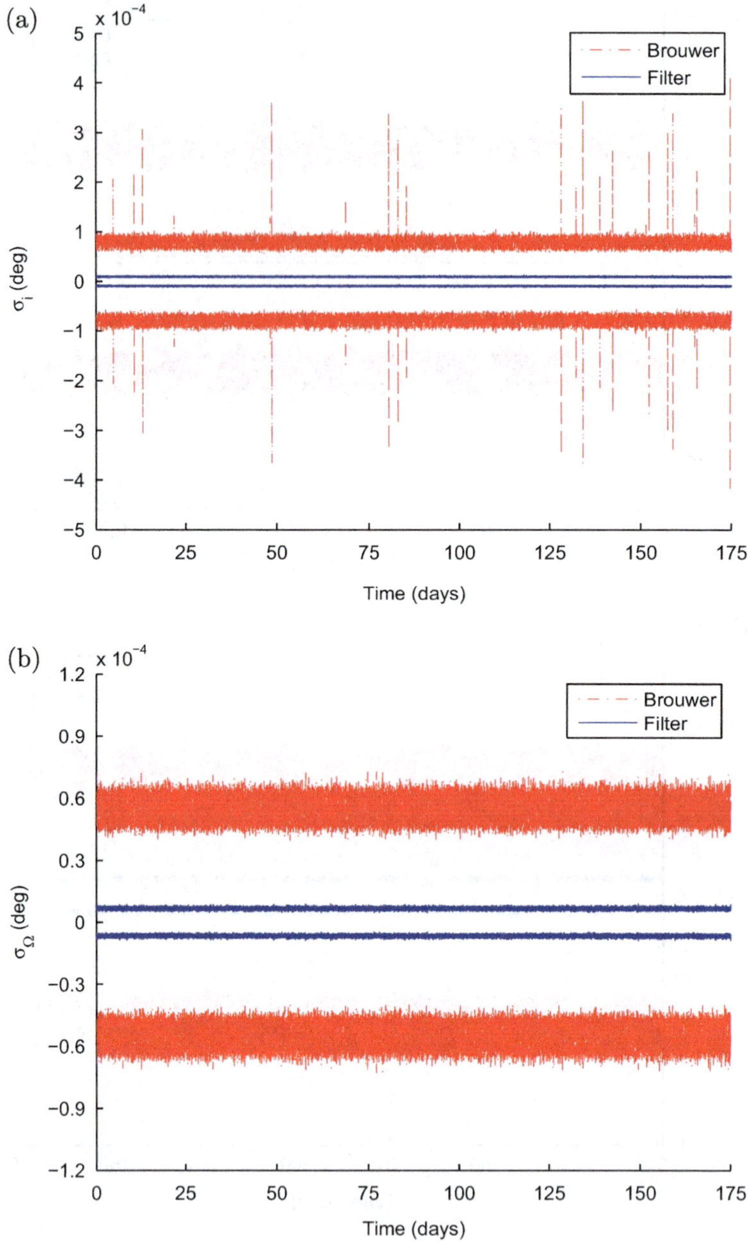

Fig. 16.9 The standard deviation of the mean (**a**) inclination and (**b**) RAAN estimation errors as obtained from 100 Monte-Carlo simulation runs, compared to the Brouwer theory

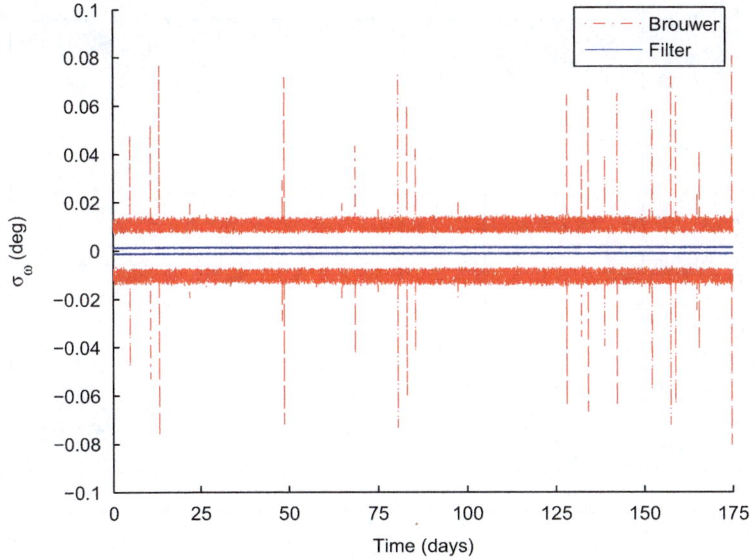

Fig. 16.10 The standard deviation of the mean argument of perigee estimation error as obtained from 100 Monte-Carlo simulation runs, compared to the Brouwer theory

Table 16.12 Standard deviation values of the mean elements estimation errors for 100 Monte-Carlo Runs

σ	Filter	Brouwer
e_a	0.6118 m	22.42 m
e_e	3.972×10^{-7}	2.994×10^{-6}
e_i	9.380×10^{-6} deg	7.877×10^{-5} deg
e_Ω	6.665×10^{-6} deg	5.486×10^{-5} deg
e_ω	0.0013°	0.0103°

To summarize, using the EKF for estimating the mean orbital elements has clear advantages over the Brouwer artificial satellite theory. It can adequately respond to thrust, and it is much less sensitive to measurement noise. In the examined scenarios, the filter provides a sub-meter 1-σ estimation accuracy of the mean semimajor axis, which is almost two orders of magnitude better than the Brouwer-based estimation.

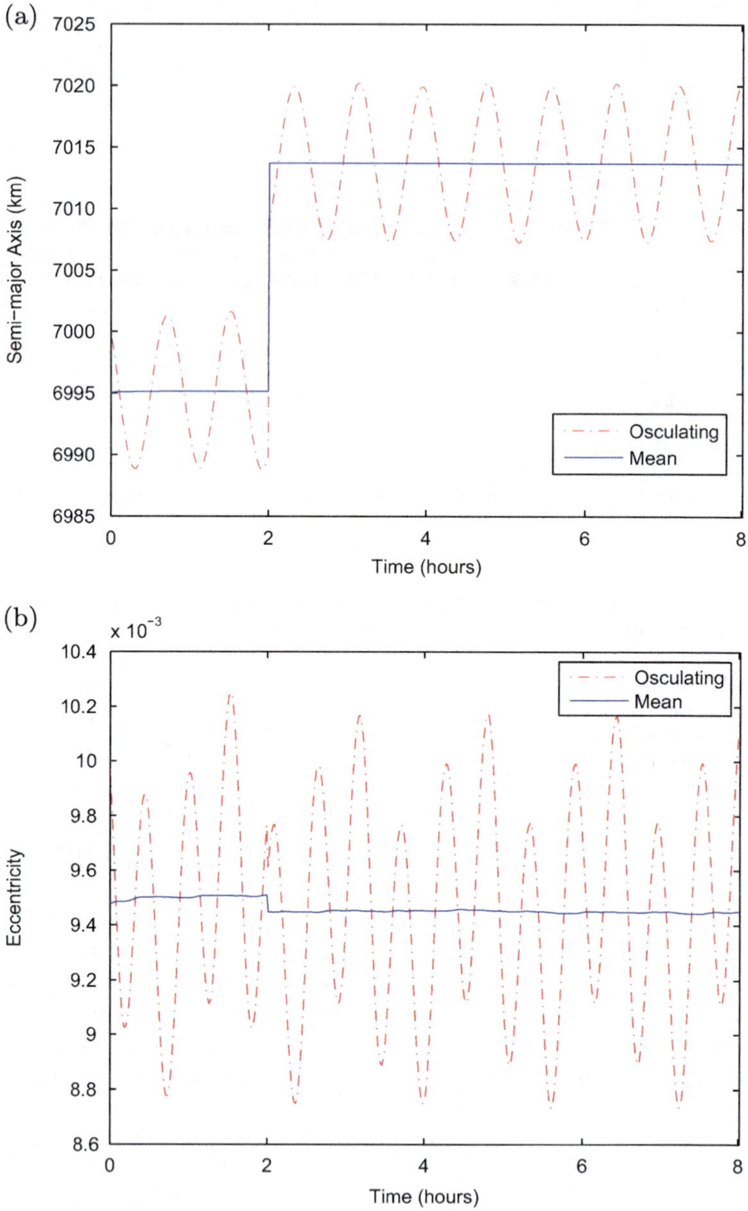

Fig. 16.11 Estimated mean (**a**) semimajor axis and (**b**) eccentricity in the presence of a single impulsive maneuver

(a)

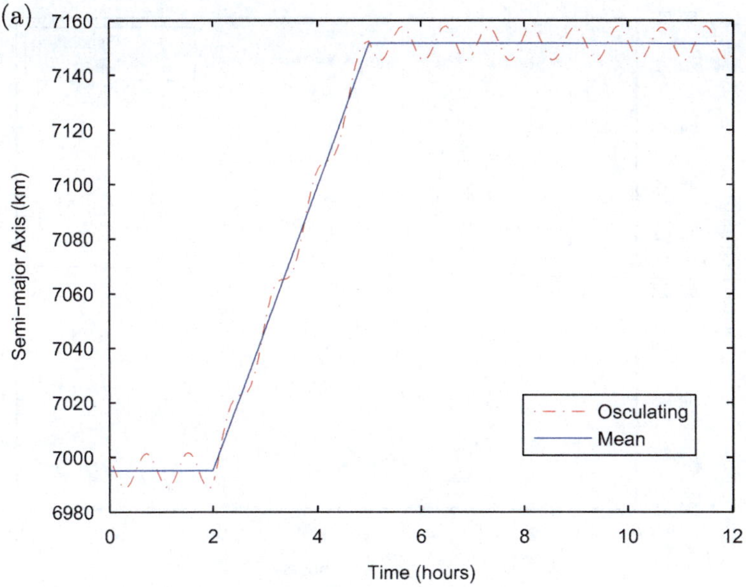

(b)

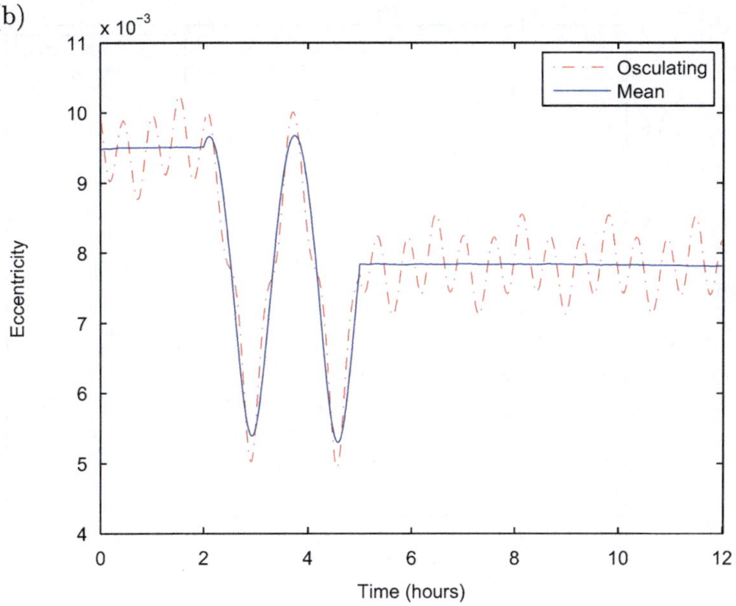

Fig. 16.12 Estimated mean (**a**) semimajor axis and (**b**) eccentricity under continuous constant-magnitude thrust

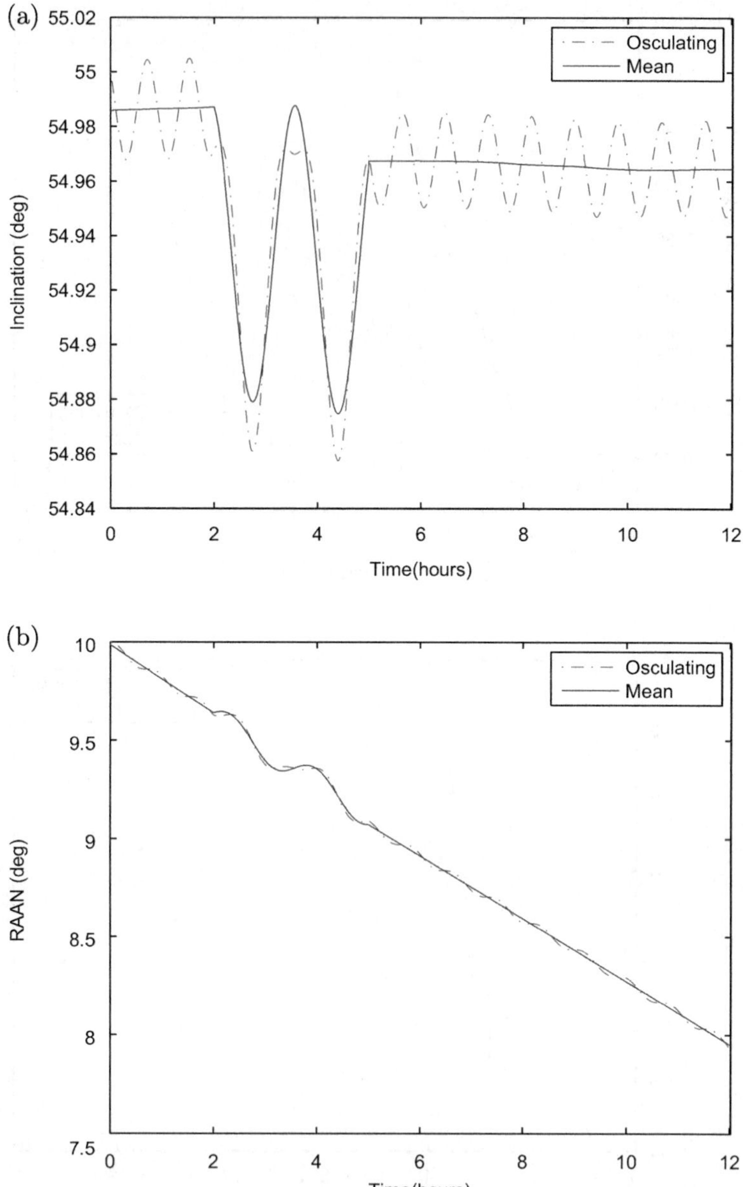

Fig. 16.13 Estimated mean (**a**) inclination and (**b**) RAAN under continuous constant-magnitude thrust

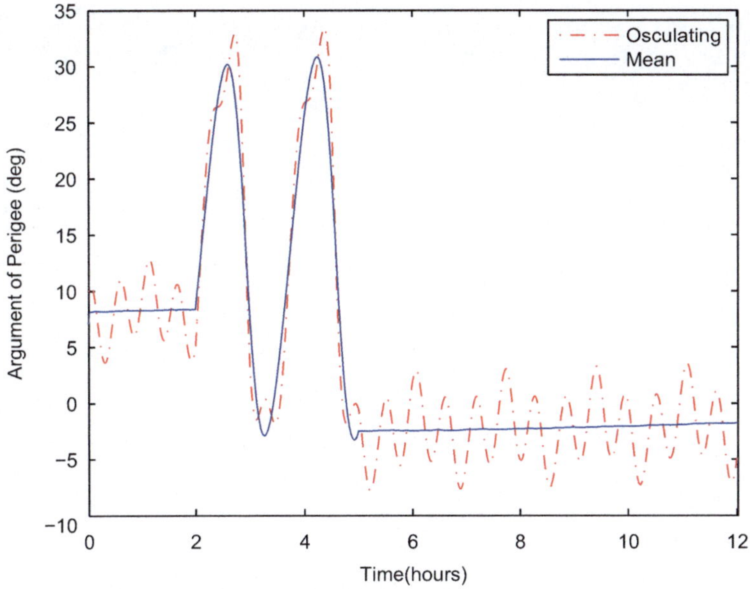

Fig. 16.14 Estimated mean argument of perigee under continuous constant-magnitude thrust

References

Busse, F.: Precise formation-state estimation in low Earth orbit using carrier differential GPS. PhD thesis, Stanford University (2003)

Carpenter, L.: Planetary perturbations in Chebyshev series. Tech. Rep. NASA TN D-3168, National Aeronautics and Space Administration, Washington, DC (1966)

Deprit, A., Pickard, H., Poplarchek, W.: Compression of ephemerides by discrete Chebyshev approximations. Navigation **26**(1), 1–11 (1979)

Grewal, M., Andrews, A.: Applications of Kalman filtering in aerospace 1960 to the present [historical perspectives]. IEEE Control Syst. **30**(3), 69–78 (2010)

Hildebrand, F.B.: Introduction to Numerical Analysis. McGraw-Hill, New York (1956)

Kalman, R.: A new approach to linear filtering and prediction problems. J. Basic Eng. **82**(1), 35–45 (1960)

Kalman, R., Bucy, R.: New results in linear filtering and prediction theory. J. Basic Eng. **83**(3), 95–108 (1961)

Lee, D.J., No, T.S., Choi, S.W., Lee, S.R., Kim, H.J., Alfriend, K.T.: Precise ephemeris reconstruction using the Clohessy-Wiltshire frame and multiple sequential compressions. J. Guid. Control Dyn. **26**(5), 781–785 (2003)

Noaa, N., Usaf, U.: Standard atmosphere, 1976. US Government Printing Office, Washington, DC (1976)

Schaub, H., Alfriend, K.T.: Impulse feedback control to establish specific mean orbit elements of spacecraft formations. J. Guid. Control Dyn. **24**(4), 739–745 (2001)

Schaub, H., Junkins, J.L.: Analytical Mechanics of Space Systems. American Institute of Aeronautics and Astronautics, Reston (2003)

Stengel, R.F.: Optimal Control and Estimation. Dover Publications, New York (1994)

Zhong, W., Gurfil, P.: Mean orbital elements estimation for autonomous satellite guidance and orbit control. J. Guid. Control Dyn. **36**(6), 1624–1641 (2013)

Chapter 17
Space Debris

17.1 Introduction

Since 1957, more than 4500 satellites have been launched and over 1000 active satellites are in orbit. The intensive use of the space environment created a large amount of man-made objects in orbit around the Earth, which no longer serve a useful purpose. These objects are commonly referred to as *space debris*. Approximately 20,000 objects larger than 10 cm are known to exist; 500,000 particles between 1–10 cm in diameter and about 10 million debris smaller than 1 cm have been estimated to exist.

Most space debris are a result of used rocket stages, anti-satellite tests, defunct or dead satellites, different explosions, and spacecraft erosion or fragmentation. From 1957 to 2003, there were 186 space object explosions that produced thousands of debris (Davidov et al. 2005). The smaller debris source could be dust from solid rocket motors, bolts and other objects released during the satellite deployment and fragmentation of larger debris. All these space debris accumulate and reside in orbit for very long periods, because there is no simple way to remove large amounts of debris from high-density debris zones, although this possibility has been investigated (Cerf 2013).

The space debris density varies in the different regions as a function of space activity. The LEO and GEO regions are highly utilized and rich in space debris because they cover almost 90% of space activities (Wright 2007). The LEO region is up to 1000 km in altitude, entailing high collision probabilities for future space flights with 70% of catalogued artificial space debris (Davidov et al. 2005). There is a drastic growth of LEO usage since these orbits are well adapted to Earth observation. The GEO debris-rich zone is at a 200 km distance from the GEO altitude, which is utilized for communication satellites.

P. Gurfil, P.K. Seidelmann, *Celestial Mechanics and Astrodynamics: Theory and Practice*, Astrophysics and Space Science Library 436,
DOI 10.1007/978-3-662-50370-6_17

Most of the debris are tracked, mapped, correlated, and cataloged by international agencies. The larger debris are tracked by the US Space Surveillance Network (SSN) using radar and optical sensors that are able to detect debris larger than 10 cm in LEO and 1m in GEO (Sánchez Ortiz et al. 2001). Many models have been developed to detect and identify the space debris by estimating physical characteristics. Other models estimate the space debris trajectories using algorithms for orbit determination (Farnocchia et al. 2010) based on information from a network of optical sensors or a radar system (Milani et al. 2012).

Several models provide a description of the environment in terms of the distribution of debris flux as a function of size, altitude, and inclination. For example, ORDEM 2000 and SIMPLE are used for LEO (Liou et al. 2001; Ananthasayanam et al. 2006); MASTER 99 is the ESA orbital debris model (Sdunnus et al. 2001); and there are many evolutionary models that are designed to predict the future debris environment and verify correlation between measurements and prediction models (Landgraf et al. 2004).

Over the years, the debris have been dispersed due to fragmentation. Several days after an explosion, for example, due to the distribution in speeds of the particles, the debris will spread out along the initial orbit. After a longer period, Earth's oblateness will cause the debris to rotate around Earth's axis. If the debris cloud is on a polar orbit, it will create a shell around Earth. Debris in orbits near the equatorial plane will spread out into a band around the equator.

Space debris pose a risk to continued reliable use of space-based services and operations. The risk of damage may vary as a function of the debris size. Small debris impacts cause erosion and may severely damage the solar panels or optical devices such as telescopes and star trackers (Carpenter et al. 2008). Shielding can protect against these smaller particles, but will increase the production and launching cost. Even very small debris (less than 1 cm in diameter) may cause fatal impact to the spacecraft, since they move at a speed of a few kilometers per second. Major debris can destroy the spacecraft or severely damage the structure due to the large kinetic energy.

There are known orbital collisions that made satellite inoperable. The first one is probably Olympus-1, which collided with a meteor in August 1993. In 1996, the French satellite Cerise had its stabilization arm damaged by a piece of an Ariane rocket. It was the first official registered space object collision. In March 2006, Russian Express-AM11 communication satellite was hit by an unknown object. In February 2009, Iridium 33, an operational communication satellite in LEO, collided over Siberia with Cosmos 2251, a defunct Russian satellite. The two satellites were completely destroyed. It was the first collision between two satellites in orbit. In addition to these fatal collisions, there are many events of dangerous approaches of large debris to the manned space stations or space shuttles. Some of them have been documented to pass at only 300 m away from the MIR station.

Catastrophic impact with large debris is considered as a primary risk for space missions. Many studies have been focused on calculation of the collision probability. According to statistical studies, the average time between collisions of two major debris is 7–8 years for the next decades, with a catastrophic collision every 12–14 years (Wright 2007). These predictions could be even worse, since in some areas the number of collisions is so high that it will produce more debris than the atmospheric drag could remove. It is known as the *Kessler syndrome*. Davidov et al. (2005) predict that even if the number of debris in LEO remains unchanged, the number of fragments will increase at least fivefold and the number of collisions will grow up to 40–60 per year.

Mitigation measures include recommendations and international consensus to minimize or prevent the creation of new debris and measuring the existing ones, as well as implementing operational procedures such as utilizing orbital regimes with less debris, using shielding to protect from smaller debris, and adopting specific spacecraft maneuvers to avoid collisions with debris.

The large number of objects in Earth orbits makes it impossible to examine possible collisions with every catalogued space debris due to the huge computational effort that would be required. Only high risk conjunction events can be considered. If the estimated collision risk exceeds a pre-defined threshold, a collision avoidance maneuver is required. Sánchez-Ortiz et al. (2006) suggest a number of evasive maneuver strategies to reduce the collision risk. Warning messages are being sent for possible collisions.

17.2 SGP4 Propagator and TLE

As we have seen throughout this book, motion can be described using the position and velocity (**r** and **v**), or orbital elements (e.g. classical, Delaunay). The North American Aerospace Defense Command (NORAD) provides information about orbiting objects in a format called *Two Line Elements* (TLE). The Center for Space Standards and Innovation (CSSI) website CelesTrak provides a catalog (SATCAT) of these TLE. When dealing with space debris, TLE are often the only source of information.

The TLE were created to work with the *Simplified General Perturbations* series of propagators. The most common propagator is SGP4 (Hoots et al. 2004). The TLE set contains the following orbital information: Spacecraft name and catalog number; epoch; eccentricity; mean motion; RAAN; inclination; argument of perigee; mean anomaly; the first and second derivatives of the mean motion; a drag parameter $B^* = B\rho_0/2$, where ρ_0 is the atmospheric density and B is the ballistic coefficient, $B = C_D S/m$, C_D is the drag coefficient, S is the cross-sectional area, and m is the mass.

TLE are generated using SGP4 for use in the SGP models only. The major advantage of the TLE and SGP4 are their availability. SGP4 was implemented on a number of computational platforms.

Another advantage of SGP4, being an analytical propagator, is its speed. The major disadvantage of both the propagator and the format is accuracy. A generated TLE may contain inaccuracies of hundreds of meters in position and mm/sec in velocity (Foster Jr 2001). The SGP4 propagation error is also quite large, tens of kilometers after a few days (Knowles 1995).

TLE can be converted into position and velocity by feeding the TLE data into the SGP4 propagator. Let $\mathbf{x} = [x_1, x_2, \ldots x_6]^T$ be the state vector of the satellite, i.e. $\mathbf{x} = [\mathbf{r}^T, \mathbf{v}^T]^T$, and let P be the orbital propagator. Denote by $\mathbf{y} = [y_1, y_2, \ldots, y_6]^T$ the set of six TLE i, ω, Ω, M, n, and e. Then

$$\mathbf{x} = P(\mathbf{y}) \tag{17.1}$$

Inverting the problem, that is, calculating the TLE from a known state, is not as straightforward. If a collision avoidance maneuver is planned, one needs to calculate the influence of a velocity impulse $\Delta\mathbf{v}$ on the orbit (see discussion of impulsive maneuvers in Sect. 14.3). To that end, first transform the TLE state to $\mathbf{r}$ and $\mathbf{v}$. Then, after adding the impulse $\mathbf{v}^+ = \mathbf{v} + \Delta\mathbf{v}$, transform the new state into a new TLE set. This problem was solved by Lee (2002). Assuming B^* is not changed due to the maneuver, the TLE can be calculated in an iterative manner. First, expand P into a Taylor series about $\mathbf{y}^i$,

$$P(\mathbf{y}) = P\left(\mathbf{y}^i\right) + \frac{\partial P}{\partial y_1^i}\left(y_1 - y_1^i\right) + \frac{\partial P}{\partial y_2^i}\left(y_2 - y_2^i\right) + \ldots \frac{\partial P}{\partial y_6^i}\left(y_6 - y_6^i\right) + \mathcal{O}\left(y^2\right) \tag{17.2}$$

In matrix form,

$$\left(\mathbf{x} - \mathbf{x}^i\right) = \mathbf{M}\left(\mathbf{y} - \mathbf{y}^i\right) \tag{17.3}$$

where

$$\mathbf{M} = \begin{bmatrix} \frac{\partial P_1}{\partial y_1^i} & \frac{\partial P_1}{\partial y_2^i} & \cdots & \frac{\partial P_1}{\partial y_6^i} \\ \vdots & \vdots & \ddots & \vdots \\ \frac{\partial P_6}{\partial y_1^i} & \frac{\partial P_6}{\partial y_2^i} & \cdots & \frac{\partial P_6}{\partial y_6^i} \end{bmatrix} \tag{17.4}$$

The iterations are preformed as follows:

$$\mathbf{y}^{i+1} = \mathbf{y}^i + \mathbf{M}^{-1}\left(\mathbf{x} - \mathbf{x}^i\right) \tag{17.5}$$

Since the partial derivatives cannot be calculated analytically, central finite differencing is used (see Chap. 9). The inverse function requires the inverted matrix to be nonsingular. In addition, the matrix inversion may be slow. Using the Moore-Penrose pseudo inverse (Penrose 1955) is faster and more robust.

17.3 Sizing the Debris

An important part of assessing the *probability of collision* between a satellite and a debris is knowing the size of the objects. The size of each satellite is known. However, when dealing with debris, the size of the object is not always known. The size of objects can be found in American, European or Russian catalogs. None of the catalogs is readily accessible to the general public. CelesTrak offers a catalog containing some radar cross sections (RCS) of objects.

To transform the RCS to the true area of the object, the wavelength with which the measurement was taken must be known. One may use the TLE to make an educated guess as to the radar that made the measurement. Knowing the bandwidth of the radar, one can assess the wavelength used. Another option is to use the power law approximation (Badhwar and Anz-Meador 1989)

$$S = 0.5712\,(RCS)^{0.7666} \tag{17.6}$$

17.4 Time of Closest Approach

Usually, when calculating the risk of collision, an operating satellite is tested against a catalog of objects. A geometric filter is used to sift out the objects with which there is no chance of approach. Then, the minimum distances among the satellite and the hazardous objects need to be calculated. The operator is interested both in the minimum distance between the satellite and the object, and in the *time of closest approach* (TCA). The most straightforward way of calculating these two parameters is by using the propagator to build a "truth table" that would tell where the two orbiting objects were at any moment. This method is computationally expensive.

Alfano (1994) suggested a method called the *Alfano-Negron Close Approach Software* (ANCAS). This method is widely used today and can be implemented on any propagator. Following is an explanation of the mechanism of ANCAS.

Let the relative position between two orbiting objects be defined as $\mathbf{r}_d$, the relative velocity is $\mathbf{v}_d = \dot{\mathbf{r}}_d$, and the acceleration $\mathbf{a}_d = \ddot{\mathbf{r}}_d$. Alfano (1994) demonstrates ANCAS on an integrator with J_2 to calculate the relative distance, velocity and acceleration. Since SGP4 only provides position and velocity, the acceleration is derived using forward finite differencing. A distance function is defined as

$$g = \mathbf{r}_d \cdot \mathbf{r}_d \tag{17.7}$$

It follows that

$$\dot{g} = 2\dot{\mathbf{r}}_d \cdot \mathbf{r}_d \tag{17.8}$$

and

$$\ddot{g} = 2(\ddot{\mathbf{r}}_d \cdot \mathbf{r}_d + \dot{\mathbf{r}}_d \cdot \dot{\mathbf{r}}_d) \tag{17.9}$$

A close approach will occur when g obtains a local minimum, i.e. when $\dot{g} = 0$ and $\ddot{g} > 0$. To approximate the time for which $\dot{g} = 0$, a cubic spline is fitted to the data set of time points. The spline is a set of cubic polynomials, i.e. for every given pair of t_i and t_{i+1}, a cubic is fitted. The cubic is defined on $\tau = [0, 1]$. The transformation between τ and t is

$$t(\tau) = t_i + \tau \Delta t \tag{17.10}$$

The cubic is formulated as

$$C(\tau) = a_0 + a_1\tau + a_2\tau^2 + a_3\tau^3, \; 0 \le \tau \le 1 \tag{17.11}$$

where

$$\begin{cases} a_0 = \dot{g}(t_i) \\ a_1 = \ddot{g}(t_i)\Delta t \\ a_2 = -3\dot{g}(t_i) - 2\ddot{g}(t_i)\Delta t + 3\dot{g}(t_{i+1}) - \ddot{g}(t_{i+1})\Delta t \\ a_3 = 2\dot{g}(t_i) + \ddot{g}(t_i)\Delta t - 2\dot{g}(t_{i+1}) + \ddot{g}(t_{i+1})\Delta t \end{cases} \tag{17.12}$$

The roots of Eq. (17.11) are found, and if a minimum is identified, the value of $\mathbf{r}_d$ is calculated using three quintic polynomials fitted between $\mathbf{r}_d(t_i)$ and $\mathbf{r}_d(t_{i+1})$. For instance, if $\mathbf{r}_d = \left[r_x, r_y, r_z\right]$, then for r_x

$$Q_{r_x}(\tau) = b_0 + b_1\tau + b_2\tau^2 + b_3\tau^3 + b_4\tau^4 + b_5\tau^5, \; 0 \le \tau \le 1 \tag{17.13}$$

where

$$\begin{cases} b_0 = r_x(t_i) \\ b_1 = \dot{r}_x(t_i)\Delta t \\ b_2 = 0.5\ddot{r}_x(t_i)\Delta t^2 \\ b_3 = -10r_x(t_i) - 6\dot{r}_x(t_i)\Delta t - 1.5\ddot{r}_x(t_i)\Delta t^2 \\ \qquad +10r_x(t_{i+1}) - 4\dot{r}_x(t_{i+1})\Delta t + 0.5\ddot{r}_x(t_{i+1})\Delta t^2 \\ b_4 = 15r_x(t_i) + 8\dot{r}_x(t_i)\Delta t + 1.5\ddot{r}_x(t_i)\Delta t^2 \\ \qquad -15r_x(t_{i+1}) + 7\dot{r}_x(t_{i+1})\Delta t - \ddot{r}(t_{i+1})\Delta t^2 \\ b_5 = -6r_x(t_i) - 3\dot{r}_x(t_i)\Delta t - 0.5\ddot{r}_x(t_i)\Delta t^2 \\ \qquad +6r_x(t_{i+1}) - 3\dot{r}_x(t_{i+1})\Delta t + 0.5\ddot{r}_x(t_{i+1})\Delta t^2 \end{cases} \tag{17.14}$$

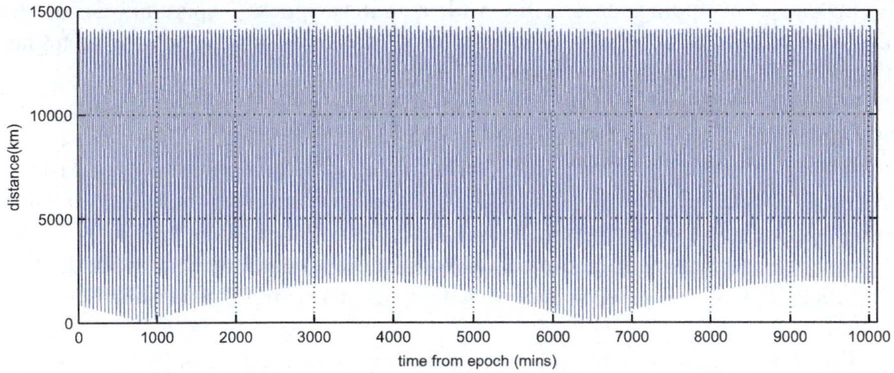

Fig. 17.1 Example of inter-satellite distance evolution under perturbations

The approximated minimum distance would be

$$r_{\min} = \sqrt{Q_{r_x}^2 + Q_{r_y}^2 + Q_{r_z}^2} \tag{17.15}$$

The advantage of this method is computational speed.

Alfano (1994) chose a uniform time step. Choosing a uniform time-step makes sense since $\mathbf{r}_d$ is usually evaluated using some form of a numerical integrator. The error in position using this method could be as large as hundreds of meters, and the error in time tenths of a second. The accuracy of the method depends on the time step chosen. Alfano (1994) stated that there was no need to iterate the splining or the root finding. Due to the inaccurate nature of ANCAS, this method can be used to perfrom initial calculations before preforming more precise ones.

Another way of looking at the TCA problem is treating it as an optimization problem, i.e. finding $\min_t(\mathbf{r}_d)$. Though the problem seems simple, and may have analytic solutions in the two-body case, when taking into account the gravitational harmonics and drag, the distance function becomes multi modal. To illustrate this phenomenon, Fig. 17.1 shows a sample evolution of distance between two orbiting objects. The graph was produced using the SGP4 propagator.

17.5 Probability of Collision

The simplest way of avoiding conflicts in space is keeping a safe distance from all threatening objects. For instance, an operator may define a safety distance of 15 km. When debris are expected to be within a smaller distance, a maneuver is initiated to keep the required distance. This approach is wasteful because the determination of the satellite orbit is inaccurate, and the determination of the debris orbit is even less accurate.

Instead of wrapping the satellite with a "safety sphere", operators define the maximum allowed probability of collision; should a collision probability be higher than the threshold, a maneuver would take place.

Many formulations for the collision probability were developed (Patera 2003; Alfano 2005; Foster and Estes 1992; Klinkrad 2006; Chan 2008), most deal with short-term conjunctions. Several criteria exist for the definition of short-term conjunction (Chan 2008; Coppola 2012). Methods of calculating long-term collision probability are also known (Chan 2008; Alfano 2006; Patera 2003; McKinley 2006). Plainly put, the conjunction time will be short if the relative speed between the two objects is high. Usually, when dealing with space debris in LEO, the relative speed is high.

The most common assumption when dealing with orbiting objects is that the object is spherical. Representing the collision probability with non-spherical objects was also studied, but because there is little to no information about the debris rotational state, using the spherical model makes sense. Additional assumptions are that the covariances of the two objects are uncorrelated, and the probability density functions are zero-mean and Gaussian. Under these assumptions Chan (2008) showed that the covariances can be combined into a single covariance located in the center of one of the objects, the primary.

Let $\mathbf{C}_{3D}$ be the combined covariance. The probability density function of an object in a position $\mathbf{r}_d$ relative to the primary is

$$p(\mathbf{r}_d) = \frac{1}{\sqrt{(2\pi)^3 \det(\mathbf{C}_{3D})}} \exp\left(-\frac{1}{2}\mathbf{r}_d^T \mathbf{C}_{3D}^{-1} \mathbf{r}_d\right) \tag{17.16}$$

The secondary object passes through the $\mathbf{C}_{3D}$ ellipsoid and sweeps a volume through it. The probability of collision would be the three-dimensional integral on Eq. (17.16) with the volume of the swept volume as the domain of integration,

$$P_c = \iiint\limits_V p(\mathbf{r}_d)dx\,dy\,dz \tag{17.17}$$

Because the covariances are combined, the secondary object's size must reflect the combined size of both objects. The secondary object is, therefore, a sphere with the radius being the sum of radii of both objects,

$$R = R_1 + R_2 \tag{17.18}$$

The solution of the 3D integral in Eq. (17.17) is quite difficult. Hence, a simplifying assumption is made. According to this assumption, if the relative velocity is large, then the relative motion is rectilinear. If the motion is rectilinear, then the volume swept by the secondary object is approximately a tube. A plane normal to the relative velocity is formed. The combined covariance and the volume traced by the secondary object are then projected onto that plane, as shown in Fig. 17.2.

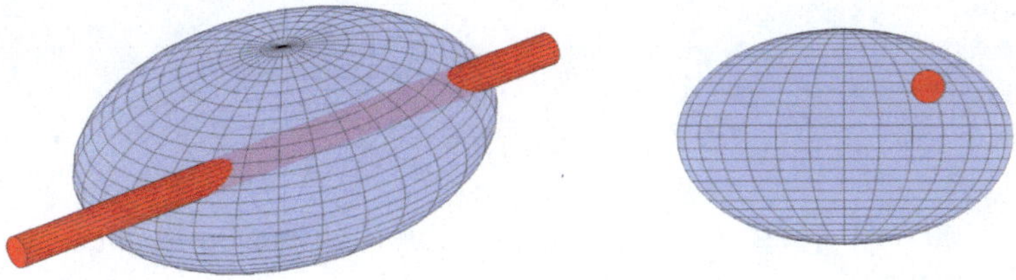

Fig. 17.2 Description of a 3D encounter and its projection to 2D

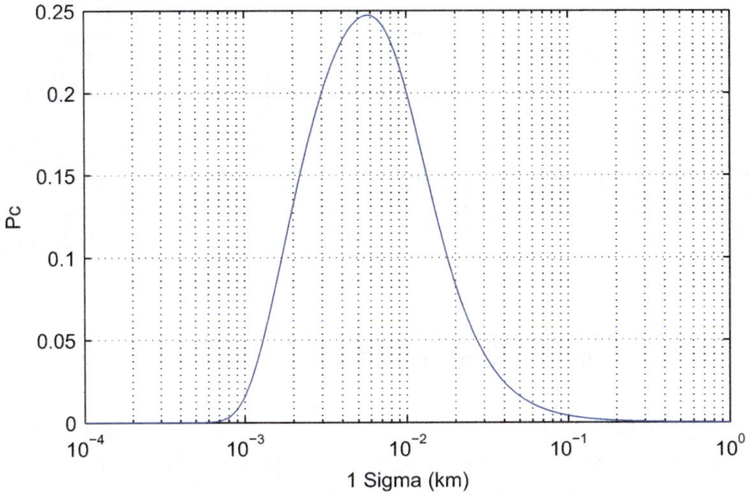

Fig. 17.3 Probability as a function of uncertainty of two objects of size 20 cm^2

Thus, the integral in Eq. (17.17) is reduced to two dimensions, and the probability of collision becomes

$$P_c = \frac{1}{2\pi\sqrt{\det(\mathbf{C}_{2D})}} \int_{-R}^{R} \int_{-\sqrt{R^2-x^2}}^{\sqrt{R^2-x^2}} \exp\left(-\frac{1}{2}\mathbf{r}^T\mathbf{C}_{2D}^{-1}\mathbf{r}\right) dy\, dx \tag{17.19}$$

where $\mathbf{C}_{2D}$ is the projected covariance, and $\mathbf{r}$ is the relative position, rotated to 2D coordinates. We can either numerically evaluate the double integral in Eq. (17.19), use a method that transforms it into a series (Chan 2008), preform a one-dimensional integral over a line (Patera 2005), or use a series of error functions (erf) (Alfano 2007).

The methods introduced so far relied on covariance matrices. Klinkrad (2006) shows that the probability does not grow as the covariance grows (see Fig. 17.3).

For some relative position, there exists a covariance matrix which makes the argument of the exponent equal to -1. Let $\mathbf{C}^*$ be that covariance. If $\mathbf{C}^* = k\mathbf{C}_{2D}$, then the argument will be -1 when $k = \sqrt{\frac{1}{2}\mathbf{r}^T\mathbf{C}_{2D}^{-1}\mathbf{r}}$, and then we get the maximum probability

$$P_c = \frac{R^2}{\exp(1)\sqrt{\det(\mathbf{C}_{2D})\mathbf{r}^T\mathbf{C}_{2D}^{-1}\mathbf{r}}} \tag{17.20}$$

Alfano (2005) provides an approximation for P_c using error functions in the following manner.

Let $r = R/\|\mathbf{r}_d\|$; assuming the covariance grows linearly, and parameterizing it using r, the following formulation is derived:

$$P_c = \frac{1}{2}\left\{\operatorname{erf}\left[\frac{r+1}{2\sqrt{r}}\sqrt{-\ln\left(\frac{1-r}{1+r}\right)}\right] + \operatorname{erf}\left[\frac{r-1}{2\sqrt{r}}\sqrt{-\ln\left(\frac{1-r}{1+r}\right)}\right]\right\} \tag{17.21}$$

The value of P_c computed using Eq.(17.21) and the one computed using Eq. (17.20) is the same up to the 5th decimal place. Using the last formulation proves to be easier since no need for covariances or the propagation thereof is required.

17.6 Calculating the Required Δv

Once the calculation of probability is complete, the operator must decide whether or not to maneuver, and what type of maneuver to preform. By defining an *accepted collision probability level* (ACPL) the operator can decide whether a maneuver is required. If the calculated probability is higher than the ACPL, the satellite must be maneuvered. The most efficient maneuver would be an along-track maneuver. Sánchez-Ortiz et al. (2006) suggest two strategies for assessing the along-track maneuver. The long-term maneuver, based on an along-track separation, would take place at the point of the collision several revolutions before it occurs. The short-term maneuver takes place at the location opposite the expected collision, and is based on a radial separation. To calculate both maneuvers, Sánchez-Ortiz et al. (2006) define an *allowed minimum miss distance* (AMMD). The AMMD is not derived directly from the ACPL, but is rather selected in a manner that would satisfy the ACPL. The short-term maneuvers are far more expensive than the long term.

The required Δv for long-term maneuvers is calculated as follows. Let v_1 be the known speed of the satellite at the collision, d_{min} be the AMMD, and T_1 be the orbital period. Then

$$\Delta T = \frac{d}{n_{rev} v_1} \tag{17.22}$$

and

$$T_2 = T_1 + \Delta T \tag{17.23}$$

where n_{rev} is the number of orbits before the collision at which the maneuver is preformed, and T_2 is the new period. From Eq. (17.23), we can derive the speed v_2 at the collision point,

$$\Delta v = v_2 - v_1 \tag{17.24}$$

This method is approximate. It is useful for assessing how much fuel will be required and is usually used together with statistical information about the flux of debris. Just as in the case of ANCAS, this method tends to produce solutions that are larger than the actual required maneuver.

To find a more accurate maneuver, Patera and Peterson (2003) suggested using an iterative method. This method uses the gradient of the probability of collision with respect to the maneuver direction, and Newton-Raphson iterations to find the optimum. This method assumes that the covariances are small and so the maneuver amplitude would be small as well.

References

Alfano, S.: Determining satellite close approaches, part ii. J. Astronaut. Sci. **42**(2), 143–152 (1994)

Alfano, S.: A numerical implementation of spherical object collision probability. J. Astronaut. Sci. **53**(1), 103 (2005)

Alfano, S.: Addressing nonlinear relative motion for spacecraft collision probability. In: AAS/AIAA Astrodynamics Specialist Conference, Keystone, CO., vol. 6760 (2006)

Alfano, S.: Review of conjunction probability methods for short-term encounters. In: AAS Specialist Conference, Sedona, AZ, Paper 07–148 (2007)

Ananthasayanam, M., Anilkumar, A., Rao, P.S.: A new stochastic impressionistic low Earth model of the space debris scenario. Acta Astronaut. **59**(7), 547–559 (2006)

Badhwar, G.D., Anz-Meador, P.D.: Determination of the area and mass distribution of orbital debris fragments. Earth Moon Planets **45**(1), 29–51 (1989)

Carpenter, J., Wells, A., Abbey, A., Ambrosi, R.: Meteoroid and space debris impacts in grazing-incidence telescopes. Astron. Astrophys. **483**(3), 941–947 (2008)

Cerf, M.: Multiple space debris collecting mission – debris selection and trajectory optimization J. Optim. Theory Appl. **156**(3), 761–796 (2013)

Chan, F.K.: Spacecraft Collision Probability. Aerospace Press, Beijing (2008)

Coppola, V.: Evaluating the short encounter assumption of the probability of collision formula. In: Proceedings of the AAS/AIAA Spaceflight Mechanics Meeting, Charleston, SC, USA (2012)

Davidov, V., Kulik, S., Mikhailov, M., Chekalin, S., Yakovlev, M., Bulinin, Y.: Measures undertaken by the russian federation for mitigating artificial space debris pollution. In: 4th European Conference on Space Debris, vol. 587, p. 53 (2005)

Farnocchia, D., Tommei, G., Milani, A., Rossi, A.: Innovative methods of correlation and orbit determination for space debris. Celest. Mech. Dyn. Astron. **107**(1–2), 169–185 (2010)

Foster, J., Estes, H.S.: A parametric analysis of orbital debris collision probability and maneuver rate for space vehicles. NASA JSC 25898 (1992)

Foster, Jr. J.L.: The analytic basis for debris avoidance operations for the international space station. In: Space Debris, vol. 473, pp. 441–445 (2001)

Hoots, F.R., Schumacher, Jr. P.W., Glover, R.A.: History of analytical orbit modeling in the US space surveillance system. J. Guid. Control Dyn. **27**(2), 174–185 (2004)

Klinkrad, H.: Space Debris: Models and Risk Analysis. Springer, Berlin (2006)

Knowles, S.H.: A comparison of geocentric propagators for operational use. AAS Preprint, pp. 95–429 (1995)

Landgraf, M., Jehn, R., Flury, W.: Comparison of EISCAT radar data on space debris with model predictions by the master model of ESA. Adv. Space Res. **34**(5), 872–877 (2004)

Lee, B.S.: NORAD TLE conversion from osculating orbital element. J. Astron. Space Sci. **19**, 395–402 (2002)

Liou, J.C., Matney, M., Anz-Meador, P., Kessler, D., Jansen, M., Theall, J.: The new NASA orbital debris engineering model ORDEM2000. In: Space Debris, vol. 473, pp. 309–313 (2001)

McKinley, D.P.: Development of a nonlinear probability of collision tool for the Earth observing system. In: AIAA/AAS Astrodynamics Specialist Conference and Exhibit, p. 6295 (2006)

Milani, A., Farnocchia, D., Dimare, L., Rossi, A., Bernardi, F.: Innovative observing strategy and orbit determination for low Earth orbit space debris. Planet. Space Sci. **62**(1), 10–22 (2012)

Patera, R.P.: Satellite collision probability for nonlinear relative motion. J. Guid. Control Dyn. **26**(5), 728–733 (2003)

Patera, R.P.: Calculating collision probability for arbitrary space vehicle shapes via numerical quadrature. J. Guid. Control Dyn. **28**(6), 1326–1328 (2005)

Patera, R.P., Peterson, G.E.: Space vehicle maneuver method to lower collision risk to an acceptable level. J. Guid. Control Dyn. **26**(2), 233–237 (2003)

Penrose, R.: A generalized inverse for matrices. Math. Proc. Camb. Philos. Soc. **51**(3), 406–413 (1955)

Sánchez Ortiz, N., Belló Mora, M., Graziano, M., Pina Caballero, F., Sánchez Pérez, J., Klinkrad, H.: Collision risk assessment. In: Space Debris, vol. 473, pp. 447–453 (2001)

Sánchez-Ortiz, N., Belló-Mora, M., Klinkrad, H.: Collision avoidance manoeuvres during spacecraft mission lifetime: Risk reduction and required ΔV. Adv. Space Res. **38**(9), 2107–2116 (2006)

Sdunnus, H., Bendisch, J., Klinkrad, H.: The ESA MASTER'99 space debris and meteoroid reference model. In: Space Debris, vol. 473, pp. 299–307 (2001)

Wright, D.: Space debris. Phys. Today **60**(10), 35–40 (2007)

Chapter 18
People, Progress, Prospects

18.1 People and Progress

The developments and progress in celestial mechanics and astrodynamics can in most cases be tied directly to the scientists who contributed to the ideas and advancements. Some of those people are identified here.

Aristotle (384–322 BC) was a very influential philosopher and thought the Earth was fixed in space, and all motions must be perfect, hence, circular.

Aristarchus (310–230 BC) measured the diameter of the Earth, the distances to the Sun and Moon, and concluded that the Sun was much bigger than the Earth and, therefore, should be at the center of the solar system. However, Aristotle's philosophy was more generally accepted.

Hipparchus (150 BC) discovered the precession of the equinox and, hence, the motion of all stellar positions.

Ptolemy (140 AD) produced the Almagest, which included a section on cosmology. He placed the Earth at the center of the universe, the stars on a sphere rotating daily around the Earth, and the Sun, Moon, and planets on different spheres rotating with uniform motions. Smaller spheres were attached to the larger ones to explain complex and retrograde motions. Ptolemy recognized the variation in apparent solar time and developed mean solar time for computation of ephemerides.

Roger Bacon (1200) proposed that theories must be based on observations, which are repeatable. This was the beginning of the renaissance period.

Nicolaus Copernicus (1473–1543) introduced the hypothesis of the solar system with the Sun at the center. His "De Revolutionibus Orbium Coelestium" was only published after his death.

Tycho Brahe (1546–1601) built for that time the finest astronomical observatory, "Uranienborg", on the island of Hven near Copenhagen. He made accurate observations of the Sun, Moon, planets, and stars for over 20 years. After being forced to leave Denmark, he became the astronomer of the Imperial Court in Prague.

P. Gurfil, P.K. Seidelmann, *Celestial Mechanics and Astrodynamics: Theory and Practice*, Astrophysics and Space Science Library 436,
DOI 10.1007/978-3-662-50370-6_18

Galileo Galilei (1564–1642) pointed a telescope at the sky and observed the planetary phases and satellites of Jupiter, confirming Copernicus's hypothesis of the heliocentric solar system.

Johanes Kepler (1571–1630) used Tycho Brahe's observations, particularly those of Mars, to develop his three laws of planetary motion. This required abandoning circular motions and believing in Tycho's observations.

Isaac Newton (1642–1727) used Kepler's laws to develop the basic principles of the law of universal gravitation and a second law of motion, that the acceleration of an object is due to the force applied to move it.

Ole Roemer (1644–1710) developed the transit circle for accurate astrometric observations and measured the speed of light based on observations of the satellites of Jupiter.

Edmund Halley (1656–1742) paid for the publication of Newton's "Principia", predicted the return of a comet that now bears his name, and discovered proper motion of fixed stars from observations of Arcturus and Sirius.

Leonard Euler (1707–1783) developed a lunar theory, which was the basis for a lunar ephemeris, published in 1767 and used for navigation.

James Bradley (1693–1762) was able to detect aberration and nutation from stellar observations.

Joseph Lagrange (1736–1813) announced the possibility of triangular libration points in the Sun-Jupiter system and predicted the possible existence of asteroids. He developed the method of variation of parameters.

William Herschel (1738–1822) discovered in 1781 the planet Uranus, and later two of its Moons, Titania and Oberon, and two moons of Saturn, Mimas and Enceladus. He discovered the existence of infrared radiation. He observed many binary and multiple star systems and nonstellar objects. He was a musician and composer of twenty four symphonies and other musical pieces.

Pierre Simon de Laplace (1749–1827) developed a perturbation theory to study the stability of the solar system and concluded that the solar system was stable. This proved to be not quite correct. He introduced the potential function known as Laplace's equation. He developed a lunar theory and wrote the five volumes of "Mécanique Céleste".

Carl Friedrich Gauss (1777–1855) was a brilliant mathematician. He developed Gauss's method of orbit computation to predict the future location of newly discovered minor planet, Ceres. His theory of the motion of celestial bodies introduced the Gaussian gravitational constant and an influential treatment of the method of least squares.

Fredrick Bessel (1784–1846) determined the positions of over 50,000 stars, was the first to use parallax to determine the distance to a star, 61 Cygni, deduced that Sirius and Procyon had unseen companions, the first correct claim of unseen companions. Bessel developed what are known as Bessel functions, for the solution of certain differential equations.

Peter Andreas Hansen (1795–1874) published his Tables of the Moon in 1857, which was believed at the time to be a complete lunar theory. However, within a decade Newcomb found deviations between the tables and observations. Hansen developed a planetary theory for the Jupiter-Saturn perturbations.

George Biddell Airy (1801–1892) was Astronomer Royal from 1835–1881 and his stellar observations were the basis for the Greenwich meridian becoming the prime meridian. He oversaw the reduction and publication of a large number of planetary and lunar observations. He determined the mean density and the polar and equatorial radius of the Earth, which established the Airy Geoid, still in use in the United Kingdom.

William Rowan Hamilton (1805–1865) reformulated Newtonian mechanics into what is called Hamiltonian mechanics. He is the inventor of quaternions.

Charles-Eugene Delaunay (1816–1872) developed the lunar theory and the infinite series converged very slowly, but it led to further development of functional analysis and computer algebra. With Ferrel he determined that the tides retard the rotation of the Earth.

U. J. J. Leverrier (1811–1877) used the perturbations of Uranus by an unknown planet to predict the location of the planet and Neptune was discovered as he predicted by J. F. Encke and L. d'Arrest in 1846. Bessel's star catalog was critical for the discovery of Neptune.

J. C. Adams (1819–1892) predicted the location of an unknown planet at the same time as Leverrier. Unfortunately for him, the planet Neptune was observed at Cambridge Observatory, but not recognized as moving, and Airy did not follow up on the prediction at Greenwich. Adams found the secular acceleration of the mean motion of the Moon.

Wilhelm Struve, Thomas Henderson (1838–1840) made stellar observations sufficiently accurate to prove there was parallactic motions, so the Earth was moving around the Sun.

Simon Newcomb (1835–1909) introduced a standardized reference system and planetary theories that were used from 1900 to 1984. He was a leading astronomer in the US for the last 30 years of his life.

George W. Hill (1838–1914) wrote many papers on celestial mechanics and developed the general theories for Jupiter and Saturn that were used for their ephemerides over the first half of the twentieth century.

Seth Carlo Chandler (1846–1913) performed research on positional astronomy and discovered the Chandler Wobble in polar motion.

Albert A. Michelson (1852–1931) used the Michelson-Morley experiment to measure the speed of light and introduce interferometry.

Henri Poincaré (1854–1912) wrote many scientific papers and more than 30 books. He established the concept of nonintegrable dynamical systems. This affected the concept of stability in the solar system. He showed the existence of deterministic chaos, where the three-body system may become chaotic, i.e. unpredictable.

E. W. Brown (1866–1938) developed the lunar theory used during the first half of the twentieth century.

P. W. Cowell (1870–1949) developed a method of numerical integration for orbit computations.

Albert Einstein (1879–1955) developed the theories of special and general relativity.

Willem de Sitter (1872–1934) investigated the variability of the Earth rotation. With Einstein he authored a paper arguing that there might be large amounts of matter that do not emit light. He developed the concept of de Sitter Space and de Sitter Universe, a solution of general relativity in which there is no matter and a positive cosmological constant.

H. Spencer Jones (1890–1960) investigated the variability of Earth rotation. He determined the solar parallax from the observations of 433 Eros and was an expert on astronomical constants. He was Astronomer Royal from 1933 to 1956.

Andre-Louis Danjon (1890–1967) designed the impersonal astrolabe, known as the Danjon astrolabe, which improved the accuracy of fundamental optical astrometry. He suggested the establishment of Ephemeris Time long before G. M. Clemence, but that was not known by Clemence.

George E. Lemaitre (1894–1966) introduced the concepts of expansion of the universe, Hubble's law, Hubble's constant, and the Big Bang theory.

Y. Hagihara (1897–1979) wrote a nine-book text on celestial mechanics.

Dirk Brouwer (1902–1966) developed a theory for artificial satellite motions and was a leader in the development of astrodynamics after Sputnik. He was head of the astronomy department at Yale and educated and attracted many of the future leaders in celestial mechanics. Brouwer, Clemence, Eckert, and Herget were the leaders in the field of celestial mechanics at the advent of the space program and astrodynamics.

Wallace Eckert (1902–1971) was a pioneer in the use of punched card equipment and computers for astronomy. He introduced their use for computing the almanacs at the US Naval Observatory during World War II. He developed the computer used for the integration of the ephemerides of the outer planets that were used internationally from 1960–1984. He developed a modern lunar theory.

Andrey N. Kolmogorov (1903–1987) investigated probability theory and stochastic processes. He is known for the Kolmogorov-Arnold-Moser theorem and the Kolmogorov complexity theory.

Boris Garfinkel (1904–1999) developed an artificial satellite theory.

Clyde W. Tombaugh (1906–1997) discovered Pluto in 1930 near the predictions of its location, but its mass is too small to be the basis of the predictions.

William Markowitz (1907–1998) developed the Moon camera used to measure the ephemeris second with respect to an atomic clock and determined the resulting value of the atomic second, which became the SI second.

John P. Vinti (1907–1990) developed an artificial satellite theory and his lecture notes were published as "Orbital and Celestial Mechanics".

G. M. Clemence (1908–1974) was a pioneer in the use of punched card equipment and computers for celestial mechanics. He proposed the introduction of Ephemeris Time. He was Scientific Director of the US Naval Observatory and developed a general planetary theory for Mars.

Paul Herget (1908–1981) was a pioneer in the use of punched card equipment and computers for celestial mechanics and developed the computational methods used for the Mercury, Gemini, and Apollo missions and artificial satellite orbits. He was director of Cincinnati Observatory and the Minor Planet Center after World War II.

Samuel Herrick (1911–1974) taught astrodynamics at UCLA and wrote books on Astrodynamics.

G. A. Chebotarev (1913–1975) studied minor planets, and was a Russian leader in the field and wrote a book on celestial mechanics.

Raynor L. Duncombe (1917–2013) worked with Eckert, Clemence and Herget with punched card computers and early orbit determination of artificial satellites. He was director of the US Nautical Almanac Office and taught in the Aerospace and Engineering Mechanics Department at the University of Texas for thirty six years.

John V. Breakwell (1917–1991) made contributions in astrodynamics and related areas of optimization, control theory, and differential games. One of his major contributions was his mentoring of many students who have made numerous contributions to astrodynamics.

Victor Szebehely (1921–1997) is best known for his work on the three-body problem and his book "Theory of Orbits: The Restricted Problem of Three Bodies". He was well known for his entertaining lectures.

Richard H. Battin (1925–2013) was responsible for developing the guidance and navigation concepts and software for the Apollo on-board flight computers. He taught astrodynamics at MIT and wrote several books on astrodynamics.

Andre Deprit (1926–2006) was one of the pioneers in the development and use of Lie Series and algebraic manipulation in the development of orbit theories.

18.2 Future Prospects: Exoplanets

A summary of celestial problems already solved, that can and should be solved more completely, and still waiting to be solved, is given by Brumberg (2013). We focus our discussion of future prospects on the topic of exoplanets. This remarkable field holds much promise for both astrodynamics and celestial mechanics. Moreover, it could answer some of the most fundamental questions of existence.

18.2.1 History

An *exoplanet*, also called extrasolar planet, is a planet outside our solar system. Objects with true masses below the limiting mass for thermonuclear fusion of deuterium (currently 13 Jupiter masses for objects of solar metallicity) that orbit stars, or stellar remnants, are planets (no matter how they formed). The minimum mass required for an exosolar object to be considered a planet should be the same as that used in our solar system. Substellar objects with true masses above the limiting mass for thermonuclear fusion of deuterium are *brown dwarfs*. Free floating objects in young star clusters with masses below the limiting mass for thermonuclear fusion of deuterium are not planets, but are *sub-brown dwarfs*.

The sixteenth century Italian philosopher Giordano Bruno and the eighteenth century scientist Isaac Newton proposed that stars similar to the Sun should likewise be accompanied by planets. There is a long history of searches for planets and life somewhere other than on Earth (Dick 1999, 2001). Hence, the search for life, or traces of past life, on planets in our solar system continues. Also the search for radio signals from other solar systems, such as SETI, has a long history going back to Drake (2009). There is the resulting *Drake equation* concerning the probability of radio signals from outside our solar system. There is also the *Fermi paradox*, which is basically "if there are so many of them, where are they?" (Prantzos 2013).

The first real discovery of an exosolar planet was from a radio astronomy observation of a planet around a pulsar by Wolszczan and Frail (1992). There have been many searches using astrometry, where the periodic motion of a star due to its motion about the center of mass of the star and a planet could be detected, just as binary stars are detected, but with the much smaller mass of the planet. None of the claimed detections proved to be real.

Campbell et al. (1988) used radial velocity observations to suggest that a planet was orbiting Gamma Cephei. The observations were at the limits of instrumental capacities and the discovery remained in doubt until 2003, when improved techniques confirmed its existence (Hatzes et al. 2003). The breakthrough in the discovery of exoplanets came from very accurate radial velocity measures, where the periodic variations in the observed radial velocities revealed the existence of a planet going around the star (Mayor and Queloz 1995). Their discovery was followed by a number of other exoplanets by Marcy and Butler (1996).

The detection of a planet passing in front of a star by the photometric measure of the stellar magnitude, which dips and returns, has resulted in over 2700 planet detections with the Kepler spacecraft (Dunham et al. 2014; Batalha et al. 2013).

Current information about the status of exoplanets and catalogs of exoplanets can be found online.[1]

18.2.2 Observations

18.2.2.1 Direct Imaging

Since the star is usually a million times brighter than the planet and causing a glare, it is necessary to block the light from the star so the planet can be seen. An occulting technique is necessary to block the light from the star. The larger the planet and the

[1] www.exoplanets.org; www.exoplanet.eu; www.exoplanetarchive.ipac.caltech.edu; www.openexoplanetcatalogue.com.

farther from the star the better. Infrared observations are better than visual, since the planets are hot and emit infrared radiation. Specially designed direct imaging instruments, Gemini Planet Imager, VLT-SPHERE, and SCExAO will image gas giants. In images taken by the Keck and Gemini telescopes, four giant planets at distances between 14.5 and 68 AU, periods between 50 and 450 years, have been detected around star HR 8799 (Sudol and Haghighipour 2012). The direct imaging of exoplanets is described in more detail by Dotson et al. (2010); Biller (2013); Kostov and Apai (2013). However, most observations of exoplanets will be made by indirect means.

18.2.2.2 Radial Velocities

When a planet orbits a star, the two bodies move around the system's center of mass. The changes in the stars radial velocity, its motion with respect to the Earth, can be detected from the displacements in the star's spectral lines due to the Doppler effect. Surveys began 25 year ago with a few hundred bright stars. The measurements must be very accurate, about 1 m/s or less. Now planets with a few Earth masses with close in orbits are being detected. A disadvantage of this method is that only a lower limit on the mass can be set. The details of the technique and the software and methods of analyzing the observations are given by Dotson et al. (2010); Ma and Ge (2012); Baluev (2013).

18.2.2.3 Transits

When a planet passes in front of a star, the brightness of the star drops by a small amount. The amount of the dimming depends on the size of the planet. Repeated observations of the transits indicate the period of the planet. Variations in the times of the transits can indicate perturbations by additional planets. Sometimes the shape of the decrease in the brightness can indicate the presence of an atmosphere. *Transit Timing Variations* (TTV) can be detected from the mutual perturbations in multiplanet systems. Also *Transit Duration Variations* (TDV) can be measured as a means of detecting multiple planets. However, this technique is subject to a large rate of false positives, and, thus, confirmation from another method is usually necessary.

The first exoplanet discovery with the transit technique was of OGLE-TR 56 b by the photometric survey *Optical Gravitational Lensing Experiment* (OGLE)(Udalski et al. 1993). The CoRoT space mission, launched in 2006, has discovered planets starting in 2007. The Kepler mission, launched in 2009, has been extremely successful in detecting planets by the transit technique, and has a large list of candidate planets requiring independent confirmation. The Canadian MOST mission has succeeded in observing transits. Future exoplanet discoveries can be expected from the Transiting Exoplanet Survey Satellite (Ricker et al. 2014), the Super WASP exoplanet transit survey (Smith 2014), and Gaia (Voss et al. 2013).

18.2.2.4 Astrometry

By measuring the position of a star and changes in the position of the star over time, the parallax and proper motion of the star can be determined. If the proper motion of the star is not linear, but has a periodic signature included, this may be due to the gravitational influence of a planet. The size of the periodic motion is dependent on the ratio of the mass of the star and planet. Astrometric satellites, like Hipparcos and Gaia, provide the accuracy for such detections, but their limited observational periods reduce the probability of such detections. Combining lengthy ground-based observing data with both Hipparcos and Gaia observations may be successful in detecting planets. The astrometric detection and characterization of exoplanets is described by Dotson et al. (2010); Sahlmann et al. (2013); Sozzetti (2012).

18.2.2.5 Microlensing

When the gravitational field of a star acts as a lens, magnifying the light of a distant background star, this is *microlensing*. If the lensing star has planets orbiting it, they cause detectable anomalies in the magnification over time. The microlensing method is most sensitive to detecting planets about 1- 10 AU away from a Sun-like star. Unfortunately, the detection of such a planet does not tell much about its characteristics. The searches for and detection of exoplanets by microlensing are described by Brown (2014); Zakharov (2012); Sumi (2012).

18.2.3 Types of Exoplanets

The indirect detection of exoplanets requires observations of at least half of the orbital period of the planet, so shorter period planets are the first to be detected. Also, the more massive planets will produce a larger signal, and will be detected first. Hence, the first exoplanets were close in large planets, called *Hot Jupiters* (periods less than 10 days). They are only found around 0.5% to 1% of Sun-like stars. With time, smaller planets with longer periods were detected, going down in size to Neptune size and smaller. As expected, the planets were similar to the gas giants of the solar system with rocky cores and gaseous atmospheres. Surveys over decades at Keck Observatory show that 10.5% of G and K dwarf stars have one or more giant planets (0.3 to 10 Jupiter masses) with orbital periods of 2 to 2000 days (distances of 0.03 to 3 AU). Planets between the Earth and Neptune in size are common in exosolar systems, but absent in our solar system. 15% of Sun-like stars have one or more planets which satisfy

$$M \sin i = 3 - 30 M_E$$

where M denotes the exoplanet's mass, M_E is Earth's mass and i is the inclination between the normal to the planet's orbital plane and the line of sight from the Earth to the star. These planets are orbiting within 0.25 AU, and by extrapolation, another 24% have planets with

$$M \sin i = 1 - 3M_E$$

The goal, of course, is to get down to the Earth size. Knowing the mass of a planet does not specify its size. A planet of Earth mass will have a size dependent on its composition and atmosphere. There are super Earths with a rock/iron core surrounded by about 3% H_2 gas by mass; water world planets, consisting of a rock/iron core, with a water ocean and atmosphere that contribute 50% of the mass; a mini Neptune composed of rock/iron, water, and H/He gas (Howard 2013).

18.2.4 Orbit Determinations

With the observations being made by indirect methods, all the orbital parameters cannot be determined. From radial velocity measurements, at best the semimajor axis, eccentricity, longitude of periastron, and time of periastron can be determined. The inclination and longitude of ascending node cannot be determined. The combination of radial velocities and transit observations of an exoplanet permits the determination of the inclination of the orbit and, thus, overcoming the $\sin i$ indetermination in the mass determined from the radial velocity, and also to determine the radius of the planet. Furthermore, the combined observations can significantly improve the eccentricity value (Beaugé et al. 2012).

The mass of transiting planets can be determined from follow-up radial velocity measures, if the star is bright enough. Masses can be determined from precise timings for multiple planet systems from the deviations from strict periodicities due to gravitational perturbations, given sufficient observations.

The orbital inclination is normally the angle between the planet's orbital plane and a plane of reference for the star and its planetary system. However, for exoplanets, the inclination is usually between the normal to the planet's orbital plane and the line of sight from the Earth to the star. So most inclinations are near 90°. Generally, when there are multiple planets around a star, the planets' orbits are nearly in the same plane. However, Hot Jupiters tend to have orbital planes very misaligned with their planet star's rotation, some even have retrograde orbits.

Exoplanets have been discovered much closer and farther from stars than those in our solar system. Exoplanets have orbital periods of hours to thousands of years. Most exoplanets discovered with short periods, 20 days or less, have near circular orbits. However, giant planets with longer orbital periods tend to have eccentric orbits, greater than 0.2.

There is a peculiar pattern to the obliquities of stars with Hot Jupiters. The obliquities are apparently random about a critical temperature of about 6250 K, but cooler systems are mostly aligned. Thus, in-situ formation is unlikely for Hot Jupiters due to the insufficient protoplanetary disk close to the star. Rather the planets likely formed in the disk at several AUs, and were gravitationally perturbed into orbits with random inclinations and high eccentricities. They were captured at about 0.05 AU by dissipation of orbital energy in tides raised on the planet.

18.2.5 Planetary Systems

As the period of observing has increased, more systems with multiple planets around a single star, i.e. planetary systems, have been discovered. Planetary systems can be resonant, when the orbital periods of the planets are in integer ratios. Kepler-223 has four planets in an 8:6:4:3 orbital resonance. In interacting systems, the planets are close enough to perturb the orbital parameters of each other.

Buchhave et al. (2014) finds that the exoplanet solar systems can be divided into three types. One, those around the most metallic stars are dominated by gas giants. Two, those around the least metallic stars have mostly rocky planets, usually larger than those in our solar system. Third, stars of medium metallicity have gas dwarfs around them. These have rocky cores with a thick atmosphere of hydrogen and helium.

18.2.6 Habitable Zone

The habitable zone is the region around a star where the temperature allows liquid water to exist on the planet. The heat from stars depends on their size and age, so the habitable zone is at different distances. The planet can have an atmosphere which influences its ability to retain heat. Dessert planets have very little water and water vapor, so they can be closer to the star. Also, the lack of water means less ice to reflect heat, so its outer edge of the habitable zone is further out. Rocky planets with a thick atmosphere could maintain surface water further from a star. If subsurface temperatures are considered, the habitable zone extends much further from the star. Hence, the habitable zone depends on the characteristics of both the star and planet.

Confirmed discovered planets in the habitable zone include super-Earths Kepler-22 b, Kepler-62 e, Kepler-62 f, Kepler-69 c, Gliese 667 Cc, Gliese 667 Ce, and Gliese 667 Cf. Two planets orbiting a red dwarf, Gliese 163, have been discovered. Kepler 186 f is an Earth-sized planet in the habitable zone of a red dwarf.

18.2.7 *Observing Program*

The SOFIA 2.5 m telescope on a Boeing 747 will make optical and near infrared photometric and spectrophotometric observations during planetary transits and eclipses (Angerhausen et al. 2012).

References

Angerhausen, D., Huber, K.F., Mandell, A.M., McElwain, M.W., Czesla, S., Madhusudhan, N., Morse, J.A.: Occultation spectrophotometry of extrasolar planets with SOFIA. Proc. Int. Astron. Union **8**(S293), 435–441 (2012)

Baluev, R.V.: Planetpack: A radial-velocity time-series analysis tool facilitating exoplanets detection, characterization, and dynamical simulations. Astron. Comput. **2**, 18–26 (2013)

Batalha, N.M., Rowe, J.F., Bryson, S.T., Barclay, T., Burke, C.J., Caldwell, D.A., Christiansen, J.L., Mullally, F., Thompson, S.E., Brown, T.M., et al.: Planetary candidates observed by Kepler. III. analysis of the first 16 months of data. Astrophys. J. Suppl. Ser. **204**(2), 24 (2013)

Beaugé, C., Ferraz-Mello, S., Michtchenko, T.A.: Multi-planet extrasolar systems detection and dynamics. Res. Astron. Astrophys. **12**(8), 1044 (2012)

Biller, B.: Detecting and characterizing exoplanets with direct imaging: Past, present, and future. Proc. Int. Astron. Union **8**(S299), 1–11 (2013)

Brown, R.A.: Faint detection of exoplanets in microlensing surveys. Astrophys. J. **788**(2), 192 (2014)

Brumberg, V.: Celestial mechanics: Past, present, future. Sol. Syst. Res. **47**(5), 347–358 (2013)

Buchhave, L.A., Bizzarro, M., Latham, D.W., Sasselov, D., Cochran, W.D., Endl, M., Isaacson, H., Juncher, D., Marcy, G.W.: Three regimes of extrasolar planet radius inferred from host star metallicities. Nature **509**(7502), 593–595 (2014)

Campbell, B., Walker, G., Yang, S.: A search for substellar companions to solar-type stars. Astrophys. J. **331**, 902–921 (1988)

Dick, S.J.: The Biological Universe: The Twentieth Century Extraterrestrial Life Debate and the Limits of Science. Cambridge University Press, Cambridge (1999)

Dick, S.J.: Life on Other Worlds: The 20th-Century Extraterrestrial Life Debate. Cambridge University Press, Cambridge (2001)

Dotson, R., Seager, S., et al.: Exoplanets. University of Arizona Press, Tucson (2010)

Drake, F.D.: Project Ozma. Phys. Today **14**(4), 40–46 (2009)

Dunham, E.W., Borucki, W., Jenkins, J., Batalha, N., Caldwell, D., Mandushev, G.: The discovery of extrasolar planets via transits. In: American Astronomical Society Meeting Abstracts# 223, vol. 223 (2014)

Hatzes, A.P., Cochran, W.D., Endl, M., McArthur, B., Paulson, D.B., Walker, G.A., Campbell, B., Yang, S.: A planetary companion to γ Cephei A. Astrophys. J. **599**(2), 1383 (2003)

Howard, A.W.: Observed properties of extrasolar planets. Science **340**(6132), 572–576 (2013)

Kostov, V., Apai, D.: Mapping directly imaged giant exoplanets. Astrophys. J. **762**(1), 47 (2013)

Ma, B., Ge, J.: A new multi-band radial velocity technique for detecting exoplanets around active stars. Astrophys. J. **750**(2), 172 (2012)

Marcy, G.W., Butler, R.P.: A planetary companion to 70 Virginis. Astrophys. J. Lett. **464**(2), L147 (1996)

Mayor, M., Queloz, D.: A Jupiter-mass companion to a solar-type star. Nature **378**(6555), 355–359 (1995)

Prantzos, N.: A joint analysis of the Drake equation and the Fermi paradox. Int. J. Astrobiol. **12**(03), 246–253 (2013)

Ricker, G.R., Vanderspek, R.K., Latham, D.W., Winn, J.N.: The Transiting Exoplanet Survey Satellite mission. In: American Astronomical Society Meeting Abstracts, vol. 224 (2014)
Sahlmann, J., Lazorenko, P., Mérand, A., Queloz, D., Ségransan, D., Woillez, J.: Astrometric detection of exoplanets from the ground. In: SPIE Optical Engineering+ Applications, International Society for Optics and Photonics, pp. 88,641B–88,641B (2013)
Smith, A.: The SuperWASP exoplanet transit survey. Contrib. Astron. Obs. Skalnaté Pleso **43**, 500–512 (2014)
Sozzetti, A.: Astrometry and exoplanet characterization: Gaia and its Pandora's box. In: Orbital Couples: Pas de Deux in the Solar System and the Milky Way, Observatoire de Paris, vol. 1, pp. 25–31 (2012)
Sudol, J.J., Haghighipour, N.: High-mass, four-planet configurations for HR 8799: constraining the orbital inclination and age of the system. Astrophys. J. **755**(1), 38 (2012)
Sumi, T.: Current and future of microlensing exoplanet search. Proc. Int. Astron. Union **8**(S293), 10–19 (2012)
Udalski, A., Szymanski, M., Kaluzny, J., Kubiak, M., Krzeminski, W., Mateo, M., Preston, G., Paczynski, B.: The optical gravitational lensing experiment. discovery of the first candidate microlensing event in the direction of the Galactic Bulge. Acta Astronaut. **43**, 289–294 (1993)
Voss, H., Jordi, C., Fabricius, C., Carrasco, J., Masana, E., Luri, X.: Exoplanetary transits as seen by Gaia. In: Highlights of Spanish Astrophysics VII, vol. 1, pp. 738–743 (2013)
Wolszczan, A., Frail, D.A.: A planetary system around the millisecond pulsar PSR 1257+ 12. Nature **355**(6356), 145–147 (1992)
Zakharov, A.: Exoplanet searches with gravitational microlensing. In: 39th COSPAR Scientific Assembly, vol. 39, p. 2242 (2012)

Bibliography

Baker, R.M., Makemson, M.W.: An Introduction into Astrodynamics. Academic Press, New York (1967)
Baker, R.M.: Astrodynamics, Applications and Advanced Topics. Academic Press, New York (1967)
Bate, R.R., Mueller, D.D., White, J.E.: Fundamentals of Astrodynamics. Dover Publications, New York (1971)
Battin, R.H.: An Introduction to the Mathematics and Methods of Astrodynamics, 2nd edn. AIAA Education Series. AIAA, New York (1999)
Belbruno, E.: Capture Dynamics and Chaotic Motions in Celestial Mechanics. Princeton University Press, Princeton (2004)
Brouwer, D., Clemence, G.M.: Methods of Celestial Mechanics. Academic Press, New York and London (1961)
Brown, E.W.: An Introductory Treatise on the Lunar Theory. Dover Publications, New York (1960)
Brown, E.W., Shook, C.A.: Planetary Theory. Dover Publications, New York (1964)
Brumberg, V.A.: Essential Relativistic Celestial Mechanics. Adam Hilger, Bristol, Philadelphia, New York (1991)
Brumberg, V.A.: Analytical Techniques of Celestial Mechanics. Springer, New York (1995)
Brumberg, V.A., Brumberg, E.V.: Celestial Mechanics at High Eccentricities. Gordon and Breach Science Publishers, New York (1999)
Chao, C.-C.: Applied Orbital Perturbations and Maintenance. AIAA Educational Series. AIAA Press, New York (2005)
Chebatarov, G.A.: Analytical and Numerical Methods of Celestial Mechanics. American Elsevier Publishing Company, New York (1967)
Cook, A.: The Motion of the Moon. Adam Hilger, Bristol and Philadelphia (1988)
Danby, J.M.A.: Fundamentals of Celestial Mechanics. Willman-Bell, Richmond (1962,1988)
Duncombe, R.L., Szebehely, V.G. (eds.): Methods in Astrodynamics and Celestial Mechanics. Academic Press, New York, London (1966)
Escobal, P.R.: Methods of Orbit Determination. Wiley, New York, London, Sydney (1965)
Escobal, P.R.: Methods of Astrodynamics. Wiley, New York (1968)
Fitzpatrick, P.M.: Principles of Celestial Mechanics. Academic Press, New York and London (1970)
Hagahari, Y.: Celestial Mechanics. Vol. 1 and 2. Massachusetts Institute of Technology Press, Cambridge (1970–1972); Vols. 2–5, Japan Society for Promotion of Science, Tokyo (1974–1976)
Herget, P.: The Computation of Orbits (1948). Published privately by the author

P. Gurfil, P.K. Seidelmann, *Celestial Mechanics and Astrodynamics: Theory and Practice*, Astrophysics and Space Science Library 436,
DOI 10.1007/978-3-662-50370-6

Herrick, S.: Astrodynamics: Orbit Determination, Space Navigation, Celestial Mechanics, Vol. I. Van Nostrad Reinhold, London (1971)

Herrick, S.: Astrodynamics: Orbit Correction, Perturbation Theory, Integration, Vol. II. Van Nostrad Reinhold, London (1972)

Kaula, W.M.: Theory of Satellite Geodesy. Blaisdell Publishing, Waltham (1969)

King-Hele, D.: Theory of Satellite Orbits in a Atmosphere. Butterworths, London (1964)

Kovalevsky, J.: Introduction to Celestial Mechanics. Springer, New York. D. Reidel Publishing, Dordrecht-Holland (1961)

Kurth, R.: Introduction to the Mechanics of the Solar System. Pergamon Press, New York, London, Paris, Los Angeles (1959)

Lagrange, J.L.: Collected Works. Gauthier-Villars, Paris (1873)

Laplace, P.S.: Traité de Mécanique Céleste, Vols. 1–3. Duprat, Paris (1799–1802); Vol 4, Courrier, Paris (1805)

Meyer, K.R., Hall, G.R., Offin, D.: Introduction to Hamiltonian Dynamical Systems and The N-body Problem. Springer, New York (2009)

McCuskey, S.W.: Introduction to Celestial Mechanics. Addison-Wesley Publishing Company, Reading, Palo Alto, London (1963)

Moulton, F.R.: An Introduction to Celestial Mechanics. Dover Publications, New York (1970)

Plummer, H.C.: An Introductory Treatise on Dynamical Astronomy. Dover Publications, New York (1960)

Poincaré, H.: Les Methods Nouvelles de la Mécanique Céleste. Gauthier-Villars, Paris (1892–1899)

Poincaré, H.: Leçons de Mécanique Céleste (1905–1910)

Poincaré, H.: New Methods of Celestial Mechanics, 3 volumes. American Institute of Physics, New York (1993)

Pollard, H.: Mathematical Introduction to Celestial Mechanics. Prentice-Hall, Englewood Cliffs (1966)

Prussing, J.E., Conway, B.A.: Orbital Mechanics. Oxford Press, New York (1993)

Roy, A.: The Foundations of Astrodynamics. The Macmillan Company, New York (1965)

Roy, A.: Orbital Motion. Adam Hilger, Bristol and Philadelphia (1988)

Smart, W.M.: Celestial Mechanics. Longmans, Green and Co., London, New York, Toronto (1953)

Soffel, M.H.: Relativity in Astrometry, Celestial Mechanics and Geodesy. Springer, Berlin, Heidelberg, New York, London, Paris, Tokyo (1989)

Sternberg, S.: Celestial Mechanics. W. A. Benjamin, New York (1969)

Sterne, T.E.: An Introduction to Celestial Mechanics. Interscience Publishers, New York (1960)

Szebehely, V.: Theory of Orbits. Academic Press, New York (1967)

Szebehely, V., Mark, H.: Adventures in Celestial Mechanics. Wiley, New York (1998)

Tisserand, F.: Traité de Mécanique Céleste, 4 volumes. Gauthier-Villars, Paris (1889–1896)

Vallado, D.A.: Fundamentals of Astrodynamics and Applications. College Custom Series. The McGraw-Hill Companies, New York (1997)

Vinti, J.P., Der, G.J., Bonavito, N.L.: Orbital and Celestial Mechanics, Progress in Astronautics and Aeronautics, vol. 177. American Institute of Aeronautics and Astronautics, Washington, DC (1998)

Watson, J.C.: Theoretical Astronomy. Dover Publications, New York (1964)

Winter, A.: The Analytical Foundations of Celestial Mechanics. Princeton University Press, Princeton (1941)

Index

P. Gurfil, P.K. Seidelmann, *Celestial Mechanics and Astrodynamics: Theory and Practice*, Astrophysics and Space Science Library 436,
DOI 10.1007/978-3-662-50370-6